Carpentry
& Joinery

Job Knowledge

Carpentry & Joinery

Book 1
Job Knowledge

Peter Brett

Nelson Thornes
a Wolters Kluwer business

Text © Peter Brett, 1981, 1985, 2004, 2005
Illustrations © Nelson Thornes Ltd 1981, 1985, 2004, 2005

First edition of *Carpentry and Joinery for Building Craft Students* Bks 1 and 2 published in 1981 by:
Hutchinson Education
First edition of *Carpentry and Joinery for Advanced Craft Students: Site Practice* published in 1985 by:
Hutchinson Education

Second edition of *Carpentry and Joinery* Bk 1 (incorporating material from *Site Practice*) published in 2005 by:
Nelson Thornes Ltd
Delta Place
27 Bath Road
CHELTENHAM
GL53 7TH
United Kingdom

07 08 09 / 10 9 8 7 6 5 4

A catalogue record for this book is available from the British Library

ISBN 978 0 7487 8501 8

Illustrations by Peters and Zabransky
Page make-up by Florence Production Ltd, Stoodleigh, Devon

Printed and bound in Slovenia by Korotan Ljubljana

Contents

Introduction

This set of two books by Peter Brett is intended to form a complete reference work for carpenters and joiners at all levels. They provide an in depth knowledge of the basic methods and principles of the craft which are essential to the new entrant. These are extended to cover the more specialist topics that may be encountered during their working life.

- *Book 1 Job Knowledge* covers the introductory skills and an appreciation of the materials used in the woodworking industry. This is followed up with coverage of the background knowledge that is required to undertake a wide range of practical activities. The final chapters give an overview of the building industry and its controls, which are desirable for understanding and progression.

- *Book 2 Practical Activities* is divided into the specific areas of work, which may be undertaken by the carpenter and joiner. The introductory topics in each chapter are covered in depth, supported by bulleted step-by-step text and illustrations, which are intended to guide the reader through the task. This is reinforced by a wider coverage of each topic area to aid understanding and enable the transfer/application of skills to other tasks.

The works are presented in a clear, concise and factual style that is fully integrated with numerous illustrations and photographs. Cross-references and essential key facts are included at the foot of each page, where appropriate as an aid to full comprehension.

Assessment activities are included in each chapter to aid learning and enable readers to evaluate their understanding of the knowledge required. See www.nelsonthornes.com/carpentry

These two books combined will prove to be invaluable to both the new entrant undergoing training and qualified craftspeople who are undertaking a particular task for the first time or who wish to extend their knowledge of craft processes and the interrelationship with activities of the building industry as a whole. Additionally, they are ideal background reading for people studying for a career in the building industry or who are already employed at various levels within the industry.

Peter Brett has written numerous books on carpentry and joinery and has extensive experience in the industry and in education and training. He has worked with both the City and Guilds of London Institute and the Construction Industry Training Board in the development and assessment of awards in Carpentry and Joinery.

Carpentry and Joinery
by Peter Brett

Book 1: Job Knowledge

1. Timber technology
2. Care and use of hand tools
3. Woodworking joints
4. Interpretation of drawings
5. Portable powered hand tools
6. Machine utilisation
7. Setting out, marking out and levelling
8. Applied calculations
9. Applied geometry
10. Building controls
11. Building principles
12. Building procedures

Book 2: Practical Activities

1. Timber ground and upper floor construction
2. Flat and pitched roof construction
3. Partition walls
4. Timber frame construction and timber engineered components
5. Stairs
6. Doors, frames and linings
7. Windows
8. Mouldings and trim
9. Units, fitments, seating and miscellaneous items
10. Casings, false ceilings and panelling
11. Temporary construction and site works
12. Repairs and maintenance

Sources of technical information

It is rarely possible to absorb and remember all the information required for a particular job; even if it were, it is essential that this information be regularly updated to take account of changes in legislation and technology.

Throughout your working life from student or trainee through to craftsperson and beyond to various other positions in the industry, you will have to make decisions and solve problems. To do this effectively, various sources of information will have to be consulted.

Specialist information and procedures can be obtained from textbooks. However as some of the information contained in them is date sensitive, they will only be truly up to date at time of writing the publication. Other sources may be used in addition to cross-reference, check, compare and further extend your knowledge.

'Browsing the Internet' via a computer is an excellent means of accessing other sources of information: simply type in the website address of the company or organisation into a web browser and you will be connected to their website.

By typing in **www.nelsonthornes.com/carpentry** you will be connected to a website associated with this book. This site contains a study guide along with assessment activities and answers for each chapter in this book and its companion.

Try some of the following sites:
- Building Regulations: www.odpm.gov.uk
- British Standards: www.bsi-global.com
- Building Research Establishment: www.bre.co.uk
- Construction training and careers: www.citb.org.uk and www.city-and-guilds.co.uk
- Government publications: www.tso.co.uk
- Health and safety: www.hse.gov.uk
- Timber and sheet material information: www.trada.co.uk
- Woodworking trade associations: www.bwf.org.uk
- Building materials and components: www.buildingcentre.co.uk

If you don't know the exact website address of the organisation you are looking for, or you simply wish to find out more information on a subject, you could use a 'search engine' to find the web pages. Search engines use 'key words' to find information on a subject. Enter a key word or words such as doors, windows, stairs or strength grading or timber, etc. or the name of a company/organisation, and it searches the Internet for information about your key words or name. You are then presented with a list of relevant websites that you can click on, which link you to the appropriate information pages.

Acknowledgements

The author wishes to acknowledge the following:

- Stanley for the handtool photographs.
- Bosch, Trend, Hitachi, DeWalt, Makita, Hilti and Paslode for the power tool photographs.
- Dominion Machinery Co. Ltd and Wadkin plc for the machinery photographs.
- Rentokil Property Care, and Science Photo Library, for the dry rot photographs Figs 1.52 and 1.53.
- Elizabeth Whiting Associates for the wet rot photograph, Fig. 1.56.
- TRADA for the map and board manufacturing processes, Figs 1.7, 1.85, 1.88.
- Australian National University for the MDF manufacturing process, Fig. 1.90.
- The CSCS card Fig. 10.28 is copyright CITB.
- Other images from Corel 376 (NT) (electric plane, 5.29, mitre saw, Fig. 5.24, electric jigsaw, Fig. 5.21), Library/Ingram (NT) (rubber mallet, Fig. 2.55), Photodisc 28 (NT) (wooden shafted claw hammer Fig. 2.55), Stockbyte 30 (NT) (club hammer, Fig. 2.63), Instant Art Signs (NT) (hazard signs Fig. 10.25).

This Carpentry & Joinery series is dedicated to my Mum and Dad, who gave me the initial encouragement to commence a career in the Construction Industry.

Finally 'all the best for the future' to those who use this book and I trust it provides you with some of the help and motivation required to progress in the Construction Industry.

Peter Brett

Chapter One

Timber technology

This chapter covers the work of site carpenters and bench joiners. It is concerned with the characteristics, uses and limitations of materials used by all woodworkers. It includes the following:

- identification and properties of softwoods, hardwoods, wood-based board materials, fixings and adhesives
- identification of defects in materials
- seasoning and preservative treatment of timber
- enemies of wood, their recognition and eradication.
- the specification and grading of timber
- board material and veneers
- screws, nails and adhesives.

There are many different types of wood and board materials available to the woodworker, each having their own specific qualities, properties and uses.

In many circumstances you will be told what particular type of wood or board to use by the client or your employer. You will have to identify the required material and select from stock the most suitable pieces with regards to finish and defects.

Alternatively you may be asked to recommend a suitable material for use in a given situation. This will involve having knowledge of different woods, their properties, how it has been sawn, stored or manufactured and any seasoning and treatment processes.

How a tree grows

Trees grow by adding a new layer of cells below the bark each growing season. In temperate climates, such as Europe, this is mainly during the spring and summer months, whereas, in tropical climates growth may be almost continuous or be linked to their dry and rainy seasons.

All trees have three common parts as illustrated in Figure 1.1:

- **Roots** absorb ground water and diluted minerals (unenriched sap) from the soil via their system of fine hair-like ends. In addition, roots act as an anchor to secure the tree in the ground. The depth and spread of the root system will depend on the type of tree and ground conditions. However, the radius spread of the roots will often exceed that of the crown.
- The **trunk** is the main stem of the tree from which timber is cut. It conducts unenriched sap up from the roots to the crown, through the sapwood. Once enriched in the leaves it travels back down again through the inner bark to the cambium layer.

■ The **crown** comprises the branches and leaves of a tree, often termed the canopy. It is here that the unenriched sap is processed into food for growth. This processing is called **photosynthesis.** The green substance in each leaf, called **chlorophyll**, absorbs daylight energy, to convert a mixture of carbon dioxide from the air and unenriched sap from the roots into carbohydrates (sugars and starches). During this process oxygen is given off as a waste product.

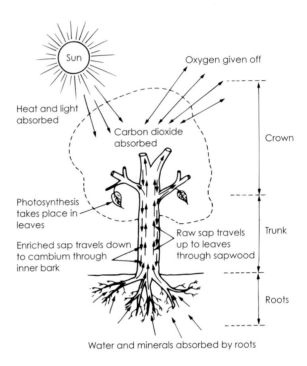

▶ **Figure 1.1** The growth of a tree

Tree trunk cross-section

An understanding of the various internal parts of the trunk is desirable as they can affect the way in which timber is cut and used, see Figure 1.2.

■ *Bark.* The often-corky outer layer of the trunk and branches. It serves to protect the inner parts from the elements, mechanical abrasion, fungal and insect attack.

■ *Bast, also termed inner bark.* It transports the enriched sap down from the crown to all growing parts of the tree. The outer layers of bast progressively become inactive as a food carrier and forms a new layer of bark, with the existing bark scaling off as the tree increases in girth (circumference of the trunk) and height.

■ *Cambium.* The thin layer of living cells under the bast. These cells divide during the growing season to form both new sapwood cells and new bast cells.

■ *Sapwood.* The newly formed outer layers of growth which convey the rising unenriched sap up to the crown. As the sapwood contains foodstuffs, timber cut from it is considered to be more prone to insect and fungi attack, unless treated with a preservative.

■ *Heartwood.* The inner, more mature part of the trunk, which no longer conveys the sap. Its main function is to give strength to the tree. Timber cut from the heartwood is often darker in colour and more durable than sapwood.

■ *Pith, also termed medulla.* It is the first growth of the tree as a sapling and it often decays as the tree ages, creating a defect in cut timbers.

■ *Rays, also termed medullary rays,* as they appear to radiate from the pith outwards towards the bark. They are groups of food storage cells. Some timbers are specially cut to show the rays on the planed surface, as a decorative effect.

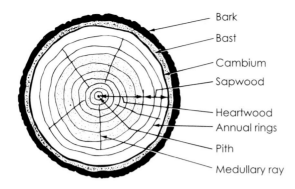

Bark

Bast

Cambium

Sapwood

Heartwood

Annual rings

Pith

Medullary ray

◀ **Figure 1.2** Cross-section of a tree trunk

Growth rings

Growth rings also termed **annual rings**. In temperate climates such as
Europe, tree growth takes place mainly during the spring and summer
months. This produces a ring of new growth. By counting the number of
rings you can tell how long the tree has been growing, see Figure 1.3.

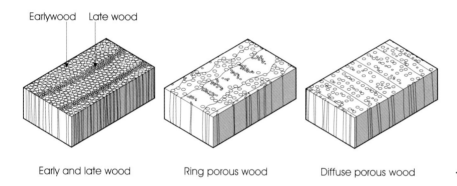

Earlywood   Late wood

Early and late wood   Ring porous wood   Diffuse porous wood

◀ **Figure 1.3** Growth rings

In tropical countries tree growth can be almost continuous or linked to other
climatic conditions. On some occasions it is virtually impossible to
distinguish between each year's growth.

In trees where growth rings are visible, it is normally possible to see that the
ring in fact consists of two rings; **early wood** (spring growth) with thin-
walled cells for maximum food transfer and appearing lighter in colour; **late
wood** (summer growth) with thicker walled cells. As growth is slowing, they
appear darker than sapwood.

Tree types

Trees can generally be divided into two groups:

■ **Softwood trees** having narrow or needle leaves (conifers) see
Figure 1.4.

■ **Hardwood trees** having broad leaves (broadleaf trees) see
Figure 1.5.

This grouping is only partly to do with the hardness of the timber but is
based on the tree's cellular structure. These terms can be misleading, as
there are exceptions. However most hardwoods are hard and heavy; likewise
most softwoods are less hard and lighter in weight. Balsa wood, classified as
a hardwood, is much softer and lighter in weight than say pitch pine or
larch, which are classified as softwoods, but are in fact much harder and
heavier than many hardwoods. These are extreme comparisons, but serve to
emphasise that the simple view, that all hardwoods are hard and all
softwoods are soft, cannot be relied upon. Generally, a clear difference can
be seen in the standing tree.

▲ **Figure 1.4** General
characteristics of softwood trees

3

Tree shapes

General identifying characteristics of **softwood trees**:

- a trunk that is very straight and cylindrical with an even taper;
- a crown that is narrow and pointed;
- needle-like leaves;
- a bark that is coarse and thick;
- seeds are borne in cones;
- evergreen, i.e. they do not drop all their leaves at once in the autumn.

General identifying characteristics of **hardwood trees**:

- an irregular, less cylindrical trunk that very often has little taper.
- a crown that is wide, rounded and contains large heavy branches.
- broad leaves
- the bark varies widely and can be very smooth and thin to very coarse and thick ranging from white to black in colour.
- they have covered seeds, e.g. berries, acorns and stoned fruits or nuts.
- they are mainly deciduous, e.g. they shed their leaves in winter.

The trunks of forest or dense woodland trees will be longer and relatively free of side branches, as the crown attempts to reach up above other adjacent trees in competition for light, as illustrated in Figure 1.6.

Sources of timber supply

There are hundreds of species of trees available for use in the United Kingdom; very few of these are in fact British or Home grown; the majority being imported from various sources worldwide.

The map illustrated in Figure 1.7 gives the main sources of supply for the range of timbers that are most frequently used.

It can be seen that the main source of supply for commercial softwoods is in the northern hemisphere, which extends across the arctic and sub-arctic regions of Europe and North America down to the south-eastern United States of America. Climate is the most important factor governing which species of hardwood grows where. Deciduous broadleaf trees are native to the temperate northern hemisphere, where they will shed their leaves and become dormant during the colder months of the year. Broadleaf trees in the

▲ **Figure 1.5** General characteristics of hardwood trees

Forest grown hardwood   Forest grown softwood

▲ **Figure 1.6** Tree shapes when grown close together

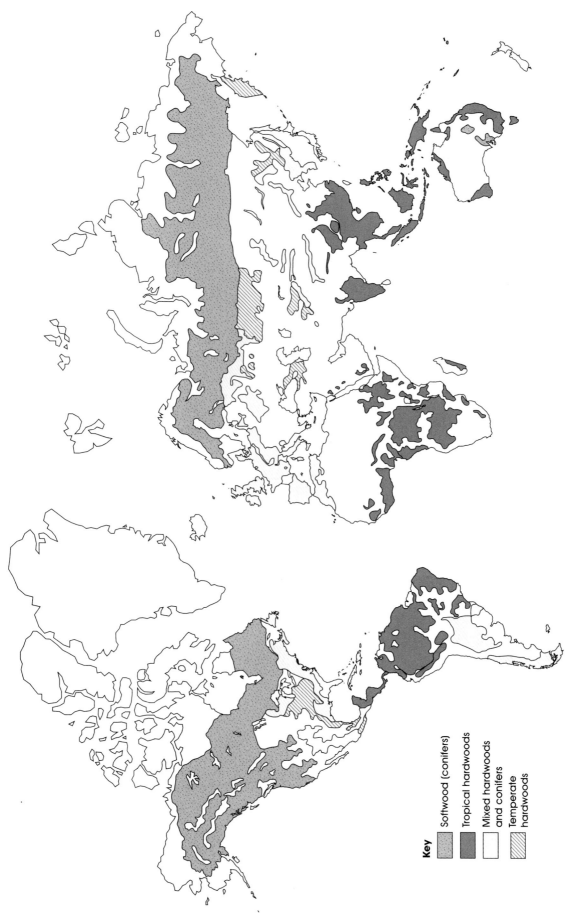

Key

Softwood (conifers)

Tropical hardwoods

Mixed hardwoods
and conifers

Temperate
hardwoods

▲ **Figure 1.7** Timber-growing regions around the world

tropics and southern hemisphere are mainly evergreens, where they may keep their leaves for two or more years.

The map also shows the mainly temperate areas, which support a mixture of evergreen softwoods and deciduous hardwoods.

The structure of wood

Apart from the differences in the general identifying characteristics, there are further important differences between the cellular structures of the two types of trees.

Wood structure

Softwood

Softwood trees have a simple structure, with only two types of cell, tracheids and parenchyma, see Figure 1.8.

- *Tracheids* are box-like cells, which form the main structural tissue of the wood and, as well as giving the tree its mechanical strength, they also conduct the rising sap. The sap passes from one tracheid to another through thin areas of the cell's wall known as pits. Tracheids are formed throughout the trees growing season, but the rapid spring growth produces a wide band of thin-walled cells (called early wood). These cells conduct sap but provide little strength. It is the summer growth of thick-walled cells (called late wood) which provides most of the tree's mechanical strength

- *Parenchyma* cells are food-storing cells, which radiate from the centre of the tree. These are also known as ray parenchyma or medullary rays. Resin ducts or pockets are also found in softwood but perform no useful function.

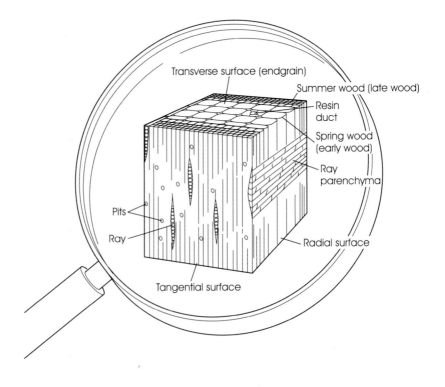

▲ **Figure 1.8** Magnified softwood cube

Slow grown softwoods are considered to be stronger than fast grown softwoods, see Figure 1.9. This is because the slow grown timber contains more thick-walled, strength giving tracheids than the fast grown timber.

Hardwood structure

Hardwood trees have a more complex structure than softwood consisting of three types of cell, fibres, parenchyma and vessels or pores, see Figure 1.10.

- Fibres are the main structural tissue of the wood, giving it its mechanical strength.
- Parenchyma cells perform the same function as they do in softwood, i.e. store food.
- Vessels or pores are sap-conducting cells.

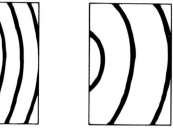

Slow grown: strong   Fast grown: weak

▲ **Figure 1.9** Fast and slow grown softwood

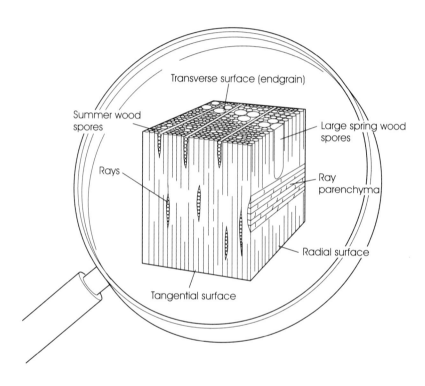

◀ **Figure 1.10** Magnified hardwood cube

The sap cells are circular or oval in section and can occur in the annual ring in one of two ways:

1. Uniform size cells spread or diffuse fairly evenly throughout the year's growth. Trees of this type are known as *diffuse porous hardwoods*.
2. A definite ring of large cells in the early spring growth, with smaller cells spread throughout the growing season. These are known as *ring porous hardwoods*.

Fast grown, ring porous hardwoods are considered to be stronger than slow grown, ring porous hardwoods, see Figure 1.11. This is because there is less room for the strength-giving fibres between the pores in the slow grown timber.

Properties and uses of timber

Some of the properties and uses of a range of timbers used in the UK are contained in Table 1.1. They are listed in alphabetical order under their common name followed by their botanical name, country of origin, timber type and colour, main properties and common uses. An explanation of the headings used in the table follows.

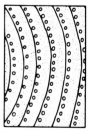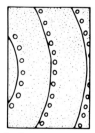

Slow grown: weak   Fast grown: strong

▲ **Figure 1.11** Fast and slow grown hardwood

Table 1.1 Properties of commercial timbers

Name, Species and Country of Origin	Type and Colour	Density kg/m³	Durability Class	Treatability	Moisture Movement	Working Qualities	Availability	Price Range	Characteristics	Structural	Cladding	Flooring	External Joinery	Internal Joinery	Fitments/furniture	General remarks
Afrormosia Pericopsis elata West Africa	Hardwood Light Brown	710	2	ED	Small	Med.	CITES 2	High	Med. fine texture. Straight to interlocked grain.		*		*	*	*	Avoid damp contact with ferrous metals. Darkens in colour on exposure to daylight.
Ash, American Fraxinus spp USA	Hardwood Grey-brown	670*	5	E	Med.	Med.	Reg.	Med.	Straight grained with a coarse texture					*	*	Also used for tool handles.
Ash, European Fraxinus excelsior Europe	Hardwood White to light brown	710*	5	ME	Med.	Good	Reg.	Med.	Straight grained with a coarse texture	*				*	*	Has good bending properties. Also used for plywood and sports goods.
Beech Fagus sylvatica Europe	Hardwood Whitish brown to yellowish brown	720	5	E	Large	Good	Reg.	Low to Med.	Fine texture Straight grained			*		*	*	Often steamed to produce a light pink colour. Also used for plywood.
Birch, American Betula spp North America	Hardwood Light to dark reddish brown	640 to 710	5	ME	Large	Good	Reg.	Low	Fine texture Straight grained			*		*	*	Also used for plywood and turnery.
Birch, European Betula pendula, Betula pubescens North America	Hardwood White to light brown	670	5	ME	Large	Good	Limited	Low	Fine texture Straight grained			*			*	Also known as silver birch and white birch. Also used for plywood and turnery.
Cedar (western red) Thuja plicata North America	Softwood Reddish-brown	390	2	ED	Small	Good	Reg.	Med.	Straight grained with a coarse texture		*					Weathers to a silver-grey. Avoid damp contact with ferrous metals. Aromatic.
Chestnut (sweet) Castanea sativa Europe	Hardwood Yellowish-brown	560	2	ED	Large	Good	Limited	Med.	Med. texture Straight or spiralled grain				*	*		Avoid damp contact with ferrous metals. Also used for fencing.
Douglas fir Pseudotsuga menziesii North America	Softwood Light reddish-brown	530	3	ED	Small	Good	Reg.	Med.	Med. texture Straight grained	*		*	*	*	*	Also known as British Columbian pine & Oregan pine. Available in long clear lengths. Avoid damp contact with ferrous metals.
Ekki Lophira alata West Africa	Hardwood Dark red to brown	1070	2	ED	Large	Difficult	CITES 2 Limited	Low to Med.	Coarse texture	*		*				Available in large sectional sizes for heavy construction and marine work.
Elm, European Ulmus spp Europe	Hardwood Light Brown	560*	4	D	Med.	Med.	Limited	Med.	Coarse texture with irreg. growth rings.	*				*	*	Also used for turnery and boat building.
Greenheart Ocotea rodiaei Guyana	Hardwood Greenish to brown	1040	1	ED	Med.	Difficult	CITES 2 Var.	Med.	Fine texture	*						Available in large sectional sizes for heavy construction and marine work.
Hemlock, Western Tsuga heterophylla North America	Softwood Pale brown	500	4	D	Small	Good	Reg.	Low to Med.	Fine texture Straight grained			*	*	*	*	Can be used externally if preservative treated. Also known as British Columbian hemlock & Pacific hemlock.
Idigbo Terminalia ivorensis West Africa	Hardwood Yellow	560	3	ED	Small	Med.	Var.	Med.	Med. texture		*	*	*	*	*	Avoid damp contact with ferrous metals. Stains yellow in contact with water.

Species / Colour / Origin	Density	Class	Durability	Movement	Working	Availability	Strength	Texture / Grain							Notes
Iroko, Milicia excelsa — Hardwood Yellow brown, West Africa	660	2	ED	Small	Med. to difficult	CITES 2 Reg.	Low to Med.	Med. texture Interlocking grain	*	*	*	*		*	May contain stone deposits. Often used as an alternative to teak.
Jelutong, Dyera costulata — Hardwood White to yellow, South East Asia	470	4	E	Small	Good	Reg.	Med.	Fine texture straight grained		*		*	*	*	May contain resin ducts. Also used for pattern making.
Keruing, Dipterocarpus spp — Hardwood Pinkish to dark brown, South East Asia	740*	3	D	Med. to large	Difficult	Reg.	Low	Med. texture			*	*			Often extrudes resin.
Mahogany, African, Khaya spp — Hardwood Reddish brown, West Africa	530*	3	ED	Small	Med.	Reg.	Low to Med.	Med. texture Interlocking grain	*	*	*	*	*	*	Variations in character particularly density. Also used in boat building.
Mahogany, American, Swietenia macrophylla — Hardwood Reddish brown, Central & South America	560*	2	ED	Small	Good	CITES 2 Reg.	High	Med. texture		*	*	*	*	*	Often used in high class furniture, cabinet-making and boat building
Makore, Tieghemella heckelii — Hardwood Pinkish to dark red, West Africa	640	1	ED	Small	Med.	Var.	Med.	Fine texture		*	*	*	*	*	Fine dust may be irritant. Also used in boat building and for plywood.
Maple, rock, Acer saccharum, Acer nigrum — Hardwood Creamy white, North America	740	4	D	Med.	Med.	Reg.	Med.	Fine texture straight grained		*		*		*	Excellent hard wearing properties.
Meranti, Lauan Shorea spp — Hardwood Med. to dark red-brown, South East Asia	660*	4	ED	Small	Med.	Limited	Low to Med.	Med. texture Slight interlocking grain	*	*	*	*	*	*	Can be wide variation in characteristics. Often used for plywood.
Oak, American red & white, Quercus spp — Hardwood Pale yellow to mid-brown, North America	790	4	D	Med.	Med.	Reg.	Med.	Med. texture straight grained		*	*		*	*	Also used for cooperage.
Oak, European, Quercus robur, Quercus petraea — Hardwood Yellowish brown, Europe	720*	2	ED	Med.	Med. to difficult	Var.	Med. to High	Med. to coarse texture straight grained	*	*	*	*	*	*	Hard wearing. Avoid damp contact with ferrous metals. Also used for cooperage.
Oak, Japanese, Quercus mongolica — Hardwood Pale yellow, Japan	670	3	ED	Med.	Med.	Very Limited	High	Med. texture straight grained		*	*		*	*	Lighter in colour than European.
Obeche, Triplochiton scleroxylon — Hardwood Creamy white to pale yellow, West Africa	390	5	D	Small	Good	Reg.	Low to Med.	Fine even texture grain can be interlocked			*		*	*	Very stable often used as core stock for plywood.
Parana Pine, Araucaria angustifolia — Softwood Golden brown with red streaks, South America	550	5	ME	Med.	Good	Reg.	Low to Med.	Fine texture mainly straight grained		*	*	*	*	*	Distortion may occur during drying.
Ramin, Gonystylus spp — Hardwood White to pale yellow, South East Asia	670	5	E	Large	Med.	CITES 2 Var.	Low/ Med.	Med. texture straight grained		*	*		*	*	Also ideal for mouldings.
Redwood, European, Pinus sylvestris — Softwood Pale yellow-brown to red-brown, Europe	510	4	ED	Med.	Med.	Reg.	Low	Med. texture	*	*	*	*	*	*	Should be preservative treated for external use. Also known as Scots pine, yellow deal & red pine.

Name / Species / Origin / Description	Density		Durability	Sizes	Working	Availability	Movement	Texture / Grain							Notes
Rosewood, Hardwood Med. to dark brown with black streaks. Dalbergia spp, India, South America	870*	1	ED	Small	Med.	CITES I	High	Med. texture straight grained	*				*	*	Highly decorative, but rarely available as a raw material. Fine dust may be irritant. Also used for turnery.
Sapele, Hardwood Reddish brown. Entandrophragma cylindricum, West Africa	640	3	D	Med.	Med.	Reg.	Med.	Med. texture interlocking grain	*		*		*	*	Has a tendency to distort.
Spruce, Sitka, Softwood Pinkish brown. Pinus sitchensis, North America, UK	450	5	D	Small	Good	Reg.	Low	Coarse texture usually straight grained	*				*	*	Also use for packaging and pallets
Sycamore, Hardwood White or yellowish white. Acer pseudoplatanus, Europe	630	5	E	Small	Good	Limited	Med.	Fine texture usually straight grained					*	*	Good turning properties. Also known as plane & great maple.
Teak, Hardwood Golden brown often with darker markings. Tectona grandis, India, Burma, Thailand	660	1	ED	Small	Med.	Reg.	Med.	Med. texture normally straight but sometimes wavy grain			*	*	*	*	Extremly resistant to chemicals. Surface can have an oily feel. Also used in boatbuilding.
Tulipwood, Hardwood Yellow to olive brown with pink stripes. Liriodendran tulipifera, North America	510	4		Med.	Good	Limited	Med.	Fine texture Straight grained					*	*	Also known as canarywood, American poplar and American whitewood.
Utile, Hardwood Reddish-brown. Entandrophragma utile, West Africa	660	3	ED	Med.	Med.	Reg.	Med.	Med. texture interlocked grain	*		*	*	*	*	Tendency to blunt cutting edges.
Walnut, African, Hardwood Yellowish-brown may have dark streaks. Lovoa trichilioides, West Afica	560	4	ED	Small	Med.	Var.	Med.	Med. texture	*			*	*	*	
Walnut, American, Hardwood Dark brown. Juglans nigia, North America	660	3	ED	Small to Med.	Good	Var.	Med. to High	Coarse texture normally straight but sometimes wavy grain					*	*	Also used for gunstocks & sports goods. Also known as American black walnut.
Walnut, European, Hardwood Grey-brown with darker streaks. Juglans regia, Europe	670	3	D	Med.	Good	Limited	High	Coarse texture straight to wavy grain			*		*	*	Also used for gunstocks and turnery.
Wenge, Hardwood Dark brown with black veins. Millettia laurentii, Central & East Africa	880*	2	ED	Small	Good	Limited	Med.	Coarse texture			*	*	*	*	Also used for turnery.
Whitewood, European, Softwood White to pale yellowish brown. Picea abies, Europe	470	4	D	Med.	Good	Reg.	Low	Med. texture	*		*	*	*	*	Should be preservative treated for external use. Also known as European spruce.
Yew, Softwood Orange-red-brown. Taxus baccata, Europe	670		D	Small to Med.	Difficult	Very limited	High	Med. texture					*	*	Distinct dark heartwood with lighter sapwood and a growth pattern that makes the wood very decorative.

* after density indicates that it may vary by 20% or more.

Species. The timbers are listed under their common name; where a timber is also known by another name this is indicated. Their geographical origin is sometimes included in these names for further identification. To enable formal scientific identification every plant also has a **botanical name**. This consists of two parts: the first its **genus** or generic name and the second its **species** or specific name. Trees with similar genera (plural of genus) are grouped into families and related families are put together into orders, see Figure 1.12.

- **Softwoods** belong to the botanical group *Gymnospermae* (plants that bear exposed seeds). Their order is Coniferae (cone bearing). A family example is Pinaceae (pine family): examples of genus are Pinus (pine), Picea (spruce) and Tsuga (hemlock). Examples of Pinus species are silvestris (European redwood or Scots pine) and palustris (American pitch pine).

- **Hardwoods** belong to the botanical group *Angiospermae* (flowering plants). Their order is *Dicotyledonae* (containing seed-bearing ovaries that develop after fertilisation into fruit or nuts). A family example is Fagoceae (beech family). Examples of genus are Fagus (beech), Quercus (oak) and Castanea (chestnut). Examples of Quercus species are petraea (European oak), mongolica (Japanese oak) and rubra (American red oak).

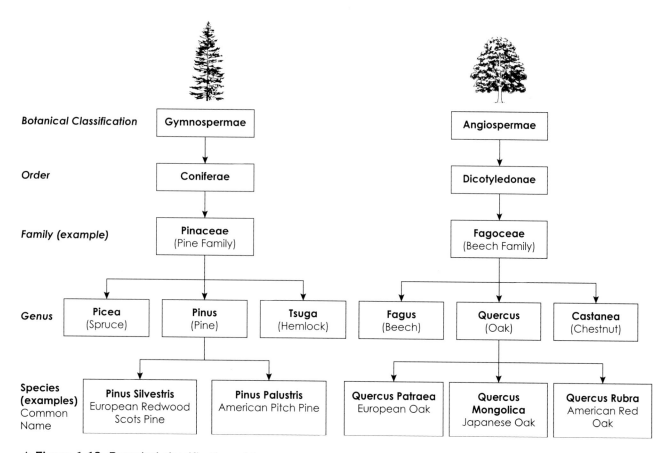

▲ **Figure 1.12** Botanical classification of trees

Only a few examples of family, genus and species are given. There are of course many more, the pine family alone has several hundred different species.

Sometimes timbers from a different genus or species having similar properties are sold together as a mixed parcel under the same common name, such as hem-fir, which is imported to the UK from North America, the hem being an abbreviation of hemlock or western hemlock and the fir being Amabilis fir.

In some cases it is not possible to give a commercial timber an individual botanical name, because it may consist of several similar species. In these cases the timber may be sold under its generic name followed by 'spp', for example oak in general may be referred to as Quercus spp.

Country of origin. This column in Table 1.1 lists the main areas of supply in the world for the most commonly used commercial timbers.

Colour. This normally relates to the heartwood, sapwood in general being much paler. The colour can vary considerably not only in the same species, but even in timber from the same tree. In use the colour will change due to light exposure (most will darken, though some do lighten) and the application of finishes (even clear finishes tend to darken the colour). In addition, unprotected timber used externally will discolour and may develop mould staining.

Density. This can vary depending on species and moisture content. Average density in kilograms per cubic metre (kg/m^3) is given for timbers with a moisture content of 15%. In many cases the density can vary by as much as 20% or more. Typically 0.5% of the stated weight can be added for every 1% increase in moisture content.

Durability. This is the natural durability of the timber and refers to the heartwood's resistance to fungal decay in use. The sapwood of most timber is classed as not durable or slightly durable and should not be used in exposed situations without preservative treatment. Each timber is put into one of five classes:

- Class 1: very durable (VD)
- Class 2: durable (D)
- Class 3: moderately durable (MD)
- Class 4: slightly durable (SD)
- Class 5: not durable (ND)

Treating with an appropriate preservative can increase the durability of timbers that are not naturally durable.

Treatability. This is a measure of the permeability or impregnability in relation to preservative penetration. Each timber is put into one of four levels of treatability:

- easy
- moderately easy
- difficult
- extremely difficult.

Complete preservative penetration is readily obtained in timbers classed as easy, whereas little more than a surface coating is possible with timbers that are classed as extremely difficult. However in general there is an inverse relationship between natural durability and treatability, e.g. timbers that are not durable tend to be more easy to treat and vice versa.

Moisture movement. This refers to the dimensional changes that occur when seasoned timbers are exposed to changes in atmospheric conditions particularly humidity. All timbers are classified into one of three groups: small (S), medium (M) and large (L).

This is not directly related to the initial shrinkage that occurs when green timber is seasoned to the equilibrium moisture content of its intended environment. The moisture movement of timbers used for structural purposes or those used in controlled humidities is not usually significant. However, species in the small movement class should be selected for use in situations where stability in varying humidities is required, such as intermittently heated buildings.

Working qualities. This refers to the ease of working, sawing, planing and fixing etc. All timbers are classified into one of three groups: good, medium and difficult.

Extra care is required when machining timbers classed as difficult, in order to produce an acceptable finished surface.

Availability. This will vary from time to time; some may not be readily available or only in limited quantities as a solid timber but can be obtained as a veneered face on sheet material. Timbers are listed under three general terms: regular, variable and limited.

Trading in certain species of timber is controlled by CITES, the Convention on International Trade in Endangered Species of Wild Flora and Fauna, which lists certain species into three grades or appendices that are regularly reviewed and amended:

- Appendix 1 lists species that are threatened with extinction, in which all trade is prohibited.
- Appendix 2 lists species, which are likely to become threatened if trade in them is not strictly controlled. An export certificate issued by the exporting country and an import permit issued by the UK must accompany any export.
- Appendix 3 lists species that have been noted by the country to which they are indigenous as being at risk. Export is permitted, but listing provides a means of controlling the amount.

As a result of CITES certain timbers will not be available or may become scarce. In most cases timber suppliers will be able to recommend similar alternatives that have been obtained from well-managed sustainable sources.

Price range. The price of timber is dependent on world market conditions, the method of purchase and the grade of timber required, but as a general guide prices are indicated as:

- High (over £1000/m^3)
- Medium (between £550 and £1000/m^3)
- Low (up to £500/m^3)

Characteristics. Comments on particular features such as grain, texture and figure:

- **Grain** refers to the general direction of the longitudinal cells relative to the axis of the tree.
- **Texture** refers to the size and distribution of the cells, which can be revealed by touch or reaction on machining. *Fine textured timber* has small, closely spaced cells; *coarse textured timber* has large cells; *even textured timber* has little variation in the size of cells and little difference between the early and latewood; *uneven textured timber* has considerable variation in the size of cells or a distinct contrast between early and latewood.
- **Figure** refers to the ornamental markings seen on the cut surface of timber.

Suitability. This depends on the timber's natural appearance, strength and durability, etc. A general guide to the typical uses of each timber is given in Table 1.1 but the list is not comprehensive as many timbers, given the right conditions or treatment, are suitable for other uses.

Conversion and seasoning of wood

The conversion of wood is the sawing up or breaking down of the tree trunk into various sized pieces of timber for a specific purpose.

Methods of conversion

Theoretically a tree trunk can be sawn to the required size in one of two ways, as illustrated in Figure 1.13, by sawing in a tangential or radial direction

The terms 'radial' and 'tangential' refer to the cut surfaces in relation to the growth rings of the tree. Both methods have their own advantages and disadvantages.

In practice very little timber is cut either truly tangentially or radially because there would be too much wastage in both timber and manpower.

To be classified as either a tangential or a radial cut the timber must conform to the following standards:

- **Tangential** (Figure 1.14). Wood is converted so that the annual rings meet the wider surface of the wood over at least half its width at an angle of less than 45°.
- **Radial** (Figure 1.15). Wood is converted so that the annual rings meet the wider surface of the wood throughout its width at an angle of 45° or more.

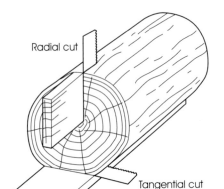

▲ **Figure 1.13** Timber conversion

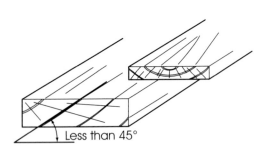

▲ **Figure 1.14** Tangential

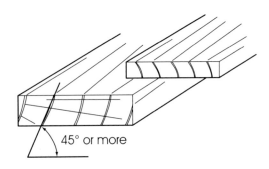

▲ **Figure 1.15** Radial

There are various methods of conversion to produce wood to these standards:

- **Through-and-through sawing** also known as *slash* or *slab sawing,* see Figure 1.16. This is the simplest and cheapest way to convert wood, with very little wastage. Approximately two-thirds of the boards will be tangential and one-third (the middle boards) will be radial. The majority of boards produced in this way are prone to a large amount of shrinkage and distortion.

- **Tangential sawing** is used when converting wood for floor joists and beams since it produces the strongest timber, see Figure 1.17. It is also used for decorative purposes on woods, which have distinctive annual rings, e.g. pitch pine and Douglas fir, because it produces *crown figuring* also known as *flame figuring* or *fiery grain* showing as arched top features on the face of the board, as illustrated in Figure 1.18.

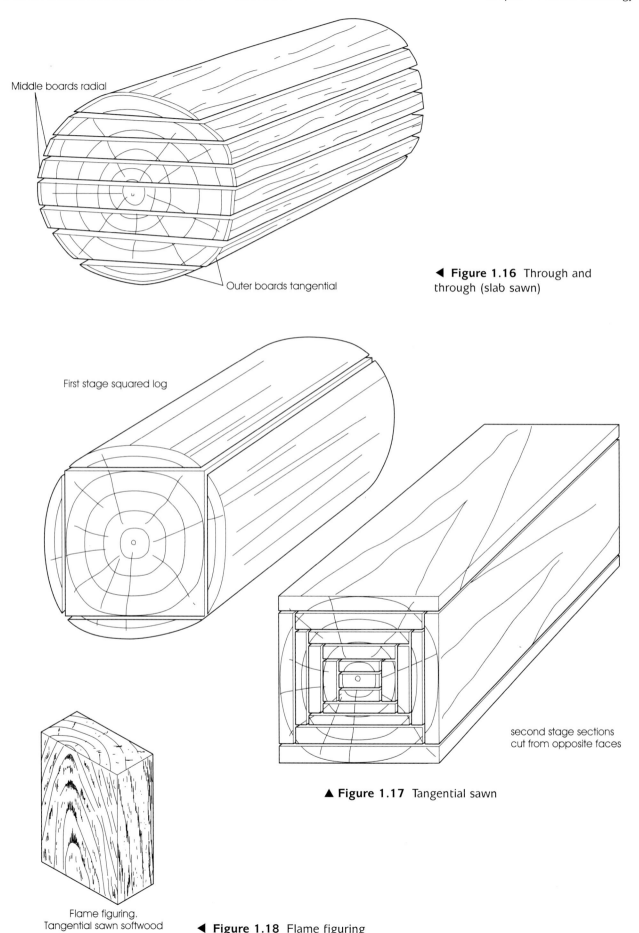

Middle boards radial

Outer boards tangential

◀ **Figure 1.16** Through and through (slab sawn)

First stage squared log

second stage sections cut from opposite faces

▲ **Figure 1.17** Tangential sawn

Flame figuring. Tangential sawn softwood

◀ **Figure 1.18** Flame figuring

■ **Quarter sawing** is the most expensive method of conversion, although it produces the best quality wood and is ideal for joinery purposes, see Figure 1.19. This is because the boards are radial sawn and have very little tendency to shrink or distort. In timber where the medullary rays are prominent such as Oak, the boards will have a silver figured finish; other timbers with interlocking grain such as African mahogany will show a stripe or ribbon figure. Both of these are illustrated in Figure 1.20.

■ **Boxed heart sawing.** When the heart of a tree is rotten or badly shaken the heart may be boxed in to keep the defect within the section. Both quarter sawn and tangential sawn timber may be produced using this method as illustrated in Figure 1.21.

■ **Rift sawn timber** falls between true tangential and radially sawn, having annual rings forming an angle between 30º and 60º to the wider face (Figure 1.22).

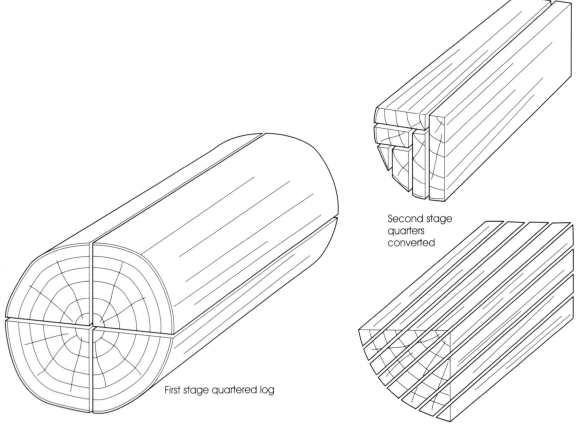

Second stage quarters converted

First stage quartered log

▲ **Figure 1.19** Quarter sawn

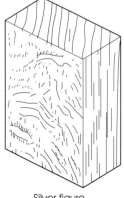

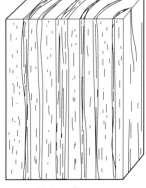

▶ **Figure 1.20** Decorative figure in quarter-sawn timber

Silver figure in species with prominent rays

Ribbon figure in species with interlocking grain

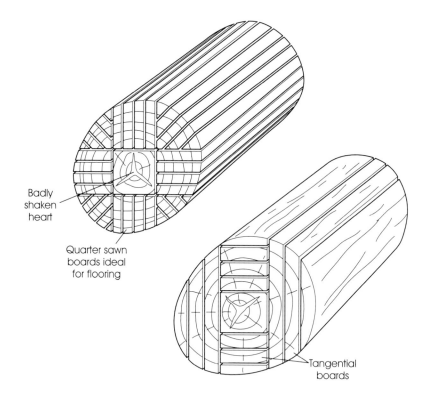

Badly shaken heart

Quarter sawn boards ideal for flooring

Tangential boards

◀ **Figure 1.21** Boxed heart conversion

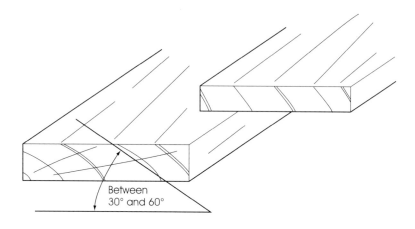

Between 30° and 60°

◀ **Figure 1.22** Rift sawn

Seasoning wood

Seasoning refers to the controlled drying by natural or artificial means of converted timber sections. There are many reasons why seasoning is necessary, the main ones being:

■ To ensure the moisture content of the timber is below the dry rot safety line of 20%.

■ To ensure that any shrinkage takes places before the wood is used.

■ Dry wood is easier to work with than wet wood.

■ Using seasoned wood, the finished article will be more reliable and less likely to split or distort.

■ In general, dry wood is stronger and stiffer than wet wood.

■ Wet wood will not readily accept glue, paint or polish.

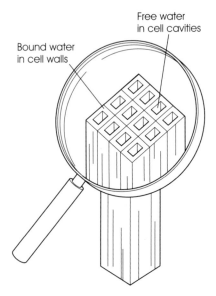

Bound water in cell walls

Free water in cell cavities

▲ **Figure 1.23** Highly magnified end grain section to show two forms of moisture

Moisture content

Moisture occurs in wood in two forms as illustrated in Figure 1.23:

■ *Free water* in the cell cavities
■ *Bound water* in the cell walls

When all of the free water in the cell cavities has been removed, the **fibre saturation point (FSP)** is reached. At this point the wood normally has a **moisture content (MC)** of between 25% and 30%.

Shrinkage only occurs when the moisture content of the timber is reduced below the fibre saturation point. The amount of shrinkage is not the same in all directions. The majority of shrinkage takes place tangentially, i.e. in the direction of the annual rings. Radial shrinkage is approximately half that of tangential shrinkage, while shrinkage in length is virtually non-existent and can be disregarded, see Figure 1.24. The results of shrinkage depend on the method of conversion used and is illustrated in Figure 1.25.

Equilibrium moisture content (EMC). Timber should be dried out to a moisture content approximately equal to the surrounding atmosphere in which it will be used. This moisture content is known as the equilibrium moisture content and, providing the moisture content and temperature of the air remains constant, the wood will remain stable and not shrink or expand. But in most situations the moisture content of the atmosphere will vary to some extent and sometimes this variation can be quite considerable, see Figure 1.26.

Wood fixed in a moist atmosphere will absorb moisture and expand. If it is then fixed in a dry atmosphere the bound moisture in the cells of the wood would dry out and the timber would start to shrink. This is exactly what happens during seasonal changes in the weather. Therefore all wood is subject to a certain amount of moisture-related movement and this must be allowed for in all construction and joinery work.

Calculating moisture content. Moisture content of wood is expressed as a percentage. This refers to the weight of the water in the wood compared to the dry weight of the wood, see Figure 1.27. In order to determine the average MC of a stack of wood, select a board from the centre of the stack, cut off the end 300 mm and discard it as this will normally be dryer than

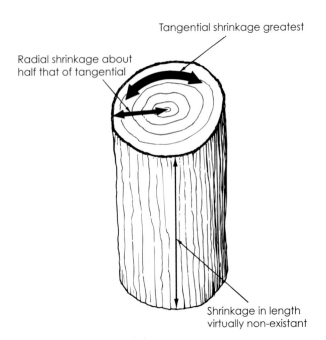

Tangential shrinkage greatest

Radial shrinkage about half that of tangential

Shrinkage in length virtually non-existant

▲ **Figure 1.24** Shrinkage

Even shrinkage

'Diamond' shrinkage

'Taper' shrinkage

'Cup' shrinkage

▲ **Figure 1.25** Results of shrinkage

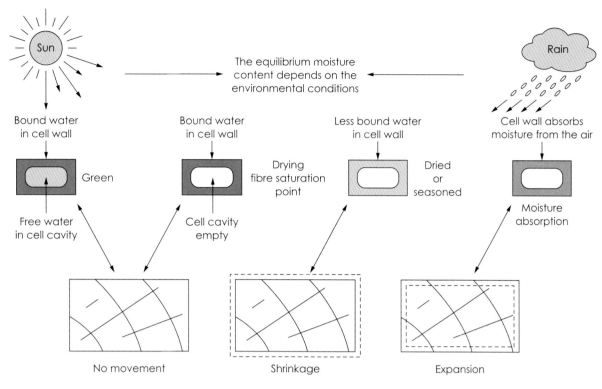

▲ **Figure 1.26** Fibre saturation and moisture equilibrium

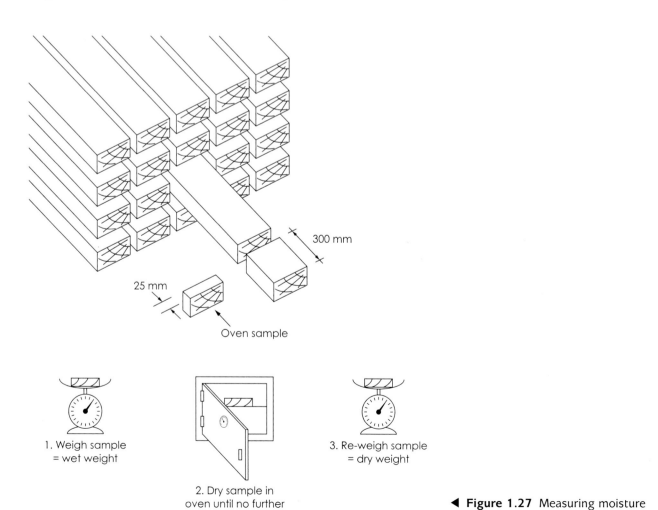

◀ **Figure 1.27** Measuring moisture content of timber

sections nearer the centre. Cut off a further 25-mm sample and immediately weigh it. This is the wet weight of the sample. Place this sample in a small drying oven and remove it periodically to check its weight. When no further loss of weight is recorded, take this to be the dry weight of the sample.

The moisture content of a piece of wood can then be found by using the following formula:

$$\text{Moisture content MC\%} = (\text{Wet weight} - \text{dry weight}) \times 100 \div \text{Dry weight}$$

Example

Wet weight of sample 50g
Dry weight of sample 40g
Moisture content $= (50 - 40) \times 100 \div 40 = 25\%$

An alternative way of finding the moisture content of wood is to use an electric **moisture meter** as illustrated in Figure 1.28. Although not as accurate, it has the advantage of giving an on-the-spot reading and it can also be used for determining the moisture content of wood already fixed in position. The moisture meter measures the electrical resistance between the two points of a twin electrode, which is pushed into the surface of the wood. Its moisture content can then easily be read off a calibrated dial, or LCD display.

▲ **Figure 1.28** Moisture meter

The ideal moisture content for timber fixed in relationship to various locations is illustrated in Figure 1.29. There is also a relationship between method of conversion, shrinkage and moisture content, which is illustrated in Figure 1.30. It can be seen that radial sawn timber shrinks less than tangentially sawn timber.

Seasoning process

Wood may be seasoned in one of two ways:

■ by natural means (air seasoning);
■ by artificial means (kiln seasoning).

Air seasoning

In this method the wood is stacked in open-sided, covered sheds, which protect the wood from rain whilst still allowing a free circulation of air.

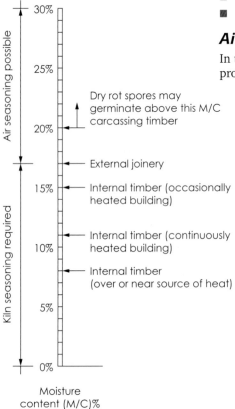

▲ **Figure 1.29** Moisture content for various locations

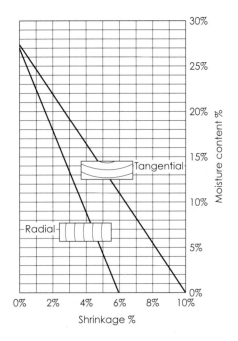

▲ **Figure 1.30** Shrinkage vs moisture content

In Britain, a moisture content of between 18% and 20% can be achieved in a period of two to twelve months, depending on the size of the wood.

A typical stack for air seasoning should conform to the following, see Figure 1.31:

■ Brick piers and timber joists keep the bottom of the stack well clear of the ground and ensure good air circulation underneath.

■ The boards are laid horizontally, largest at the bottom, smallest at the top, one piece above the other. This reduces the risk of timber distorting as it dries out.

■ The boards on each layer are spaced approximately 25 mm apart.

■ **Piling sticks** or 'stickers' are introduced between each layer of timber at approximately 600 mm intervals to support the boards and allow a free air circulation around them. The piling sticks should be the same type of wood as that being seasoned, otherwise staining may occur.

■ The ends of the boards should be painted or covered with strips of wood to prevent them from drying out too quickly and splitting.

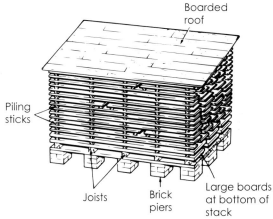

▲ **Figure 1.31** Timber stack or shed

Both softwoods and hardwoods can be seasoned in the same air-seasoning sheds. When the boards have been converted using the through-and-through method, they are often stacked in the same order as they were cut from the log, known as a **boule**, as illustrated in Figure 1.32.

Kiln seasoning

The majority of wood for internal use is kiln seasoned as this method, if carried out correctly, is able to safely reduce the moisture content of the wood to any required level without any danger of degrading (or causing defects). Although wood can be completely kiln seasoned, sometimes when a sawmill has a low kiln capacity, the timber is air seasoned before being placed in the kiln for final seasoning. The length of time the timber needs to stay in the kiln normally varies between two days and six weeks according to the type and size of wood being seasoned.

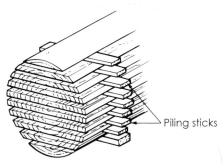

▲ **Figure 1.32**

There are two main types of kiln in general use: compartment kiln and progressive kiln. The **compartment kiln** is normally a brick or concrete building in which timber is stacked, see Figure 1.33. The timber will remain stationary during the drying process, while the conditions of the air are adjusted to the correct levels as the drying progresses. The timber should be stacked in the same way as that used for air seasoning.

The drying of wood in a compartment kiln depends on three factors:

■ *Air circulation* and extraction, which is controlled by fans.

■ *Humidity* (moisture content of the air). Steam sprays are used for raising the humidity in the early stages of the process to avoid rapid drying and distortion.

■ *Heat,* which is normally supplied by heating coils through which the vapour flows.

Air movement, humidity and heat are applied and regulated along the whole length of the compartment.

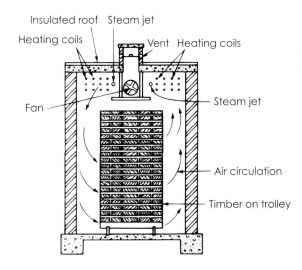

▲ **Figure 1.33** Compartment kiln

The **progressive kiln** can be thought of as a tunnel full of open trucks containing wood, which are progressively moved forward from the loading end to the discharge end. The drying conditions in the kiln become progressively more severe so that loads at different distances from the loading end are at different stages of drying.

Progressive kilns are mainly used in situations where there is a need for a continuous supply of wood of the same species and dimensions.

Drying schedules are available for different types of timbers. These set out the drying conditions required for a given size and type of timber. Although all types of timber require different conditions for varying lengths of time, the drying process in general involves three stages, these being:

- Wet timber stage: controls set to high steam, low heat.
- Timber drying stage: controls set to reduce steam, increase heat.
- Timber removal stage: controls set to low steam, high heat.

The seasoned timber can then be removed from the kiln almost dry.

Developments in seasoning

Other methods of seasoning are in limited use or are being developed, including the following:

- *Chemical or salt seasoning* where the timber is treated with hygroscopic salts before air or kiln seasoning. This encourages the moisture in the inner core to move outward while at the same time preventing the surface layers drying prematurely.

- *Press drying* can been used for the rapid seasoning of very permeable timber. It involves pressing the timber between two metal plates across which an electric potential is applied. This raises the temperature of the moisture up to boiling, when it escapes as steam.

- *Microwave energy* can be used to season timber. The centre core of the timber is excited by the microwave energy; at the same time cool air is circulated over its surface. This creates a temperature difference, which causes the moisture in the warmer centre core to move to the cooler surfaces.

- *Dehumidifying equipment* can be used in kilns to aid the seasoning process. Basically this involves forcing completely dried air through timber stacked in a drying chamber. This absorbs a great deal of moisture from the timber. The wet air is then dehumidified by a refrigeration and heating technique before being recirculated through the chamber. This process is repeated until the required moisture content is achieved.

- *PEG treatment* of timber can be carried out in place of seasoning, although only really suitable for small samples used for turning and carving work. It involves soaking green timber in a PEG (polyethylene glycol) solution. The PEG replaces the bound moisture in the cell walls and when set prevents any moisture movement.

- *Second seasoning* is rarely carried out nowadays, but refers to a further drying of high-class joinery work after it has been machined and loosely framed up, but not glued or wedged. The framed joinery is stacked in a store that has a similar moisture content to the building where it will finally be fixed. Should any defects occur during this second seasoning, which can last up to three months, the defective component can easily be replaced.

- *Water seasoning* is not seasoning, as we understand it, at all. It refers to logs which are kept under water before conversion in order to protect them from decay. This process is also sometimes used to wash out the sap of some hardwoods that are particularly susceptible to attack by the Lyctus beetle.

- *Conditioning* the prior storage of joinery and timber trim in the area where they are to be fixed, so that an equilibrium moisture content can be achieved before fixing; also the practice of brushing water into the mesh side of hardboard about twenty-four hours before fixing so that the board expands before fixing and subsequently shrinks on its fixings when installed. If this were not done expansion could take place, which would result in the board bowing or buckling.

Seasoning and conversion defects

Timber is subject to many defects, which should, as far as possible, be cut out during its conversion. These defects can be divided into two groups: timber processing/seasoning defects and natural defects

Timber processing/seasoning defects

Distortions. These may be the result of an inherent weakness in the timber or bad conversion, but often they develop during seasoning through poor stacking or uneven air circulation (Figure 1.34). Returning the timber to the kiln and subjecting it to a high humidity followed by restacking, placing heavy weights on top of the load and then reseasoning to a suitable drying schedule may straighten the timber out. Often the results of this reversal are of a temporary nature, the distortion soon returning.

- **Bowing** is a curvature along the face of a board, and often occurs where insufficient piling sticks are used during seasoning.
- **Springing** is a curvature along the edge of the board where the face remains flat. It is often caused through bad conversion or curved grain.
- **Winding** is a twisting of the board and often occurs in wood which is not converted parallel to the pith of the tree.
- **Cupping** is a curvature across the width of the board and is due to the fact that wood shrinks more tangentially than it does radially.

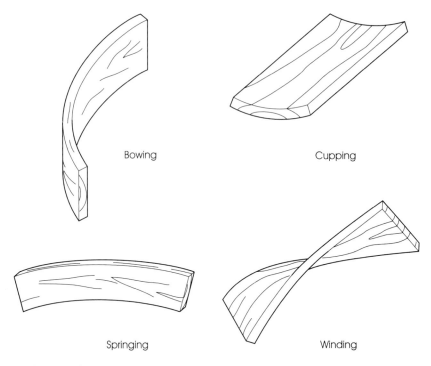

Bowing

Cupping

Springing

Winding

▲ **Figure 1.34** Distortions

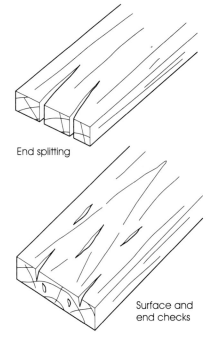

End splitting

Surface and end checks

▲ **Figure 1.35** Fissures

Shaking or splitting. These are **fissures** (separation of wood fibre) which develop along the grain of a piece of wood, particularly at its ends, and are the result of the surface or ends of the wood drying out too fast during seasoning (Figure 1.35). Termed **splits** where the fissure extends through the board from side to side and **checks** where seen on one face or at the end grain but do not extend through to the other side. Small surface checks often close up or, if shallow may be removed completely by planing. The term 'fissure' also includes resin pockets when they appear on the face or edge of a board.

Collapse. This is also known as **washboarding** and is caused by the cells collapsing through being kiln dried too rapidly (Figure 1.36). Collapse can

▲ **Figure 1.36** Collapse

rarely be reversed but in certain circumstances prolonged high-humidity treatment is successful.

Case hardening. This is also the result of too rapid kiln drying (Figure 1.37). In this case the outside of the board is dry but moisture is trapped in the centre cells of the wood. This defect is not apparent until the board is resawn when it will tend to twist and bind on the saw blade. Case hardening can sometimes be remedied if the timber is quickly returned to the kiln and given a high-humidity treatment followed by reseasoning. A simple test for case hardening can be carried out on an end grain sample about 25 mm thick; the centre is cut out to form a two-pronged 'U' shape; case hardening is indicted by the prongs closing up.

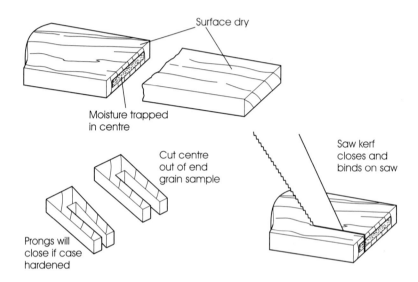

▶ **Figure 1.37** Case hardening

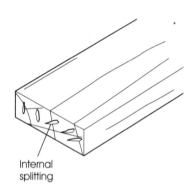

Internal splitting

▲ **Figure 1.38** Honeycombing

Honeycombing is characterised by internal splitting and may occur when the inner core of case-hardened timber subsequently dries out (Figure 1.38). No reversal of this defect is possible.

Stick staining is the result of using a different species of wood for the piling sticks to that being seasoned (Figure 1.39). It can sometimes be removed when boards are processed, but total success depends on the depth of penetration.

Natural defects

Shakes are caused by a separation of the wood fibres along the grain developed in the standing tree, on felling or prior to seasoning (Figure 1.40). The shake is formed as a result of stress relief, causing a longitudinal crack radiating from the heart and spreading through the diameter of the trunk:

- *Heart shakes* are cracks along the heart of a tree and are probably due to over maturity.
- *Star shakes* are a number of heart shakes that form an approximate star.
- *Radial shakes* are cracks along the outside of the log and are caused by the too rapid natural drying of the outside of the log before it is converted or by heavy felling onto hard ground
- *Cup shakes or ring shakes* are a separation between the annual rings and are normally the result of lack of nutriment. It is also said to be caused by the rising sap freezing during early spring cold spells.

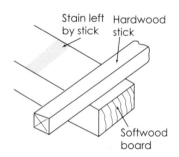

▲ **Figure 1.39** Stick staining

Waney edge is where the bark is left on the edge of converted wood (Figure 1.41).

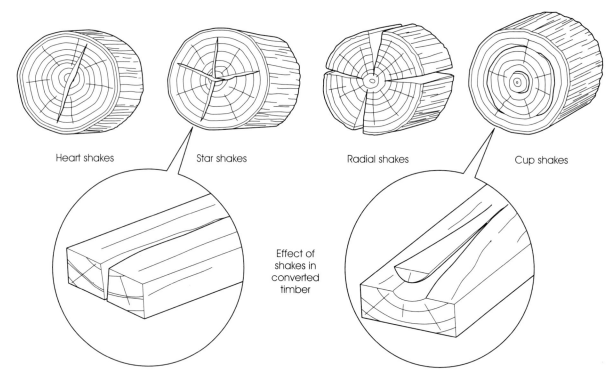

Heart shakes　　　Star shakes　　　Radial shakes　　　Cup shakes

Effect of
shakes in
converted
timber

▲ **Figure 1.40** Shakes

Knots are the end sections of branches where they grow out of the trunk
(Figure 1.42). Knots can be considered as being either *live knots*, i.e. knots
which are firm in their socket and show no signs of decay (also known as
sound or *tight knots*), or *dead knots*, i.e. knots which are separated from the
surrounding wood by the bark of the branch, have become loose in their
sockets or show signs of decay. The presence of knots on the surface of a
piece of wood often causes difficulties when finishing because of the
distorted grain that knots cause. Dead loose knots are also a potential
danger to the machinist as they may be picked up by the cutters and ejected
rapidly towards the operator.

Most **timber grading systems** use the size, location, and number of knots as
an indication of quality (Figure 1.43):

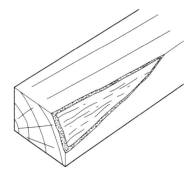

▲ **Figure 1.41** Waney edge

- *Knot size:* the larger the knot the greater the defect and the lower the
  quality.

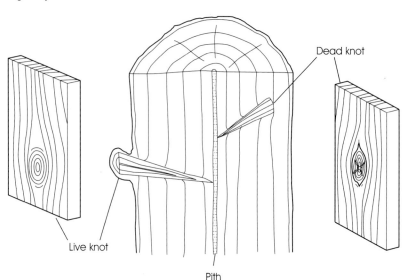

Dead knot

Live knot

Pith

◀ **Figure 1.42** Knots

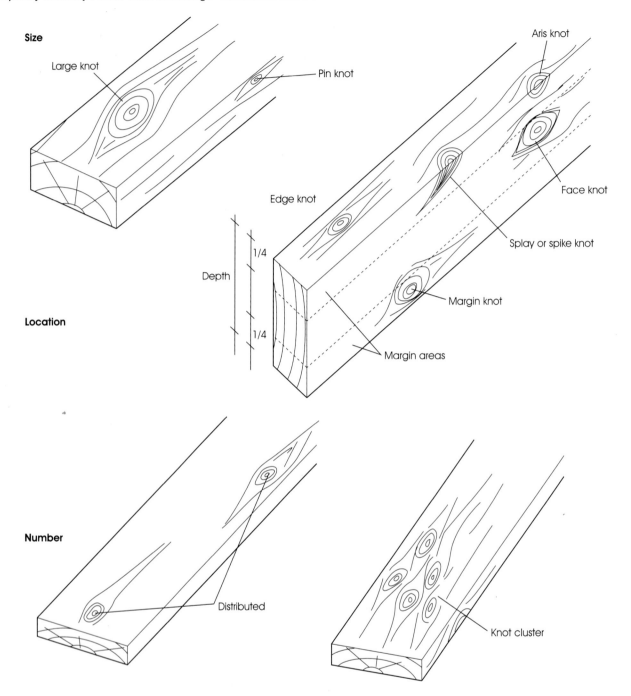

▲ **Figure 1.43** Knots and grading

■ *Knot location:* different names are given to knots depending on their position in a section of timber. Margin knots (those that appear in the upper and lower quarters of a section) reduce the strength of timber greater than those in the middle section.

■ *Number of knots and their distribution:* generally the greater the distance between knots the better the quality.

Although mainly considered to be a defect, knots are sometimes used to provide a decorative feature, e.g. knotty pine cladding.

Upsets. This is a fracture of the wood fibres across the grain also known as **thunder shake** (Figure 1.44). Lightning damage sometime during its growth can cause upsets but the severe jarring the tree receives when being felled is the main cause. This is a serious defect, most common in mahogany and is not apparent until the wood has been planed.

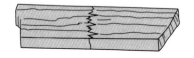

▲ **Figure 1.44** Upsets

Sloping grain is where the grain does not run parallel to the edge of the board and is often caused by bad conversion or by bent logs (Figure 1.45). When the sloping grain is pronounced, the defect is called **short graining**. This seriously affects the strength of the wood and it should not be used for structural work.

Reaction wood is produced in timber that has had to grow in a leaning position either on a slope or against strong prevailing winds (Figure 1.46). The tree will attempt to produce extra growth on the trunk to counteract the force of gravity or wind force. Any extra growth will result in an eccentric appearance around the pith on the trunk cross-section. Softwoods put the extra growth on the compressed (lower or squashed) side of the trunk and is therefore termed **compression wood**. Hardwoods put the extra growth on the tension (upper or stretched) side and is termed **tension wood**. Reaction wood is prone to serious distortion during seasoning, its grain is often woolly and applied finishes may appear patchy.

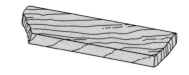

▲ **Figure 1.45** Sloping grain

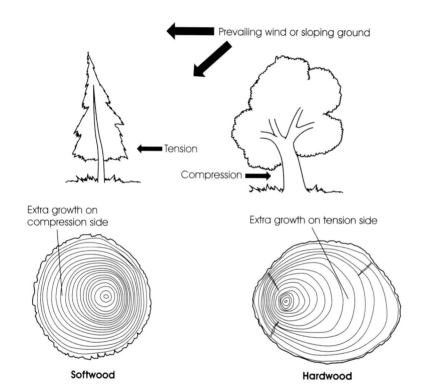

◀ **Figure 1.46** Reaction wood

Sap staining is also known as *blue sap stain* or *bluing*. This often occurs in felled logs while still in the forest. It can also occur in damp wood that has been improperly stacked close together without piling sticks or with insufficient air circulation. The stain is the result of a harmless fungus feeding on the contents of the sapwood cells. This fungus causes no structural damage. It is only normally considered as a defect in timber that requires a polished or clear varnish finish.

Protection of timber and joinery

The seasoning of timber is a reversible process. As stated previously, timber will readily absorb or lose moisture in order to achieve equilibrium moisture content. Care must therefore be taken to ensure the stability of the required moisture content during transit and storage.

Transit and storage

Conditions during transit and storage are rarely perfect, but by observing the points in the following checklist, materials wastage, damage and subsequent drying-out defects can be reduced to a minimum. Long-term preservation of timber and components is also essential. Figure 1.47 refers.

Checklist

1. *Plan all deliveries of timber and joinery* to coincide with the work programme, in order to prevent unnecessarily long periods of site storage.

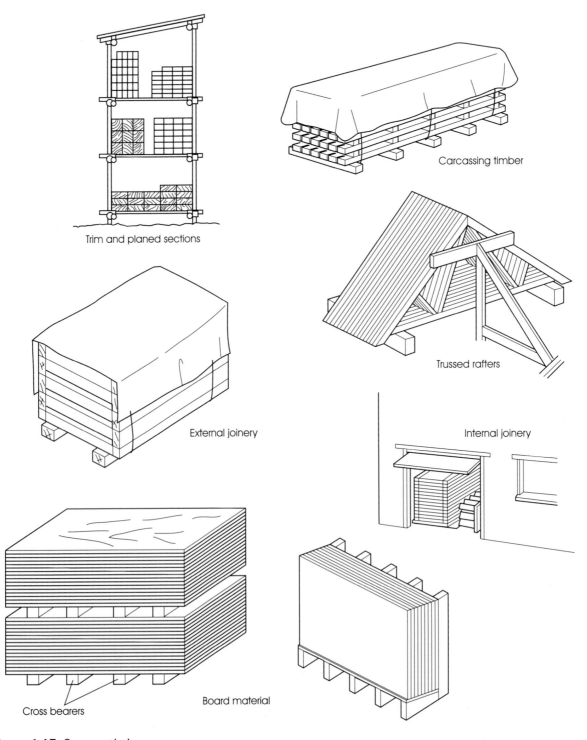

Trim and planed sections

Carcassing timber

External joinery

Trussed rafters

Internal joinery

Cross bearers

Board material

▲ **Figure 1.47** Storage timber

2. *Prepare suitable storage areas* in advance of deliveries:
   - Carcassing timber and external joinery should be stacked on bearers clear of the ground using piling sticks between each layer or item, and covered with waterproof tarpaulins as shown in Figure 1.47. The stack must be covered to provide protection from rainwater, snow and direct sunlight, etc. Care must be taken to allow a free air circulation through the stack thereby preventing problems from condensation that could form under the covering.
   - Trussed rafters can be racked upright and covered with a waterproof tarpaulin. Alternatively they could be laid flat on bearers.
   - Internal joinery items should be stacked in a dry, preferably heated store, using piling sticks where required. As far as practically possible the conditions in the store (humidity, temperature etc.) should be equal to those in which the material is to be used.
   - Where the materials are stored in the building under construction it should be fully glazed, heated and ventilated. The ventilation of the building is essential to prevent the build-up of high humidities, which could increase the moisture content of the timber.
   - All types of board material must be kept flat and dry. Ideally they should be stacked flat in conditions similar to internal joinery. Piling sticks can be used to provide air circulation between each sheet. Alternatively, where space is limited, board material can be racked on edge .

3. *Ensure timber and joinery items are protected* during transit. The supplier should make deliveries in closed or tarpaulin-covered lorries. Priming or sealing should preferably be carried out before delivery to the site or, where this is not possible, promptly thereafter. Many joinery suppliers now protect their products by vacuum sealing them in plastic coverings directly after manufacture.

4. *Build to the work programme.* Do not allow carcassing work to stand exposed to the weather for any longer than necessary:
   - Ensure glazing and roof tiling is complete before laying the flooring.
   - It is advisable to protect the floor after laying by completely covering with polythene sheeting or building paper.
   - Prevent moisture absorption from 'wet trades' by drying out the building before introducing kiln-dried timber and joinery components.

5. *Handle with care.* Careless or unnecessary, repeated, handling can cause extra costs through damaged material and even personal injury. A little planning and care can avoid both. (See Figure 1.48.)

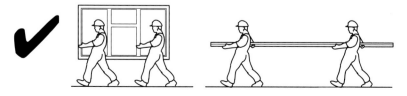

▲ **Figure 1.48** Handle with care

Enemies of timber

Timber, including structural, non-structural and timber-based manufactured items are the targets for biological attack, weathering, insect damage and fire damage. Given the right conditions, an attack by one or more of these agents is almost inevitable in the long term. Book 2, pages 411–419 refer.

There are two main types of fungi that cause biological decay in building timbers commonly known as dry rot and wet rot.

Biological attack – dry rot

Dry rot is the more serious type and is also more difficult to eradicate than wet rot. It is caused by a fungus (*serpula lacrymans*) that feeds on the cellulose found mainly in sapwood. This causes timber to:

■ lose strength and weight;
■ develop cracks in brick-shape patterns;
■ become so dry and powdery that it can easily be crumbled in the hand.

The appearance of a piece of timber after an attack of dry rot is shown in Figure 1.49

▲ **Figure 1.49** Timber after dry rot attack

Two initial requirements for an attack of dry rot in timber are:
■ Damp timber (i.e. timber with a moisture content above 20%).
■ Bad or non-existent ventilation (i.e., no circulation of air).

Given these two conditions and a temperature above freezing, an attack by dry rot is practically certain.

Stages of attack. An attack of dry rot occurs in three stages as illustrated in Figure 1.50:

■ **Stage 1.** The microscopic spores (seeds) of the fungus are blown about in the wind and are already present in most timbers. Given the right conditions, these spores will germinate on damp timber and send out hyphae (fine hair-like rootlets), which bore into the timber surface.
■ **Stage 2.** The hyphae branch out and spread through and over the surface of the timber forming a mat of cotton-wool-like threads called mycelium. It is at this stage that the hyphae can start to penetrate plaster and brickwork in search of new timber to attack. The hyphae are also able to conduct water and this enables them to adjust the water content of the new timber to the required level for their continued growth. Once the mycelium becomes sufficiently prolific a fruiting body (sporophore) will start to form.

■ **Stage 3.** The fruiting body, which is like a large fleshy pancake, with a white border and an orange-brown centre, starts to ripen (Figure 1.51). When fully ripe, the fruiting body starts to discharge into the air millions of rust-red spores which are ready to begin the process elsewhere.

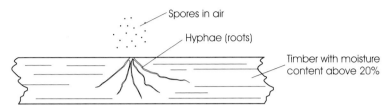

Stage 1: spores land on damp timber and send out hyphae

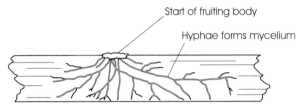

Stage 2: hyphae branch out and form mycelium and a fruiting body starts to grow

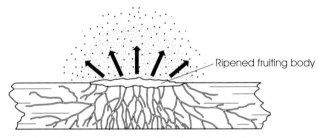

Stage 3: fruiting body ripens and starts to eject millons of spores into the air

◀ **Figure 1.50** Stages of a dry rot attack

◀ **Figure 1.51** Dry rot fruiting body

31

Prevention. As the two main factors for the growth of dry rot are damp timber and bad ventilation, by paying attention to the following points an attack of dry rot can be prevented:

- Always keep all timber dry (including before fixing into the building).
- Always ensure good ventilation. All constructional timbers should be placed so as to allow a free circulation of air around them.
- Always use well-seasoned timber.
- Always use preservative-treated timbers in unfavourable or vulnerable positions.

Recognition. In the early stages there is little evidence of a dry rot attack on the surface of the timber. Figure 1.52 shows the outward signs of an advanced dry attack to a timber suspended floor. However, often it's not until panelling, skirting boards or floorboards are removed that the full effect of an attack may be realised, as shown in Figure 1.53.

▶ **Figure 1.52** Advanced dry rot

▶ **Figure 1.53** Extensive dry rot exposed on removal of floor boards

When dry rot is suspected a simple test is to probe the surface of the timber with a small penknife blade. If there is little resistance when the blade is inserted there is a good possibility that dry rot is present. In addition to the other results of dry rot previously mentioned, a damp musty smell can also be taken as an indication of the presence of some form of fungal attack.

Eradication. By following the procedures in the order given, an attack of dry rot can be successfully eradicated:

1. Increase the ventilation and cure the cause of the dampness, which may be one or a combination of any of the following, see Figure 1.54:

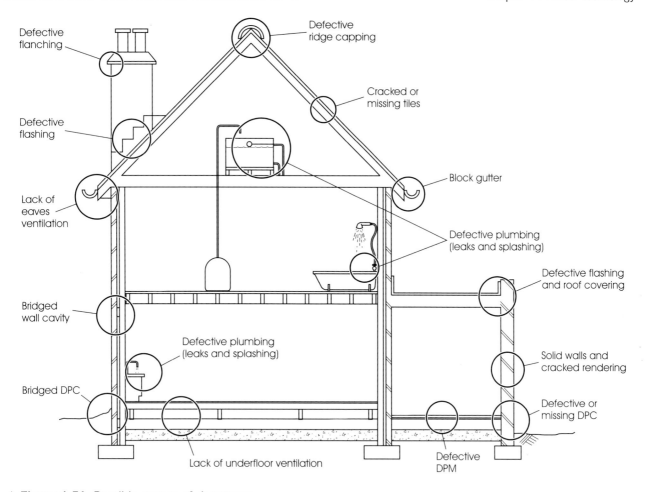

Defective flanching

Defective ridge capping

Cracked or missing tiles

Defective flashing

Block gutter

Lack of eaves ventilation

Defective plumbing (leaks and splashing)

Defective flashing and roof covering

Bridged wall cavity

Defective plumbing (leaks and splashing)

Solid walls and cracked rendering

Bridged DPC

Defective or missing DPC

Lack of underfloor ventilation

Defective DPM

▲ **Figure 1.54** Possible causes of dampness

- Cracked or missing tiles.
- Defective flashings to parapet walls and chimneys, etc.
- Defective drains and gullies.
- Defective, bridged or non-existent damp-proof course.
- Defective plumbing, including leaking gutters, downpipes, radiators, sinks, basins or WC, etc.
- Blocked, or an insufficient number of, air bricks.

2. Remove all traces of the rot. This involves cutting away all the infected timber and at least 600 mm of apparently sound wood beyond the last signs of attack, since this may also have been penetrated by the hyphae, see Figure 1.55.

3. All affected timber including swept-up dust, dirt and old wood shavings, etc. must be sealed in airtight polythene bags and arrangements made for their incineration (contact your local authority for information). This prevents spreading and kills hyphae and spores.

4. Strip the plaster from the walls at least 600 mm beyond the last signs of hyphae growth.

5. Clean off all brickwork with a wire brush and sterilise the walls by playing a blowlamp flame over them until the bricks are too hot to touch. While still warm brush or spray the walls with a dry rot fungicide. Apply a second coat when the first is dry.

6. Treat all existing sound timber with three coats of a dry rot preservative. This can be applied with brush or spray.

7. Replace all timber which has been taken out with properly seasoned timber, which has also been treated with three coats of dry rot

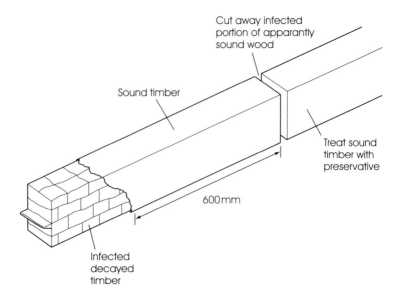

Cut away infected portion of apparently sound wood

Sound timber

Treat sound timber with preservative

600mm

Infected decayed timber

▶ **Figure 1.55** Cut-away infected timber

preservative, or timber which has been pressure impregnated with a preservative. All fresh surfaces, which have been exposed by cutting or drilling, must be re-treated with two flood coats of preservative applied with a brush.

Biological attack – wet rot

Wet rot, also known as **cellar rot**, is the result of the fungus Coniophora puteana, which is mainly found in wet rather than damp conditions such as:

- cellars;
- neglected external joinery;
- ends of rafters;
- under leaking sinks or baths;
- under impervious (waterproof) floor coverings.

Recognition. The timber becomes considerably darker in colour and has longitudinal cracks (along the grain). Very often the timber decays internally with a fairly thin skin of apparently sound timber remaining often hidden by paint on the surface. Use a penknife or bradawl to test suspect surfaces The hyphae of wet rot, when apparent, are yellowish but quickly turn to dark brown or black. Fruiting bodies, which are rarely found, are thin, irregular in shape and olive-green in colour. The spores are also olive-green. Figure 1.56 shows the results of wet rot in the rafters of a roof.

Eradication. Wet rot does not normally involve such drastic treatment as dry rot, as wet rot does not spread to adjoining dry timber. All that is normally required to eradicate an attack of wet rot is to cure the source of wetness. Where the decay has become extensive, or where structural timber is affected, some replacement will be necessary (see Book 2, page 430). Non-structural timber may be treated with a wet rot wood hardening fluid.

Weathering

In addition to wet rot, exterior timber is also subject to the effects of the weather. Exposure to sunlight can cause bleaching, colour fading and movement. Exposure to rainwater will cause swelling and distortion. Exposure to freezing causes moisture in the timber to expand, thus assisting in the break-up of the surface layers. Weathering results in repeated swelling and shrinkage of the timber or joinery item, often causing distortion, surface splitting and joint movement leading to a breakdown in the paint or other finish. Water will penetrate into the splits and joints, accelerating deterioration.

▲ **Figure 1.56** Wet rot in a roof

Weather conditions must be good when performing painting and repair work. As the main paint finish is intact there will be little ventilation to the actual surface of the timber to evaporate any moisture: the timber will therefore remain damp, almost inevitably leading to an attack by a wet rot fungi.

Wood-boring insects

The majority of damage done to building timber in the United Kingdom can be attributed to five species of insect or **woodworm** as they are more commonly called. These are illustrated in Figure 1.57. In addition to these five main types of wood-boring insects, several other species, such as the bark borer, the pinhole borer and the wharf borer may occasionally be found in building timbers, although they are only normally found in forests or timber yards and, in the case of the wharf borer, in water-logged timber.

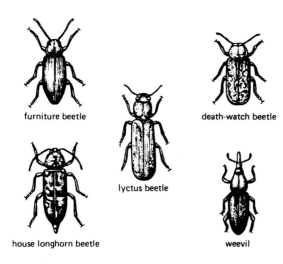

furniture beetle

death-watch beetle

lyctus beetle

house longhorn beetle

weevil

◄ **Figure 1.57** Wood-boring insects

Another outdoor pest – the wood-eating wasp – may disfigure softwood outbuildings and fences and reduce water-shedding properties.

Recognition of woodworm. It is easy to distinguish between an attack of wood-boring insects and other forms of timber decay by the presence of characteristic flight holes that appear on the surface. Also, when a thorough inspection is made below the surface of the timber, the tunnels or galleries bored by the larvae will be found. The adult beetles of the different species can be readily identified but, as these only live for a short period in the summer, the identification of the species is generally carried out by a diagnosis of the flight holes, bore dust and the type and location of the timber attacked. See Table 1.2.

Life cycle. The life cycle of all the wood-boring insects is a fairly complex process, but it can be divided up into four distinct stages as illustrated in Figure 1.58:

- ■ **Stage 1: Eggs.** The female insect lays eggs during the summer months, usually on the end grain or in the cracks and shakes of the timber, which afford the eggs a certain amount of protection until they hatch.
- ■ **Stage 2: Larvae.** The eggs hatch into larvae, or *woodworm,* between two and eight weeks after being laid. It is at this stage that damage to the timber is done. The larvae immediately start to bore into and eat the timber. The insects, while boring, digest the wood and excrete it in variously shaped pellets or bore dust, which is often used to identify the particular species of insect that is attacking the timber. The destructive stage can last between six months and ten years, according to the species.
- ■ **Stage 3: Pupae.** During the early spring, some larvae hollow out a pupal chamber near the surface of the timber, in which it can change into a pupa or **chrysalis**. The pupa remains inactive for a short period of time

Table 1.2 Characteristics of wood-boring insects

Species	Actual size	Bore dust	Location and timber attacked
Furniture beetle (*Anobium punctatum*) This is the most common wood-boring insect in the British Isles. Its life cycle is usually two to three years, with adult beetles emerging during the period between May and September. After mating the females usually lay between twenty and forty eggs each.	beetle flight holes	Small, gritty pellets which are egg-shaped under magnification	Attacks both hardwoods and softwoods, although heart-wood is often immune. Commonly causes a considerable amount of damage in structural timber, floorboards, joists, rafters and furniture
Death-watch beetle (*Xestobium rufavillosum*) Rarely found in modern houses, its attack being mainly restricted to old damp buildings, normally of several hundred years old, e.g. old churches and historic buildings. Named the death-watch because of its association with churches and the characteristic hammering noise that the adults make by hitting their heads against the timber. This hammering is in fact the adult's mating call. Its life cycle is between four and ten years, with adult beetles emerging during the period between March and June. Females normally lay between forty and seventy eggs each.		Coarse, gritty, bun-shaped pellets	Attacks old oak and other hardwoods. Can occasionally be found in softwoods near infested hardwoods. Mainly found in large-sectioned structural timber in association with a fungal attack
Lyctus or powder-post beetle In the British Isles there are four beetles in this group which attack timber. The most common is *Lyctus brunneus*. This species is rarely found in buildings, as it attacks mainly recently felled timber before it has been seasoned. Therefore it is only usually found in timber yards and storage sheds. Its life cycle is between one and two years, but is often less in hot surroundings. Adult beetles emerge during the period between May and September. Females normally lay 70–220 eggs each.		Very fine and powdery	Attacks the sapwood of certain hardwoods, e.g. oak, ash, elm, walnut, etc., normally before seasoning. But it has been known to attack recently seasoned timber. An attack is considered unlikely in timber over ten to fifteen years old.
House longhorn beetle (*Hylotupes bajulus*) This is by far the largest wood-boring insect found in the British Isles. It is also known as the Camberley beetle for its attacks are mainly concentrated around Camberley and surrounding Surrey and Hampshire areas. Its life cycle is normally between three and ten years, but can be longer. The adult beetles emerge during the period between July and September, with females laying up to 200 eggs each.		Barrel-shaped pellets mixed with fine dust	Attacks the sapwood of softwoods, mainly in the roof spaces, e.g. rafters, joists and wall plates, etc. Owing to its size and long life cycle, very often complete collapse is the first sign of an attack by this species. Complete replacement of timber is normally required
Weevils (*Euophryum confine* and *Pentarthrum huttoni*) These are mainly found in timber which is damp or subject to a fungal attack. Unlike other wood-boring insects, the adult weevils as well as the larvae bore into the timber and cause damage. Its life cycle is very short, between six and nine months. Two life cycles in one year is not uncommon. Adult beetles can be seen for most of the year. Females lay about twenty-five eggs each.		Small, gritty egg-shaped pellets, similar to bore dust of furniture beetles but smaller	Attacks both damp or decayed hardwoods and softwoods. Often found around sinks, baths, toilets and in cellars

in a mummified state. It is during this period that it undergoes the transformation into an adult insect.

■ **Stage 4: Adult insects.** Once the transformation is complete, the insect bites its way out of the timber, leaving its characteristic flight hole. This is often the first external sign of an attack.

Once the insect has emerged from the timber, it is an adult and capable of flying. The adult insect's sole purpose is to mate. This usually takes place within twenty-four hours. Very soon after mating, the male insect will die, while the female will search for a suitable place in which to lay her eggs, thus completing one life cycle and starting another. The female insect will normally die within fourteen days.

Prevention. The wood-boring insects feed on the cellulose and starch which is contained in all timber, both heartwood and sapwood, although sapwood is usually more susceptible to insect attacks. The only sure way of preventing an attack is to poison the food supply by pressure-treating the timber before it is installed in the building with a suitable preservative.

Eradication. By following the procedure in the order given, an attack of wood-boring insects can be successfully eradicated:

1. Open up the affected area, for example, take up floorboards or remove panelling and carefully examine all structural timber.
2. Remove all badly affected timber. Note: the affected timber, which has been removed including swept-up dust, dirt and old wood shavings etc. must be sealed in airtight polythene bags for incineration.
3. Strip off any surface coating or finish on the timber, for example, paint, polish or varnish, as the fluid used to eradicate the woodworm will not penetrate the surface coating.
4. Thoroughly brush the timber in order to remove all surface dirt and dust.
5. Replace all damaged timber with properly seasoned timber that has been pressure-impregnated with a preservative. All fresh surfaces that have been exposed by cutting or drilling must also be retreated with two flood coats of preservative applied by brush.
6. Apply a proprietary woodworm killer, by brush or spray, to all timber, including that which is apparently unaffected. Pay particular attention to joints, end grain and flight holes. Apply a second coat of the fluid as soon as the first has been absorbed. Floorboards must be taken up at intervals to allow thorough coverage of the joists and the underside of the floorboards. Care must be taken to avoid the staining of plaster, particularly when treating the ceiling joists and rafters in the loft.
7. In most cases it is possible to completely eradicate an attack by one thorough treatment as outlined in steps 1 to 6, but to be completely sure inspections for fresh flight holes should be made for several successive summers. If fresh holes are found, retreat timber with a woodworm killer.

Treatment of large-section timber

To eradicate an attack of wood-boring insects in large-section structural timber, deeper preservative penetration is required. A timber injection system is available. This involves drilling holes into the timber at intervals, inserting a nozzle and then pumping preservative under pressure into the timber. This method is also suitable for the in-situ treatment of external joinery, etc.

Alternatively, a paste preservative can be used for structural timbers. The paste, like thick creamy butter, is spread over all the exposed surfaces of the affected timber. Over a period of time the paste releases toxic chemicals which penetrate deep into the timber.

Note: Neither of these methods impart structural strength to the timber; therefore replacement of badly affected timbers is still required.

Stage 1

female lays eggs in shakes, etc.

Eggs

Stage 2

Eggs hatch and larvae start boring into the timber

Larvae

Stage 3

Larvae change into pupae near the surface of the timber

Pupae

Stage 4

Adult beetles emerge from flight/holes and mate

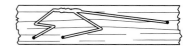

Adult insects

▲ **Figure 1.58** Life cycle of wood-boring insects

Treatment of furniture

When a wood-boring insect has attacked timbers inside a building, it is almost certain that the furniture will also be affected. Therefore, to successfully eradicate the attack, the furniture must also be treated using the following stages in the order given.

1. Remove all dirt and dust, then inject a proprietary woodworm killer into the flight holes. A special aerosol can with tube, or nozzled injector bottles, are available for this purpose.
2. Apply two coats of a proprietary woodworm killer to all unfinished surfaces, that is, all surfaces which are not painted, polished or varnished.
3. Make inspections for fresh flight holes for several successive summers. Repeat if required.

Preservative pre-treatment of timber

All timbers, especially their sapwood, contain food on which fungi and insects live. The idea behind timber preservation is to poison the food supply by applying a toxic liquid to the timber.

There are three main types of timber preservative available:

■ *Coal tar oils* are derived from coal and are dark brown or black in colour. They are fairly permanent, cheap, effective and easy to apply. However they are only for external use and should not be used internally as they are inflammable and possess a strong lingering odour. They should never be used near food as odour will contaminate. The timber once treated will not accept any further finish: it cannot be painted or glued. Its main use is for the treatment of external timber such as fences, sheds, telegraph poles, etc. **Note:** Due to is carcinogenic risk (substance that can produce cancer) coal tar creosote is not available for public use and strict controls are in place for both professional use of the preservative and timber treated with it.

■ *Water-soluble preservatives* are toxic chemicals, which are mixed with water. Unitl recently CCA (copper, chrome and arsenic) was the main type in general use. Following a European directive CCA preservative is now restricted to professional industrial use. Due to the arsenic content it can no longer be used for domestic, residential or salt-water marine environments. In these situations an alkaline copper preservative, which is arsenic free can be used as an alternative. They are suitable for use in both internal and external situations and are odourless and non-flammable. Their main disadvantage is that being water based they tend to swell the timber and will cause corrosion of metals until drying out is complete. However after re-drying following treatment, the wood can be readily painted and glued. As the toxic chemicals are water-soluble some of the types available are prone to leaching out when used in wet or damp conditions.

■ *Organic solvent preservatives* consist of toxic chemicals (chloronaphthalenes, metallic naphtanates and pentachlorophenol), which are mixed with a spirit that evaporates after the preservative has been applied. This is an advantage because the moisture content of the timber is not increased and thus does not cause swelling or have a corrosive effect on metals. The use and characteristics of these types of preservative are similar to those of water-soluble preservatives, but with certain exceptions. Some of the solvents used are flammable so care must be taken when applying or storing them. Some types also have a strong odour. In general, organic solvent preservatives are the most expensive type to use but are normally considered to be superior because of their durable preservation properties.

Methods of application

To a large extent it is the method of application rather than the preservative that governs the degree of protection obtained. This is because each method of application gives a different depth of preservative penetration. The greater the depth of penetration the higher the degree of protection (Figure 1.59).

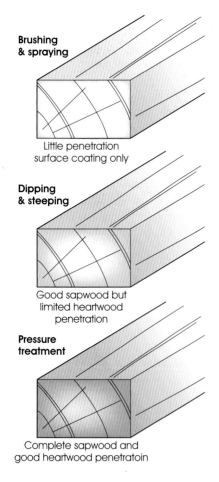

Brushing & spraying

Little penetration surface coating only

Dipping & steeping

Good sapwood but limited heartwood penetration

Pressure treatment

Complete sapwood and good heartwood penetratoin

▲ **Figure 1.59** Depth of preservative penetration is dependent on method of application

Preservatives can be applied using a number of methods but all of these can be classed in two groups.

Group 1. Non-pressure treatment (Figure 1.60) where the preservative is applied by brushing, spraying or dipping:

- *Brushing and spraying* can be used for most types of preservative but the effect is very limited as only a surface coating is achieved.
- *Dipping or immersing* the timber into a tank of preservative gives a greater depth of penetration than brushing or spraying, which can be further increased by a process known as steeping. The depth of penetration is dependant on the length of time that the timber is immersed.
- *Steeping* involves heating the preservative with the timber in the tank, also known as the hot and cold method. As the preservative is heated, the air in the cells expands and escapes as bubbles to the surface. On cooling, the preservative gets sucked into the spaces left by the air, giving a fairly good depth of penetration, making steeping by far the best non-pressure method.
- *Diffusion.* Freshly felled green timber is often immersed in a tank of preservative and on removal covered for a period of time to prevent drying and allow the preservative to diffuse into the timber.

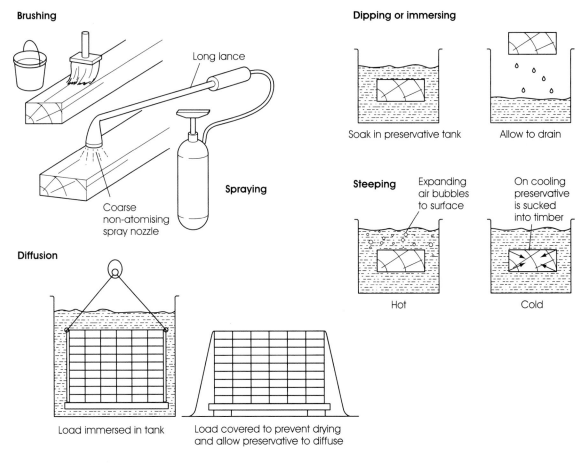

▲ **Figure 1.60** Non-pressure preservative treatment

Group 2. Pressure treatment (Figure 1.61) is the most effective form of timber preservation as almost full penetration is achieved. Suppliers with specialist compressed air or vacuum equipment carry out this process:

- *Empty-cell process.* The timber is placed in a sealed cylinder. The air in the cylinder is then subjected to pressure, which causes the air in the timber cells to compress. At this stage preservative is run into the cylinder and the pressure increased further. This forces the preservative

▶ **Figure 1.61** Pressure preservative treatment

into the timber. The pressure is maintained at this high level until the required amount of penetration is achieved. The pressure is then released and the surplus preservative is pumped back into a storage container. As the air pressure is reduced, the compressed air in the cells expands and forces out most of the preservative, leaving only the cell walls coated.

- *Full-cell process.* The timber is placed into the sealed cylinder as before but this time, instead of compressing the air, it is drawn out. This creates a vacuum in the cylinder, as well as a partial one in the cells of the timber. At this stage the preservative is introduced into the cylinder. When the cylinder is full, the vacuum is released, and the preservative is sucked into the timber cells by the partial vacuum. This method is ideal for timbers which are to be used in wet locations, e.g. marine work, docks, piers, jetties, etc. as water cannot penetrate into the timber cells because they are already full of preservative.

- *Double vacuum.* The principles of this method are similar to the full cell process except that a final vacuum is applied to draw out the excess preservative, leaving empty cells with a surface coating.

Preservative safety

- Always follow the manufacturer's instructions with regard to use and storage.
- Avoid contact with skin, eyes or clothing.
- Avoid breathing in the fumes, particularly when spraying.
- Keep away from food to avoid contamination.
- Always wear personal protective equipment (PPE): barrier cream or disposable protective gloves; eye protection; dust mask or a suitable respirator.
- Do not smoke or use near a source of ignition.
- Ensure adequate ventilation when used indoors.
- Thoroughly wash your hands before eating and after work with soap and water or an appropriate hand cleanser.
- In the case of accidental inhalation, swallowing or contact with the eyes, medical advice should be sought immediately.

Fire protection of timber

When timber is exposed to a flame (Figure 1.62), its temperature will not increase above 100°C until the majority of its moisture has evaporated. This causes a check in the spread of flame. When all of the moisture has been dried out of the timber, the temperature will rise to approximately 300°C centigrade, causing a chemical change to take place. Flammable gases are given off which immediately ignite, causing the timber to burn and gradually disintegrate leaving charcoal.

Research has proved that the charring rate is little affected by the intensity of the fire, thus enabling the strength carrying capacity of a structural timber member to be calculated after a known period of exposure to a fire. Structural timber members can be designed oversize with a layer of

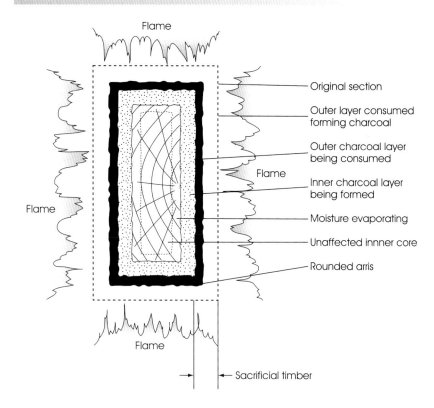

Flame

Original section

Outer layer consumed
forming charcoal

Outer charcoal layer
being consumed

Flame

Inner charcoal layer
being formed

Moisture evaporating

Unaffected innner core

Rounded arris

Flame

Flame

Flame

Sacrificial timber

◀ **Figure 1.62** Effect of fire on timber

(sacrificial) timber to allow for the expected loss in dimensions after a particular period of fire.

Fire resisting treatment

No amount of treatment can make timber completely fireproof, but treatment can be given to increase the timber's resistance to ignition and, to a large extent, stop its active participation in a fire.

There are two types of treatment that can be used:

- Impregnation with a fire resisting or flame retardant chemical, which when heated in a fire gives off a vapour that will not burn. Typical chemicals include aqueous solutions of monammonium phosphate, ammonium chloride and boric acid. Preservatives can also be included in the mix to provide a dual protection. These can be applied by steeping, but pressure impregnation is considered the most effective. Such treatment may significantly reduce the timber's strength due to cell wall damage if the timber is not correctly dried after treatment.
- Surface coating with an intumescent paint or varnish, which on heating in a fire bubbles and expands to form an insulating layer that cuts off the fire's oxygen supply. This does not adversely affect the timber's strength properties.

Purchasing timber

The terms 'wood' or 'timber' are taken to mean sawn or planed pieces in their natural or processed state. Different sections are available depending on their intended end use (Figure 1.63).

Log

This normally refers to a debarked tree trunk. Main uses:

- Hardwoods are often shipped in this form for conversion in United Kingdom saw mills.
- Poles or posts for agricultural, garden, dock and harbour work.

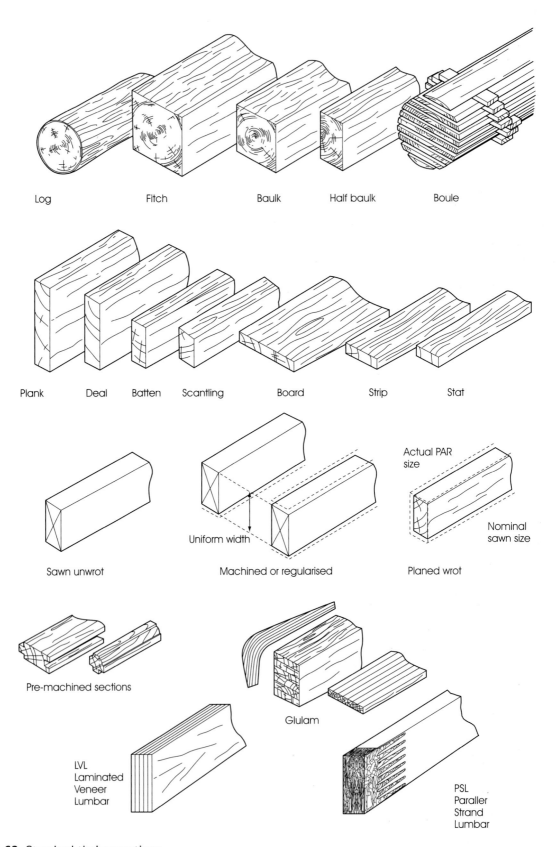

Log   **Fitch**   **Baulk**   **Half baulk**   **Boule**

Plank   **Deal**   **Batten**   **Scantling**   **Board**   **Strip**   **Stat**

Sawn unwrot

Uniform width

Machined or regularised

Actual PAR size

Nominal sawn size

Planed wrot

Pre-machined sections

Glulam

LVL Laminated Veneer Lumbar

PSL Paraller Strand Lumbar

▲ **Figure 1.63** Standard timber sections

Terminology:

- **Flitch.** A large square log.
- **Baulk.** A squared log over 100 mm × 125 mm.
- **Half baulk.** Any baulk sawn through its section in half. Main uses:
  - reconversion into smaller sections;
  - dock and harbour work, shoring, hoarding, beams and posts in heavy structural work.
- **Boule.** A through-and-through sawn log re-stacked into its original log form. Each piece having waney edges. Main uses:
  - mainly hardwood for re-conversion into smaller sections;
  - rustic cladding and garden furniture.

Sawn or planed sections

Sawn wood may be termed as unwrot (unworked) and planed as wrot (worked). These may be further defined as:

- **Plank.** 50 mm to 100 mm in thickness and 250 mm or more in width.
- **Deal.** 50 mm to 100 mm in thickness and 225 mm to 300 mm in width. Also used to refer to a species of timber e.g. red deal for redwood.
- **Batten.** 50 mm to 100 mm thick and 100 mm to 200 mm in width. Also a term loosely used in the trade for timber up to 25 mm by 50 mm in section and in certain parts of the country, to refer to a scaffold board.
- **Scantling.** 50 mm to 100 mm thick and 50 mm to 125 mm in width.
- **Board.** Under 50 mm thick and over 100 mm in width.
- **Strip.** Under 50 mm thick and under 100 mm in width.
- **Slat or slatting.** 25 mm and under in thickness and 100 mm and under in width.
- **Die squared.** A baulk dimension stock wood that has been converted to standard sizes.

Main uses:

- Sawn and regularised/machined sections for carcassing, first-fixing and other general construction work. Also for joinery, furniture and cabinetwork for reconversion and further processing.
- Planed sections; finished sections in carpentry; first and second fixing; also in joinery, furniture and cabinetwork for reconversion and/or further processing.

Pre-machined timber

These are planed sections that have been further processed for a particular end use. Main uses:

- Carpentry flooring, cladding, skirting, architrave and window boards, etc.
- Joinery sections for window, door and other components.
- Picture frames and other general moulding for furniture and cabinetwork.

Fabricated timber

- *Glue-laminated timber*, also known as **glulam**, is timber formed by gluing a number of small sections together, to form large cross-sections or longer lengths. Main uses:
  - structural work especially where large sections, long spans or shapes are required;
  - joinery, furniture and cabinetwork, especially for unit construction, shelving, counter tops, tabletops and other worktops.
- *Laminated veneer lumber (LVL)* is made from thin sheets of rotary cut veneers (like plywood veneers but thicker). The veneers are first glued together to provide the required thickness and then sawn into a range of structural sectional sizes.

■ *Parallel strand lumber (PSL)* is made from long strands of timber (similar to those used for OSB orientated strand board) bonded together with a synthetic resin in a heated press to form a continuous structural size section. These are then cross-cut into standard lengths for end use as joist and beams, etc.

Available sizes

Sawn softwoods are available in a wide range of **preferred sectional sizes** (the most usual sizes available in Europe) as illustrated in Figure 1.64, which shows the basic sizes (sawn size) at a moisture content of 20%.

The *permitted deviations* (difference between actual size and target size) for sawn timber are:

–1 mm to +3 mm for thicknesses and widths up to 100 mm

–2 to +4 mm for thicknesses and widths over 100 mm

The sum of permitted deviations is referred to as the **tolerance.**

Regularised or machined softwoods. Sawn or planed on one or both edges to give a consistent depth for structural use (such as floor joists) have a **target size** (the desired size after a production process) of 3 mm less than the sawn size for sections up to 150 mm deep and 5 mm less for those over 150 mm in depth (Figure 1.65).

The *permitted deviations* for regularised or machined timber are:

–1 mm to +1 mm for depths up to 100 mm

–1.5 mm to +1.5 mm for depths over 100 mm

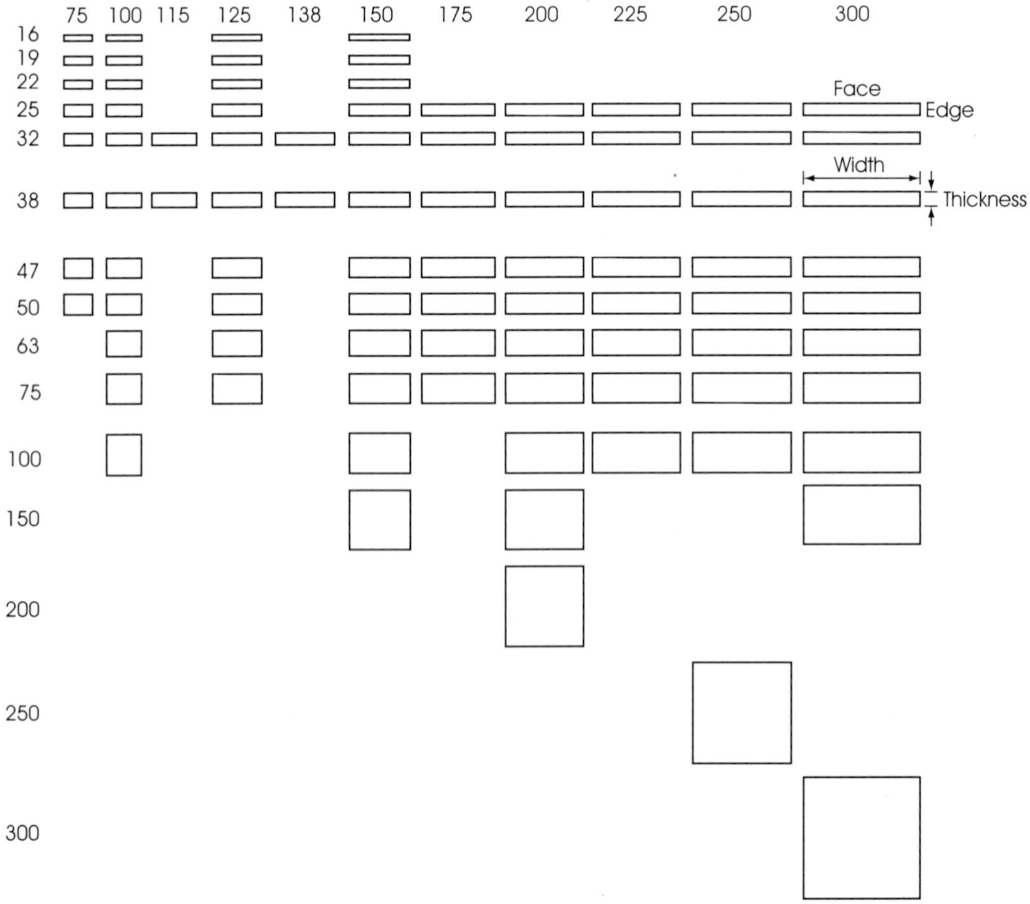

▲ **Figure 1.64** Typical range of sawn softwood sizes

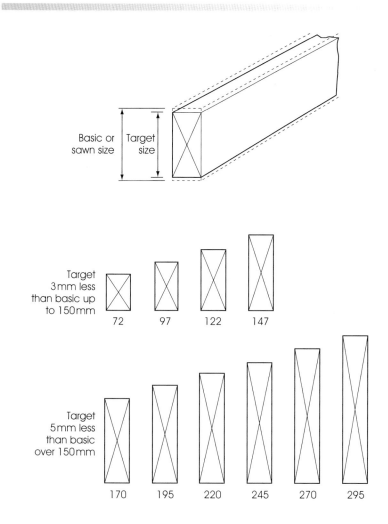

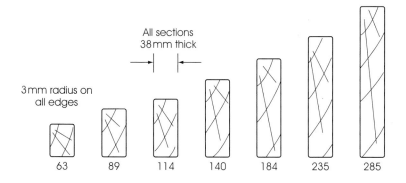

◀ **Figure 1.65** Target sizes for structural softwoods

Surfaced timber known as CLS or ALS (Canadian or American Lumber sizes/stock) is a planed timber with rounded arrises of up to 3 mm radius for ease of handling, which is used for structural purposes including joists, timber frame panels and stud partitioning. Available in a thickness of 38 mm and widths of 63 mm, 89 mm, 114 mm, 140 mm, 184 mm, 235 mm and 285 mm (Figure 1.66).

◀ **Figure 1.66** Range of widths for surfaced timber (ALS/CLS)

Planed or **prepared timber** is sawn timber that has been machined for its full length and width on at least one face to provide a smooth surface. Prepared timber may also have been cut to length. Normally referred to by its **Ex** (out of) nominal or sawn size. The following reductions apply to the sawn section:

3 mm less than the basic sawn section up to 150 mm

5 mm less than the basic sawn section over 150 mm

The *permitted deviation* for planed timber is −0 mm to +1 mm. Various terms associated with planed timber are illustrated in Figure 1.67.

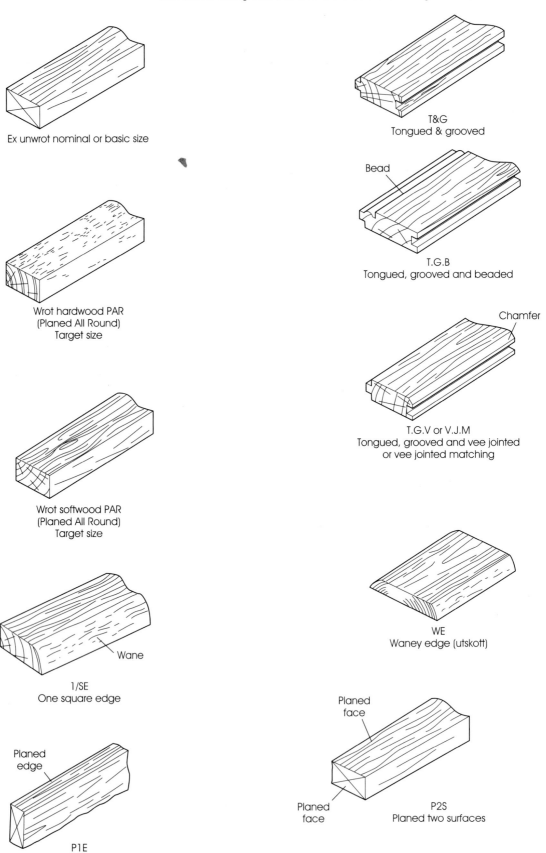

▲ **Figure 1.67** Selection of timber terms and abbreviations

Effect of moisture content

The previous sectional sizes relate to timber with a moisture content of 20%. The change in sectional size due to changes in moisture content can be calculated using the following:

0.25% increase in width and thickness for each 1% increase in moisture content over 20% and up to 30% (no further increase in size occurs above 30%)

0.25% reduction in width and thickness for each 1% reduction in moisture content below 20%

Standard sizes

Softwood lengths are in 300 mm increments starting at 1.8 m up to 7.2 m, although lengths above 5.7 m are limited and lengths of 6 m and over may not be available without finger jointing (Figure 1.68). No minus tolerance in length is permitted; however, plus tolerances can be agreed at time of order.

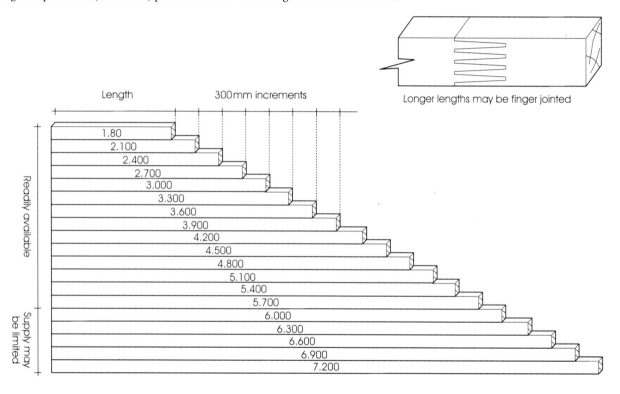

▲ **Figure 1.68** Available softwood lengths

Hardwood sections are quoted in metric or imperial thickness depending on the source (Figure 1.69).

$\frac{1}{2}$ inch or 13 mm

$\frac{3}{4}$ inch or 19 mm

1 inch or 25 mm

$1\frac{1}{4}$ inch or 32 mm

$1\frac{1}{2}$ inch or 38 mm

2 inch or 50 mm

$2\frac{1}{2}$ inch or 63 mm

3 inch or 75 mm

4 inch or 100 mm

Thicker sections may be available rising in 1 inch or 25 mm stages. Widths are in accordance with the grade, normally 3 inches or 75 mm and wider,

47

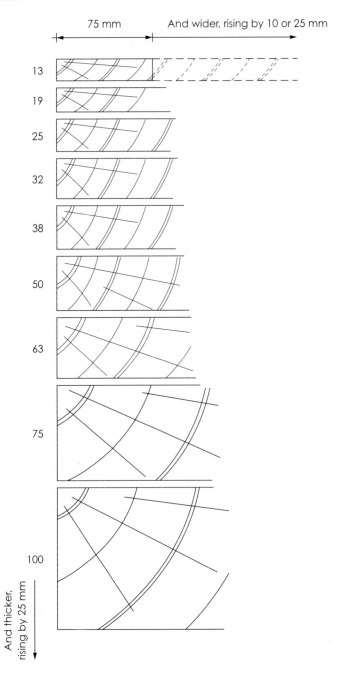

75 mm And wider, rising by 10 or 25 mm

13
19
25
32
38
50
63
75
100

And thicker, rising by 25 mm

▲ **Figure 1.69** Standard hardwood sawn sizes

rising in increments of 10 or 25 mm. The width of waney edge pieces are measured at three points along the narrow face including half the wane. The stated width is the average of the three measurements (Figure 1.70).

Hardwood lengths are in 100 mm increments and may be described as either (Figure 1.71):

- *Normal lengths:* 1.8 m and above rising in 100 mm increments.
- *Short lengths* or *shorts:* 1.7 m and less falling by 100 mm increments.

Grading of wood

Wood is available graded to give an indication of its appearance or strength. Softwoods and hardwoods are graded in different ways and the rules governing grading systems vary from country to country.

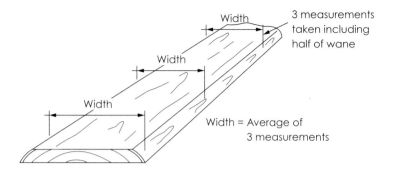

Width
Width
Width
3 measurements taken including half of wane
Width = Average of 3 measurements

◀ **Figure 1.70** Measuring waney-edged boards

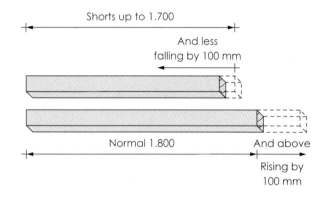

Shorts up to 1.700

And less falling by 100 mm

Normal 1.800

And above

Rising by 100 mm

◀ **Figure 1.71** Standard hardwood lengths

Softwood appearance grading

Softwood from *European sources* are generally graded into six numbered groups, I, II, III, IV, V and VI, according to the amount of defects they contain (Figure 1.72). First being the best quality through to sixth being the poorest quality.

■ Grades 'I, II, III and IV' are rarely available separately. They are normally marketed together as U/S or unsorted, containing a mixed batch of grades I to IV. This is the normal grade specified for joinery work.

■ The 'V' grade is available separately and is used for general non-structural construction and carcassing work.

■ The 'VI' grade is a low quality wood used mainly for packaging. It is available separately and is often known as 'Utskott', a general term meaning waney edged boards.

■ Other parcels of timber may be described as 'saw felling quality', this being a mixed batch of grades I to V sold without further sorting, but the majority will consist of IV and V grades with very little of I, II and III grades. Some suppliers market this as 'V and better' or 'fifths and better'.

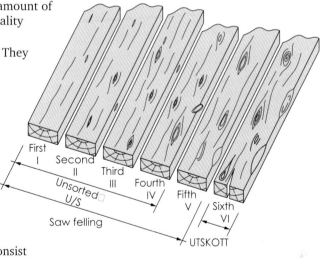

First I
Second II
Third III
Fourth IV
Fifth V
Sixth VI
Unsorted U/S
Saw felling
UTSKOTT

▲ **Figure 1.72** Softwood appearance grades

The *Russian Federation* uses a similar system. But only have five numbered grades I, II and III being marked as unsorted, IV and V being equivalent or slightly better than the European V and VI.

Canadian and North American grading rules use the term 'clear' for their best grade, being virtually free of defects. These are equal to or better than European I grade. 'Selected merchantable' and 'merchantable' are the middle grades with 'commons' being the lowest grade at the equivalent of the European grade VI.

Wood graded in other parts of the world may use other grading rules and terms; check with the supplier with regards suitability for your intended end use.

Hardwood appearance grading

Hardwoods are obtained from a far wider range of sources than softwoods, resulting in more varied grading systems. Often a grading will not be applied and the wood is sold for a specific end use by mutual agreement between the buyer and seller.

Most hardwood grades are based on the usable area of clear wood that is obtainable from a board, which is known as the cutting system (Figure 1.73). The greater the area of clear wood the higher the grade.

- The best grades 'first and seconds' (FAS) provide long, wide, clear cuttings of at least 83.4% of the board from the worst face. The other face will be at least the same or better. They are ideally suited for furniture, high quality joinery and mouldings.
- 'First and seconds one face' (FAS 1F) and selects grade the board from both faces. The best face must meet the requirements for FAS, the other meet No. 1 common requirements. The difference between FAS 1F and selects is the cuttings minimum length and width requirements. Typically 2440 mm by 150 mm for FAS 1F and 1830 mm long by 100 mm for selects.
- No. 1 commons which provide clear cuttings of medium length and width from at least 66.7% of the boards are typically at least 900 mm by 75 mm or 600 mm by 100 mm.
- No. 2. commons will provide clear cuttings of short length and narrow width from at least 50% of the board area. Typically at least 600 mm by 75 mm.

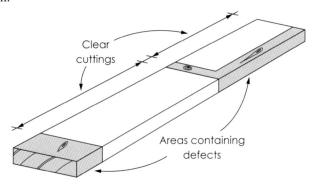

▶ **Figure 1.73** Hardwood appearance grading based on cutting system

Clear cuttings

Areas containing defects

Both common grades are an economical option for smaller joinery components, furniture, framework and a range of smaller wooden items.

Parcels of timber may contain a mixture of grades. No. 1 commons may be mixed with FAS 1F boards and sold as 'No 1 commons and better'. The supplier will identify the expected percentage of better grades.

Shipping marks. Exporting saw mills mark the cut ends of every piece of wood with their own private grade in the form of a shipping mark. They may be stamped, stencilled or branded using a combination of letters, symbols and colours. Knowledge of these can identify the timber's place of origin and its grade (Figure 1.74). Details may be obtained by consulting a directory of shipping marks.

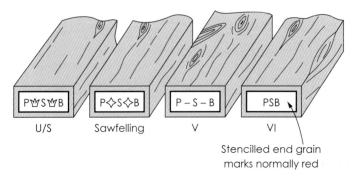

P�†S☆B   P◇S◇B   P – S – B   PSB
U/S   Sawfelling   V   VI

Stencilled end grain marks normally red

▶ **Figure 1.74** Typical shipping marks and grades

Strength grading

Both softwoods and hardwoods are available in strength grades (formerly known as stress grades). There are two methods used for strength grading (Figure 1.75):

■ *Visual strength grading.* Each piece is inspected and either rejected or given a grade, taking into account the number, size and positioning of defects and the slope of grain.

■ *Machine strength grading.* Each length is passed through a machine, which deflects the piece to test its stiffness. Based on the stiffness a grade is automatically marked on the face or each piece of wood.

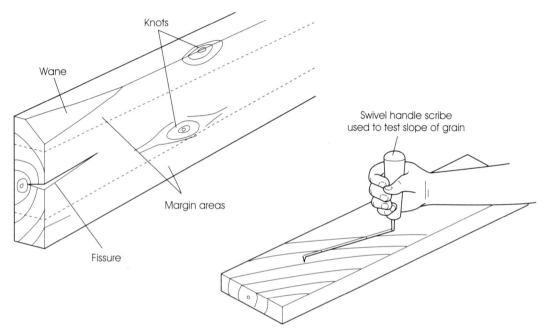

Visual strength grading based on defects and slope of grain

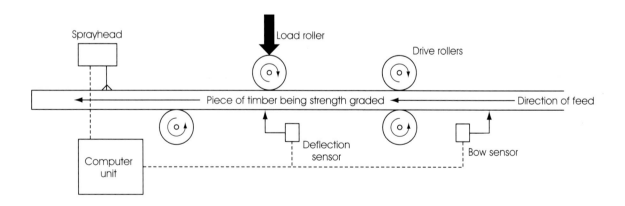

Machine strength grading based on amount of deflection under load

▲ **Figure 1.75** Strength grading

Grades are class 'C' for softwoods: C14, C16, C18, C22, C27, C30, C35, and C40. Hardwood grades are class 'D': D30, D35, D40, D50, D60, and D70. The numbers indicate a characteristic value of strength and stiffness. The higher the number, the stronger the piece. Typical strength grading marks are illustrated in Figure 1.76.

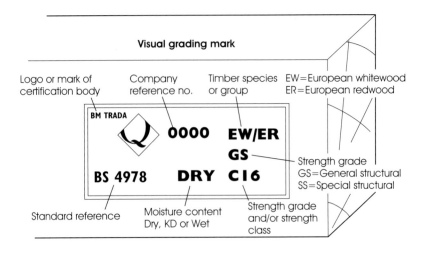

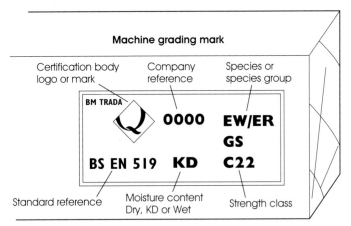

▶ **Figure 1.76** Typical strength grading marks

Moisture content of structural timber is vitally important as it has a direct effect on strength, dimensions and durability. Three service classes are defined, depending on the end use; they are used in design to determine the allowable stresses for structural timber:

■ Service Class 1. Characterised by a moisture content corresponding to a temperature of 20°C and the relative humidity of the surrounding air only exceeding 65% for a few weeks each year. In such conditions most timber will attain average moisture content not exceeding 12%. Mainly used for internal heated locations.

■ Service Class 2. Characterised by a moisture content corresponding to a temperature of 20°C and the relative humidity of the surrounding air only exceeding 85% for a few weeks each year. In such conditions most timber will attain average moisture content not exceeding 20%. Mainly used for internal unheated locations.

■ Service Class 3. Characterised by higher moisture contents than Service Class 2. Mainly used for external locations.

Specifying timber for joinery

Timber for joinery use is normally specified to be from one of the higher sawn timber appearance grades. However a further grading may be used in joinery specifications. This is applicable to timber after it has been incorporated into an item of joinery or into individual joinery components and is therefore inappropriate for the purchase of timber.

Five classes for joinery timber are specified J2, J10, J30, J40 and J50, which are based on the allowable size of knots and other characteristics. A distinction is made between the visible and concealed faces of installed joinery. A concealed face is one that is not visible once the item of joinery is installed, being concealed by other parts or elements of construction (such as the rear face of a door lining that is fixed against a wall). Faces that are only seen when the item is open are still considered to be visible faces, as are painted surfaces. Defects present in commercial grades may be eliminated by selective cutting or positioning during the production of individual joinery components.

The **joinery classes** are based primary on the size of knots, being expressed as a maximum knot size and maximum percentage of knot compared with the width or thickness of the finished piece as shown in Table 1.3. J2 grade is the highest or best grade and J50 is the lowest or worst grade. The distribution of knots or knot clusters over 10 mm diameter is limited to no closer than an average of 150 mm measured over the length of the piece. Other defects are permitted to a limited extent in grades lower than J2 and may be made good with a plug of matching timber or a filler.

Table 1.3 Classes of timber for joinery

| | Joinery Class | | | | |
	J2	J10	J30	J40	J50
Maximum Knot size	2mm	10mm or 30%	30mm or 30%	40mm or 40%	50mm or 50%

Example:

25 mm diameter knot in 100 mm board is 25% of face, thus classed as J30

100 mm

25 mm

Information required for purchasing wood

This will vary depending on type of timber, end use and supplier, but will typically include:

- **Species.** The common name followed in some cases by the botanical name.
- **Grade.** The appearance grade or strength grade.
- **Amount.** A combination of cross-sectional size and volume, area or length (Figure 1.77).
- **Cross-sectional size.** The thickness first followed by the width e.g. 50 mm × 225 mm. Hardwoods may only be available in a stated thickness and minimum width e.g. 50 mm × 100 mm wide and up.
- **Volume or area or length.** Larger quantities of both hardwoods and softwoods are purchased by the cubic metre (m^3). Hardwoods may still be specified by the cubic foot.
  - **Area.** Floor boarding and cladding is purchased by the square metre (m^2).
  - **Length.** Pre-machined mouldings, skirtings and architraves etc. are purchased by the running metre (m/run).
- **Finish.** State the type of finish or processing required: sawn, regularised, machined/planed or moulded.
- **Moisture content.** State the moisture content or the method of seasoning if required.
- **Preservative treatment.** State the type of treatment if required.
    ### Example
    European Redwood, U/S, 50 × 225 mm, 2m³, PAR, KD.

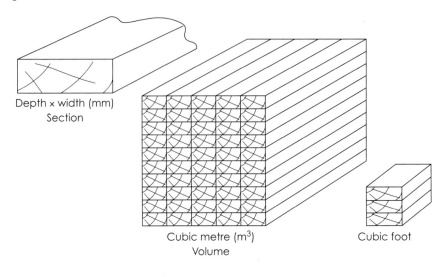

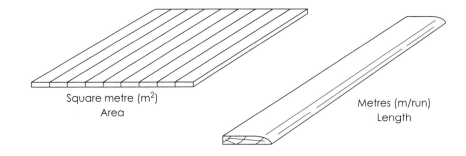

▶ **Figure 1.77** Methods of purchasing wood

Boards and veneers

There is a wide range of wood-based board materials available to the woodworker. These generally fall into three main categories:

- laminated boards
- particle boards
- fibre boards.

Laminated boards

Plywood is a laminated board material made from thin sheets of wood, termed construction veneers or plies. Its panel size, stability and ease of use make it an ideal material for both construction work (flooring, cladding and formwork) and interior joinery and carcass work (panelling, furniture and cabinet construction).

Plywood is normally constructed with an odd number of layers, with their grain direction alternating across and along the sheet to counter the movement in the wood. It is glued together to form a strong board, which will retain its shape and not have a tendency to shrink, expand or distort.

The number of plies varies according to the thickness of the finished board, three being the minimum (Figure 1.78). Whatever the number of plies, the construction is symmetrical about the centre ply or centre line of the board, so that both sides are equally balanced to resist warping. Where an even number of plies are used the central two plies are bonded with their grain running parallel to each other, to act as a single layer thereby maintaining the balance of the board achieved with an odd number of plies.

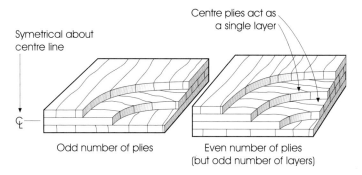

▲ **Figure 1.78** Symmetrical construction

Plywood sizes. Plywood is manufactured in a range of sizes, varying in thickness from 3 mm to 30 mm in increments of approximately 3 mm the width of the board is typically 1.22 m but 1.52 m boards are also available (Figure 1.79).

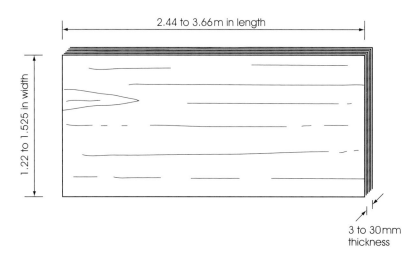

◀ **Figure 1.79** Typical plywood sizes

The length of the most common board is 2.44 m, but boards up to 3.66 m are manufactured. The grain direction of the face plies normally runs the longest length of the board. However short grain boards are also available. Thus when specifying plywood it is standard practice for the grain direction to run parallel to the first stated dimension. Therefore a 2.44 m × 1.22 m board will have long grain and a 1.22 m × 2.44 m board will have a short grain (Figure 1.80).

Appearance grading

Plywood is graded according to the appearance of its outer faces, each face being assessed separately. In common with timber, the grading rules for plywood vary widely from country to country:

Northern Europe/Russia with permitted defects:

Grade A: practically defect free

Grade B: a few small knots and minor defects

Grade BB: several knots and well-made plugs

Grades C or WG: all defects allowed; only guaranteed to be well glued

It is rarely necessary for the face and back veneer to be of the same grade; most manufacturers offer a wide combination of face and back veneer grades: A/A A/B A/C B/BB B/C, etc. The back veneer may also be described as a balancer where it will not be seen and need not be from the same

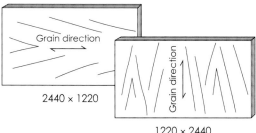

▲ **Figure 1.80** Plywood sizes and grain direction

Typical label information

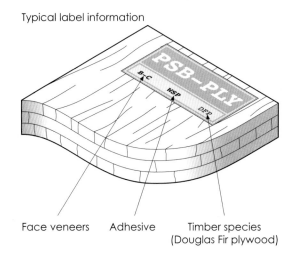

▶ **Figure 1.81** Typical plywood labelling

Face veneers   Adhesive   Timber species (Douglas Fir plywood)

species of wood. These boards may be graded as A/bal or B/bal. Some suppliers of veneered boards term A/A or A/B boards as double-sided (DS) and A/bal or B/bal boards as single-sided (SS). Typical labelling details are shown in Figure 1.81.

Plywood and other panels for use in construction have to comply with the European Construction Products Directive (CPD) and may have a CE label on them (Figure 1.82)

Canada/America gradings are as follows with the approximately equivalent Northern Europe/Russia grades shown:

> Grade G2s (good two sides) equivalent to A/A
> Grade G/S (good one side/solid reverse) equivalent to A/B
> Grade G1s (good one side) equivalent to A/C
> Grade Solid 2s (Solid two sides) equivalent to B/B
> Grade Solid 1s (solid 1 side) equivalent to B/C
> Grade Sheathing equivalent to C/C

The performance of plywood in use is dependent not only on the quality of the plies used, but also by the type of **bonding adhesive** used in manufacture:

- INT, interior grade will not withstand humidity or dampness.
- MR, moisture resistant and BR, boil resistant, are suitable for use under normal conditions, but will not withstand continuous exposure to extreme conditions.

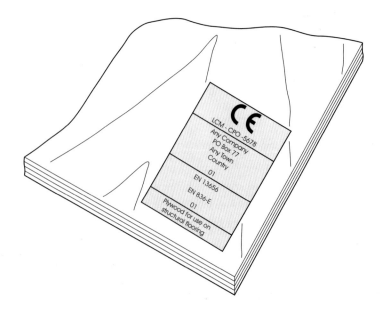

▶ **Figure 1.82** CE marking of plywood and other panels

■ WBP plywoods, weather and boil proof, are suitable for use under any conditions.

See Adhesives later in this chapter for further details.

Types of plywood

Plywood boards are manufactured in many parts of the world, and thus the species of wood used varies considerably dependent on their origin. The core and face plies may be from the same wood throughout or of different species.

■ Softwood boards are normally from Douglas fir.
■ Hardwood boards using light coloured temperate woods are normally from either Birch or Beech.
■ Tropical hardwood boards using a variety of darker (red) woods, including Gaboon, Lauan and Meranti.

The main types of plywood available are (Figure 1.83):

■ *Three-ply,* which consists of three equal thickness layers. Ideal for drawer bottoms and cabinet backs.
■ *Stout-heart*, which also consists of three layers, but the middle layer is thicker.
■ *Multi-ply,* which is the name given to any plywood which has more than three layers. Softwood boards are used for formwork and structural work, whilst hardwood boards are used for furniture and cabinet construction.
■ *Drawer-side ply* is the exception to the alternating cross-grained construction as all plies are in the same direction. Normally made from birch in a thickness of 12 mm, it is used in place of solid timber for drawer construction.
■ *Four-ply and six-ply* have their central plies bonded together with their grain in the same direction. Mainly manufactured from softwood for structural use.
■ *Decorative-ply*, made from multiply, which has been faced with selective crown or quarter cut matching veneers. Used for panelling, furniture and cabinetwork. To maintain the stability of the board, a balancer, often of a lesser quality veneer, must be applied to the reverse face of the board. These may be graded as A/A, A/B, A/C or A/bal (balancer). Boards graded as A/A or A/B may be termed two-sided and A/C or A/bal as one-sided.

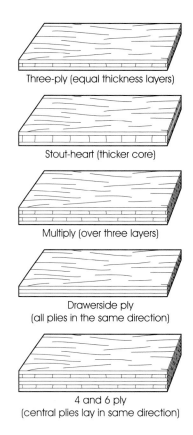

Three-ply (equal thickness layers)

Stout-heart (thicker core)

Multiply (over three layers)

Drawerside ply
(all plies in the same direction)

4 and 6 ply
(central plies lay in same direction)

▲ **Figure 1.83** Types of plywood

Other laminated boards

Laminboard and **blockboard** are similar to ply, using a layered construction with the grain direction alternating. They differ from plywood in that the core is made from strips of solid timber (Figure 1.84). These are faced on both sides with one or two plies. The width of the strips vary with each type of board. Laminboard has strips that are up to 8 mm. In blockboard the strips are up to 25 mm in width. Both forms of laminated board are available in the same panel sizes as plywood. Thicknesses vary between 12 mm and 38 mm.

Both are used as a core material for veneering, panelling, partitioning, door-blanks, furniture and cabinetwork. Laminboard is superior to blockboard, as there is less likelihood of the core strips distorting and showing through on the surface of the board – a defect termed 'telegraphing'.

Particle boards

Wood particleboards are manufactured using small wood chips or flakes impregnated with a synthetic resin and bonded together under pressure (Figure 1.85). Softwood forest thinnings and waste from sawmills and other wood manufacturing processes are mainly used. Other materials are also used including 'shivers' from the flax plant (also used in the clothing industry to make linen) to produce a particleboard known as flaxboard.

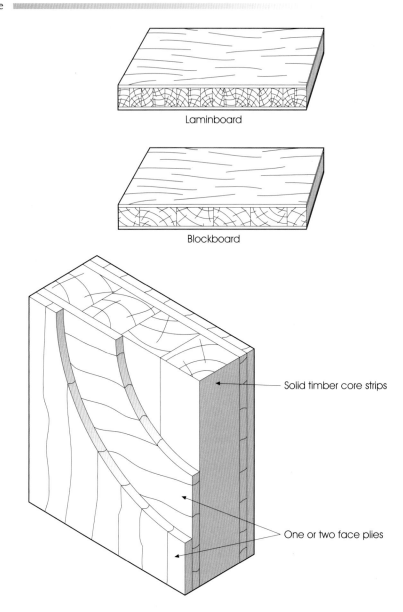

Laminboard

Blockboard

Solid timber core strips

One or two face plies

▶ **Figure 1.84** Solid core laminated boards

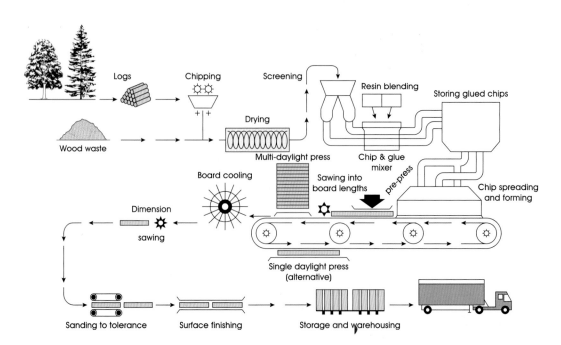

▶ **Figure 1.85** Particle board manufacturing process

Single-layer

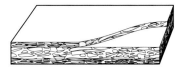

Three-layer

Graded density

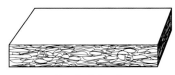

Extruded

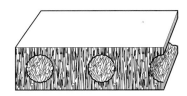

Extruded tubular core

Board sizes. Particleboard is manufactured in a range of sizes, varying in thickness from 2.5 mm to 38 mm with extruded tubular core boards increasing to 50 mm or more. The standard imperial eight by four board is still readily available as 2440 mm × 1220 mm. The fully metric equivalents are 2.4 m × 1.2 m boards based on 600 mm increments. Lengths range from 1.83 m up to 3.66 m. 600-mm wide boards are made mainly for flooring use.

Chipboard

Various types exist (Figure 1.86):
- *Single-layer chipboard* is made from a mass of similar size wood particles, pressed to form a flat, relatively coarse surface board. This is suitable as a core board for veneering, or plastic laminating, but not painting.
- *Three-layer chipboard* consists of a low-density core of large particles, sandwiched between two outer layers of finer, higher density particles. The outer layers contain a higher proportion of resin. This produces a very smooth, even, surface that is suitable for painting. Extra care is required when sawing layered boards, as compared to single layered, since they tend to split or delayer at the cut edge.
- *Graded density chipboard* has a core of coarse particles and faces of very fine particles. But, unlike the layered boards there is a gradual transition from the coarse core to the fine surface. These are ideal for use in furniture and cabinet construction.
- *Extruded boards* are formed by forcing a blend of wood particles and resin through a die, resulting in a continuous length of board to the required width and thickness. Most particles are at right angles to the board's face, thus reducing strength. The holes in *tubular core boards* are formed by heating coils used in the curing process, which enables much thicker boards to be made, as well as reducing the overall weight. Main uses are for partitioning, door blanks and core stock for veneers and melamine foils.
- *Decorative-faced chipboard;* normally a single layer, graded density or extruded board faced with a wood veneer, plastic laminate or a thin melamine foil (*MFC, melamine faced chipboard*). The veneered boards are supplied sanded ready for polishing. No further surface finishing is required for the plastic laminated or melamine foil faced boards. These are used extensively for furniture, shelving and cabinet construction. The plastic laminate-faced boards often have a thicker core and a post-formed rounded edge for use as kitchen worktops.

If exposed to excess moisture, most chipboard considerably in thickness, loses strength and does not subsequently recover on drying. However stronger, moisture resistant grades, tinted green in colour are made for flooring and use in damp conditions. In general chipboards strength increases with its density.

Other particle boards:
- *Cement-bonded particleboards (CBPB)* are bonded with ordinary Portland cement (OPC) or with a magnesium-based cement to give a high density board that has a high durability as well as good sound and fire resistant properties. They can be used in dry, humid or external conditions.
- *Oriented strand board (OSB)* (Figure 1.87) is manufactured from large softwood strands bonded with resin and wax in three layers. The strands in each layer are oriented in the same direction, with face layers being at right angles to the core layer, much like three-core plywood. Main uses are as floor and roof coverings, wall sheathing, site hoardings, formwork and packing cases.
- *Flake-board or wafer-board* consists of wood flakes or shavings, often hardwood. The flakes are laid horizontal and overlap each other, with their grain direction orientated randomly. These have a similar range of uses to OSB.

▲ **Figure 1.86** Types of chipboard (particle board)

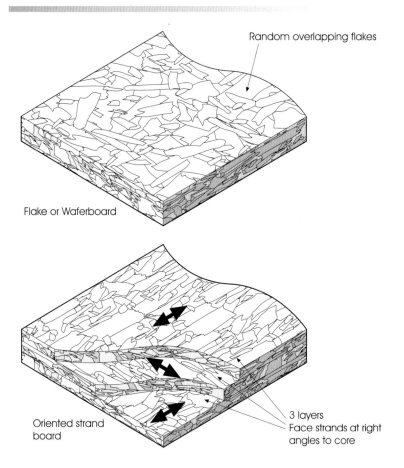

Random overlapping flakes

Flake or Waferboard

Oriented strand board

3 layers
Face strands at right angles to core

▶ **Figure 1.87** Other particle boards

Fibreboards

These are manufactured from pulped wood and other vegetable fibres, pressed into boards of the required thickness. Boards of varying density are produced depending on the pressure applied and the process used (Figure 1.88).

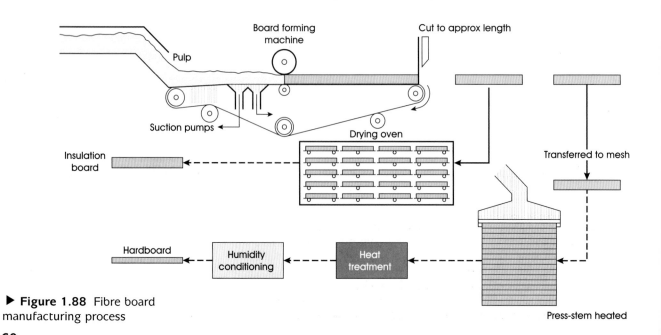

Board forming machine

Cut to approx length

Pulp

Suction pumps

Drying oven

Insulation board

Transferred to mesh

Hardboard

Humidity conditioning

Heat treatment

Press-stem heated

▶ **Figure 1.88** Fibre board manufacturing process

The main types of fibreboard available (Figure 1.89) are as follows.

Hardboard. This is a high density fibreboard made from wet wood fibres pressed together at high pressure and temperature. This wet process uses the natural resins in the wood fibres to bond them together.

- *Standard hardboard* has a smooth upper face; the opposite face is rough, caused by the meshed lower press plate that allows drainage of the moisture.
- *Duo-faced hardboard* has two smooth faces and is produced by further pressing standard boards between two flat plates.
- *Tempered hardboard* is a standard board that has been impregnated with oil and resin to produce a stronger board that is more resistant to moisture and abrasion.
- *Decorative hardboard* is produced in moulded, perforated, wood grained, painted and melamine foil faced boards.

Medium and soft boards:

- *Medium board* is made in the same way as hardboard, using a wet process, but lower pressure, resulting in a mid-density board.
- *Soft board or insulation board* is a low-density board, again made using the wet process, but only lightly compressed and dried in an oven. Bitumen-impregnated soft boards are available for use where increased moisture resistance is required.
- *Medium density fibreboard (MDF)* is a fibre board with two smooth faces, but unlike other fibre boards it is made using a dry process with synthetic resin adhesive being added to bond the fibres together (Figure 1.90).

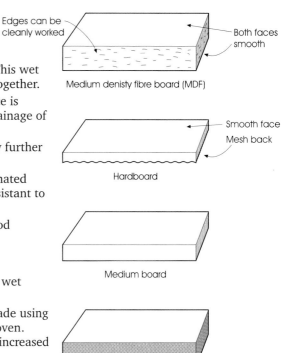

▲ **Figure 1.89** Types of fibreboard

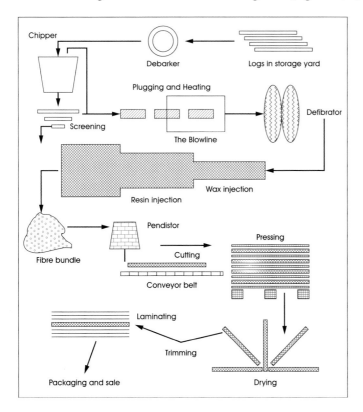

◀ **Figure 1.90** MDF manufacturing process

MDF can be worked with hand and machine tools like solid wood. It has a fine texture that allows faces and edges to be cleanly work and finished. It provides an excellent core board for veneers and plastic laminates and also melamine foil known as MFMDF (melamine-faced MDF). In addition, MDF can be painted or polished. Standard boards can be modified during manufacture to increase their moisture or fire resistance. Green-tinted boards are moisture resistant (MRMDF). Red-tinted boards are fire resistant

(FRMDF). MDF is also widely used for panelling, furniture, cabinet construction and joinery sections such as mouldings, skirting, architraves, fascia and soffit boards.

Fibreboard sizes. Hardboards are available in thickness from 1.5 mm to 1.2 mm, medium and soft boards from 6 mm to 12 mm thick and MDF from 1.5 mm to 60 mm thick. Board sizes in general vary from 2440 mm to 3660 mm in length and 1220 mm to 1830 mm in width. Smaller and larger sizes may be available for certain boards to special order.

Veneers

Veneers are very thin sheets or 'leaves' of real wood cut from a log for decorative or constructional purposes (Figure 1.91).

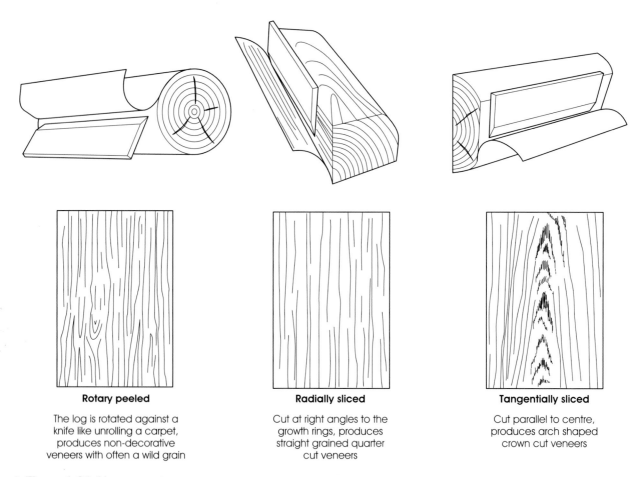

Rotary peeled

The log is rotated against a knife like unrolling a carpet, produces non-decorative veneers with often a wild grain

Radially sliced

Cut at right angles to the growth rings, produces straight grained quarter cut veneers

Tangentially sliced

Cut parallel to centre, produces arch shaped crown cut veneers

▲ **Figure 1.91** Veneer cutting

Constructional veneers are normally rotary peeled. The whole log is put into a giant lathe. The log is rotated against a knife, which runs the whole length of the log to peel off a continuous sheet of veneer. A widely varying grain pattern is produced which is not normally considered decorative. Thus these veneers are mainly used for plywood manufacture.

Decorative veneers are sliced either radially or tangentially to provide either quarter cut or crown cut **figuring**. These leaves are used for facing plywood and other decorative boards normally using book or slip **matching** for commercial grade boards. Boards of a higher quality, which are often referred to as classic or architectural grades are normally veneered to order, when it is possible to specify other methods of matching the veneer leaves (Figure 1.92). These veneered boards will then be graded in the same way as plywood by the quality of the veneer used on each face e.g. A/A or A/B, etc.

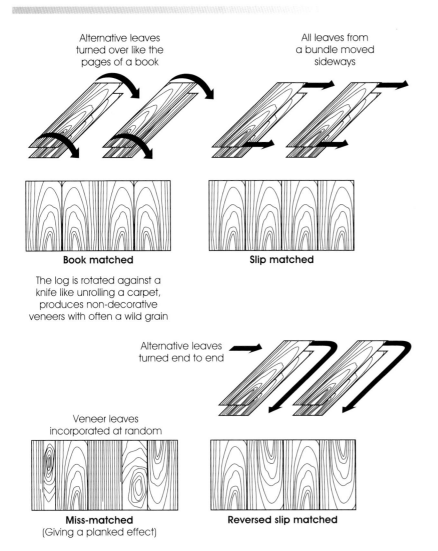

Alternative leaves
turned over like the
pages of a book

All leaves from
a bundle moved
sideways

Book matched

Slip matched

The log is rotated against a
knife like unrolling a carpet,
produces non-decorative
veneers with often a wild grain

Alternative leaves
turned end to end

Veneer leaves
incorporated at random

Miss-matched
(Giving a planked effect)

Reversed slip matched

◀ **Figure 1.92** Types of veneer matching

Most decorative veneers are sliced at 0.6 mm thick. Veneers are normally available in lengths of 2440 mm and longer. Leaf widths of tangential sliced veneers, range from about 225 mm and 600 mm depending on the species. Radial sliced leaves tend to be much narrower.

Once sliced and dried the veneers are kept in multiples of four leaves and bundled into parcels of 16 to 32 leaves for matching purposes. Each bundle or parcel is taped together and re-assembled in consecutive order into its log form, ready for use.

More **exotic figured veneers** are available by slicing from the butt, burr or crotch of a tree (Figure 1.93). These are highly prized for furniture and small wooded articles. These exotic pieces are normally much shorter in length or may be of an irregular shape. Burrs for example are irregular and typically from 150 mm × 100 mm to 1000 mm × 450 mm in size. Figure 1.94 illustrates typical methods of matching these exotic veneers on a board.

Laminated plastic

This is a thin synthetic plastic sheet made from layers of craft paper bonded and impregnated with a resin. A plain colour, patterned or textured paper or real wood veneer is used for the upper decorative face, which is coated with a clear coat of melamine formaldehyde to provide a matt, satin or gloss hard-wearing hygienic finish (Figure 1.95).

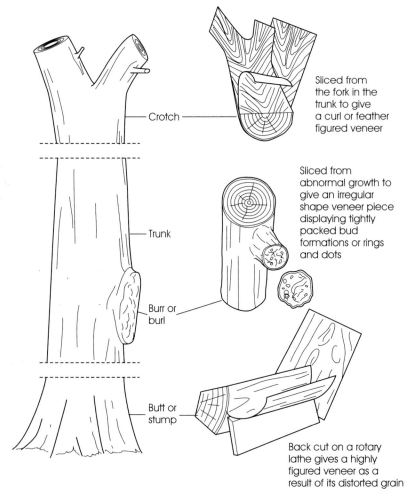

Crotch

Sliced from
the fork in the
trunk to give
a curl or feather
figured veneer

Trunk

Sliced from
abnormal growth to
give an irregular
shape veneer piece
displaying tightly
packed bud
formations or rings
and dots

Burr or
burl

Butt or
stump

Back cut on a rotary
lathe gives a highly
figured veneer as a
result of its distorted grain

▲ **Figure 1.93** Exotic figured veneers

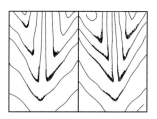

Book matched curls (normally
on cut panels or short grain boards)

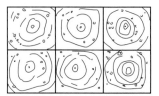

Centre matched burrs
(small cut panels)

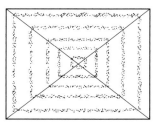

Quarter or diamond matched
(small cut panels)

▲ **Figure 1.94** Matching of
exotic veneers

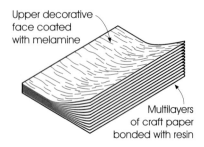

Upper decorative
face coated
with melamine

Multilayers
of craft paper
bonded with resin

▲ **Figure 1.95** Laminated plastic

Plain non-face sheets are also available for use as backing laminates.
Thickness ranges from 0.6 mm to 1.5 mm. Lengths range from 1830 mm to
3660 mm and widths from 600 mm to 1830 mm. The thinner sheets are used
for post-forming (bending around curved edges) and other curved work and
the thicker ones for flat surfaces. Colour core sheets are also available in a
range of colours that don't show a dark or black line at the edge; as are
thicker double-faced boards with the core built up from many layers of craft
paper for use in damp locations.

Purchasing board material

Board material is normally sold by the sheet, but may be priced by either the
square metre (m²), 10 square metres (10 m²) or by the individual sheet.
Typical information required when purchasing board material is:

- thickness of board
- type of board
- board finish
- size of board; quote longest grain size first where applicable
- number of boards.

> **Example**
>
> 18 mm, MDF, Crown cut, Cherry veneer A/B, 2440 × 1220 mm, 25 No. off

Environmental considerations

Wood and wood-based products are widely recognised as ecologically beneficial materials when compared to other structural materials.

- Conversion of timber into planks or sheet materials requires comparatively little energy and produces few harmful waste products. Future supplies are assured from well-managed sustainable sources.
- The active process of growing and replacing trees is a means of controlling or even reducing the amount of carbon dioxide in the atmosphere.
- Greater use of wood products can further improve the 'greenhouse effect' due to the high levels of insulation achievable in timber buildings, which in turn require less energy to heat or keep cool.

A comparison of the energy requirements to produce structural beams of equal performance is illustrated in Figure 1.96.

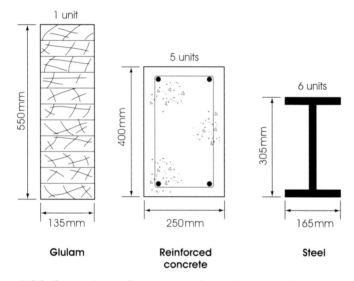

▲ **Figure 1.96** Comparison of energy requirements to produce structural beams of equal performance

Fixings

Popular methods of fastening or fixing timber products include screws, nails and adhesives.

Wood screws

Wood screws are used for joining wood to wood; the clamping force they provide make a strong joint that can be easily dismantled. Screws are also used to attach items of ironmongery, such as hinges, locks and handles.

Most screws are made of steel; they may be treated to resist corrosion and case hardened for additional strength. Chrome-plated or black japanned steel screws are used for decorative purposes. Brass screws are more decorative and stainless steel screws are more resistant to corrosion. Both brass and stainless steel screws can be used with acidic hardwoods such as oak, which are stained by steel screws.

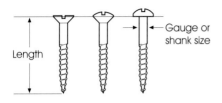

▲ **Figure 1.97** Screw size data

Screw sizes

The length of the part that enters the wood and the diameter of the shank are used to specify the size of a screw (Figure 1.97). Metric shank sizes are in millimetres and imperial sizes are indicated by a gauge number A typical range of standard screw sizes in both metric and imperial is illustrated in Figure 1.98. There are no direct equivalents between the shank sizes of metric and imperial screws. However, the following are close equivalents:

No. 4 = 3.0 mm
No. 6 = 3.5 mm
No. 8 = 4.0 mm
No.10 = 5.0 mm
No.12 = 6.0 mm

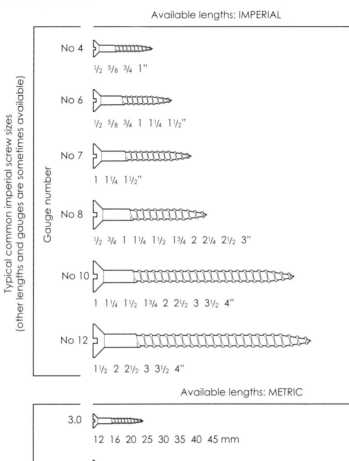

▶ **Figure 1.98** Standard screw sizes

Screw threads and heads

Conventional **woodscrews** are threaded for about 60%, of their length, the plain shank acts as a dowel, and the larger head holds the two pieces of wood or the item of ironmongery in place (Figure 1.99).

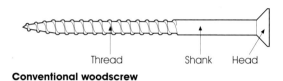

Thread   Shank Head

Conventional woodscrew

Thread   Shank Head

Twin-threaded screw

◀ **Figure 1.99** Woodscrew details

Twin-threaded wood screws have a coarser twin thread that provides a stronger fixing in wood-based boards such as chipboard and MDF compared with conventional woodscrews. More of the screw length is threaded and the steeper pitch enables the screw to be driven in more quickly. In addition the shank is narrower to reduce the risk of splitting.

Screw heads. Both conventional and twin-threaded screws are available with a variety of screw heads (Figure 1.100):

- *Countersunk head screw,* a flat-topped screw with a tapered bearing surface for use where the head is required to finish flush with the surface. It fits into a countersunk hole formed into either the timber or the item of ironmongery.
- *Round head screw,* a dome head screw with a flat bearing surface used to fix sheet material and un-countersunk ironmongery.
- *Raised head screw,* a slightly rounded head screw with a tapered bearing surface. For use on exposed fixings in conjunction with cups or for fixing surface ironmongery such as handles and bolts.
- *Bugle head screw,* a flat-topped screw with a stronger taper to the bearing surface than countersunk screws. This allows them to be driven without the need for countersinking. Mainly used as drywall screws for fixing plasterboard.

Screw slots. The recess that is cut into screw heads is designed to accommodate the tip of a screwdriver (Figure 1.101):

- *Slotted screws* have a single groove slot machined right across their head, for a straight tipped screwdriver.
- *Cross-head screws* have two intersecting slots designed to accept the tip of a matching cross-head screwdriver. The majority of screws are now of the cross-head type to minimise slipping, especially when using power screwdrivers. In theory a different pattern cross-head screwdriver is required for different brands of screws.
- *Clutch-head screws, security or thief-proof screws.* They can be inserted with a standard straight tipped screwdriver, but the tip rides out of the slot when turned in an anti-clockwise direction for removal.
- *Screw cups* enhance the appearance of screws used for surface fixings (Figure 1.102). They also increase the bearing area and protect the countersinking from wear, in situations that require screws to be removed, such as glazing beads and access panels. They are available as either recessed or surface mounted, for use with countersunk or raised head screws.
- *Screw covers* are used to conceal the screw head, for a neater finish whilst still indicating the screws position.

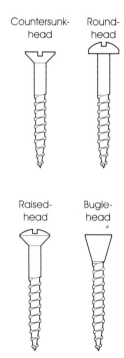

Countersunk-head   Round-head

Raised-head   Bugle-head

▲ **Figure 1.100** Screw heads

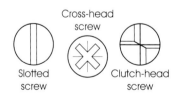

Cross-head screw

Slotted screw   Clutch-head screw

▲ **Figure 1.101** Screw slots

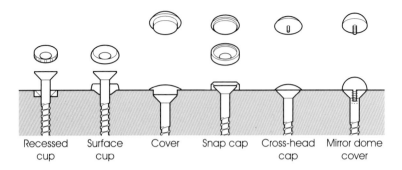

Figure 1.102 Screw cups and covers

- *Snap caps* consist of a plastic dome that snaps over the rim of a matching screw cup, after the countersunk screw has been inserted.
- *Cross-head cover caps* are a moulded plastic cover that has a spigot on the underside to locate into the head of cross head screws.
- *Mirror screw covers* are brass, chrome or stainless steel, often domed, covers, which have a threaded spigot that screws into the threaded hole in the screw-head of special countersunk screws. They are intended to cover the head of screws used to fix mirror glass.

Using screws

The length of a screw should be approximately three times the thickness of the wood being secured, but to avoid through penetration or dimpling the backing surface, the screw point should finish a minimum of 3 mm short (Figure 1.103)

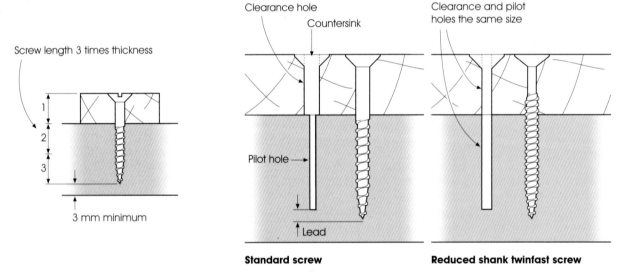

▲ **Figure 1.103** Screw length guide; clearance and pilot holes

Depending on the type of screw being used a clearance hole and countersinking should be drilled in the piece of wood being secured. A smaller pilot hole is required in the piece to which it is secured. With smaller screws into softwood the pilot hole may be made with a bradawl. Pilot holes for larger screws and all screws in hardwoods and edge screwing into sheet materials should be drilled. The pilot hole should be shorter than the full depth of the screw. The undrilled depth is called the lead. The clearance hole for twin fast screws can be the same size as the pilot hole because they have a reduced shank gauge.

Nails and staples

Nailing is the simplest way of joining pieces of wood together and securing other materials to wood. If carried out correctly, nailing can result in a strong, lasting joint. A typical range of nails and fasteners is illustrated in Figure 1.104.

■ *Wire nails* are also known as French nails and normally available in lengths from 12 mm to 150 mm. They are round in section and the larger sizes normally have a chequered head to reduce the possibility of the hammer slipping while the nail is being driven in. The top part of the shank is ribbed to give the nail extra grip in the timber. These are general-purpose carpentry nails which can be used for all rough work where the presence of the nail held on the surface of the timber is not important such as most carcassing, first fixing and formwork.

■ *Oval nails*, as the name implies, are oval in section and are available in lengths from 12 mm to 150 mm. They have a small head, which can be punched below the surface. This is an advantage when fixing timber that has to be painted because the hole when filled is not visible.

■ *Cut nails* are cut from mild sheet steel and are square in section, which gives a good grip. They are normally available up to 100 mm in length. The cut nail can be used for general construction work, but nowadays it is mainly used for fixing timber to block-work, walls, etc. They can also be used for nailing into the mortar joints of green, (freshly laid) brickwork.

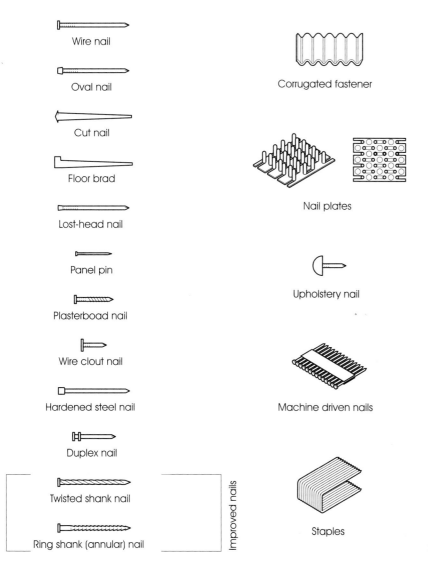

◀ **Figure 1.104** Typical range of nails and fasteners

- *Floor brads.* These are similar to the cut nail, but they are lighter in section. They are available in lengths up to 75 mm and are used for the surface fixing of floorboards.
- *Lost head nails* are similar to the wire nail, but they have a small head, which can be easily punched below the surface of the timber without making a large hole. Hence the name 'lost head'. They are also used for fixing timber that has to be painted and also for the secret fixing of tongued and grooved flooring boards. They are available in lengths up to 75 mm.
- *Panel pins* are a much lighter version of the lost-head nail and are used for fixing fine mouldings around panels, etc. They are normally available in lengths between 12 mm and 50 mm.
- *Improved nails* are wire nails with a ringed or twisted shank that can be hammered in easily, but have the holding power of a screw. They can be used for all construction work where extra holding power is required. They are available in lengths up to 75 mm.
- *Plasterboard nails* are galvanised to prevent them rusting, and they have a roughened shank for extra grip. They are available in lengths up to 40 mm and can be used to fix plasterboard and insulating board to timber ceiling joists, studwork and battens, etc.
- *Wire clout nails* are a short galvanised wire nail with a large head. Their main use is for fixing roofing felt and building paper, etc. They are available in lengths between 12 mm and 25 mm.
- *Hardened steel nails* are similar to the lost head nail but are made from a specially hardened zinc-plated steel for making hammered fixings directly into brickwork and concrete without plugging. They are available in lengths up to 100 mm. Goggles must always be worn when driving these nails.
- *Duplex nails* are a wire nail with a double head. The lower head is driven to the surface for maximum holding power whilst leaving the upper head protruding for easy withdrawal. Mainly used for formwork and other temporary fixings.
- *Corrugated fasteners* are a corrugated bright steel strip used to reinforce mitres and butt joints in rough framing work. The fastener is placed across the joint line and driven flush with the surface. They are available in sizes from 6 mm to 25 mm deep.
- *Nail plates* are spiked or flat metal plates used for securing framework joints, normally made from zinc plated or galvanised steel. These plates are positioned or pressed into the wood. Flat plates are secured with improved nails.
- *Upholstery nails* are small domed head, decorative nails used for fixing upholstery fabric and trim. Normally 12 mm long and available in brass, bronze and chrome.
- *Machine-driven nails* are wire nails and pins packed in strips or coils for use in various types of pneumatic nailers. A wide range of types and sizes are available, depending on the manufacturer, for use in structural work right through to fine pins for fixing finishing trim.
- *Staples,* normally supplied in strips, temporarily bonded together. They are used as an alternative to pneumatically driven nails for fixing plywood, fibreboard, plasterboard, insulation board and plastic sheeting.

Using nails

Nails are driven in with an appropriate hammer (Figure 1.105). Use a Warrington or pin hammer for fine work and a claw hammer for heavier work. Always keep the hammer face clean, by rubbing it with a fine abrasive paper. A dirty hammer face tends to slip on the nail head, damaging the work piece and bending the nail. Keep an eye on the nail and check the angle as it enters the wood. To avoid damage and bending, the hammer shaft at the moment of impact should be at right angles to the nail.

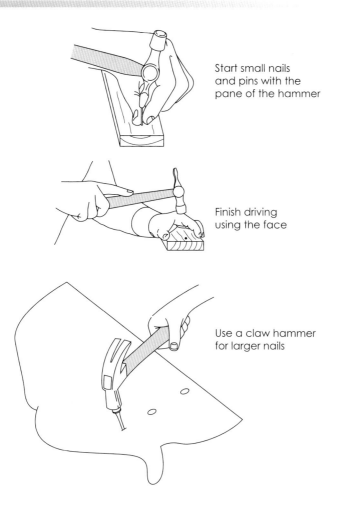

Start small nails
and pins with the
pane of the hammer

Finish driving
using the face

Use a claw hammer
for larger nails

◀ **Figure 1.105** Driving nails

Methods of using nails are listed below and illustrated in Figure 1.106:

- Always nail the thin piece to the thick piece.
- Use a length of nail that is about $2\frac{1}{2}$ to 3 times the thickness of the wood it is being driven through. This gives approximately two thirds of the nail to provide the holding power.
- When joining two thin pieces, use a nail length 4 to 6 mm longer than the combined thickness of the pieces. This allows the protruding end to be clenched over for strength.
- Where extra strength is required always dovetail or skew nail. Using this method prevents the nails from being pulled out or working loose.
- Where oval or rectangular section nails are used, the widest dimension must be parallel to the grain of the timber. Their use in the opposite direction, across the grain, will normally result in the timber splitting.
- When nailing near the end of a piece of timber, the timber has a tendency to split. In order to overcome this, the point of the nail should be tapped with a hammer to blunt the point before the nail is used. The point of a nail tends to part the fibres of the timber and therefore split it, while the blunted end tends to tear its way through the fibres, making a hole for itself.
- Pilot holes may be required or specified to receive wire nails in structural joints. These pilot holes should be up to 80% of the nail shank diameter, in order to prevent splitting
- Pilot holes are essential when nailing hardwoods.
- Stagger nails across the grain: do not nail in the same grain line more than once, as this will split the wood. Nails in surfaces to be painted should be punched just below the surface ready for filling.

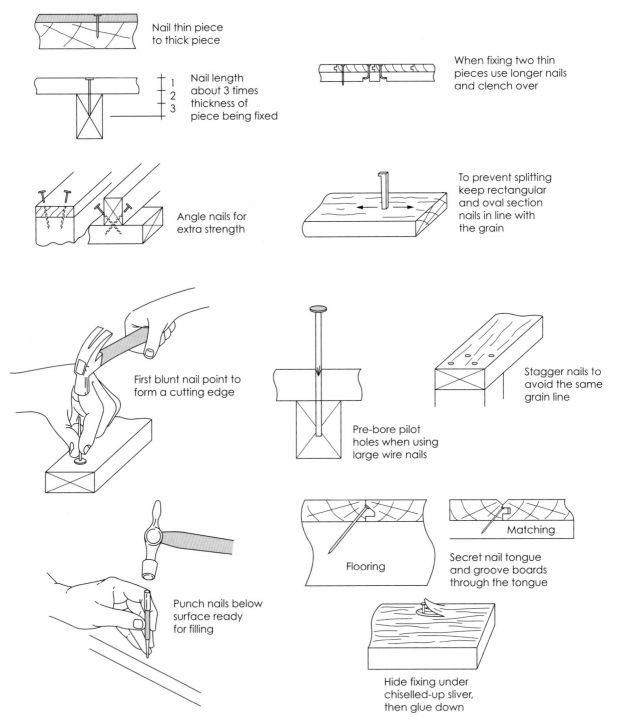

Nail thin piece to thick piece

Nail length about 3 times thickness of piece being fixed

When fixing two thin pieces use longer nails and clench over

Angle nails for extra strength

To prevent splitting keep rectangular and oval section nails in line with the grain

First blunt nail point to form a cutting edge

Pre-bore pilot holes when using large wire nails

Stagger nails to avoid the same grain line

Punch nails below surface ready for filling

Flooring

Matching

Secret nail tongue and groove boards through the tongue

Hide fixing under chiselled-up sliver, then glue down

▲ **Figure 1.106** Using nails

- Boards with tongued and grooved joints can be secret nailed through the tongue. Each subsequent board effectively hides the nail head.
- For general surface nailing, nail heads can be concealed by chiselling up a sliver of wood and nailing into the recess. The sliver is then glued back down to hide the nail head.

Mechanical fasteners

In addition to screws and nails there are wide range of specialist fasteners and fixings used in the construction industry for both structural work and fixing purposes. The chart shown in Table 1.4 illustrates a typical range of fasteners and fixings and summarises their main characteristics.

Table 1.4 Mechanical Fasteners

Application	Suitable fixing device	Installation procedure	
Screw fixings in brickwork, blockwork or concrete	Plastic plugs: *Yellow* for gauge 6–8 (3.5–4mm shank) screws using a 5mm drill; *Red* for gauge 8–10 (4–5mm shank) screws using a 6mm drill; *Brown* for gauge 10–12 (5–8mm shank) screws using a 7mm drill.	1 Drill correct size hole in background material 2 Remove dust from hole using the head of a wire nail or blow out bulb (*don't blow out the dust using your mouth*) 3 Insert plug into hole 4 Insert screw through the item to be fixed and place point of screw into plug 5 Drive screw with screwdriver to secure item to the background	
Light duty fixing in thin material, e.g. plasterboard ceilings and walls, hollow core doors and cavity blocks, etc.	Nylon toggle anchor *Note*: This type of anchor can also be used as a normal plug in thicker materials.	1 Drill hole through material and fold anchor 2 Insert anchor into hole and tap flush with material 3 Expand anchor using toggle key 4 Insert wood screw or self-tapping screw and tighten	
Fixing mineral wool, glass wool or expanded polystyrene insulation to brickwork, concrete or timber, etc.	Insulation fastener (made from polypropylene)	1 Drill hole through insulation into base material 2 Insert fastener into hole and gently tap home	
Rapid plug and screw through fixings into brickwork, concrete, etc.	Hamma screw (combined nylon plug and zinc plated screw)	1 With item to be fixed in position, drill through item into base material 2 Insert hamma screw into hole 3 Drive hamma screw home with a sharp blow from a hammer. *Note*: The hamma screw can be removed with a screwdriver if required	
Fixings for suspended ceilings and other items hanging from concrete ceilings	Steel ceiling suspension anchor	1 Drill hole in concrete ceiling 2 Gently tap anchor into hole 3 Expand anchor by giving the ring bolt a sharp pull with a claw hammer	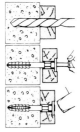

Table 1.4 *continued*

Application	Suitable fixing device	Installation procedure
Medium to heavy duty through-fixing to concrete of machinery, door frames and steel sections, etc.	Zinc-plated stud anchor	1 Position item to be fixed and drill through fixing holes into concrete 2 Remove dust from hole using blow-out bulb 3 Insert anchor into hole and gently tap home 4 Expand anchor by tightening nut
Fixing door and window frames that have not been 'built in', to concrete, brickwork or blockwork	Galvanized steel sleeve expansion/ frame anchor	1 With frame in position drill hole through frame and into wall 2 Insert anchor and tap through frame into wall with setting tool 3 Adjust position of frame if required and tighten screw *Note*: Plastic sleeved versions are also available
Medium to heavy duty bolt fixing of machinery and steel sections, etc. to concrete and solid brickwork	Steel expansion anchor	1 Drill hole in base material. Remove dust from hole using blow-out bulb 2 Insert anchor into hole and tap home using setting tool. This expands anchor which will then be ready to accept a suitable bolt
Heavy duty through-fixing of machinery and structural steelwork, etc. to concrete	Heavy duty steel expansion anchor	1 With item in position, drill through fixing holes into concrete. Remove dust from hole using blow-out bulb 2 Insert anchor into hole and tap home 3 Expand anchor by tightening bolt
Self-drilling, medium/ heavy duty bolt for fixing machinery and steel sections, etc. to concrete	Self-drilling steel expansion anchor	1 Insert anchor into a rotary hammer drill and bore into concrete. Remove dust from hole using blow-out bulb 2 Place taper plug into anchor 3 Insert anchor into hole and drive home using tool. This will expand plug 4 Move tool sharply downwards to snap off anchor. The anchor will then be ready to accept a suitable bolt

Table 1.4 *continued*

Application	Suitable fixing device	Installation procedure	
Heavy duty fixings into concrete, especially when subject to vibration	Chemical resin anchor	1 Drill hole into concrete and clean out using blow-out bulb. Insert resin cartridge 2 Using a rotary hammer drill and adaptor, drive anchor rod into resin cartridge 3 Remove tool and then adaptor 4 Allow resin to harden before using anchor. This takes between ten minutes at 20 degrees and over, and up to five hours at –5 degrees	

Woodworking adhesives

There is currently a vast range of adhesives available for use in the building industry. It must be remembered that each adhesive has its own specific range of uses and that no one adhesive will satisfactorily bond all materials.

Primarily an adhesive must be capable of sticking or adhering to timber. Adhesion results from the formation of a large number of very small chemical bonds between the adhesive and the timber when these are brought into contact. In addition, for the formation of a strong-glued joint, it is essential that the adhesive penetrates the timber surface and keys into the porous layers below. This is known as mechanical adhesion.

The most important factors affecting the depth of penetration when gluing are as follows:

■ The amount of pressure applied to the joint (this forces the adhesive into the timber surface).
■ The viscosity (thickness) of the adhesive (thinner adhesives penetrate more easily than thicker adhesives).

Curing and cohesion

Apart from penetrating and adhering to the timber an adhesive must also have strength within itself (cohesion). Only solids have high cohesive strength; therefore every adhesive after application must be able to change from its liquid state to a solid state. This change takes place during the setting and curing of the adhesive in one of the following ways:

■ *By loss of solvent*, which takes place either by the evaporation of the solvent as with contact adhesives or by its absorption into the timber, as with emulsion adhesives.
■ *By cooling*. Some adhesives are applied hot or in a molten condition and subsequently solidify on cooling. This method has the advantage that a very fast set is achieved, as with animal and hot melt adhesives.
■ *By chemical reaction*, which is brought about by the addition of a hardener or catalyst or by the application of heat, as with synthetic resin adhesives.
■ *Combination*. Many adhesives set by using a combination of solvent loss, cooling and chemical reaction when mixed or brought into contact.

Classification of adhesives

Timber adhesives are made from either naturally occurring animal or vegetable products (glues) or from synthetic resins.

Adhesives that will set at room temperature are known as *cold setting adhesives* and those that require heating to a temperature of around 100°C are known as *hot setting adhesives*. In addition some adhesives require heating between these two ranges and are called intermediate temperature setting adhesives.

The **durability** of an adhesive is important, as it must retain its strength under the conditions it will be subjected to during its service. Durability can be tested by exposure tests over long periods, or by quick tests made over a few days, that subject the adhesive to heating, soaking, cooling and micro-organisms. Timber adhesives can be classified into one of the following durability classes.

- *Weather and boil-proof (WBP)* adhesives have a very high resistance to all weather conditions. They are also highly resistant to cold and boiling water, steam and dry heat, and micro-organisms.

- *Boil resistant (BR)* adhesives have a resistance to boiling water and a fairly good resistance to weather conditions, and are highly resistant to cold water and micro-organisms, but they will fail on prolonged exposure to the weather.

- *Moisture resistant (MR)* adhesives are moderately weather resistant and will withstand prolonged exposure to cold water but very little to hot water. They are also resistant to micro-organisms.

- *Internal (INT)* adhesives are only suitable for use in dry internal locations. They will fail on exposure to weather or in damp conditions, and are not normally resistant to micro-organisms.

The main types of **adhesives** used in the woodworking industry are illustrated in (Figure 1.107):

Types of natural adhesives

- **Animal glue** (INT) is also known as **Scotch glue**. It is made from animal hides and bones, but it is rarely used now, except in some small shops, because of the time taken to prepare and its limitations. These glues are supplied in pearl or cake form and must be broken up, soaked and heated before use.

- **Casein adhesive** (INT) is manufactured from soured, skimmed milk curds, which are dried and crushed into a powder. An alkali and certain fillers are added to the powder to give it its gap-filling properties. Its main use is for general joinery although it is inclined to stain some hardwoods, particularly oak. Little preparation is required as the powder is simply mixed in a non-metal container with a measured quantity of cold water and stirred until a smooth creamy consistency is achieved.

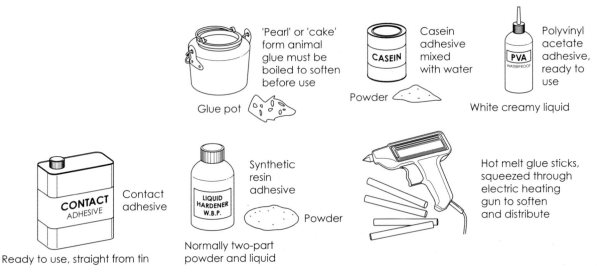

▲ Figure 1.107 Types of adhesive

Many **synthetic adhesives** require heat in order to flow. They fall into two main classes, thermoplastic and thermosetting:

- **Thermoplastic adhesives** set by heating and cooling and will soften again by reheating or by use of a solvent.
- **Thermosetting adhesives** undergo a chemical change that causes the adhesive to set and solidify on cooling. This is an irreversible change and reheating or solvents cannot soften the adhesive.

Types of synthetic resin adhesive:

- *Phenol formaldehyde (WBP)* is a two-part adhesive, the resin and the hardener, which are mixed together as required and gives a dark brown glue line. Classified as thermosetting adhesive that sets at either high temperatures or upon the addition of a catalyst. It is mainly used for exterior plywood, exterior joinery and timber engineering.

- *Resorcinol formaldehyde (WBP)* is a dark purplish-brown thermosetting adhesive that is classified as cold setting and sets by the addition of a hardener. As the resorcinol is very expensive it is often mixed with the cheaper phenol, making phenol/resorcinol formaldehyde adhesive. This has the same properties as the pure resorcinol adhesive, although a higher setting temperature is required. It is sold as a liquid to which a powder or liquid hardener is added and is mainly used for timber engineering and marine work.

- *Melamine formaldehyde (BR)* is a colourless thermosetting adhesive that sets at high temperatures and is suitable for use where the dark colour of the phenol and resorcinol adhesives are unacceptable. Also widely used in the production of plastic laminates.

- *Urea formaldehyde (MR)* is also a colourless thermosetting adhesive that will set at either high or low temperatures. Urea is often mixed with the more expensive melamine or resorcinol resin to form fortified urea formaldehyde adhesive that has an increased durability to BR class. Urea formaldehyde is available as either a one-part adhesive that is mixed with water or as a two-part adhesive with a separate hardener. Its main uses are for general joinery, furniture and plywood.

- *Polyvinyl acetate glue (INT)* commonly known as PVA or white glue is a thermoplastic adhesive, which is widely used for furniture and internal joinery. No preparation is required as this adhesive is usually supplied as a white creamy liquid in a nozzled, polythene bottle. It does not stain timber, but some types can affect ferrous metal. PVA glue is one glue that does not blunt the cutting edge of tools. Modified PVA adhesives are also available to increase gap-filling capacity, moisture resistance rated (MR) and with a surface etching capability for use on pre-finished surfaces such as melamine-faced and lacquered boards.

- *Polyurethane (MR)* an adhesive that is normally supplied in tubes and is applied using a skeleton gun. It expands and foams up on application making the use of cramps essential. Suitable for use on timber, metal, plastic, ceramics and a wide range of other porous and non-porous surfaces.

- *Aliphatic (INT)* a yellow resin glue with similar properties to PVA, but has a better initial tack and a faster setting time, which helps to reduce the cramping period. The yellow glue line is also less noticeable when used with darker timbers.

- *Contact adhesive (INT)* is available in either a solvent or water-based form. It is mainly used for bonding plastic laminates and sheet floor covering. The adhesive must be applied to both surfaces and allowed to become touch dry before being brought together. This normally takes between ten and thirty minutes depending on the make used. Once the two surfaces touch, no further movement or adjustment is normally possible, as an immediate contact or impact bond is obtained. Modified types are available which allow a limited amount of adjustment to be made after contact. Solvent-based contact adhesives must be used in a well-ventilated area where no smoking or naked lights are allowed. Water-based contact adhesives are safer to use but take longer to dry.

■ *Hot melts (INT)* are made from ethylene vinyl acetate (EVA) and are obtained in a solid form. They become molten at very high temperatures and set immediately on cooling. Small hand-held electric glue guns and automatic edging machines normally use this type of adhesive. Hot melt glue is also available in thin sheets for veneering work. The glue is placed between the veneer and board then activated with a heated domestic iron.

■ *Epoxy resins* are two-part adhesives that are useful for bonding wood, metal, glass and some plastics. The adhesive is normally supplied in two tubes or in a double-barrelled syringe, the two parts being mixed in equal proportions to a stiff paste.

■ *Cyanoacrylate adhesive* commonly known as Superglue™, although not widely used in woodworking, is useful for repairs and instant tacking of small objects, mouldings and mitres, where an instant bond is required. An amount of moisture is required for activation, which can be provided by a spray for an instant bond.

Using adhesives safely

Adhesives can be harmful to your health over time. Many adhesives have an irritant effect on contact with the skin and may result in dermatitis. Some are poisonous if swallowed, while others can result in narcosis if the vapour or powder is inhaled. These and other harmful effects can be avoided if proper **safety precautions** are taken:

■ Always follow the manufacturer's instructions.
■ Always use a barrier cream or disposable protective gloves.
■ Do not use those with a flammable vapour near a heat source.
■ Always provide adequate ventilation.
■ Avoid inhaling toxic fumes or powders.
■ Thoroughly wash your hands before eating or smoking and after work with soap and water or an appropriate hand cleanser.
■ In the case of accidental inhalation, swallowing or contact with eyes, medical advice should be sought immediately.

There are various guidelines, which are important to follow when applying adhesives (Figure 1.108). Always read the manufacturer's instructions as the following general rules may not always apply:

■ *Timber preparation.* The timber should be seasoned, preferably to the equilibrium moisture content it will obtain in use. The timber should be planed to a smooth even surface and all dust must be removed before gluing. The gluability of a timber surface deteriorates with exposure; therefore the time between preparation and gluing should be as short as possible.

■ *Adhesive preparation.* Adhesives that consist of two or more components must be mixed accurately in accordance with the manufacturer's instructions. It is normally advisable to batch the component parts by weight rather than volume.

■ *Application of adhesives.* They may be applied either by brush, roller, spray, spatula or mechanical spreader. For maximum strength a uniform thickness of adhesive should be applied to both sides of the joint. Adhesives that set by chemical reaction begin to react as soon as the components are mixed. This reaction rate is dependent mainly on the temperatures of the adhesive, timber and surrounding room or workshop. These factors must be taken into account to ensure that pot life of the adhesive is not exceeded; otherwise the strength of the joint will be affected.

■ *Assembly time.* This is the elapsed time between the application of the adhesive and the application of pressure. Some adhesives benefit in strength if they are allowed to partly set before the surfaces are brought into contact (*open assembly time*). Other adhesives require a period to thicken while the surfaces are in contact but before pressure is applied (*closed assembly time*). These times will be specified by the manufacturer and must be carefully controlled to ensure that the adhesive is squeezed out when the pressure is applied.

■ *Pressure* should be applied to the glued joint in order to: spread the adhesive uniformly; squeeze out excess adhesive and pockets of air; ensure close contact between the two adjoining surfaces. This pressure must be sustained until the joint has developed sufficient strength (*cramping period*).

■ *Temperature* will affect the speed at which strength develops in an adhesive and thus the cramping period. Check the temperature range of the adhesive before use to make sure the workshop is not too hot or too cool for the process.

■ *Curing.* This is the process that leads to the development of full strength and resistance to moisture. It starts during the cramping period and is completed while the components are in storage prior to use (conditioning period). Curing is also dependent on temperature and for many adhesives can be can be speeded up by heating.

■ *Radio frequency (RF) heating* can be used to shorten the time in cramps, but requires specialist equipment and knowledge and can be potentially dangerous. It involves sandwiching the timber component between metal plates (electrodes) that are connected to a source of radio frequency

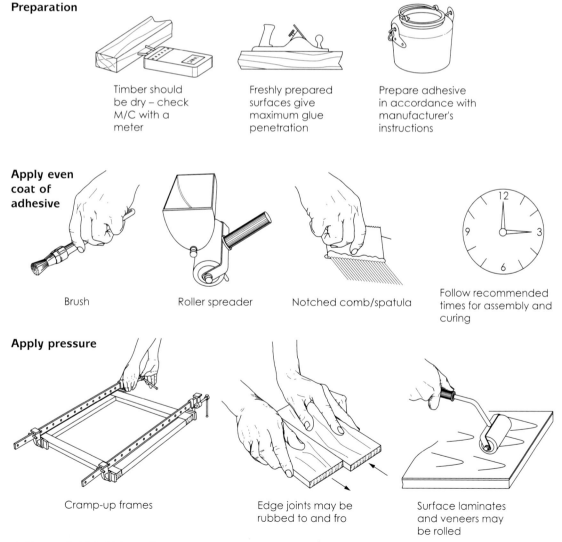

Preparation

Timber should be dry – check M/C with a meter

Freshly prepared surfaces give maximum glue penetration

Prepare adhesive in accordance with manufacturer's instructions

Apply even coat of adhesive

Brush

Roller spreader

Notched comb/spatula

Follow recommended times for assembly and curing

Apply pressure

Cramp-up frames

Edge joints may be rubbed to and fro

Surface laminates and veneers may be rolled

▲ **Figure 1.108** Using adhesives

energy. When positive voltage is applied to one electrode and a negative voltage to the other the timber molecules (minute particles) which all have a positive and negative end will tend to turn so that their charged ends face their like-charged electrodes. On reversing the voltage the molecules will tend to turn back immediately, causing internal friction and thus heat. The amount of heat is dependent on the frequency of the voltage reversals – quicker reversals generate higher temperatures. Glue lines can be cured in a matter of minutes or even seconds using this process.

■ *Storage life* refers to the period in which the materials forming the glue can be stored and still remain usable. After being stored for a certain length of time, glue becomes useless. This time varies with different types of glue. All glues should be stored in cool frost-free conditions.

■ *Pot life* refers to the time available for using the glue after it has been mixed. After a certain amount of time most glues become too thick to work with ease. Pot life and setting time depends to a large extent on the temperature. High ambient temperatures shorten life.

■ *Gap filling* refers to adhesives that are capable of filling gaps up to 1.3 mm wide without affecting the strength of the joint.

Chapter Two

Care and use of hand tools

This chapter covers the work of site carpenters and bench joiners. It is concerned with the principles of using, maintaining and sharpening hand tools. It includes the following:

- use of hand tools for measuring, marking, ripping, cross-cutting, planing, rebating and grooving, drilling and boring, recessing, driving fixings and inserting screws
- use of abrasive papers
- maintenance and sharpening of woodworking hand tools
- safe working practices when using, maintaining and sharpening woodworking hand tools
- typical requirements for workshop equipment.

The skills required to use and sharpen hand tools, are relatively easy to acquire. However they will take much patience and practice to perfect, but the rewards will be with you for life. Tool skills coupled with woodworking practices will be evident in the quality of the finished product, for better or worse!

Measuring and marking out

Accuracy is a simple matter, take care to mark the timber to size and then cut to the marked line. This way your components cannot fail to fit exactly together.

Timber cannot be accurately marked out for cutting until it has been first prepared with flat faces and square parallel edges. Face and edge marks are applied to indicate that the two adjacent surfaces are true and square to each other (Figure 2.1). It's from these surfaces that marking out progresses. The face side mark is a looped character, like a figure nine, with its leg extending to the face edge. The face edge mark is indicated by an upside down letter 'V' with its apex pointing towards the face side and joining up with the leg of the face side mark.

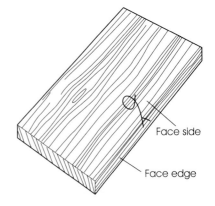

▲ **Figure 2.1** Face side and edge marks

Rules and tape measures

You should be aware of the trade terms used in measuring a piece of wood.

The **length** of a piece is taken along the grain. The **width** of a piece is the distance across the widest face from edge to edge. The **thickness** of a piece is the distance between the faces, Figure 2.2.
Rules are used to mark out and check dimensions (Figure 2.3).

The **four-fold metre rule** was traditionally the basic woodworkers' rule. They are still available in either wood or plastic and with markings in imperial or metric measurements or a combination of both. Some patterns have a chamfered edge on the first fold for more accurate marking off.

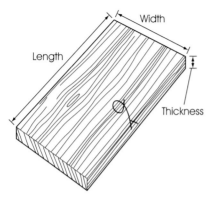

▲ **Figure 2.2** Terms used in measuring

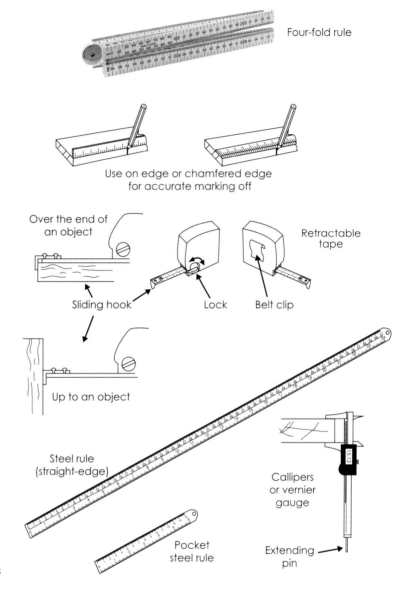

▶ **Figure 2.3** Rules and tape measures

Retractable tape measures are in common use by many woodworkers; they fit easily in a pocket or clip over the belt. The flexible steel tape is enclosed in a compact spring-loaded container. Many models can be locked in an extended position and be retracted back into its case automatically at the touch of a button. Typical tape lengths for woodworking use are between 3 and 8 metres. On the end of a tape is a hook to aid measuring long lengths. It is designed to slide in and out to compensate for its thickness when measuring up to an object or over the end of an object.
For accuracy check before every use that the slide is working and not damaged or stuck with surplus glue, etc.

Steel rules are considered essential for accurate workshop use. A 600 mm or 1 m long rule in required on the bench and a 150 mm rule for your pocket. The longer steel rules also have a dual use as a straightedge.

Dial callipers or **vernier gauges** are useful in the workshop, principally for accurate thickness measurement. The less expensive types, with either a dial scale or digital readout and capable of measuring to 0.25 mm, are most suitable for woodwork use.

Other uses for rules

To divide the width of a board into equal parts see Figure 2.4:

- ■ Place the rule diagonally across the board.
- ■ Adjust the rule to give the required number of parts.
- ■ Mark off with a pencil.

Rules can be used as an aid when 'ruling off' lines parallel to an edge, for saw cuts, etc. Position the pencil against the end of the rule; grip the rule between the thumb and first finger using the knuckle as a guide against the edge of the wood, while pulling the rule towards you. When marking parallel lines close to an edge dispense with the rule and use the tip of the first finger as a guide as illustrated in Figure 2.5. Take care with sharp corners or irregular grain to avoid cut fingers and splinters.

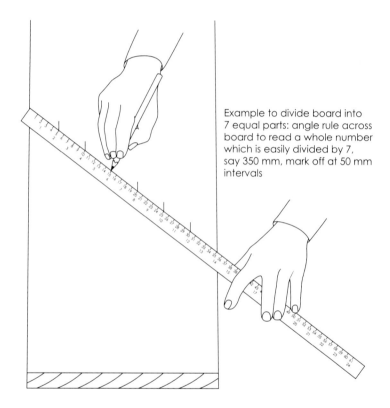

Example to divide board into 7 equal parts: angle rule across board to read a whole number which is easily divided by 7, say 350 mm, mark off at 50 mm intervals

◀ **Figure 2.4** Use of rule diagonally to divide into equal parts

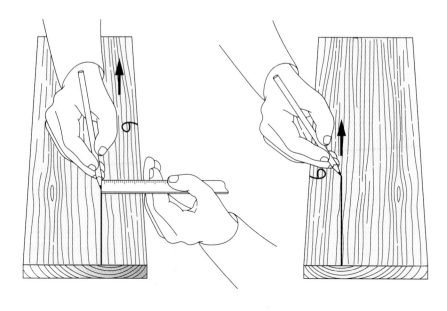

◀ **Figure 2.5** Marking out and ruling parallel lines

Squares and sliding bevels

Try squares

Try squares are used to mark lines at right angles across from the face side or face edge of a piece of wood (Figure 2.6). In addition, as their name suggests, they test pieces of wood or made-up joints etc. for square. The woodworker's try square has a steel blade, which is often 'blued' to prevent it rusting, and a rosewood stock.

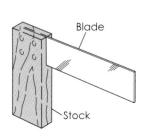

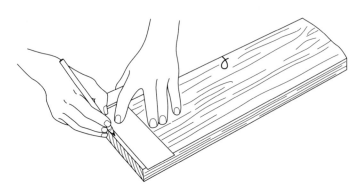

▲ **Figure 2.6** Using a try square

In use the blade of the square should rest flat on the piece with the stock pressed up firmly to the edge. Mark out using the outer edge of the blade, keeping the inner edge for testing. Most marking out is done with a pencil, however shoulder lines for joints are often marked with a marking knife to cut the surface fibres cleanly, leaving a guide for the saw to follow.

Checking for accuracy. To ensure accuracy you should periodically test your square for trueness as illustrated in Figure 2.7:

■ Place the stock against an edge known to be perfectly straight and mark a line.
■ Turn the square over and check the previously marked line.
■ If the square is 'true', they line up perfectly.
■ Any misalignment between the blade and line will result in inaccuracies.
■ Time to replace the square in case of misalignment, taking extra care in the future not to misuse it or accidentally drop it.

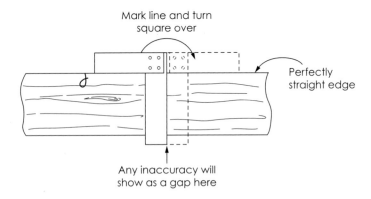

Mark line and turn
square over

Perfectly
straight edge

▶ **Figure 2.7** Checking a square
for trueness

Any inaccuracy will
show as a gap here

Checking for square. When using a square to check the squareness or flatness of a piece of timber, a good light source behind the blade is essential (Figure 2.8).

Place the stock against the face of the piece with the blade just touching the edge, slide the square along the edge, watching out for any light appearing under the blade.

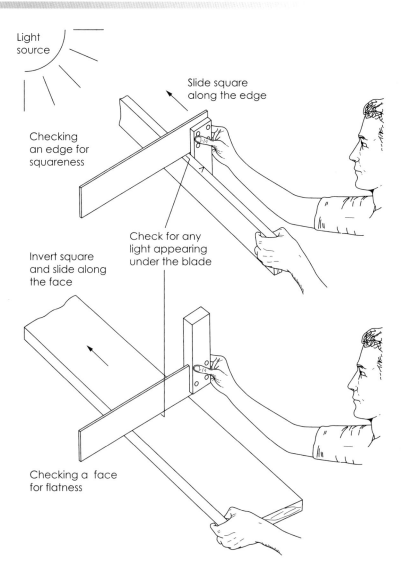

Light source

Slide square along the edge

Checking an edge for squareness

Check for any light appearing under the blade

Invert square and slide along the face

Checking a face for flatness

◄ **Figure 2.8** Using a try square to check for squareness and flatness

Flatness is checked with the blade placed on its edge across the face. Again, slide the square along the piece: any unevenness is shown up by light appearing under the blade.

Mitre square

This is similar to a try square with a steel blade and a rosewood stock but the blade of this square is set permanently at an angle of 45° to the stock, for marking out and testing mitres (Figure 2.9).

Combination square

All metal construction with a stock that can slide along the blade (Figure 2.10). Useful for ruling off and measuring depths as well as marking out and testing both right angles and mitres. In addition the blade can be removed for use as a rule. Often the stock incorporates a spirit level for checking plumb and level. Before use, ensure that the screw securing the blade to the stock is fully tight. Figure 2.11 illustrates typical uses.

▼ **Figure 2.9** Mitre square

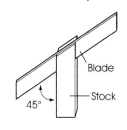

Blade

Stock

45°

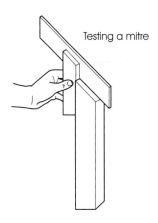

Testing a mitre

◄ **Figure 2.10**
Combination square

Sliding bevel

An adjustable square that can be set to any angle and can be used for marking out and testing angles, bevels and chamfers. It consists of a hardwood, plastic or metal stock and a sliding steel blade that is normally secured with a machine screw or wing nut (Figure 2.12).

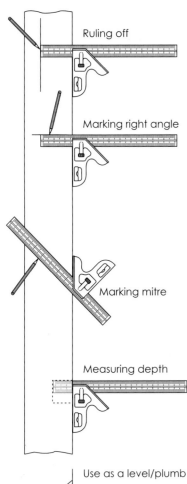

Ruling off

Marking right angle

Marking mitre

Measuring depth

Use as a level/plumb

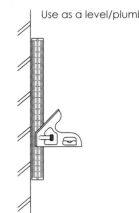

▲ **Figure 2.11** Uses of a combination square

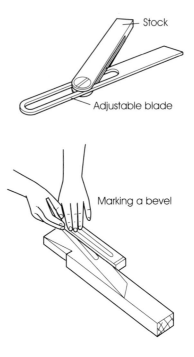

Stock

Adjustable blade

Marking a bevel

▲ **Figure 2.12** Sliding bevel

Marking knives

Marking knives (Figure 2.13) are used in preference to a pencil for accurately marking shoulder lines. This procedure is used before cutting shoulders with a tenon saw, particularly when using hardwood. The knife is sharpened like a chisel with one bevel.

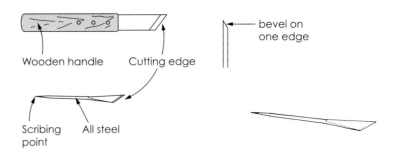

Wooden handle   Cutting edge

bevel on one edge

Scribing point   All steel

▲ **Figure 2.13** Marking knives

The flat edge is used against the blade of a try square and the bevel on the waste side of the joint.

With the knife held at 90° to the job, draw it along the outside edge of the square to give a V-line across the piece. This cleanly cuts the surface fibres to prevent tearing. Before sawing, the bevel side of the line may be enlarged with a chisel, to create a channel that acts as a guide for your saw and aids a clean start as illustrated in Figure 2.14.

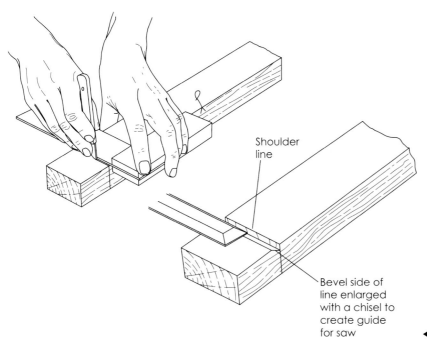

Shoulder line

Bevel side of line enlarged with a chisel to create guide for saw

◀ **Figure 2.14** Use of a marking knife

Gauges

The surest way to score cutting lines on a piece of timber parallel to an edge or end is with a gauge. This consists of a wooden stem containing the steel marking pin or blade and a wooden stock, which slides along the stem and is locked off in the required position with a turn-screw (Figure 2.15). Gauges are usually made from either beech or rosewood. Better quality gauges have brass strips inlaid across the working face of the stock to reduce wear.

- A **marking gauge** is fitted with a steel pin or spur for marking with the grain or across the end grain. Since the pin tends to tear the surface fibres it is not recommended for use across the grain.
- A **cutting gauge** is fitted with a blade in place of the pin, to cut a clean line across the face grain. It is also useful for cutting laminate strips, thin wood sections and easing the sides of grooves or rebates.
- A **mortise gauge** has two pins, the inner one is adjustable to enable the scoring of parallel lines for mortise and tenon joints.
- A **combination gauge** is dual purpose: it can perform as either a marking gauge or a mortise gauge by simply turning over.

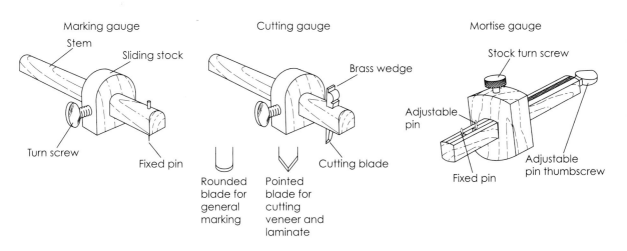

Marking gauge
Stem
Sliding stock
Turn screw
Fixed pin

Cutting gauge
Brass wedge
Cutting blade
Rounded blade for general marking
Pointed blade for cutting veneer and laminate

Mortise gauge
Stock turn screw
Adjustable pin
Fixed pin
Adjustable pin thumbscrew

▲ **Figure 2.15** Carpenter's gauges

Using gauges

A correct grip is essential to scoring accurate gauge lines. Grip the stock and stem, placing your thumb on the stem close to the pin and your index finger over the top of the stock to control the gauge's tilt. Your other fingers hold the stem and push the stock against the workpiece to be marked. Tilt the gauge with the pin trailing into the work and steadily slide the gauge forward. If the point digs into the work, push down on your index finger to adjust the tilt. (See Figure 2.16)

To centre the gauge's pin on the workpiece for halving joints, etc:

1. Approximately set the stock using a rule.
2. Mark from both faces.
3. Reset the stock so that the pin is central between the two marks.
4. Check that it is central again from both faces.
5. Fine adjustment of the gauge can be made by gently tapping the stem on the bench top. Ensure the screw that secures the stock to the stem is tight before use.

To set a mortise gauge, slacken the screw that secures the stock to the stem. Adjust the thumbscrew, so that the distance between the pins is the exact width of the chisel. Set the stock so that the pins are in the required position, check from either face for central mortises, or line up the mortise with the rebate or mould for offset mortises. Re-tighten the stock, securing the screw before use.

Cutting gauges are set up and used in the same way as a standard marking gauge. Like a marking knife its blade should be sharpened on one side only with the bevel facing the stock.

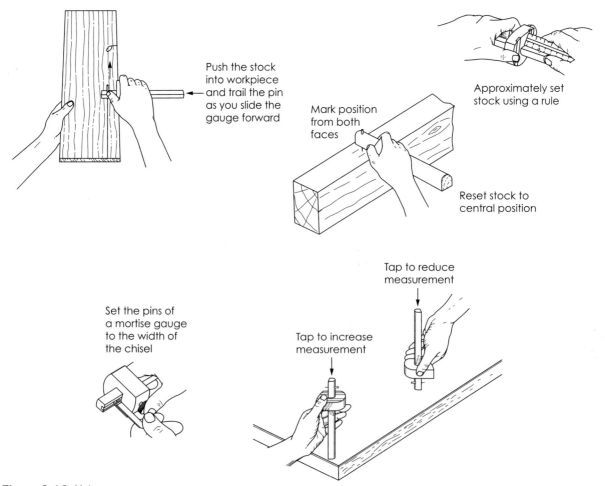

Push the stock into workpiece and trail the pin as you slide the gauge forward

Mark position from both faces

Approximately set stock using a rule

Reset stock to central position

Tap to reduce measurement

Set the pins of a mortise gauge to the width of the chisel

Tap to increase measurement

▲ **Figure 2.16** Using gauges

Saws

There is a wide variety of saws available to the woodworker; in general they can be considered in four main groups:

- handsaws
- backsaws
- frame-saws
- narrow-blade saws.

Cutting action

In order to prevent the saw blade from jamming in the timber, its teeth are 'set', that is each alternate tooth is bent outwards to make a saw-cut, or kerf, just wide enough to clear the blade (Figure 2.17).

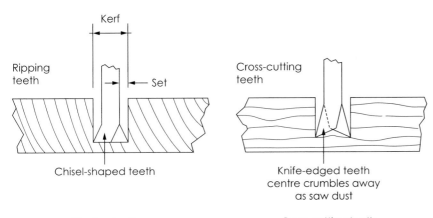

Ripping teeth

Cross-cutting teeth

◀ **Figure 2.17** Saw teeth

Ripping teeth are chisel-end teeth and the cutting action is in fact like a series of tiny chisels, each cutting one behind the other.

Cross-cutting teeth have knifepoints to sever the fibres of the timber. These points are so arranged that they cut two knife lines close together. The centre fibres between these knife lines crumble away as sawdust.

Handsaws

These are used for the preliminary cutting of components to size. They take three forms (Figure 2.18):

- The **ripsaw** is the largest, up to 750 mm long with 3–6 teeth per 25 mm (i.e. three to six teeth per inch (or 3–6 TPI) in imperial measurements). They are used for cutting along the grain only.
- The **cross-cut saw** is up to 650 mm long with 6–8 teeth per 25 mm or TPI and is mostly used for general-purpose cross-cutting to length.
- The **panel saw** is the smallest of the handsaws at around 550 mm in length, with 10 teeth per 25 mm or TPI. Generally used for cutting up sheet material.

Cross-cut and panel saws can also be used for ripping but progress will be slower.

Ripsaws are almost impossible to use for cross-cutting as their tooth shape will cause it to 'jar' and tear the fibres. Good quality handsaws have a tapered ground blade that is thicker near the teeth and get thinner towards its upper edge. This gives it added clearance in the kerf and aids to guide the saw.

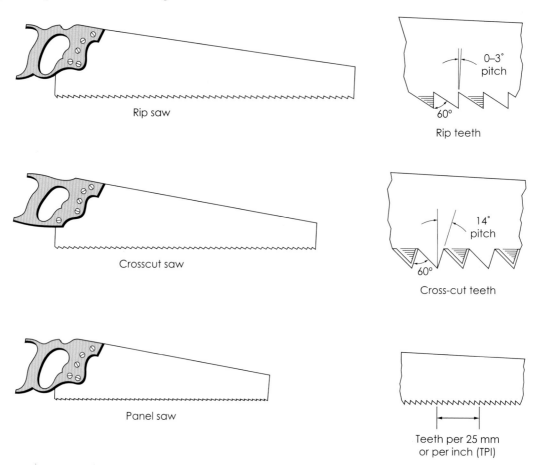

▲ Figure 2.18 Handsaws

Choosing a saw

Check the metal quality when choosing a saw to purchase: a firm tap with your knuckle should produce a nice ringing sound, not a flat thud. Try bending the blade to a 'U' shape: on release, it should return perfectly straight.

Hardpoint saws. Saw teeth will dull (become blunt) very quickly when cutting MDF and chipboard materials. Many woodworkers will purchase an inexpensive hard-point saw for such occasions. Hardpoint saws have teeth that retain their edge longer (do not dull as quick) but when they do dull they are considered as throwaway items as they cannot be resharpened.

Backsaws

These are used for bench work. They have their upper edge stiffened with a brass or steel folded strip to prevent twisting or buckling during use (Figure 2.19):

- The **tenon saw** is used for cutting shoulders of joints and general bench cross-cutting. Blade lengths range from 250 mm to 350 mm with 10–14 teeth per 25 mm (TPI). Some woodworkers find it useful to have two tenon saws, one with bevelled teeth for cross-cutting and the other with teeth re-cut square across their face for cutting along the grain.
- The **dovetail saw** is for cutting dovetails, mouldings and other delicate work. Blade lengths range from 200 to 250 mm with 16–20 teeth per 25 mm (TPI).
- The **gents' saw** is for the finest cutting of mouldings etc. Blade lengths range from 100 to 250 mm with up to 32 teeth per 25 mm (TPI).

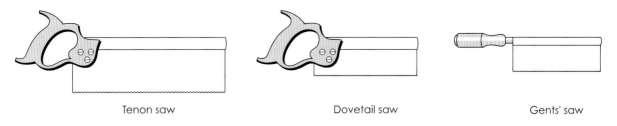

Tenon saw   Dovetail saw   Gents' saw

▲ **Figure 2.19** Backsaws

Frame saws and other saws

Frames saws have in general a fine replaceable blade which is held in tension by a frame (Figure 2.20):

- **Bow saw.** The traditional all wooden versions were once used as a general bench saw. Its blade could rotate in the frame and be used for ripping, cross-cutting and cutting curves. The modern metal frame saw with the same name, is really intended for cross-cutting carcassing timber (floor joists, etc.) especially in damp timber as the narrow blade gives less resistance than a handsaw.

- **Coping saw.** Designed to make curved cuts in wood and board material. A fret saw is similar, but has a deeper bowed frame for cutting further from the edge of a piece.

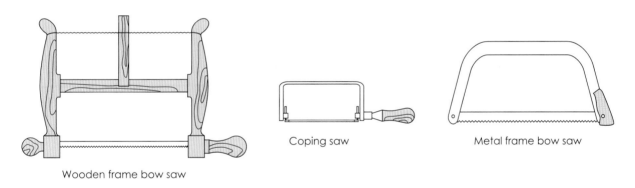

Wooden frame bow saw   Coping saw   Metal frame bow saw

▲ **Figure 2.20** Frame saws

- The **keyhole or padsaw,** as its name implies, is mainly used for cutting keyholes and other small shapes and holes away from the edge of the timber. (See Figure 2.21.) The **compass saw** is mainly used for cutting larger shapes or holes well away from the edge of the material.

- **Japanese saws.** These are becoming popular for fine woodwork. They have long teeth, a thin blade, fine kerf and generally cut on the pulling stroke rather than the pushing stroke. (See Figure 2.22.)

- **Mitre saws.** A modern frame saw set in a metal jig that can be set to cut at predetermined angles. Often used for the mitring of picture framing and fine mouldings, also for the accurate general cross-cutting of small components.

- **Hacksaws.** Frame saw used by the woodworkers to cut metal sections.

- **Floorboard saw.** The front end of this saw is curved with teeth continuing round the curve. It can be used to start a cut in the middle of a board. Its main use is for cutting heading joints, when forming access points in existing timber-boarded floors.

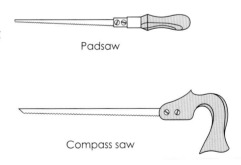

Padsaw

Compass saw

▲ **Figure 2.21** Narrow-blade saws

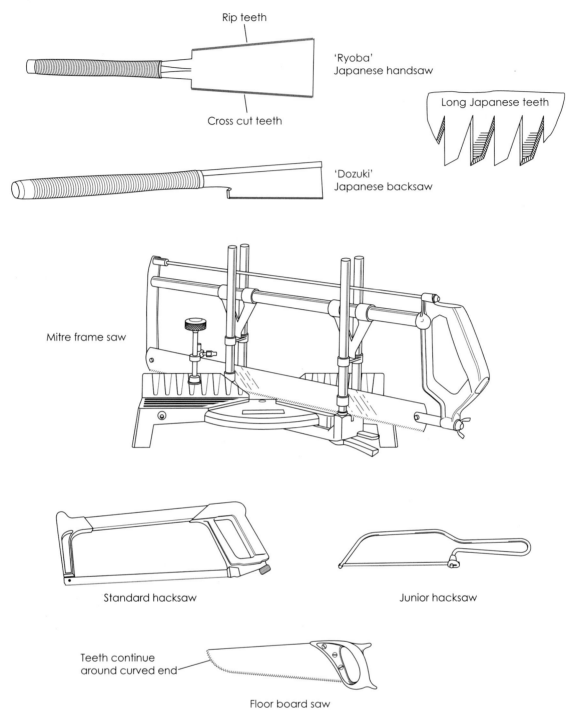

Rip teeth

'Ryoba' Japanese handsaw

Cross cut teeth

Long Japanese teeth

'Dozuki' Japanese backsaw

Mitre frame saw

Standard hacksaw

Junior hacksaw

Teeth continue around curved end

Floor board saw

▲ **Figure 2.22** Other types of saw

Method of saw use

This will vary depending on the material being cut, the direction of cut and where the work is being undertaken.

Grip the saw handle firmly. It helps to extend your index finger along the side of the handle as a guide to give you maximum control as you saw (Figure 2.23).

1. Start a cut by positioning your thumbnail on the marked line to give the saw a starting location.
2. Draw the saw backwards a few times, using short strokes and a low angle to give the saw a start.

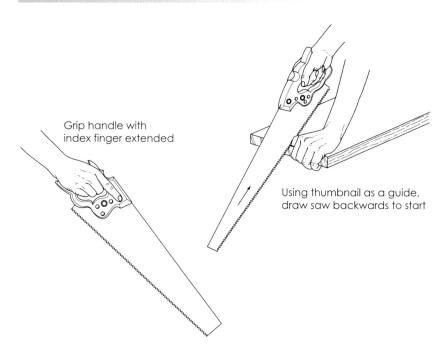

Grip handle with index finger extended

Using thumbnail as a guide, draw saw backwards to start

◀ **Figure 2.23** Method of saw use

3. Slacken your grip and continue sawing using full blade-length strokes.
4. Let the saw do the work, use a light pressure in the forward stroke and relaxing on the return. Your concentration should be on 'thinking' the saw along the line.

Ripping on stools. Use a cutting angle of about 60°. Two saw stools may be required for long lengths or wide boards. Start from beyond the first stool, continue between the stools and complete the cut beyond the second stool. If the grain closes on the blade causing binding, insert a wedge in the kerf to keep it open and apply a little candle wax to lubricate the blade (Figure 2.24).

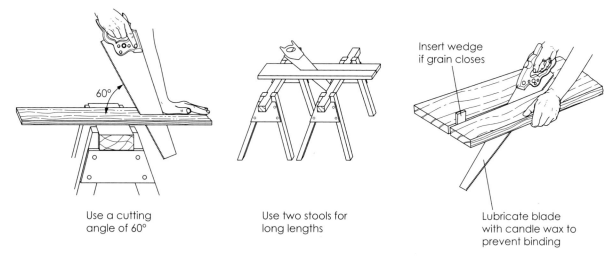

Insert wedge if grain closes

60°

Use a cutting angle of 60°

Use two stools for long lengths

Lubricate blade with candle wax to prevent binding

▲ **Figure 2.24** Ripping on stools

Cross-cutting on stools. Use a cutting angle of about 45°. Clamp the board secure with the aid of your knee. Support the overhanging end and use very gentle strokes towards the end of the cut to avoid splitting (Figure 2.25).

Ripping in the vice. Place the piece to be cut vertically in the vice when cutting the full length. Reverse the piece in the vice and saw from the other end to complete the cut.

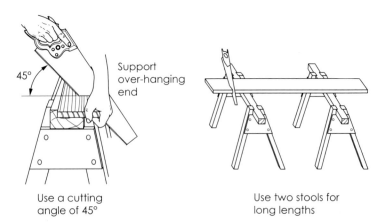

45°

Support
over-hanging
end

Use a cutting
angle of 45°

Use two stools for
long lengths

▶ **Figure 2.25** Cross-cutting on stools

When ripping joints, angle the piece in the vice, so that both the ripping line and the line across the end grain can be seen. The aim when ripping down a pencil line is to saw on the waste side, just leaving the pencil line visible on the job. With gauged lines you should saw at 'half mark', again on the waste side, taking out half the gauged line and leaving a trace of it still present (Figure 2.26).

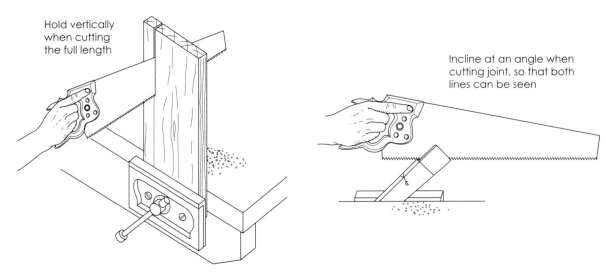

Hold vertically
when cutting
the full length

Incline at an angle when
cutting joint, so that both
lines can be seen

▲ **Figure 2.26** Ripping in the vice

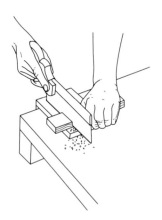

▲ **Figure 2.27** Cross-cutting on the bench

Cross-cutting on the bench. Use a bench hook held in the vice or butting over the front of the bench. This allows you to secure your job and also protect the bench (Figure 2.27).

Coping saws are often used to cut out the waste when forming dovetail joints. This is best achieved with the piece held vertically in the vice and about 50 mm protruding above the bench surface. When using coping or padsaws to cut internal shapes, pre-drill a hole on the waste side, to give a starting point (Figure 2.28).

Cutting sheet material. Use a fine tooth panel saw when cutting manufactured boards. Preferably have a second saw kept exclusively for this purpose, as the resin they contain will quickly blunt the teeth. Lay battens across two saw stools for support and place the sheet on top (Figure 2.29).

To prevent 'break-out' when cutting across the grain of plywood or when cutting faced or laminated boards, pre-score both faces of the board with a marking knife and cut on the waste side. Alternatively masking tape applied to both sides of the board and cutting through it has the same effect.

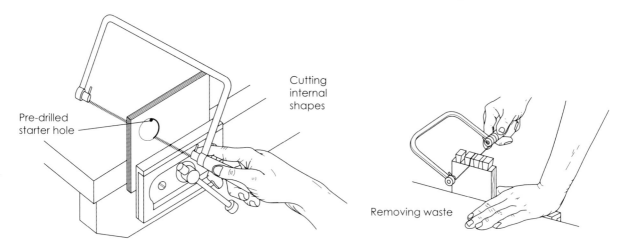

▲ **Figure 2.28** Use of coping saws

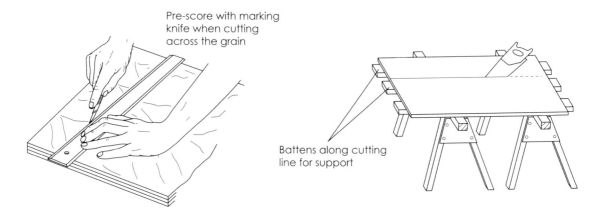

▲ **Figure 2.29** Cutting sheet material

Saw care and maintenance

Simplest and important saw care procedures include (Figure 2.30):
1. Protect blade teeth when not in use, with a plastic or timber sheath.
2. Hang up or store flat in a box to prevent distortion.
3. Lightly smear the blade with oil to prevent rusting and pitting.
4. Avoid cutting reclaimed timber as this often contains hidden nails and screws.

To keep saws working efficiently they must be regularly sharpened and set. Signs of a dull (blunt) saw are 'shiny' teeth tips, increased effort required during sawing and the saw binding in the cut.

Saw maintenance

The techniques used and equipment required for saw maintenance is similar for all types of saw. A saw in poor condition due to inaccurate sharpening or misuse (cutting through nails or screws etc.) may require all of the following operations: topping, shaping, setting and sharpening

The equipment required (Figure 2.31) for these operations include:
- A *flat file* for topping.
- A *triangular saw file*, which must be the right size for the saw to be sharpened. In general the saw tooth should rise just over halfway across the file face.
- *Saw set pliers* for setting the teeth.

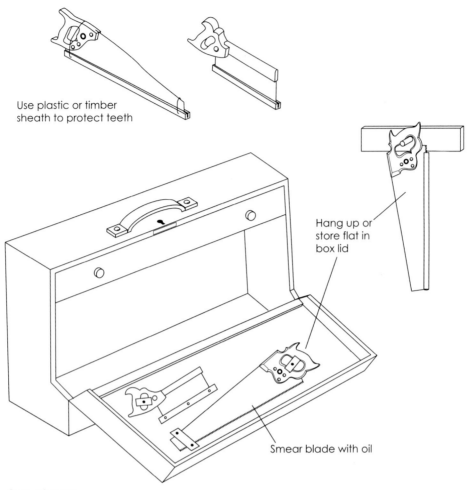

Use plastic or timber
sheath to protect teeth

Hang up or
store flat in
box lid

Smear blade with oil

▲ **Figure 2.30** Care of saws

- A *means of clamping* the saw blade firmly along the entire length, to prevent vibration during filing.
- *'Saw chops'* for backsaws and a *'saw horse'* for hand saws. These are simple wooden devices easily made by the woodworker.
- A medium grade *oilstone* or *slipstone* for side dressing.

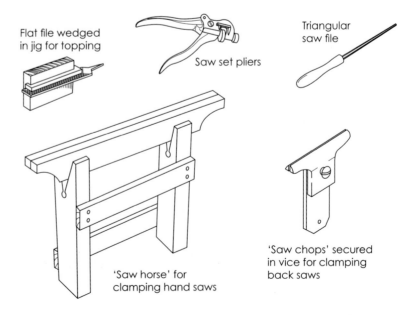

Flat file wedged
in jig for topping

Saw set pliers

Triangular
saw file

'Saw horse' for
clamping hand saws

'Saw chops' secured
in vice for clamping
back saws

▶ **Figure 2.31** Saw sharpening
equipment

Saw maintenance operations are shown in Figure 2.32:

- **Topping** is the first stage of sharpening a saw, when the teeth are at different heights due to poor sharpening in the past (known as cows and calves) or have worn unevenly, possibly by contact with metal objects. To hold the file square to the saw blade and in order to protect your fingers, wedge the file in a grooved block of timber. Lightly draw the file along the entire length of the saw. Removing just enough metal from the teeth tips to bring the larger teeth down to the same height as the smaller ones (i.e. levelling up the cows and calves) and also producing a 'shiner' (newly filed surface) on the tip of every tooth.

- **Shaping or reshaping** is the process of filing the teeth to their original size and shape. With the blade securely clamped, start filing from the handle end on the front edge of the first tooth bent away from you. Press the file into the gullet and file horizontally across at right angles to the blade. Two or three firm strokes may be required to remove half the 'shiner'. Repeat the process on each alternate gullet up to the tip of the blade. Turn the saw around and repeat the process on the other teeth not yet filed, this time removing the other half of the shiners, leaving all the teeth the same size and shape.

- **Setting** is the bending over of the teeth tips to give the blade clearance in the kerf. Adjust the saw set pliers to the number of teeth (TPI). Place the set over the blade, squeezing the handles firmly to set each alternate tooth. Turn the saw around and repeat the process to set the remaining teeth.

- **Sharpening** is putting a cutting edge on the teeth. Firstly, lightly draw a flat file along the blade, to recreate small shiners on the top of each tooth. Working from one end with the triangular file held horizontal, and

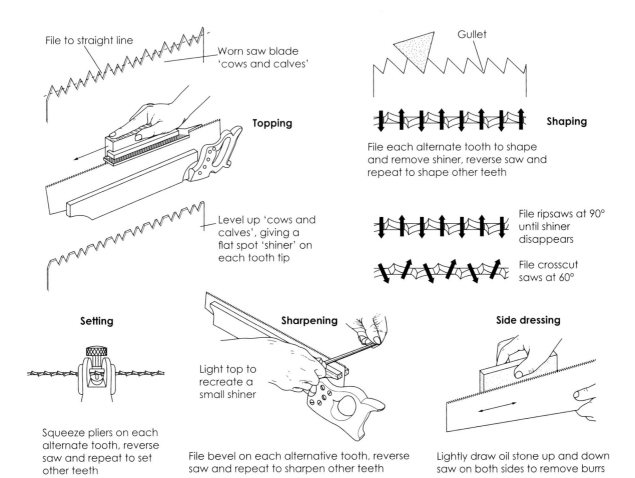

▲ **Figure 2.32** Saw maintenance

at the correct angle for the type of saw, lightly file each alternate tooth, until its shiner disappears. Filing the bevel on the front edge of one tooth also puts the bevel on the back edge of its adjacent tooth. After reaching the end, turn the saw around and repeat the process.

■ The final stage of sharpening is **side dressing,** to remove burrs and to ensure all the teeth are evenly set. Lightly draw a medium grade oilstone or slipstone, up and down the teeth on either side of the saw blade.

Although it is possible to maintain saws yourself, many woodworkers will send their saws periodically to a 'saw doctor' (specialist in saw maintenance) for a full reshape and sharpening, whilst extending the cutting life themselves by 'touching up' with a set and sharpen.

Planes and chisels

Planes

All planes cut wood by producing shavings. Some are used to produce flat, straight and true surfaces, whilst others are used to produce rebates, grooves, mouldings and curves. They may be considered in two groups: bench planes and specialist planes.

Bench planes

Various sizes, are used by most woodworkers (Figure 2.33) to dimension timber, prepare surfaces prior to final finishing and trim and fit both joints and components.

■ The **smoothing plane** is a finishing plane. It is used for smoothing up a job after the jackplane has been used, and for general cleaning up work. Its length is 250 mm.

■ The **jackplane** is mainly used for reducing timber to the required size and for all rough planing work. Its length of 375 mm enables it to be used satisfactorily for straight planing. Being an all-round, general-purpose plane it is ideal for both site and benchwork.

■ The **try plane** is the largest plane a craftsman uses. Its main use is for straight planing and levelling. This type of plane varies from 450 mm to 600 mm in length.

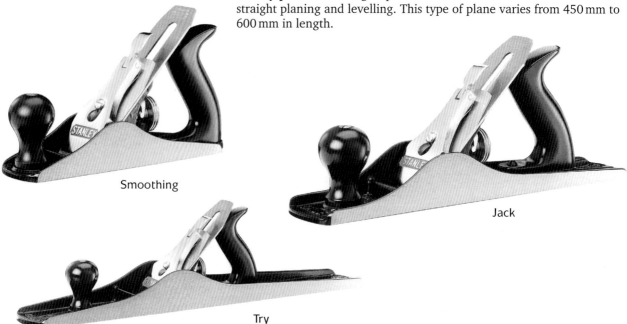

Smoothing

Jack

Try

▲ **Figure 2.33** Bench plane types

Cutting action and setting of bench planes

All bench planes are similarly constructed, with the same type of cutting unit (Figure 2.34). When planing, the thickness of the shaving and the smoothness of the finish is controlled by four main factors (Figure 2.35):

■ *The amount the cutting iron projects below the sole (bottom) of the plane*. Hold the plane up and sight along the sole. Turn the adjusting nut, so that the cutting iron projects by about 0.5 mm. Commence planing and re-adjust if required to achieve a clean, slightly transparent shaving.

■ *The alignment of the cutting iron*. This should be parallel with the sole. Sight along the sole to check that the cutting iron projects equally right across the mouth (opening in sole). If not it may be corrected by moving the adjusting lever sideways.

■ *The distance the back iron is set from the cutting iron edge*. This should be set between 0.5 mm for fine finishing work and up to 2 mm for heavier planing. To adjust, slacken the cap iron screw (preferably using a wide flat blade screwdriver, not the lever cap!) Set to the required

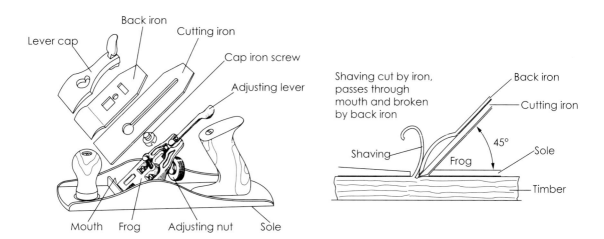

▲ **Figure 2.34** Bench plane exploded view and cutting action

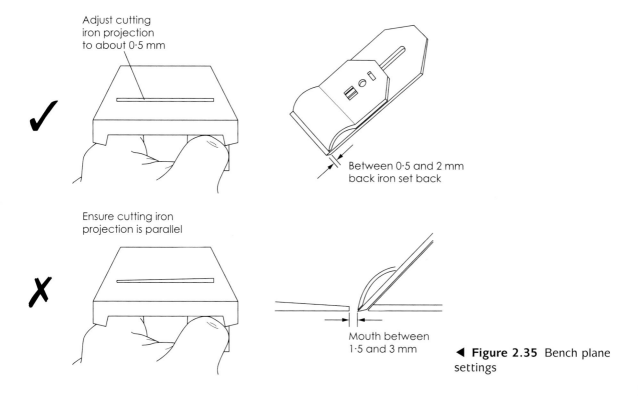

◀ **Figure 2.35** Bench plane settings

distance and retighten. At this stage it is also worthwhile checking that the back iron sits down closely on the cutting iron without any gaps, otherwise shavings will get jammed between them causing the plane to clog.

■ *The size of the mouth with the cutting iron in position.* Between 1.5 mm and 3 mm is suitable for most timbers. Use a wide mouth for heavy coarse shavings and a narrow mouth for fine work and timber with interlocking grain. The mouth size on bench planes is altered by adjusting the frog. Slacken the frog securing screws, turn the frog adjusting screw in or out to move the frog forwards or backwards. Retighten frog securing screws.

Using bench planes

To accurately plane long boards, requires the use of a long jack or try plane (Figure 2.36). A shorter smoothing plane will merely follow the existing bumps and hollows.

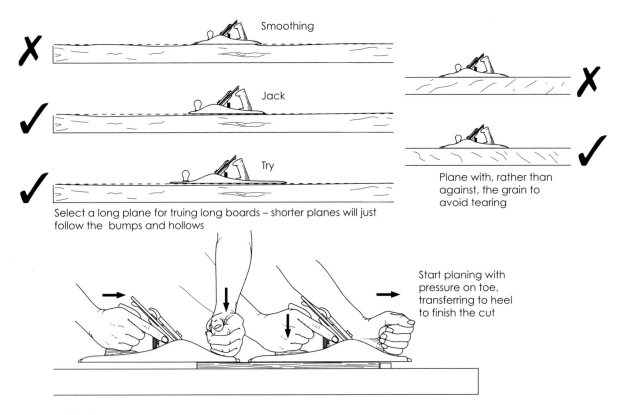

▲ **Figure 2.36** Using bench planes

Always try to plane with the grain to minimise the risk of tearing. Look at the edge of the board and aim the plane up any sloping grain. Planing down will result in tearing.

1. Begin planing with pressure on the front or toe of the plane. Planing requires pressure down on the job as well as moving forward.
2. Move forward over the job, transferring the pressure evenly over the toe and heal.
3. Finish the cut with pressure on the back or heel of the plane.

Chalk marks can be made across a board before planing to act as a guide. Chalking three or four lines across a board enables you to see where the plane takes a shaving and what part still needs to be planed.

When cleaning up large areas (Figure 2.37) such as a solid timber table top, use either a jack plane with a slightly curved cutting iron or a smoothing plane where the corners have been removed from the cutting

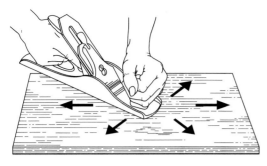
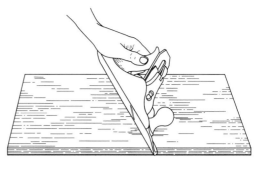

Plane wide tops diagonally | Use edge of sole to check flatness

▲ **Figure 2.37** Planing large areas

edge. This reduces the chance of creating corner tracks across the work. Use a fine setting and plane diagonally, taking care not to remove too much at the edges. Test the surface for flatness by tilting the plane to about 45° using the corner of the sole as a straightedge. Any light showing through indicates bumps and hollows for further planing.

Planing edges. Edges are normally planed square to a face side. Use your plane with the job in the centre of the sole. Steer the plane off to one side or another if you need to correct an angle. A finger curled under the sole at the front acts as a guide and assists steerage (Figure 2.38). It is often too difficult to balance a bench plane on very thin edges (10 mm or less). A shooting board can be used in these circumstances. The workpiece is supported at right angles to the plane and held against a stop at one end of the shooting board. Lay the plane on its side. Using pressure up to the workpiece and down onto the shooting board, slide the plane along.

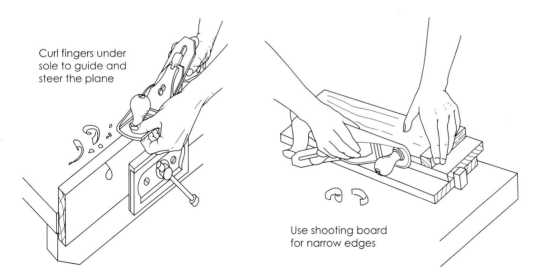

Curl fingers under sole to guide and steer the plane

Use shooting board for narrow edges

▲ **Figure 2.38** Planing edges

Framed joinery. Use a smoothing plane when cleaning up assembled framed joinery (Figure 2.39). A steady turning action towards the joints can help to prevent tearing where the grain direction changes. Tilt the plane across the joints to check for flatness, again any light showing through indicates bumps or hollows, requiring further work.

Planing end grain. Even when using a sharp set smoothing plane, planing end grain finely can often result in splintering at the corner. This can be avoided using a number of methods (Figure 2.40):

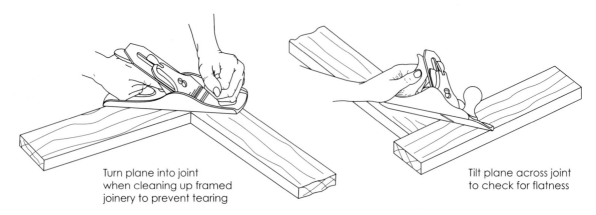

Turn plane into joint when cleaning up framed joinery to prevent tearing

Tilt plane across joint to check for flatness

▲ **Figure 2.39** Cleaning up frame joinery

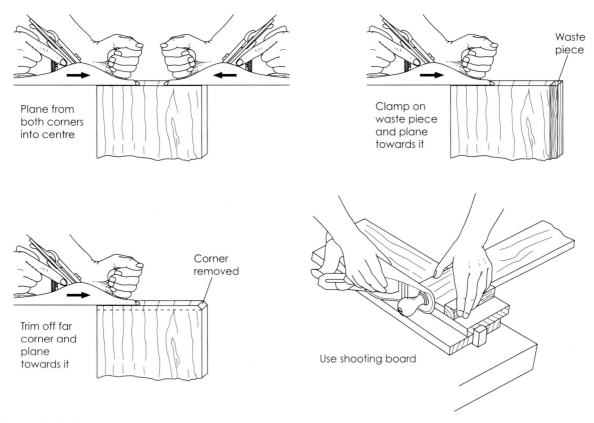

Plane from both corners into centre

Clamp on waste piece and plane towards it

Waste piece

Trim off far corner and plane towards it

Corner removed

Use shooting board

▲ **Figure 2.40** Planing end grain

- by planing from both corners into the centre of the end (not always easy to get a straight edge);
- trimming off the corner as a backing cut and planing towards it;
- clamping a piece of waste behind the corner and plane towards it; any splintering will occur in the scrap piece;
- using a shooting board.

Planing sawn timber

Preparing sawn timber by hand can be carried out using the following procedure (Figure 2.41). It is normal practice to complete each operation on every piece required for a job before moving on to the next operation.

1. Select the best faces and direction of working.

2. Use a jack plane to plane a face side straight and out of twist.

3. Winding strips can be used to sight along the length as these will accentuate the amount of twist.

4. Plane diagonally along the board if required to remove any twist.

5. Mark as face side. (Repeat operation on all other pieces required.)

6. Use a jack or try plane to plane the face edge straight and square to the face side.

7. Check periodically with a try square.

8. Mark as face edge.

9. Gauge to thickness using the stock of the gauge from the face side.

10. Plane down to the gauge lines.

11. Gauge to width from both faces, using the stock of the gauge from the face edge.

12. Plane down to gauge lines.

13. Use the blade of a try square or corner of the plane sole to check for flatness across the piece.

Select best face plane and mark as face side

Plane edge square to face and mark as face edge

Gauge the thickness from the face side

Plane to thickness down to the gauge lines

Gauge the width from the face edge

Plane to width down to the gauge lines

▲ **Figure 2.41** Preparing sawn timber

Other bench planes

■ *Corrugated sole versions of bench planes* are available (Figure 2.42). These create less friction and are used where resinous timber is regularly encountered. Lubricating the sole of standard bench planes can have a similar effect.

■ *Bench rebate plane.* This is a specialised version of the jack plane with the blade extending the full width of the sole (Figure 2.43). Also termed as a badger plane or carriage plane, it is mainly used for either forming or cleaning up large rebates. It has no fence or depth stop and is used with a guide batten pinned or cramped to the workpiece.

Corrugated sole plane for use with resinous timber

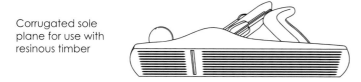

▲ **Figure 2.42** Corrugated sole bench plane

Specialist planes

With the exception of the compass plane, specialist planes have their cutting irons set with their bevel side up and no back iron. The bevel on the cutting iron acts as a back iron to break the wood shaving (Figure 2.44).

Various types of specialist planes are illustrated in Figure 2.45, along with their typical uses in Figure 2.46.

■ **Block planes** can be used for fine cleaning up and finishing work, both with the grain and for end grain. They are also useful for trimming the edges of manufactured boards and plastic laminates. Versions are available with low angled blades and adjustable mouths, to fine tune their cutting efficiency.

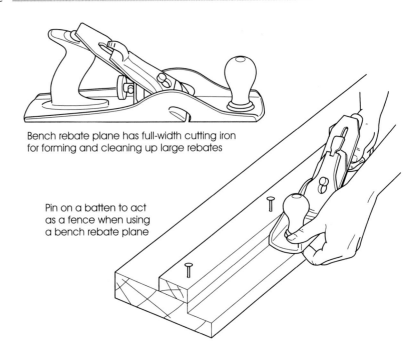

Bench rebate plane has full-width cutting iron
for forming and cleaning up large rebates

Pin on a batten to act
as a fence when using
a bench rebate plane

▶ **Figure 2.43** Bench rebate plane

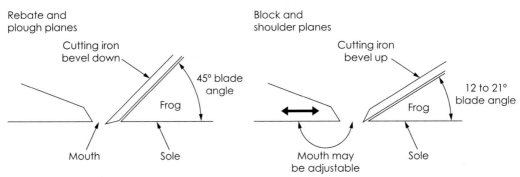

Rebate and
plough planes

Cutting iron
bevel down

45° blade
angle

Frog

Mouth   Sole

Block and
shoulder planes

Cutting iron
bevel up

12 to 21°
blade angle

Frog

Mouth may
be adjustable   Sole

▲ **Figure 2.44** Specialist planes – cutter arrangement

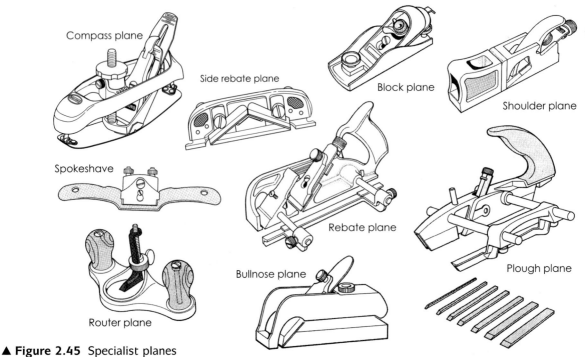

Compass plane

Side rebate plane

Block plane

Shoulder plane

Spokeshave

Rebate plane

Router plane

Bullnose plane

Plough plane

▲ **Figure 2.45** Specialist planes

- **Shoulder planes** are used for cleaning up and truing rebates and the end grain shoulders of tenon joints, etc.
- **Bull-nose planes** are shortened versions of a shoulder plane.
  Their cutting iron is mounted towards the front so that they can work close into corners, such as stopped rebates and chamfers. Some types have a removable front end, referred to as a chisel plane, which enables closer working into stopped ends.
- **Router planes** (often termed granny's tooth) are used to clean out the waste wood from housings and grooves. Various blade widths are available to suit the work in hand and the depth of cut is adjustable.

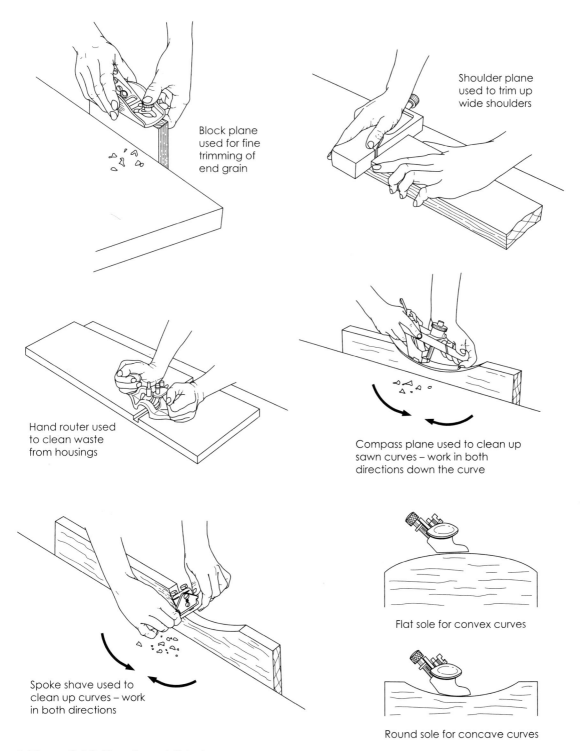

Block plane used for fine trimming of end grain

Shoulder plane used to trim up wide shoulders

Hand router used to clean waste from housings

Compass plane used to clean up sawn curves – work in both directions down the curve

Spoke shave used to clean up curves – work in both directions

Flat sole for convex curves

Round sole for concave curves

▲ **Figure 2.46** Use of specialist planes

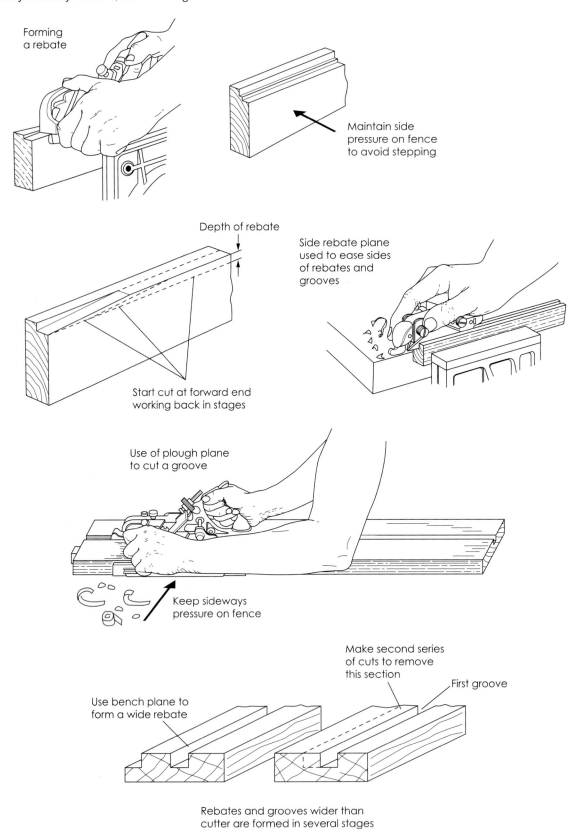

Forming a rebate

Maintain side pressure on fence to avoid stepping

Depth of rebate

Side rebate plane used to ease sides of rebates and grooves

Start cut at forward end working back in stages

Use of plough plane to cut a groove

Keep sideways pressure on fence

Make second series of cuts to remove this section

First groove

Use bench plane to form a wide rebate

Rebates and grooves wider than cutter are formed in several stages

▲ **Figure 2.46** Use of specialist planes *continued*

■ **Compass planes** are used to smooth and clean up sawn curves. Their cutting assembly is the same as bench planes, but the sole is made from a flexible steel strip. Enabling it to be adjusted to fit against internal (concave) or external (convex) curved shapes.

- **Spoke-shaves** are used for the final working and cleaning up of curved edges. The flat-soled spoke-shave version can be used for planing narrow edges and convex curves. Concave curves being worked with a round-soled spoke-shave. In use, frequent changes of direction may be required. To avoid tearing always work with the grain.

- **Rebate planes** are used to cut rebates. The size of rebate is controlled by the use of an adjustable side fence for width and a depth stop to control the maximum depth of cut. In use the fence must be firmly held against the face or edge of the work to prevent 'stepping'. The cutting iron has an additional mounting position at the front, so that it can work close into the corners of stopped rebates. A spur near the sole can be adjusted to cut the grain when planing rebates across the grain.

- **Plough planes** are used to cut plough grooves both along the grain and, with a spur, across the grain. A range of blade widths is supplied with the plane. They may also be used for rebates. Before commencing each job, set up the fence and depth stop, ensuring the fence is parallel to the plane body. As with rebate planes, the fence must be firmly held against the workpiece to prevent stepping. Rebates and plough grooves wider than the blades supplied may be formed in two or more stages.

- **Multiplanes** similar in appearance to a plough plane are available. These are supplied with an extended range of cutters, allowing rebating, ploughing and moulding operations.

The method of planing is the same for the rebate, plough and multiplane. The cut is started at the forward end of the piece and the plane gradually worked back until the final cut is along the full length.

Sharpening planes and chisels

Plane cutting irons (blades) and chisels are ground to a bevel at the 'sharp' end (Figure 2.47). An angle of 25º is the most suitable for general woodwork. A slacker angle results in a thin edge, which is easily damaged; any steeper and cutting will be difficult. A secondary bevel of 30º is honed on the very edge of the grinding bevel, to strengthen the bevel and provide a fine cutting edge.

The shape of a blade's cutting edge will depend on its use: most will be square. The corners of a smoothing plane blade are removed to prevent them digging in and ridging the surface. Jack plane irons are slightly rounded to prevent digging in and aid the quick, easy removal of shavings.

The sharpening of plane cutting irons and chisels is a two-stage process – grinding followed by honing.

Grinding is carried out on:
- *A sandstone,* which is a wet grinding process, water being used to keep the cutting iron cool and to prevent the stone becoming clogged.
- *A high speed carborundum grinding wheel,* which is a dry grinding process, although to prevent the cutting iron overheating, it can be periodically cooled in water. If the tool is allowed to overheat it will turn 'blue' and lose its hardness.
- *A coarse sharpening stone,* when access to a grinding wheel is not available

Honing is carried out on a fine sharpening stone, the fine side is for honing, the coarse side only being used when a grindstone is not available.

Stone materials

Sharpening stones are available in many materials:
- **Oilstones** such as natural Arkansas or synthetic silicon carbide. Coarse, medium and fine grades are available. In addition combination stones are available, which have two different grade stones glued back to back. Use the finer side for honing and the coarser side to occasionally replace

The secondary honing bevel strengthens the grinding angle and provides a fine cutting edge

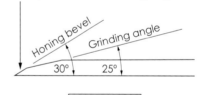

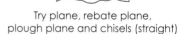

Try plane, rebate plane, plough plane and chisels (straight)

Jack plane (slightly round)

Smoothing plane (straight, corners removed)

▲ **Figure 2.47** Sharpening and shaping cutting edges

the grinding bevel. As their name suggests, oil is normally used to lubricate the stones and prevent them clogging with metal particles.

■ **Japanese waterstones** both natural or synthetic are quick cutting and available in finer grades than oilstones, but are more expensive. They are lubricated with water and it helps to build up a slurry on the wet surface prior to honing. This is realised by rubbing the surface with a chalk-like Nagura stone.

■ **Diamond stones** have a grid-like pattern of diamond particles bonded in a plastic base. They are available in various grades, but are expensive. Diamond stones can be used to flatten out worn oil and waterstones.

■ **Slipstones**, both oil and waterstones, are made in a teardrop section for honing inside ground gouges.

Sharpening process for planes and chisels

■ *Grind the iron* when required on a grinding wheel with the tool rest set at an angle of 25° (Figure 2.48).

■ *Hone the iron* on the fine side of a lubricated sharpening stone. Place the grinding bevel flat on the stone. Lift slightly and use firm to-and-fro strokes over the full width of the stone or in a figure-of-eight pattern to form the honing bevel. This avoids forming a hollow in the stone. Wide plane blades may require angling across the stone. Continue honing until a secondary bevel about 1 mm wide is formed.

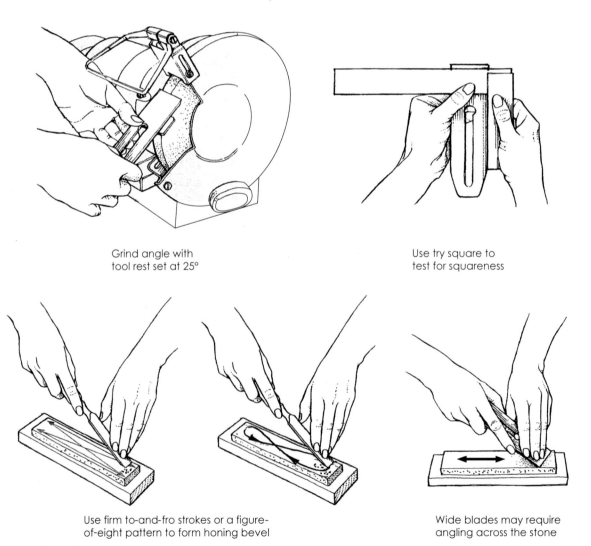

Grind angle with tool rest set at 25°

Use try square to test for squareness

Use firm to-and-fro strokes or a figure-of-eight pattern to form honing bevel

Wide blades may require angling across the stone

▲ **Figure 2.48** Sharpening procedure

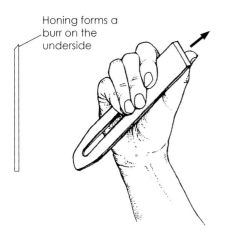

Honing forms a burr on the underside

Use your thumb to feel for the burr

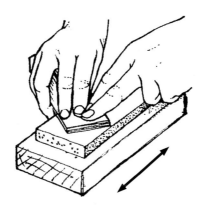

Use light to-and-fro strokes to remove burr, keep the back of the iron flat on the stone

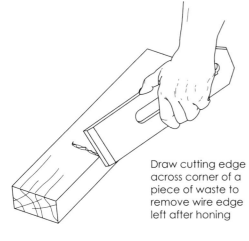

Draw cutting edge across corner of a piece of waste to remove wire edge left after honing

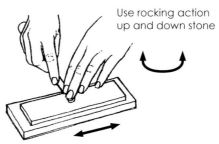

Further honing on a piece of leather will produce a super–sharp edge for working difficult grains

Scribing gouge

Bevel honed on slipstone

Use rocking action up and down stone

Burr removed on flatstone

Firmer gouge

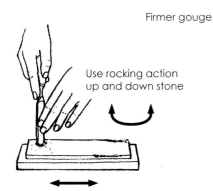

Use rocking action up and down stone

Bevel honed on flatstone

Burr removed on slipstone

◀ **Figure 2.48** Sharpening procedure *continued*

■ *Remove the burr.* The honing process forms a burr on the underside of the iron. Check for it by running your thumb over the back of the blade. Use light strokes to remove the burr with the back of the iron flat on the stone. Finally remove the thin wire edge left after honing by drawing the cutting edge across a piece of wood. When working difficult grains some woodworkers further hone a cutting iron by working the bevel on a thick piece of leather.

Chisels and gouges

These are available in a range of types, which are illustrated in Figure 2.49.

Firmer chisels are general-purpose chisels, which can be used for all types of woodwork. They have a rectangular section blade strong enough to be driven through wood with preferably the aid of a wooden mallet or alternatively the flat (side) face of the hammer. Firmer blades are available in widths rising in regular increments from 3 mm to 30 mm. 45 mm and 50 mm blades are also available in some ranges. Registered pattern chisels

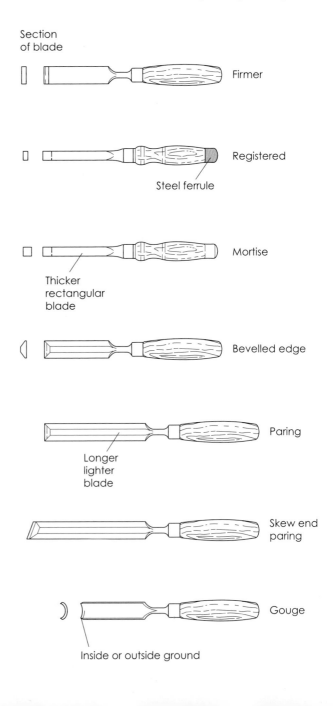

Section of blade

Firmer

Registered

Steel ferrule

Thicker rectangular blade

Mortise

Bevelled edge

Longer lighter blade

Paring

Skew end paring

Gouge

Inside or outside ground

▶ **Figure 2.49** Chisels and gouges

have a steel band (ferrule) at the end of the handle to stop the wooden handle splitting under constant hammering.

Mortise chisels are made much stronger than firmer chisels with a thicker section rectangular blade to withstand the heavy hammering and levering involved in chopping mortises. Standard widths are 6, 9 and 12 mm. Other sizes up to 25 mm wide are available in some ranges.

■ **Bevelled-edge chisels** are a lighter form of firmer chisel with bevels on the front edges of the blade. The thinner edges of this chisel enable it to be used for chiselling corners, which are less than 90º, such as dovetails, etc. They are available in widths ranging from 3 mm to 50 mm.

■ **Paring chisels** are similar to bevelled-edge chisels, but are much longer and lighter. They must be used for handwork only and never hit with a mallet. Their extra length makes them easier to control when paring either vertically or horizontally. They are available in various widths up to 50 mm. **Skew-end paring chisels** are also available for paring into awkward corners.

Gouges

These are in fact curved chisels mainly used for shaping, scribing and carving. They are available in two types:

A scribing gouge is ground and honed on its inside curvature for paring and scribing concave surfaces.

A firmer gouge is ground and honed on its outside curvature for hollowing and carving.

Using chisels and gouges

The most important point when using a chisel is to ensure that it is sharp since much of the time you will be using hand pressure. When sharpness is lost and extra pressure is required, re-hone it to restore the edge.

Paring

This is the cutting of thin slices of wood, either across the grain's length or across the end grain (Figure 2.50).

Vertical end grain paring:

1. Place the workpiece flat on the bench. A bench hook or cutting board should be used to protect the bench surface.
2. Grip the chisel handle with your thumb over the end.
3. Control the blade with the thumb and forefinger of your other hand.
4. With the chisel vertical and close to your body, use your shoulder power to apply a steady downwards force.
5. Work from the corner in towards the centre, paring off a little wood at a time.

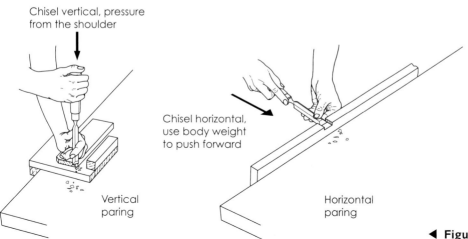

Chisel vertical, pressure from the shoulder

Chisel horizontal, use body weight to push forward

Vertical paring

Horizontal paring

◀ **Figure 2.50** Paring with a chisel

Horizontal across the grain paring:

1. Secure the workpiece to the bench top.
2. Use a tenon saw to cut the sides of the housing. Wide housings may require additional saw cuts in the waste.
3. Grip the chisel handle with your index finger extended towards the blade. Use the thumb and forefinger of your freehand to grip and guide the blade.
4. Take up a position in front of the workpiece with your forearm and the chisel parallel with the floor. Use your body weight to push the chisel forward.
5. Work from one side towards the centre taking a little wood off at a time. Turn the workpiece around and complete from the other side. Where extra force is required a mallet may be used. Never use your palm to strike the end of the handle, as it will cause personal injury in the long term.

Chopping with a chisel and mallet (Figure 2.51):

1. Secure the workpiece to the bench top.
2. Grip the chisel handle with a clenched fist. Hold the chisel vertical and strike the end of the handle squarely with the face of the mallet for maximum force.
3. For more delicate chopping hold the mallet shaft just below the head and lightly tap the chisel with the side of the mallet.

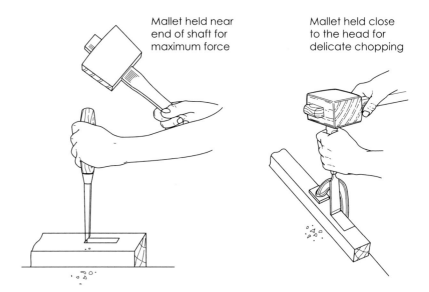

Mallet held near end of shaft for maximum force

Mallet held close to the head for delicate chopping

▶ **Figure 2.51** Chopping with a chisel and mallet

Paring and chopping with a gouge requires the same techniques used for standard chiselling:

1. Use a scribing gouge for paring internal concave, curved shoulders.
2. Use the firmer gouge for scooping out hollows in the surface of the workpiece.

Sharpening gouges

A similar procedure is used to sharpen gouges (refer to Figure 2.48). Since they are actually curved-bladed chisels, contoured grindstones and slipstones are required.

■ *Scribing gouges.* These are ground on a shaped grindstone and honed using a teardrop-shaped slipstone. The burr is removed on a standard flat stone using a rocking action as it is moved to and fro along the stone.

■ *Firmer gouges.* These can be ground and honed on standard grinding wheels and flat stones using a rocking action. A teardrop slipstone is used to remove the burr.

Hand drills and braces

With the increasing availability of power tools and especially battery-powered drills, the use of hand drills and braces is far less commonplace today. However, hand tools are a useful addition to the toolkit as they are simpler, safer and do not require mains or battery power sources.

- **Wheel brace.** Usually referred to as a hand drill, this is used both on site and in the workshop (Figure 2.52). Its main use is for boring pilot holes for screws, etc.
- **Twist drills** are used in this type of brace. They are available in a range of sizes from 1 mm to 6 mm in diameter. A **countersink bit** is also available for the wheel brace.
- **Ratchet brace.** This is mainly used for site work (Figure 2.53). The handle of the brace has a sweep of 125 mm. Sometimes it is necessary to bore holes in or near corners. In order to do this the ratchet on the brace is put into action.

Drill bits

The following are a selection of the wide range of drill bits available:

- *The Jennings bit* (6 mm to 38 mm in diameter) can be used for boring both across the grain and into end grain.
- *The centre bit* (6 mm to 32 mm in diameter) should only be used for boring shallow holes.
- *The Irwin bit* (6 mm to 38 mm in diameter) has the same range of uses as the Jennings bit.
- *The Forstner bit* (10 mm to 50 mm in diameter) is used for cutting blind or flat-bottomed holes.
- *Expanding bits* (expands 13 mm to 75 mm) are mainly used for boring large holes.
- *Countersink bits* are used to prepare holes in order to receive countersunk screws.
- *Screwdriver bits* can be used as an alternative to the hand screwdriver and gives more leverage and a quicker action.

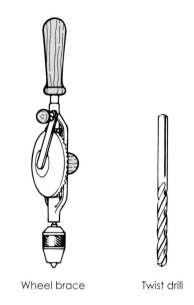

Wheel brace   Twist drill

▲ **Figure 2.52** Wheel brace and twist drill

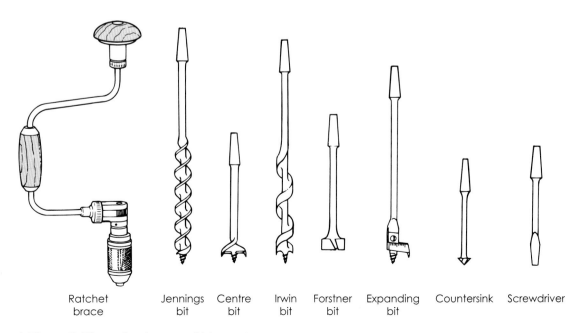

Ratchet brace | Jennings bit | Centre bit | Irwin bit | Forstner bit | Expanding bit | Countersink | Screwdriver

▲ **Figure 2.53** ratchet brace and bits

Use of hand drills

The most difficult part of drilling or boring is to start and keep the drill or bit going in the right direction. Good all round vision is essential. An assistant can be useful until the skill is mastered, as they can tell you whether the drill is leaning out of line either horizontally, vertically or parallel with the edge (Figure 2.54).

Small holes up to say 6 mm can be drilled using a twist drill in a hand or power drill.

Larger holes up to about 50 mm can be bored either using a hand brace and auger bit or a spade bit in a hand or power drill.

In order to prevent splitting out the workface when boring through holes either:

■ Bore from one face until the point of the bit appears, complete the hole from the other side placing the point of the bit into the small hole and complete the work.

■ Temporarily cramp a waste piece of wood on the back of the workpiece, and bore right through. Any splitting will then be in the waste and not the workpiece.

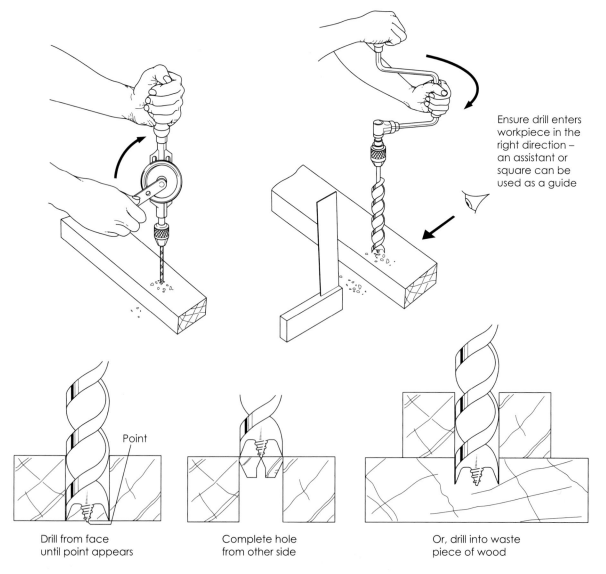

Ensure drill enters workpiece in the right direction – an assistant or square can be used as a guide

Point

Drill from face until point appears

Complete hole from other side

Or, drill into waste piece of wood

▲ **Figure 2.54** Using a drill

Fixing tools

Hammers and mallets

Hammers are available in three main types for use by the carpenter and joiner (Figure 2.55):

The **Warrington** or **cross-pein hammer** is used mainly by the joiner for shop work.

The claw hammer is a heavier hammer than the Warrington. It is used mainly for site work.

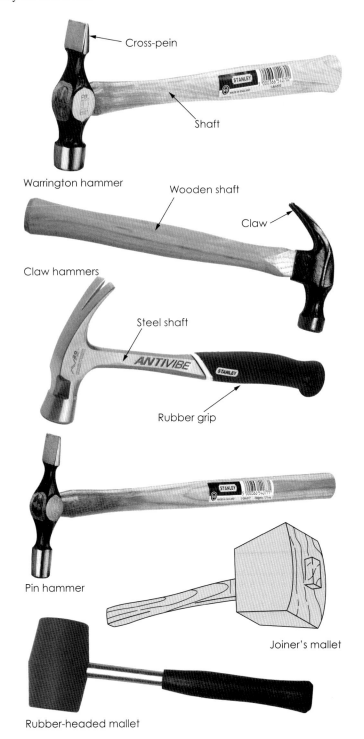

Cross-pein

Shaft

Warrington hammer

Wooden shaft

Claw

Claw hammers

Steel shaft

ANTIVIBE

Rubber grip

Pin hammer

Joiner's mallet

Rubber-headed mallet

◀ **Figure 2.55** Hammers and mallets

The **pin hammer** is a light version of the Warrington. It is used both on site and in the workshop for driving small pins.

The **mallet** is mainly a joiner's tool used for driving chisels when cutting joints and for framing up. A suitable material for the head of a mallet is beech. Rubber headed mallets are useful for frame and assembly, without fear of damaging the wood.

Hammer handles or shafts were traditionally available in hickory, beech or ash. Today, steel and glass-fibre are more common. Hickory is the best wooden handle, but the steel or glass-fibre shafted hammer is often preferred for its balance and increased strength.

Use of hammers and mallets

Observe the following instructions using Figure 2.56:

1. Grip the hammer or mallet at the end of the shaft, not near the head. Less effort and better control is achieved in this way.
2. Use the cross-pein to start small pins.
3. Use a nail punch to drive nail heads below the surface to avoid bruising the work.
4. Keep the hammerhead clean to prevent slipping and bending nails.
5. Place a scrap block of wood under a claw hammer when withdrawing nails. This protects the work piece and increases leverage.

Small bruises in the workpiece caused by a slipping hammer may be raised, by immediately soaking the bruise with water to swell the wood. The application of a hot iron will speed the raising process. Once the fibres are raised to the surface, allow it to dry and then sand off smooth.

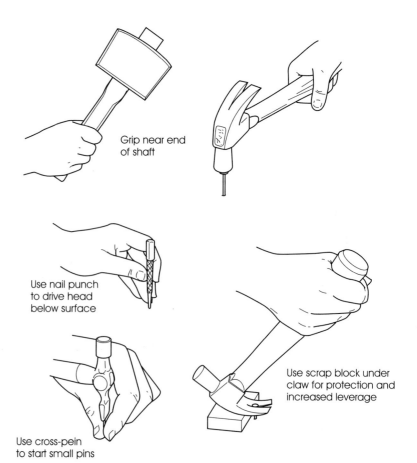

Grip near end of shaft

Use nail punch to drive head below surface

Use scrap block under claw for protection and increased leverage

Use cross-pein to start small pins

▶ **Figure 2.56** Use of hammer and mallet

Screwdrivers

There are four main types of screwdriver used by the woodworker (Figure 2.57):

- **Cabinet screwdriver.** Traditionally the cabinet pattern screwdriver has a rounded steel blade and wooden handle. The London pattern has a flat blade and flat faces on the handle. Both are available in sizes (length of the blade) ranging from 50 mm to 300 mm and various blade widths for slot-head screws.

- **Ratchet screwdriver.** This allows screws to be turned in or removed without releasing the handle. It is available in a similar range of sizes to those of the cabinet screwdriver. Both slot head and cross-head types are available.

- **Pump-action screwdriver** is popular with both carpenters and joiners as it allows quick and easy insertion and removal of screws without requiring a turn of the wrist. A range of interchangeable screwdriver slot or cross-head bits, drill bits and countersinks is often available. When in use, grip the chuck end to steady the screwdriver bit and prevent it from jumping out of the screw head and damaging the workpiece.

- **Cross-head screwdriver** is available in a range of blade lengths, tip patterns and sizes to suit the variety of cross-head screws manufactured.

Selecting a screwdriver

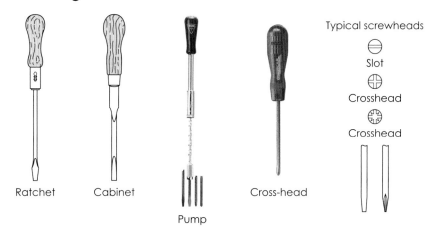

▲ **Figure 2.57** Screwdrivers

Always use the correct size of screwdriver (Figure 2.58). This ensures that:

- damage is not done to the screwdriver or screw slot;
- the screwdriver does not slip and damage the workpiece.

Slot-head screws. Match the width of the blade to the slot of the screw: too small and it will twist or chip; too large and it will tend to slip out and the protruding edges will score marks in the surrounding surface as the screw is driven home.

Cross-head screws. Different blade point patterns and sizes are available to match the gauge or screw being used. Phillips are coded as PH and Pozidrive as PZ. Both are available in the following sizes:

    Size 1: up to gauge 6

    Size 2: gauge 8 – 10

    Size 3: gauge 12 –14

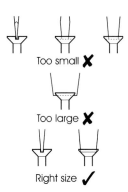

▲ **Figure 2.58** Screwdriver selection

Finishing tools and abrasives

Scrapers

Scrapers are used to remove paper-thin shavings, leaving a smooth finish, even on irregular grain (Figure 2.59).

Cabinet scrapers are a piece of hardened steel sheet, either rectangular for flat surfaces or shaped for finishing mouldings and other shaped work.
A burr is turned on the long edge of the scraper to form the cutting edge.

1. Hold the scraper in both hands using your thumbs to flex it into a curve.
2. Tilt the scraper away from you using a push forward action along the work.
3. Vary the tilt and the flex so that you are removing the required amount.

For wide surfaces use diagonal strokes in both directions before finishing parallel with the grain.

Scraper plane. Flexing a cabinet scraper can get hard on the thumb after a period. The blade of a scraper plane is held in the plane body at the required cutting angle and is flexed into a curve by an adjustment screw.
The plane is simply pushed forward. Adjust the curve until the required shaving is made. Slack curves will only produce dust. Larger curves produce thicker shavings.

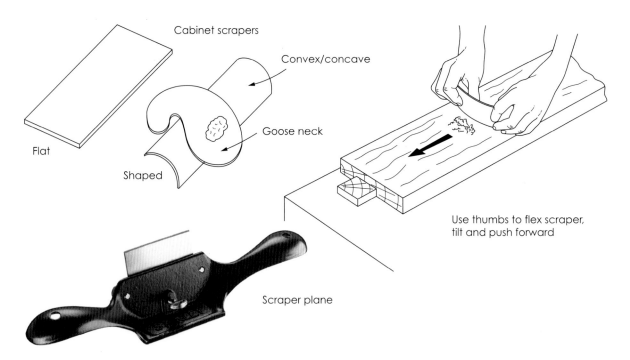

Cabinet scrapers

Convex/concave

Goose neck

Flat

Shaped

Use thumbs to flex scraper, tilt and push forward

Scraper plane

▲ **Figure 2.59** Scrapers

Sharpening a scraper

Figure 2.60 refers:

1. Secure the scraper in a vice and 'draw file' its long edges true and square. Use your fingertips to prevent the file rocking, as you pull it along the scraper.
2. Hone the filed edge by drawing an oiled slipstone over it. Use the stone on each side to remove the burr.
3. To raise a burr, lay the scraper on the bench top, with the long edge just overhanging. Stroke the face at a slight angle along the entire edge four

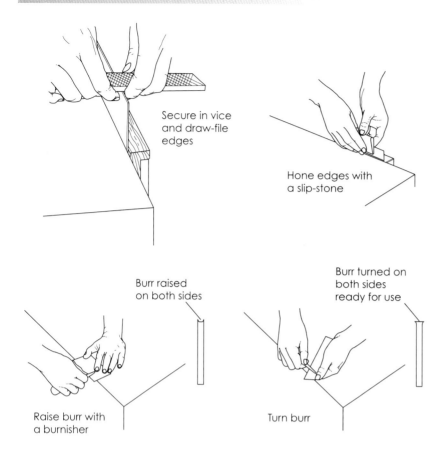

Secure in vice and draw-file edges

Hone edges with a slip-stone

Burr raised on both sides

Burr turned on both sides ready for use

Raise burr with a burnisher

Turn burr

◀ **Figure 2.60** Sharpening a scraper

or five times using a burnisher (like a toothless file) or the curved back of a gouge blade. Repeat from the other face.

4. To turn a burr, hold the scraper on end, with the burnisher or gouge held at a slight angle; make two or three firm strokes from both sides.

When the scraper starts to dull, simply raise and turn a new burr. Draw filing and honing is only required periodically when the edge has been damaged.

Sanding

This is the process of finishing the surface of wood with an abrasive paper or cloth. The method used will depend on the surface:

■ Hardwoods are normally clear finished. Planed surfaces can be scraped to remove minor blemishes and torn grain before finally rubbing down parallel with the grain with an abrasive paper.

■ Softwoods are normally painted. Minor blemishes will not show through. Machine marks should be removed with a smoothing plane, followed by rubbing down diagonally to the grain. The small scratches left as a result help to form a key for the priming paint.

Abrasive papers

A variety of abrasive materials are bonded to backing sheets. Traditionally the most popular for hand sanding wood was glass and garnet (harder than glass). Increasingly aluminium oxide (harder than garnet) and silicon-carbide (used in wet and dry paper) are used.

Most are available in standard sheet sizes of 280 mm × 230 mm. Smaller sheet sizes exist to suit standard power sander bases and also rolls of various widths.

Table 2.1 Grading of abrasive papers

Typical use	Comparison of typical abrasive grades		
Paint removal	Very coarse	50	1
		60	1/2
Preparation and levelling	Coarse	80	0
		100	2/0
General use	Medium	120	3/0
		150	4/0
		180	5/0
Finishing	Fine	220	6/0
		240	7/0
		280	8/0
Delicate finishing and denibbing between coats	Very fine	320	9/0
		360	—
		400	—

Abrasive papers are graded according to the size of particle (**grit size**). Typically termed very coarse, coarse, medium, fine or very fine. They are also graded by number, typically from 400 to 50 or 9/0 to 1, the higher the number the finer the grit (Table 2.1).

In addition to grades, abrasives are also termed as either *open-coated* or *closed-coated* depending on the spacing of the particles. The particles of close-coated abrasives are bonded closely together for fast, fine finishing. In open-coated the abrasive particles are spaced further apart. These clog up less readily and are thus better for use when finishing resinous timbers.

Hand sanding

Figure 2.61 refers:

- *Flat surfaces* should always be sanded with the abrasive paper wrapped around a sanding block to ensure a uniform surface.
- *Mouldings* may be sanded with the abrasive wrapped around a shaped block.
- *Curved surfaces* and rounded edges may be sanded without a block. Use your palm or fingertips to apply pressure to the abrasive.
- *Arrises* (the sharp external edge where two surfaces meet) are removed (de-arrised) using the abrasive wrapped around a block. The purpose of

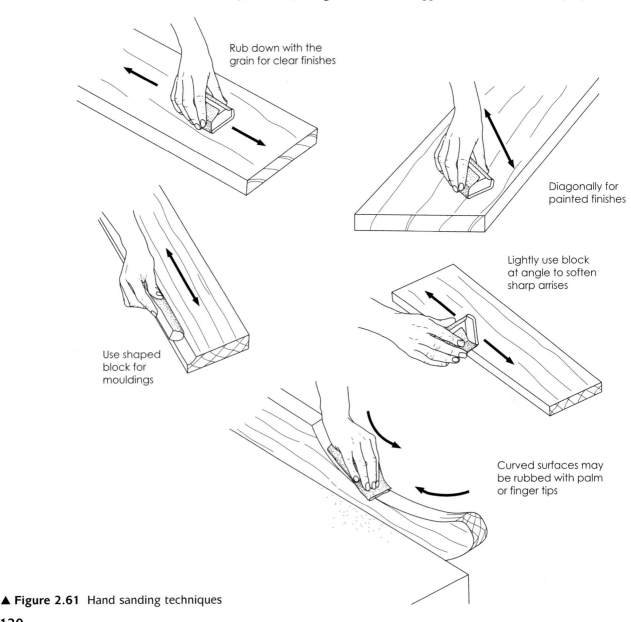

Rub down with the grain for clear finishes

Diagonally for painted finishes

Use shaped block for mouldings

Lightly use block at angle to soften sharp arrises

Curved surfaces may be rubbed with palm or finger tips

▲ **Figure 2.61** Hand sanding techniques

de-arrising is to soften the sharp edge and provide a better surface for the subsequent paint or polished finish. However take care not to overdo it, by completely rounding over a sharp corner. All that is required is a very fine chamfer. Gently run your thumb over a number of arrises and you will soon get the 'feel' for what is required.

Rasps, files and knives

See Figure 2.62:

Rasps are rough woodworkers files used for roughing out shaped work.

Metal-workers' files are used for the maintenance of tools and adjusting metal fittings, etc. They are also useful for the final finishing of edges to plastic laminates.

In use, files and rasps can become clogged and no longer cut: they should be cleaned using a **file card** or coarse wire brush.

Surform files are a development of the rasp. They have thin perforated metal blades that allow wood shavings to pass through without clogging. There is a wide range available suitable for planing and roughing-out shaped work.

Replaceable blade knives to which various blades can be fitted enabling them to be used for cutting plasterboard, plastic laminate, building paper, roofing felt and insulation board, etc. They can also be fitted with a padsaw blade for cutting small sections of timber, or a hacksaw blade for cutting metal.

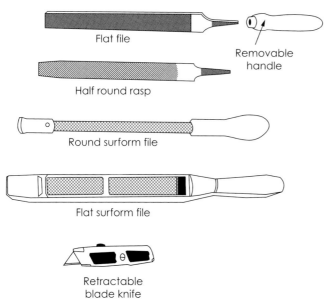

Flat file

Removable handle

Half round rasp

Round surform file

Flat surform file

Retractable blade knife

◀ **Figure 2.62** Files, rasps and knives

Miscellaneous tools

Refer to Figure 2.63:

- **Pincers** are used both on site and in the workshop for removing small nails and pins. To avoid bruising the wood, joiners place a scraper or a thin block of timber under the levering jaw.
- **Cold chisels** are steel chisels used for cutting holes or openings in brickwork, etc.
- **Plugging chisels** are also steel chisels but they are exclusively used for raking out the mortar joints in brickwork to receive wooden plugs and joist hangers, etc.
- **Nail punches** are used to set nails below the surface of the timber. They are available in a wide range of sizes to suit most nails.

121

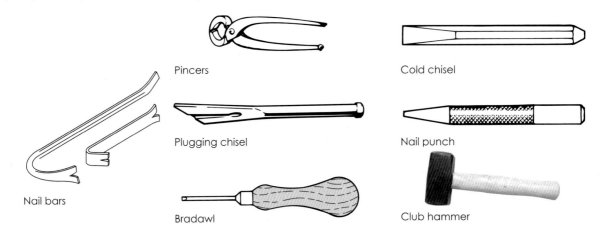

▲ **Figure 2.63** Miscellaneous tools

■ **Bradawls** are used to form starting holes for nails and small screws. Pilot holes for larger screws should always be bored with a twist drill.

■ **Club hammer,** also known as a lump hammer, is mainly used by bricklayers but is useful to the woodworker for heavy work when using cold or plugging chisels etc.

■ **Nail bar** also known as a **wrecking bar, tommy bar** or **crow bar.** Used for levering, prizing items apart and withdrawing nails.

Safety

In use, cold chisels, plugging chisels and to some extent nail punches may develop a dangerous 'mushroom' head as a result of hammering. These should be re-ground to avoid any risk of metal fragments flying off.

Workshop equipment

A large number of workshop items are required in order to produce a finished piece of work. The main range of equipments is illustrated in Figure 2.64.

Bench and stools

The ideal woodworkers bench should be 800 mm high, 3 to 4 m in length and have a well or recess running down its centre so that tools may be placed in it without obstructing the bench surface.

The bench should also have a **bench stop**, **bench peg** and a steel instantaneous-grip **vice** which opens at least 300 mm.

Most woodworkers require one or two **sawing stools** for their own use. These should be sturdily constructed with 50 mm × 50 mm legs, housed into a 75 mm × 100 mm top.

Cramps

Many types of cramp or clamp are used in the workshop:

■ **Sash cramp,** which is available in sizes from 500 mm to 2 m. This cramp is used to pull up joints, etc. when assembling and gluing up.

■ **Gee cramp,** which is available in sizes from 100 mm to 300 mm. This cramp is used for general holding and cramping jobs before and after assembly. Edging and deep throat versions are also available.

■ **Holdfast,** which is used for cramping jobs to the bench for sawing, cleaning up and finishing. The leg of the holdfast locates in a hole in the bench top.

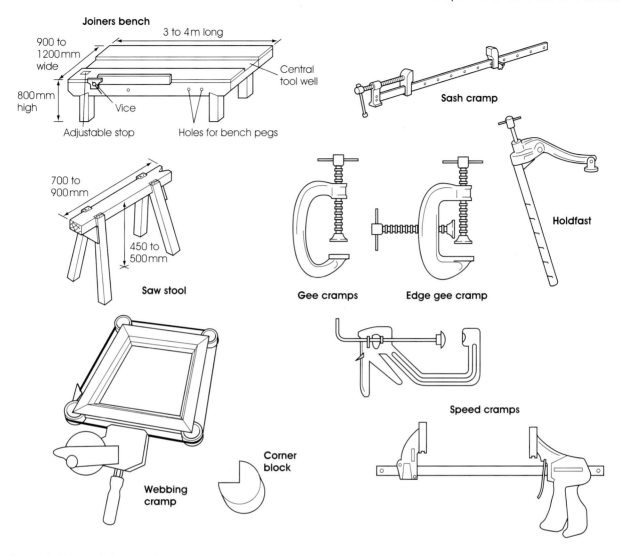

Figure 2.64 workshop equipment

- **Speed cramps** are available in a variety of types. All enable rapid single-handed cramping whilst the workpiece is held in the other hand. Mainly used in place of gee cramps.

- **Long bar, speed or ratchet cramps** may be used in place of sash cramps when assembling small frames. However it is almost impossible to apply the same amount of pressure to the work.

- **Webbing cramps** are also termed **picture frame cramps** as they are useful in the assembly of small mitred frames. Can also be used to cramp irregular shapes. They consist of a length of webbing, which is pulled taut around the workpiece by a ratchet mechanism. Corner blocks may be used to apply extra pressure at these points.

Bench equipment

Most woodworkers will make various items of wooden equipment for themselves. These may include the following items, which are illustrated in Figure 2.65:

- **Bench hook,** which is used to steady small pieces of timber when crosscutting, e.g. squaring ends, cutting to length and sawing shoulders.

- **Bench pegs,** which can be inserted in various positions along the side of the bench to support long boards, etc.

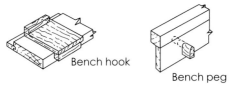

Bench hook

Bench peg

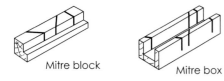

Mitre block

Mitre box

Winding strips

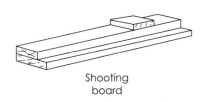

Shooting board

▲ **Figure 2.65** Bench equipment

■ **Mitre block,** which is used when cutting mitres for picture frames and other small mouldings.

■ **Mitre box,** which is used when cutting mitres on larger mouldings such as architraves and skirtings.

■ **Winding strips,** which are used to check timber and framed joinery items for wind or twist.

■ **Shooting board,** which is used as an aid when planing end grain and thin boards. An angled stop can be used when trimming mitres.

Templates

Woodworkers often make up jigs and templates in order to make the job easier. These include the following, which are illustrated in Figure 2.66.

■ Dovetail template used to mark out dovetails.

■ Mitre template used to guide the chisel when mitring or scribing joints.

■ Box square used to square lines around moulded sections.

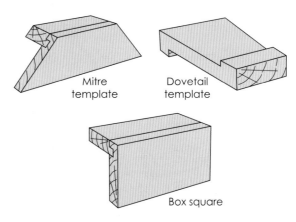

Mitre template

Dovetail template

Box square

▲ **Figure 2.66** Templates

Site equipment

See Figure 2.67:

■ **Straightedge.** A straight length of timber or an off-cut from the factory-machined edge of sheet material. Used on its own to test the alignment of spaced members or in conjunction with a spirit level for plumbing and levelling.

■ **Builder's square,** a purpose-made large square used for setting out right angles, such as partitions and walls. Can be cut from the corner of a sheet or jointed together. Use the 3, 4, 5 rule to set it out, typically make the sides 900, 1200 and 1500 mm.

■ **Spirit levels.** Traditionally having a hardwood body but now almost exclusively aluminium. Inset in the body are a number of curved tubes, partially filled with spirit and containing a bubble of air. The bubble position indicates whether a surface is truly plumb (vertical) or level (horizontal). Before use check the level for accuracy. Use the level to mark a pencil line on a wall, either vertical or horizontal. Turn the level over and re-check the line. Any discrepancy and the level is out of true. The tubes may be reset in some levels, otherwise it is time to get a new one.

■ **Tripod-mounted laser levels** can be used to project level datum points around a room.

■ **Water levels** consist of a length of hose with a glass or transparent plastic tube at each end. The water surfaces in the two tubes will give two level points. This uses the fact that water will always find its own level. It is used mainly to transfer levels around corners, etc.

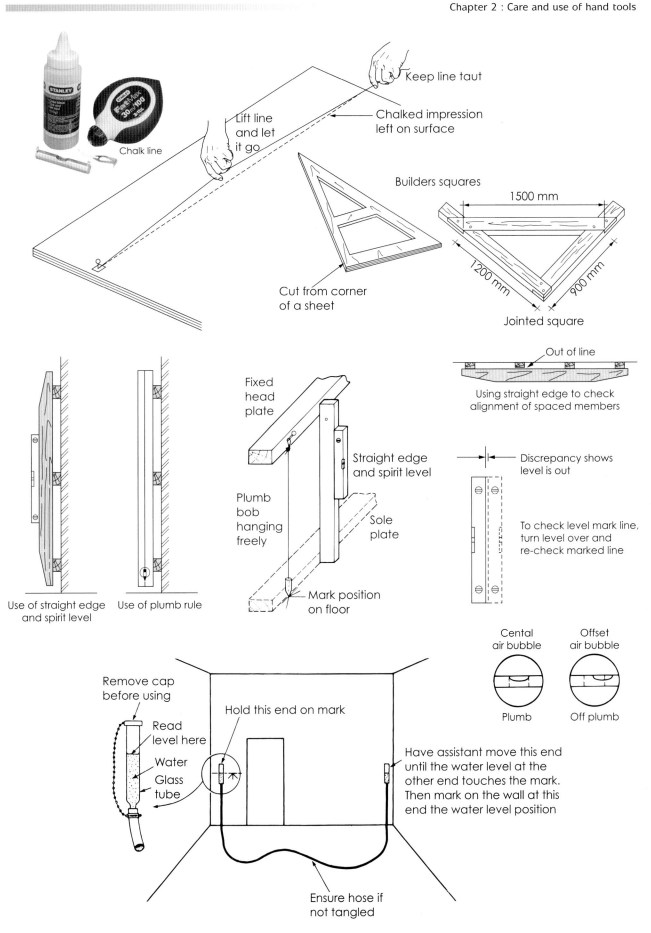

▲ Figure 2.67 Site equipment

- **Plumb bob** is a metal weight that is attached to a length of cord. When freely suspended, it will produce a true vertical plumb line. It is used to indicate drop positions or as a margin line from which other positions can be measured.

- **Plumb rule** is made using a straightedge and plumb bob. It allows the rule edge to touch the item being plumbed or tested. A hole is cut in one end of the edge to accommodate the bob and a central gauge line is used as the plumb indicator.

- **Chalk line.** A line coated with chalk used to mark straight lines on a surface. The line is stretched over a surface and snapped in the centre leaving a chalked impression. Purpose-made chalk lines are available, consisting of a container filled with coloured chalk into which the line is wound. In addition these can often double up as a plumb bob, enabling vertical lines to be chalked.

Chapter Three

Woodworking joints

This chapter covers the work of site carpenters and bench joiners. It is concerned with the principles of marking out and forming woodworking joints using a range of hand tools. It includes the following:

- ℱ   marking out and forming butt, lapped, halving, notched, mortised, bridle, edge, angle, dowelled and dovetailed joints;
- ℱ   assembly of framed joinery;
- ℱ   board joints, edging and laminating.

All woodworkers make joints. The carpenter makes joints that are normally load bearing. The joiner uses mainly framing joints for doors, windows and decorative trims. The cabinet/ furniture maker also uses framing joints, both to create flat frames and also to build up three-dimensional carcasses. In addition both the joiner and the cabinet/furniture maker may be involved with a range of jointing methods for manufactured boards.

Joint making is regarded as a measure of a woodworker's skill, since it requires the mastering of a variety of very accurate marking and cutting techniques. In training, joint making instils a 'feel' for both materials and the use of hand tools, which will not be lost in later years, even when progressing on to machines and powered hand tools.

Butt joints and mitres

Butt joints are the simplest form of joint where one piece of timber meets another end on or side on. It is not a strong joint on its own, as end grain does not take glue well and there are no interlocking parts. However they are often reinforced in some way for increased strength.

- ■   *Butt joints in length* are mainly used when joining structural timber and covering material (Figure 3.1). They must always be made over a support (wall plate or joist, etc).
- ■   *Butt joints for boxes and frames* are simple angle joints made with the end grain of one glued to the face or edge of the other (Figure 3.2).

Forming butt joints

1.   Mark the length of the parts and with a try square and marking knife, square a shoulder line across the face and edge (See Figure 3.3.)
2.   Use a bench hook to hold the piece; cut square on the shoulder line using a tenon saw.
3.   Trim the end grain to the shoulder line using a plane and shooting board to provide a smooth surface for gluing.

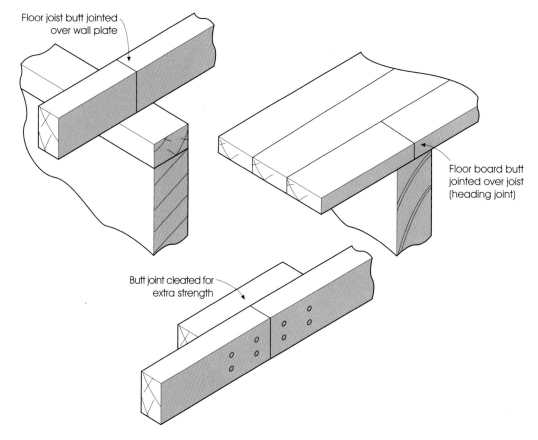

Floor joist butt jointed over wall plate

Floor board butt jointed over joist (heading joint)

Butt joint cleated for extra strength

▲ **Figure 3.1** Butt joints in length

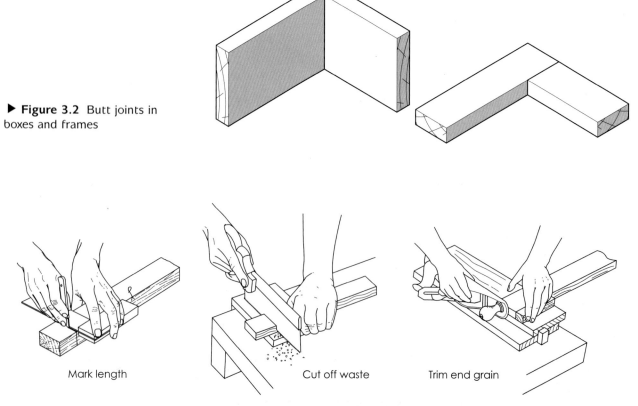

▶ **Figure 3.2** Butt joints in boxes and frames

Mark length

Cut off waste

Trim end grain

▲ **Figure 3.3** Forming a butt joint

4. Apply glue to both parts, rub together to expel excess glue and cramp together.
5. Use a damp cloth to wipe off the excess glue.

Reinforcing butt joints. As end grain does not glue well, the joint is often reinforced with nails or glue blocks (Figure 3.4). Butt joints can also be screwed or dowelled for additional strength.

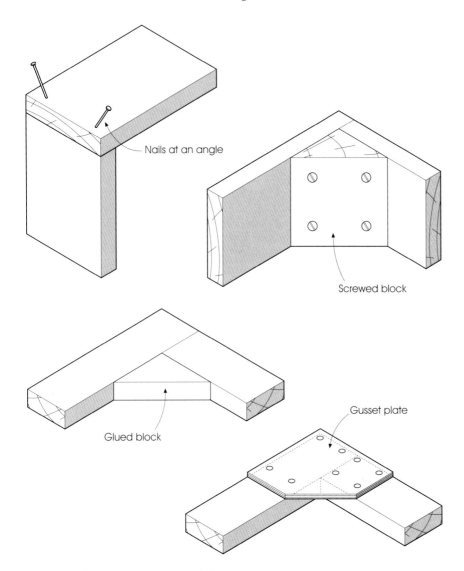

Nails at an angle

Screwed block

Glued block

Gusset plate

▲ **Figure 3.4** Reinforcing butt joints

Mitred butt joint

The use of a mitre increases the gluing area compared with a squared-end butt, avoids end grain being visible and also allows moulded sections to be joined (Figure 3.5). In addition to gluing the mitre is normally reinforced by nailing, splines or tongues.

Forming a mitred butt joint

1. Mark the cutting lines on the face or edge with a marking knife and mitre square. Square the lines onto the adjacent face or edge using a try square. See Figure 3.6.
2. Cut to the lines using a tenon saw. A mitre box may be useful to hold the workpiece and guide the saw.

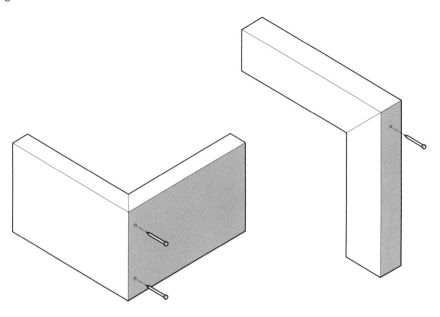

▶ **Figure 3.5** Mitre joints

3. Trim the cut mitres with a plane and shooting board. Wide mitres for boxes can be trimmed, held in a vice, using an off-cut on the back edge to prevent breakout.

4. Apply glue, nail or cramp joint together, remove excess glue with a damp cloth.

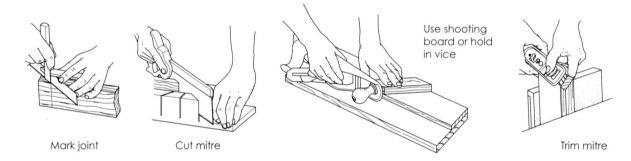

Use shooting board or hold in vice

Mark joint   Cut mitre   Trim mitre

▲ **Figure 3.6** Forming a mitred butt joint

Joint reinforcement

■ **Spline reinforcement.** This is carried out when the joint has been assembled (Figure 3.7). Use a tenon saw to cut slots across the corner. The cuts may be square to the end or angled like a 'dovetail' for additional strength. Glue thin plywood or veneer splines into the slots, trim up flush when the glue has set.

■ **Tongue reinforcement.** The grooves or slots for this must be cut before the mitre is glued and assembled (Figure 3.8). Cut the tongue from 3 mm MDF or plywood. Solid wood may also be used but for strength make it with the grain running across the width.

Cutting the groove for tongue reinforcement (Figure 3.9)

1. Use a mortise gauge to mark the parallel lines across the end grain.

2. Cut down each line with a tenon saw.

3. Chop out the waste with a chisel. Alternatively a plough plane may be used to form the groove.

4. Centre the groove when the tongue is with the mitre, but move it off centre nearer to the inside of the mitre when it runs across the mitre. This is to minimise the weakening effect of short grain.

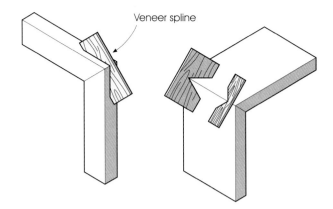

▲ **Figure 3.7** Spline-reinforced mitre joints

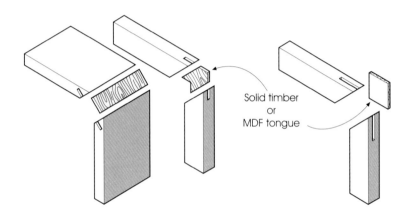

▲ **Figure 3.8** Tongue-reinforced mitre joints

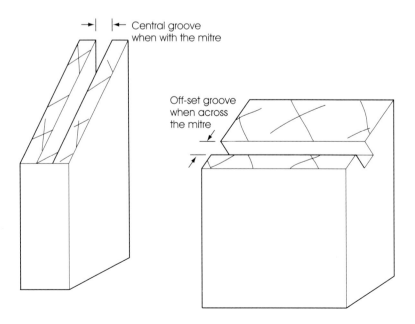

▲ **Figure 3.9** Positioning the groove for mitre reinforcement

Lap joints

In structural work the lap joint is used to lengthen timber (Figure 3.10). The two pieces overlap each other side by side, and are secured by nails, screws or bolts. Typical uses are for lengthening floor joists or pitched roof rafters. Laps must always be made over a support, (wall plate or purlin). The lap joint is also a basic corner joint used for box and carcass construction. Sometimes known as a **rebate joint** as one piece is rebated to receive the plain end of the other.

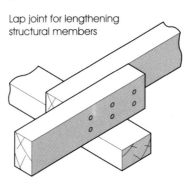

Lap joint for lengthening structural members

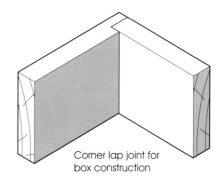

Corner lap joint for box construction

▶ **Figure 3.10** Lap joints

The rebated joint has advantages over a plain butt joint: increased gluing area, resistance given by the shoulder of inward pressure and a partial concealment of end grain. Reinforcement is given by nailing in one or both directions.

Forming a corner lap joint

Follow these steps referring to Figure 3.11:

- Cut the piece to length.
- Mark a shoulder line square across the face of the piece to be rebated and continue down both edges. The distance from the end will be the thickness of the other piece.
- Set a marking gauge to between a half to two-thirds the thickness of the rebated piece. Scribe a gauge line on the end grain and over each edge. Mark the waste with a pencil.
- Hold the rebate piece vertically in a vice and use a tenon saw to cut down the gauge line to the shoulder line.
- Using a bench hook to hold the piece, cut across the shoulder line with a tenon saw to remove the waste. (Alternatively the shoulder line may be cut first and the waste removed with a chisel by paring down the end grain.)
- Apply glue to the rebated piece, assemble parts and secure with pins in one or both directions. Remove excess glue with a damp cloth.

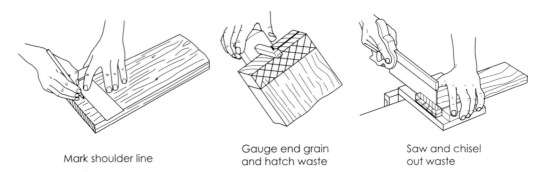

Mark shoulder line

Gauge end grain and hatch waste

Saw and chisel out waste

▲ **Figure 3.11** Forming a lap joint – box construction

Mitred lap joint

In structural work these are termed as **splayed lap joints** or **scarf joints** and used for lengthening in one line without increasing the cross-section at the joint (Figure 3.12). A long taper is cut on both pieces and are secured by nails, screws or bolts. Shoulders at right angles to the taper may be formed at either end for location.

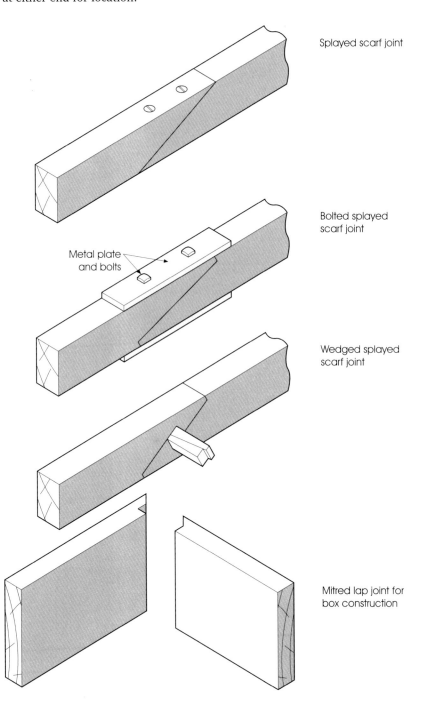

Splayed scarf joint

Bolted splayed
scarf joint

Metal plate
and bolts

Wedged splayed
scarf joint

Mitred lap joint for
box construction

◀ **Figure 3.12** Mitred lap joints

A further variation is the **wedged scarf joint,** where the taper is stepped and the joint tightened with folding wedges.

The **corner mitre lap joint** is used for box construction. It is stronger than a plain mitre due to the locating shoulder and neater than the corner lap joint due to the absence of end grain. Also termed a **birdsmouth mitre joint**.

Forming a mitred lap joint

Follow these steps referring to Figure 3.13:

■ Mark and cut the laps using a similar method to the corner lap, except that this time both pieces are lapped.

■ Mark a mitre on both pieces. Plane off the waste using a block plane for the rebated piece and a shoulder plane for the abutting piece.

■ Glue and secure the joint as before.

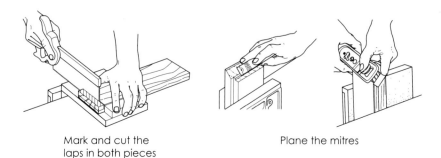

▶ **Figure 3.13** Forming a mitred lap joint

Mark and cut the laps in both pieces

Plane the mitres

Halving joints

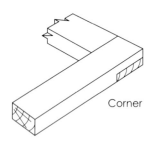

Corner

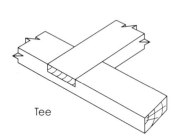

Tee

Halving joints are constructed with half of the timber cut from each piece at the point of intersection. They are used for both lengthening members and for constructing flat frames.

The **scarf halving joint** is used for lengthening structural timber, typically wall plates where the joint is supported throughout its length (Figure 3.14). Nails or screws are used to secure the joint. An improvement is the **bevelled scarf joint,** which will resist pulling stresses.

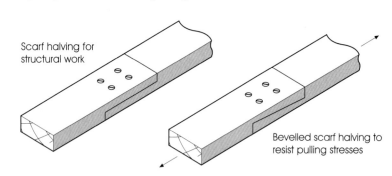

Scarf halving for structural work

Bevelled scarf halving to resist pulling stresses

▲ **Figure 3.14** Scarf halving joint

Angle halving joints are used for both structural work and flat frame construction where one piece abuts or crosses over another (Figure 3.15).

Forming a corner halving joint

Follow these steps referring to Figure 3.16:

1. Use the piece of timber to mark the width on both face sides. Use a try square to mark the shoulder lines across the face and continue down the edges.

2. Set a marking gauge to half the timber's thickness and from the face side scribe a line on the edge from the shoulder, up to the end, across the end grain and finish up to the shoulder on the other edge. Mark the waste and repeat on the other piece of timber.

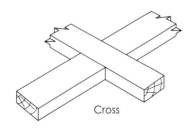

Cross

▲ **Figure 3.15** Angle halving joints

3. Position one piece of timber vertically in the vice, saw a guide cut across the end grain on the waste side (half mark of the gauge line).

4. Reposition the piece at an angle, keeping the saw in the guide cut; make an angled cut down across the end grain and down to the shoulder line.

5. Reverse the piece in the vice, position the saw in the guide cut and make a second angled cut down the other edge.

6. Reposition the piece vertically in the vice and, using the angled cuts as a guide, saw down to the shoulder line.

7. Holding the piece on a bench hook use a tenon saw to cut across the shoulder line and remove the waste.

8. Repeat the process to cut halving in the other piece.

9. Dry assemble the joint, checking for flatness and square. Clean up halving if required with a paring chisel.

10. Glue up and cramp secure with pins or screws if required.

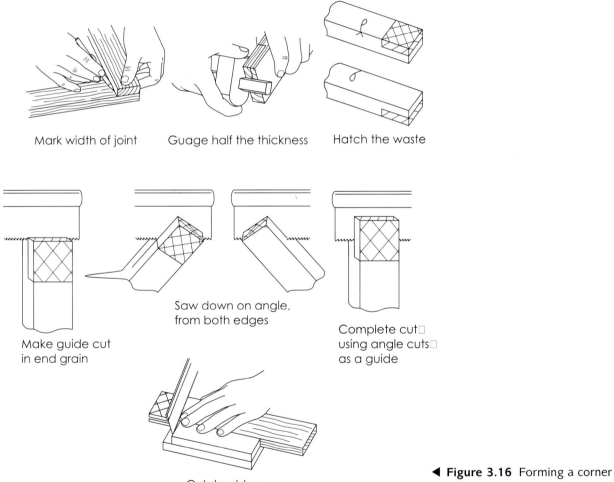

Mark width of joint   Guage half the thickness   Hatch the waste

Make guide cut in end grain

Saw down on angle, from both edges

Complete cut using angle cuts as a guide

Cut shoulders

◀ **Figure 3.16** Forming a corner halving joint

The **mitred corner halving joint** is a refined version suitable for moulded timber trim (Figure 3.17). However, as the reduced gluing area makes this a weaker joint, it should be secured from the rear face with screws.

The **single or double bevelled corner halving** joint can be used for structural work as these will resist pulling stresses on one or both directions (Figure 3.18). Marking out and cutting is the same as for standard halving joints except that the bevel is marked with a sliding bevel or template.

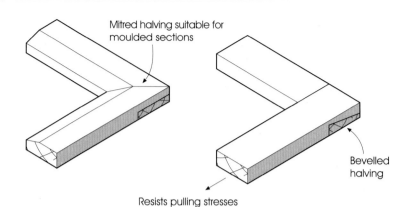

▶ **Figure 3.17** Corner halving with moulded sections

Mitred halving suitable for moulded sections

Bevelled halving

▶ **Figure 3.18** Corner halving with bevelled halving

Resists pulling stresses

Forming a tee halving joint

Follow these steps referring to Figure 3.19:

1. Mark and cut the upright member of the tee using the method described for the corner halving.
2. Use the upright member to mark the width of the cut-out in the cross-member of the tee.
3. Use a try square to mark shoulder lines across the face and down the edges.
4. With the previously set marking gauge, scribe a line on both edges between the marked shoulder lines. Mark the waste.

▶ **Figure 3.19** Forming a tee halving joint

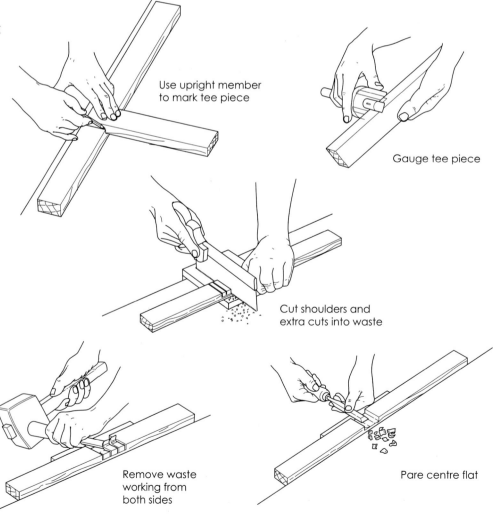

Use upright member to mark tee piece

Gauge tee piece

Cut shoulders and extra cuts into waste

Remove waste working from both sides

Pare centre flat

5. Holding the piece with the aid of a bench hook, use a tenon saw to cut down the shoulders to the gauged line. Ensure you cut inside the waste: a tight joint can be eased, a slack one can't. It is also unsightly and weak.

6. Make one or more additional saw cuts across the waste to make chiselling out easier.

7. Holding the piece in the vice, use a wide chisel and mallet to remove the waste. Work from both sides towards the centre with the chisel held at a slightly upward angle.

8. Having removed most of the waste, use the chisel flat to pare away the raised centre portion. Use the chisel vertically along the base of the shoulder line if required to cut any remaining fibres not cut by the saw. Use the side edge of the chisel to check for flatness.

9. Dry assemble the joint; pare shoulders if too tight. Check for level and square. Adjust if required.

10. Glue up and cramp. Secure with pins or screws if required.

Dovetail tee halving

These are an improvement on the standard joint as they resist pulling stresses better (Figure 3.20).

The dovetail tee joint is formed in a similar way to the standard tee (Figure 3.21):

1. Mark out and cut the piece to be inserted first using the same method as the standard joint.
2. Set a sliding bevel to the required slope and mark the pin.
3. Saw and pare away the waste.
4. Lay the dovetailed pin on the cross-member.
5. Mark the shoulder lines, square the shoulder lines down the edges and score a gauge line between them.
6. Saw shoulder lines, chisel out and assemble as described before.

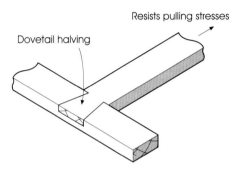

▲ **Figure 3.20** Dovetail tee halving joint

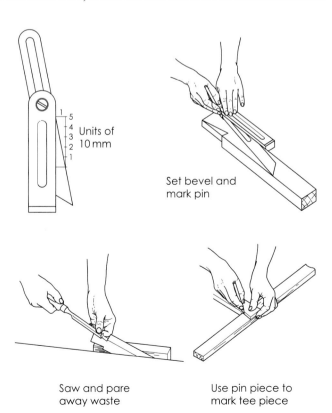

Set bevel and mark pin

Saw and pare away waste

Use pin piece to mark tee piece

▲ **Figure 3.21** Forming a dovetail tee halving joint

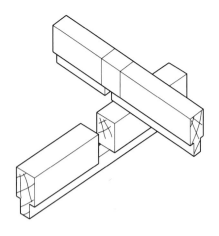

▲ **Figure 3.22** Cross-nailing joint square-edge sections

Cross-halving joints

These joints are used where horizontal rails, cross-vertical members such as cabinet front frames and glazing bars of doors or windows are required (Figure 3.22). The joint in square timber is marked out and cut using a similar method to that described for the tee halving, except as both pieces are cross-members, the shoulders are cut with a tenon saw and the waste chiselled out.

Mitred cross-halving

This joint is used for moulded glazing bars. The basic joint is a cross-halving in the centre section with the moulding mitred as illustrated in Figure 3.23.

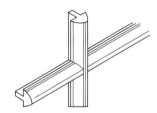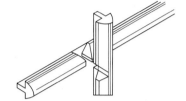

▲ **Figure 3.23** Mitred cross-halving

Forming mitred cross-halving joint

Follow these steps referring to Figure 3.24:

1. Mark the two central positions of the joint around both a box square will be useful for this stage.

2. Cut away the moulded section on both faces of each piece, down to the central section. The square cut of a mitre box can be used as a guide.

3. Place a box mitre over the section and secure in the vice. Pare away the corners of each moulding in turn using the box mitre as a guide for the final cuts.

4. Cut the cross-halving in the centre sections, making the cut-outs level with the rebate depth.

5. Dry assemble, pare mitres and shoulders as required if any are too tight. Take extra care when dry assembling and fitting as the slender remaining cross-section is easily snapped.

6. Glue up and cramp, removing excess glue with a damp cloth.

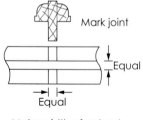

Mark joint

Make width of cut-out the same as width of central section

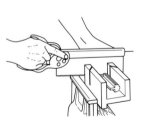

Cut away central section

Use box mitre as guide to pare waste

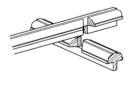

Carefully assemble joint

▲ **Figure 3.24** Forming a mitred cross-halving joint

Housing joints

Housings are grooves (also known as **trenches**) cut across the grain of a piece of timber to accommodate the square cut ends of another piece, thus forming a right-angled joint. They have a variety of uses ranging from stud partitions and stairs to fixed shelving and intermediate standards in cabinet carcasses

The through housing joint

The depth of housings should be restricted to about one-third the material's thickness (Figure 3.25). Follow these steps referring to Figure 3.26:

1. Measure and mark in pencil, the position of the groove.
2. Mark square parallel lines across the piece, equal to the thickness of the stud or shelf, etc. Continue line down the edges.
3. Score shoulder lines across the face, using a marking knife and try square.
4. Set marking gauge to one-third the thickness and scribe lines between the marks on both edges.
5. Cut down on the waste sides of the shoulder lines to the gauge lines. On wide housings make further saw cuts in the waste to ease chiselling out the waste.

◀ **Figure 3.25** Through housing joint

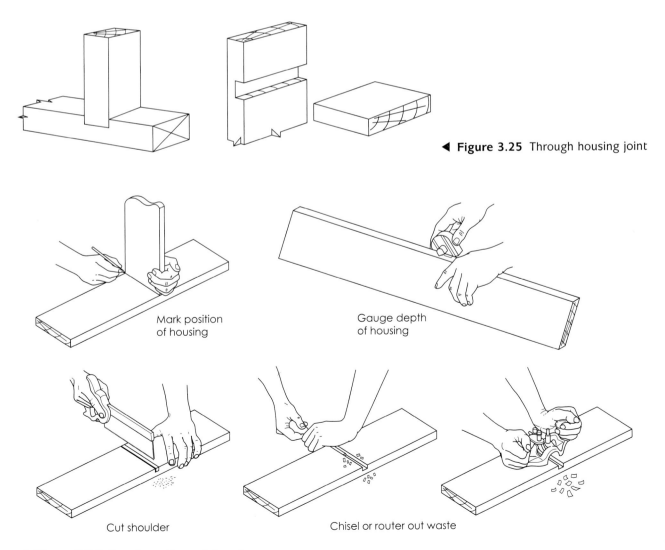

Mark position of housing

Gauge depth of housing

Cut shoulder

Chisel or router out waste

▲ **Figure 3.26** Forming a through housing

6. Chisel out the waste working from both sides towards the centre. This will help to prevent the risk of breakout.

7. Pare the bottom of the housing flat. Alternatively, for housings in wide boards, use a hand router plane to remove the waste between the sawn shoulder lines. Start off making shallow cuts, resetting the cutter deeper each time until the full depth is reached.

8. Dry assemble, easing if required, before gluing up.

The dovetail housing joint

The dovetail housing is a variation of the through housing and is used to resist endways pulling stresses (Figure 3.27). The housing may be cut with one or both sides of the groove cut at an angle.

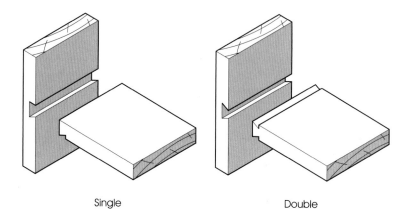

Single   Double

▶ **Figure 3.27** Dovetail housing joints

The marking and cutting of the housing is similar to the standard through housing, except that the lines down the edges are marked using the dovetail outline of the shelf angle with the shoulder lines being cut to the marked angles (Figure 3.28). An angled block may be cramped along the shoulder line to act as a guide for the tenon saw.

1. Mark the shoulder line on the tailpiece (end of shelf) equal to the depth of housing and score with a marking knife. Square the shoulder line around the edges.

2. Mark the dovetail shape from the end of the shelf towards the shoulder line, using a sliding bevel set to a suitable dovetail angle.

3. Saw along the shoulder line down to the marked angle.

4. Pare out the waste using a purpose-made block to guide the chisel.

5. Mark out the shelf housing using the shelf as a guide. Square the shoulder lines across the face.

6. Saw along the shoulder lines using an angled block to guide the saw.

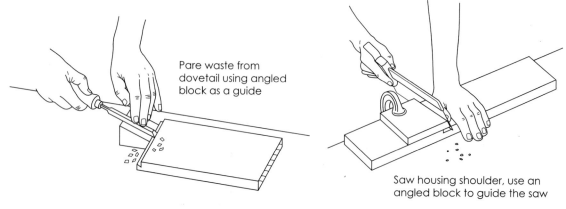

Pare waste from dovetail using angled block as a guide

Saw housing shoulder, use an angled block to guide the saw

▲ **Figure 3.28** Forming a dovetail housing joint

7. Remove the waste using a bevel edged chisel.
8. Dry assemble, easing the shoulders and bottom of the groove if required. The joint is assembled by sliding the tailpiece into the housing from one end.

Stopped housing joint

The groove of a stopped housing is stopped short of the full width of the timber for use where the member to be joined is narrower, as inset back, fixed shelves and stair strings (Figure 3.29).

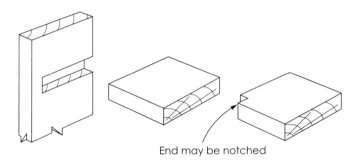

End may be notched

◀ **Figure 3.29** Stopped housing joint

Forming a stopped housing joint (Figure 3.30):

1. Mark out the groove partially across the face and down one edge, as described for the standard joint. Gauge the stopped end of the groove.
2. Bore a hole in the stopped end of the groove down to the required depth using a flat-bottomed Forstner bit.
3. Square up the hole using a chisel. Alternatively chop a cut-out at the stopped end using a chisel and mallet. Wrap a piece of masking tape around the chisel blade to indicate the required depth.

◀ **Figure 3.30** Forming a stopped housing joint

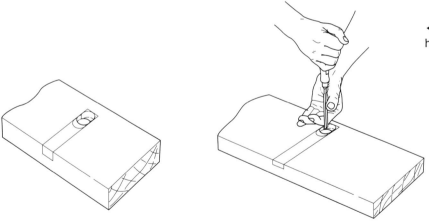

Bore holes at stopped end and square out with chisel

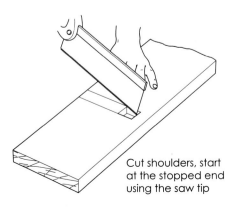

Cut shoulders, start at the stopped end using the saw tip

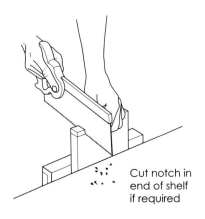

Cut notch in end of shelf if required

141

4. Starting from the cut-out at the stopped end, use the tip of a tenon saw to cut along the shoulder lines down to the required depth.
5. Remove waste from the groove with a chisel and/or a hand router.
6. Dry assemble, easing shoulders and bottom of groove if required.

Stopped housings can also be cut as dovetails where resistance to endwise pullout is required. A further variation can be made to the stopped housing, which conceals the end of the groove and masks the effects of subsequent shrinkage. This involves cutting a shoulder or notch on the front edge of the shelf.

Bare-faced housing joint

These are also termed **bare-faced tongue joints**. It is a stronger corner joint than the standard housing (Figure 3.31). Often used for joining door linings. The groove is cut in the head (horizontal member) and the single-shouldered tongue on the end of the jamb (vertical member).

Forming a bare-faced housing joint (Figure 3.32):
1. Mark the width of the vertical member on the face of the horizontal member square across the face and onto the edges.
2. Set the marking gauge to one-third the thickness. Use the gauge to mark the shoulder line across the back face and edges of the vertical piece. Then also across the end grain from the face side to mark the width of the tongue.
3. Continue the lines on either edge to meet the shoulder lines.
4. Saw down the shoulder line.
5. Cut out the waste with a chisel by paring the end grain.
6. Align the face side of the tongue with the line marked on the horizontal piece. Use the tongue to mark the width of the housing. Square the lines across the face and onto the edges.

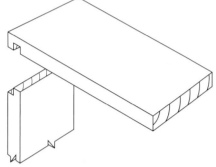

▲ **Figure 3.31** Bare-faced housing joint

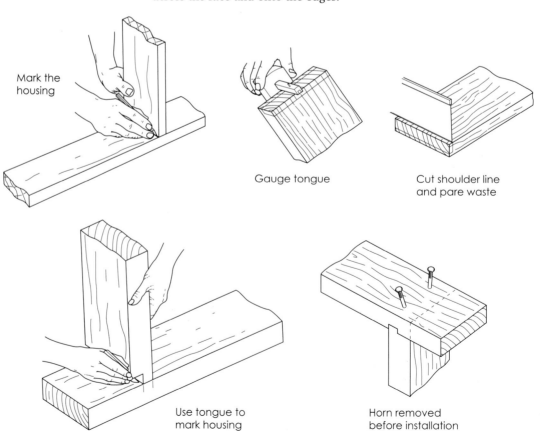

Mark the housing

Gauge tongue

Cut shoulder line and pare waste

Use tongue to mark housing

Horn removed before installation

▲ **Figure 3.32** Forming a bare-faced housing joint

7. Use the pre-set gauge to mark the depth of the housing on the edges.
8. Saw and chisel out the housing as above.
9. Dry assemble, easing if required before gluing up.

Nailing or screwing is often used to reinforce the joint. It is normal practice to leave the horizontal member long (leaving a horn) when working and assembling the joint. The horn, which helps to prevent grain splitting, can be cut off before the assembled component is installed.

Edge joints

Edge joints are used to enable narrow boards to be built up to cover large areas for floor boarding or cladding or to form wider boards for shelves, cabinet carcasses, panels, counter/worktops and table tops etc. (Figure 3.33).

The edges of the boards may be plain (butt) or shaped. The shaping provides a means of interlocking to line up the surfaces and for glued joints; it increases the gluing area for additional reinforcement.

Unglued edge joints. These are mainly used in carpentry work for flooring and cladding. Each board will expand or contract across their width, with changes in the moisture content. On shrinking, gaps will appear between the boards. Narrow boards will show a smaller gap on shrinkage than wider boards. Tongued and grooved (T&G) edge joints are normally used for flooring. The interlocking increases the strength and prevents a through gap on shrinkage.

Glued edge joints. These are used to join narrow boards together forming a wide panel for use in cabinetwork. The joints may be simply butted or reinforced with a tongue or dowels, to aid location and increase the glue line.

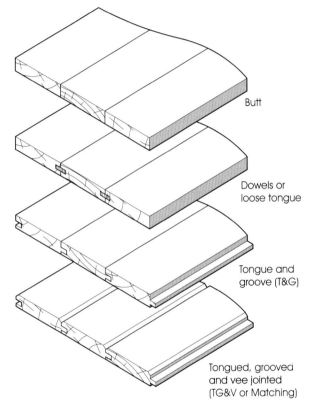

Butt

Dowels or loose tongue

Tongue and groove (T&G)

Tongued, grooved and vee jointed (TG&V or Matching)

▲ **Figure 3.33** Edge joints

Forming edge joints

Follow these steps referring to Figure 3.34:

1. Lay out boards to be joined, selecting them for colour and grain direction. When using tangential sawn timber make sure the heart side alternates, to minimise the effect of distortion.
2. Number each board and mark face and edge marks. When working with them keep numbers facing the same way.
3. Set the first two boards 1 and 2 back to back in the vice and use a try plane to true the edges. Using this method, the squareness of the edge is not critical, as they will still fit together and form a flat surface.
4. Repeat the previous stage using boards 2 and 3 again back to back. Note that board 2 will have to be rotated in order to match its unplaned edge with board 3. Continue to plane the edges of each pair of boards.

A close fitting butt joint can often be glued up without any applied pressure (cramping) termed a **rubbed joint,** see Figure 3.35:

1. Apply glue to both edges.
2. Bring the edges together and rub to and fro until they start to stiffen. This action squeezes out the excess glue and air from the joint and brings the components into close contact.

143

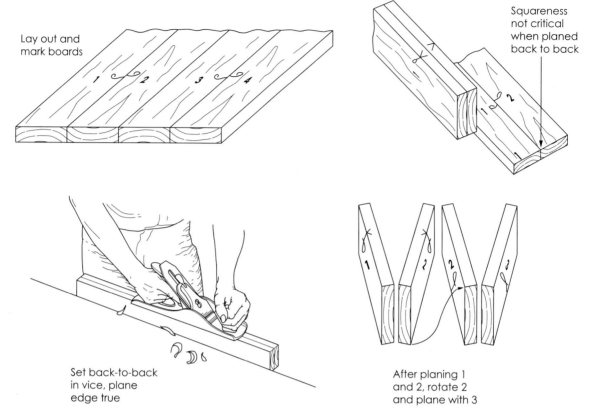

Lay out and mark boards

Squareness not critical when planed back to back

Set back-to-back in vice, plane edge true

After planing 1 and 2, rotate 2 and plane with 3

▲ **Figure 3.34** Preparing edge joints

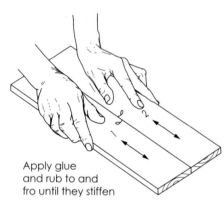

Apply glue and rub to and fro until they stiffen

▲ **Figure 3.35** Rubbed edge jointing method

3. Remove excess glue with a damp cloth and allow curing before truing up joined faces with a smoothing plane.

Cramped joint

Where more than two boards are being jointed, the use of cramps is preferable to a rubbed joint as illustrated in Figure 3.36:

■ Lay out the numbered boards on the bench. Use battens at right angles for support.

■ Dry assemble and check the boards for fit. Use two or more sash cramps, positioned about a quarter of the board's length from each end. Do not forget to use scrap pieces between the cramp heads and the board edges to prevent damage.

■ If all joints are close fitting remove the cramps. Apply glue to the jointing edges and re-cramp.

■ Tap any misaligning joints if required with a hammer and block of wood to make them flush.

■ Turn the panel over and cramp the centre. This centre cramp will pull up the joints and also helps to keep the panel flat.

■ Remove the excess glue that has squeezed out with a damp cloth.

■ Leave the panel in the cramps until the glue has cured before finally cleaning up the face of the panel with a smoothing plane.

Notched and cogged joints

These joints are used in heavy carpentry work to provide a location and ensure a uniform depth (Figure 3.37). Single joints cut in on one member allow movement in one direction. Double joints are cut in on both members to prevent movement in both directions.

The method used to form notched and cogged joints is similar to that used for housing and halving joints, except for the joint proportions and the timber section normally being much larger.

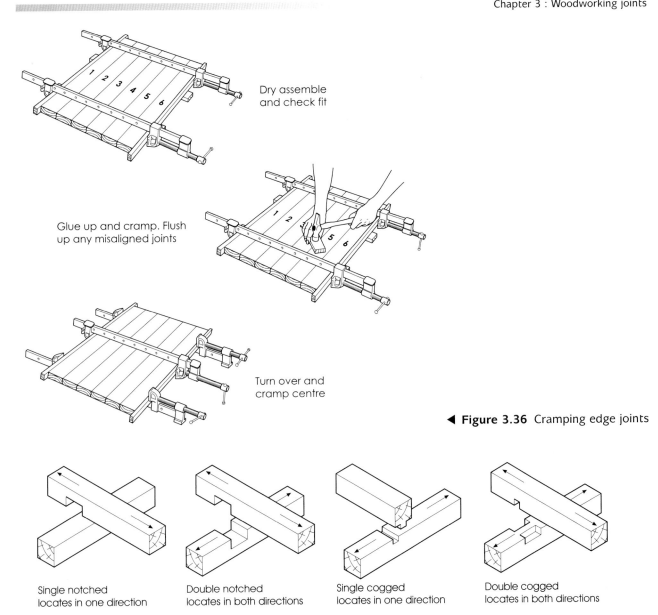

Dry assemble
and check fit

Glue up and cramp. Flush
up any misaligned joints

Turn over and
cramp centre

◀ **Figure 3.36** Cramping edge joints

Single notched
locates in one direction

Double notched
locates in both directions

Single cogged
locates in one direction

Double cogged
locates in both directions

▲ **Figure 3.37** Notched and cogged joints for structural work

Dovetail joints

Dovetail joints are designed to resist tensile or pulling forces. They are used mainly in box and drawer construction, see Figure 3.38. **Through dovetails** are used for boxes and backs of drawers. On the front of drawers, **lapped dovetails** are used because they provide a neater finished appearance.

Dovetail joints should have a pitch or slope of one in six for softwood and one in eight for hardwood (Figure 3.39). An excessive slope is weak due to the short grain, whereas an insufficient slope will tend to pull out under load. The slope can be marked out using a dovetail template or a sliding bevel (Figure 3.40).

Two alternative methods may be adopted when marking out and cutting dovetails, either:

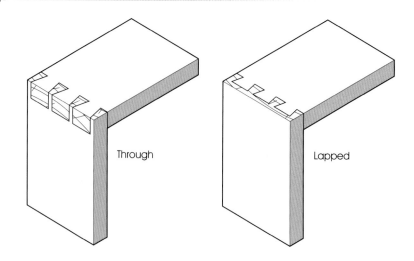

▶ **Figure 3.38** Dovetail joints

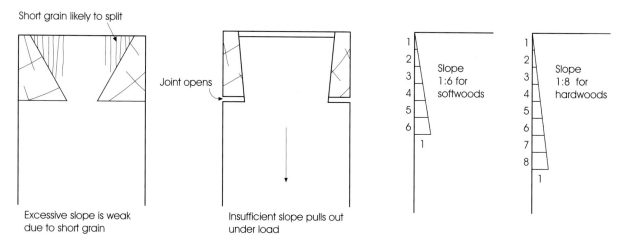

Short grain likely to split

Joint opens

Slope 1:6 for softwoods

Slope 1:8 for hardwoods

Excessive slope is weak due to short grain

Insufficient slope pulls out under load

▲ **Figure 3.39** Dovetail angles

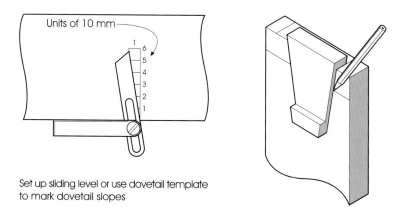

Units of 10 mm

▶ **Figure 3.40** Using bevel or dovetail template

Set up sliding level or use dovetail template to mark dovetail slopes

- mark and cut the dovetail and use the dovetail to mark out the pins and sockets, or
- mark and cut the pins and sockets, then mark the dovetails out from these.

Forming through dovetail joint

Dovetails, pins and sockets are illustrated in Figure 3.41. Dovetail joints may also be formed using a powered router and jig, see Chapter 5.

The following operations can be used when making a through dovetail joint in wide boards using the dovetail-before-pin method. The actual size and

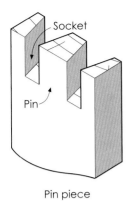

Socket

Pin

Pin piece

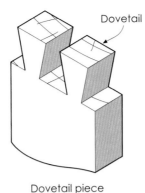

Dovetail

Dovetail piece

◀ **Figure 3.41** Dovetails, pins and sockets

number of dovetails will vary with the width of the board and type of timber. In softwoods dovetails are normally cut wider and thus fewer in number than those in hardwoods (Figure 3.42).

1. Cut square the ends of the pieces to be joined. The end grain can be trued up and smoothed with the aid of a shooting board.
2. Use a pencil and try square to mark the thickness of the board around the ends of both pieces.
3. Divide the width of the board by the number of dovetails (Figure 3.43).
4. Sub-divide each division into six equal parts, one at either end for the pin and four in the middle for the dovetail.
5. Square lines over the end grain and down to the shoulder line.
6. Use a dovetail template or a sliding bevel and pencil to mark the slope of the dovetails on both faces.
7. Hatch out the waste with a pencil to avoid later confusion.
8. Position the piece of wood in the vice at an angle so that one slope of each dovetail is vertical.
9. Use a dovetail saw to cut down one side of each dovetail, keeping just inside the waste.
10. Reposition the piece in the vice and cut down the other slope of each dovetail.
11. Position the pin piece of wood vertically in the vice. Lay the cut dovetail piece in position, lining up the edges and shoulder line.
12. Using a dovetail saw in each dovetail saw cut to mark their slope on the end grain. Square these lines down to the shoulder line on both faces. Again hatch the waste.
13. With the pin piece positioned vertically in the vice, use a dovetail saw to cut down to the shoulder line following the angle marked from each dovetail. Keep to the waste side aiming to leave it just visible.
14. Saw the corner waste from the dovetail piece.
15. Use a coping saw to remove most of the waste between both the dovetails and the pins.
16. Working from both faces, use a bevel-edged chisel to pare out the remaining waste down to the shoulder lines.
17. Partly assemble the joint to check the fit, paring out any high spots if required.
18. Clean up the inside faces of both pieces.
19. Apply glue to both halves of the joint and tap the joint together. Use a piece of waste wood to protect the surface.
20. Use a damp cloth to remove the excess glue.
21. Allow the glue to set before cleaning up the end grain with a smoothing or block plane. Work in from both edges towards the centre to avoid the end grain breaking out.

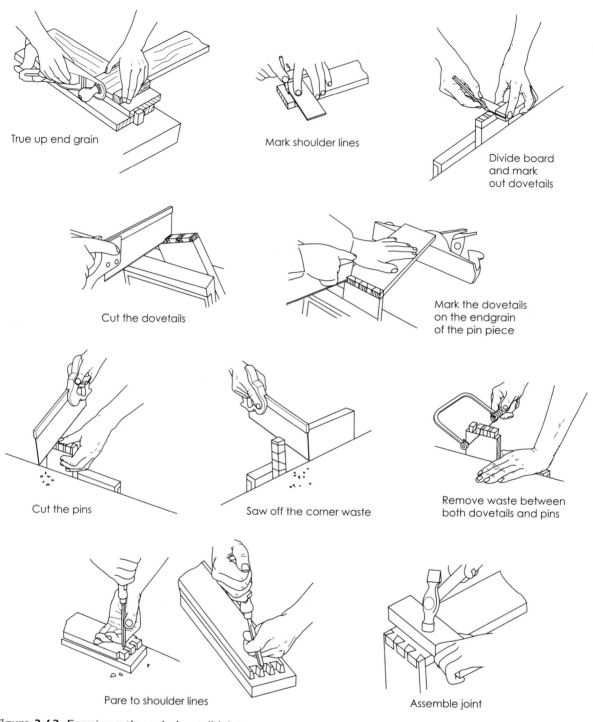

True up end grain

Mark shoulder lines

Divide board and mark out dovetails

Cut the dovetails

Mark the dovetails on the endgrain of the pin piece

Cut the pins

Saw off the corner waste

Remove waste between both dovetails and pins

Pare to shoulder lines

Assemble joint

▲ **Figure 3.42** Forming a through dovetail joint

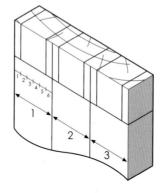
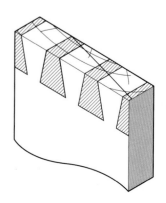

▶ **Figure 3.43** Marking out dovetails

Forming a lapped dovetail joint

This joint is formed using a similar process to that of the through dovetails, except for cutting the pin piece (Figure 3.44):

1. Cut and prepare the ends of the pieces to be joined.
2. Use a pencil and try square to mark two-thirds the thickness of the pin piece around the end of the dovetail piece.
3. Set out the dovetails and hatch the waste as before.
4. Cut dovetails with a dovetail saw using a coping saw to remove most of the waste. Pare to the shoulder line with a bevel edge chisel.
5. Use a pencil and try square to mark the thickness of the dovetail piece on the inside face of the pin piece.
6. Lay the dovetail piece on the pin piece; line up the edges and shoulder line. Pencil around the dovetails to mark the pins.
7. Square the pin lines down to the shoulder line and hatch the waste.
8. Position the pin piece vertically in the vice. Use a dovetail saw held at an angle to cut down the waste, stopping at the lap and shoulder lines.
9. Use a cramp to secure the pin piece to the bench top.
10. Chop out the waste using a chisel and mallet. Start just in from the end and work back towards the shoulder.
11. From the end of the pin piece, pare out the remaining waste and trim into the corners using a bevel-edged chisel.
12. Clean up, fit and assemble the joint as before.

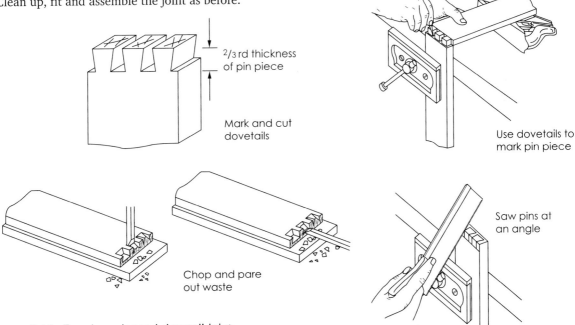

2/3 rd thickness of pin piece

Mark and cut dovetails

Use dovetails to mark pin piece

Chop and pare out waste

Saw pins at an angle

▲ **Figure 3.44** Forming a lapped dovetail joint

Mortise and tenon joints

This joint is probably the most widely used angle joint. It is used extensively for door and window construction and general framework in a variety of forms (Figure 3.45).

- **Through mortise and tenon joint.** In this joint the rectangular mortise slot goes right through the timber member.
- **Stub-mortise and tenon joint.** The mortise slot is stopped short and does not pass completely through the timber member. Also termed a **blind mortise**.

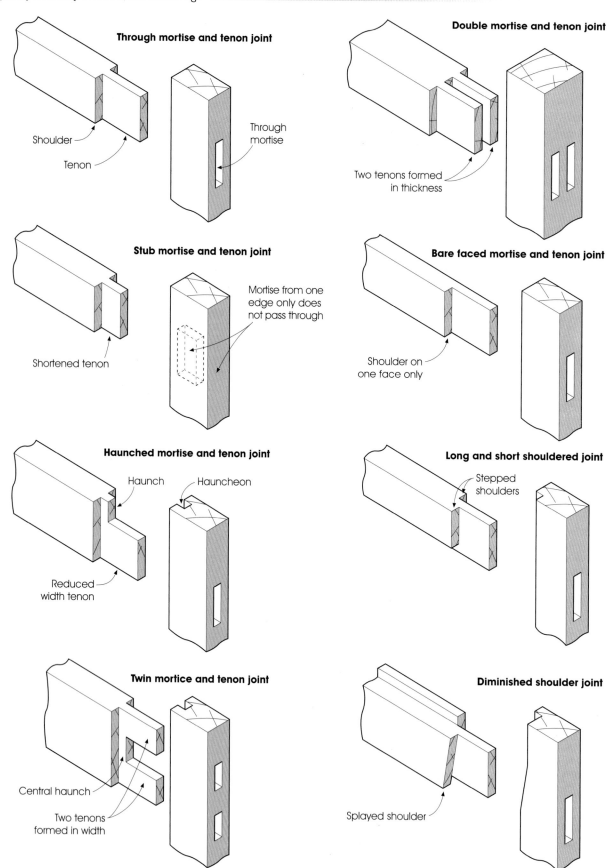

Through mortise and tenon joint

Shoulder

Tenon

Through mortise

Double mortise and tenon joint

Two tenons formed in thickness

Stub mortise and tenon joint

Shortened tenon

Mortise from one edge only does not pass through

Bare faced mortise and tenon joint

Shoulder on one face only

Haunched mortise and tenon joint

Haunch

Hauncheon

Reduced width tenon

Long and short shouldered joint

Stepped shoulders

Twin mortice and tenon joint

Central haunch

Two tenons formed in width

Diminished shoulder joint

Splayed shoulder

▲ **Figure 3.45** Varieties of mortise and tenon joint

Franked mortise and tenon joint

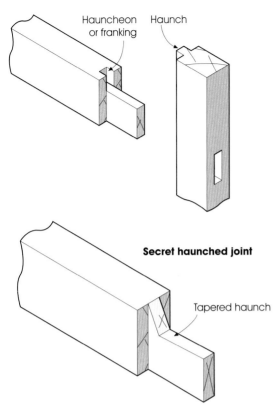

Hauncheon or franking Haunch

Secret haunched joint

Tapered haunch

Moulded frame mortise and tenon joints

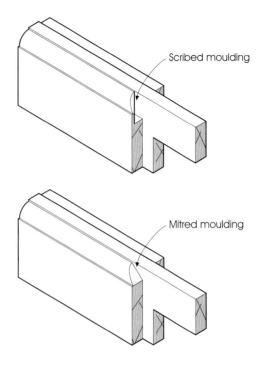

Scribed moulding

Mitred moulding

▲ **Figure 3.45** *continued*

- **Haunched mortise and tenon joint.** Here the tenon has been reduced in width leaving a shortened portion of tenon protruding called a haunch. The purpose of forming a haunch rather than cutting the reduced part of the tenon flush with the shoulder is to provide a positive location for the full width of the rail, preventing it from twisting. Haunches are used in wide members to reduce the width of the tenon or where the joint is at the end of a framed member to permit wedging up. The sinking adjacent to the mortise to receive a haunch is termed a **hauncheon**.

- **Twin mortise and tenon joint.** In this joint a haunch is formed in the middle of a wide member, creating two tenons one above the other.

- **Double mortise and tenon joint**. This has two tenons formed in the thickness of a member side by side.

- **Bare-faced mortise and tenon joint.** The tenon has a shoulder on one face only. Used for stair strings and framed matchboarded doors.

- **Long and short-shouldered joint.** Here the shoulder lines are stepped. Used for joints in rebated members.

- **Diminished shoulder joint.** One or both of the shoulders are splayed to accommodate a change in section above and below the mortise and tenon joint.

- **Franked mortise and tenon joint.** The haunch is formed on the mortised member and franking or hauncheon is cut into the tenoned member. Used where a standard haunch would remove too much timber and weaken the joint. Also known as a **sash haunch**.

- **Secret haunched joints** also known as a **table haunch joints**, have a tapered haunch which does not show on the end grain.

- **Moulded frame mortise and tenon joint.** Where the members have moulded edges it is necessary to scribe or mitre the moulding. It is normal to make the depth of the moulding the same as the rebate so that the shoulders are not stepped.

Joint proportions

The proportions of the mortise and tenon are important to the strength of the joint (Figure 3.46).

- The tenon should be one-third the thickness of the timber to be joined. If a chisel this size is not available to chop the mortise to receive the tenon, the thickness of the tenon may be adjusted to the nearest available chisel size.

- The width of the tenon should not exceed five times its thickness. This is to overcome the tendency of a wide tenon to buckle and also to reduce the effect of shrinkage.

- Where haunches are used to reduce the tenon width, they should be about one-third the width of the member and equal in length to the thickness of the tenon. Haunches between twin tenons are one-third of the member for middle rails or one quarter of the member where twin haunches are used for bottom rails.

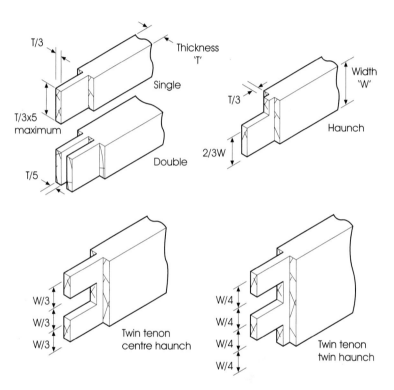

▶ **Figure 3.46** Joint proportions

Forming a through mortise and tenon joint

Follow these steps referring to Figure 3.47:

1. Use a square and pencil to mark the shoulder line for the tenon, around the member. Then score the shoulder lines with a marking knife. Where tenons are required at either end, mark out the distance between the shoulders precisely.

2. Mark the edge of the mortise on the mortise member and use the tenon member to mark its width.

3. Square lines all around the member using a pencil.

4. Set the pins of a mortise gauge to the selected mortise chisel. Set the stock of the gauge to centre the mortise on the edge of the member.

5. Using the gauge stock from the face side, score lines on both edges of the mortise member and from the tenon shoulder line on one edge over the end and back to the shoulder line on the other edge.

6. Cramp the mortise member to the bench or secure it in the vice. Position the mortise chisel in the middle of the mortise holding it vertically with the cutting edge at right angles to the gauged lines.

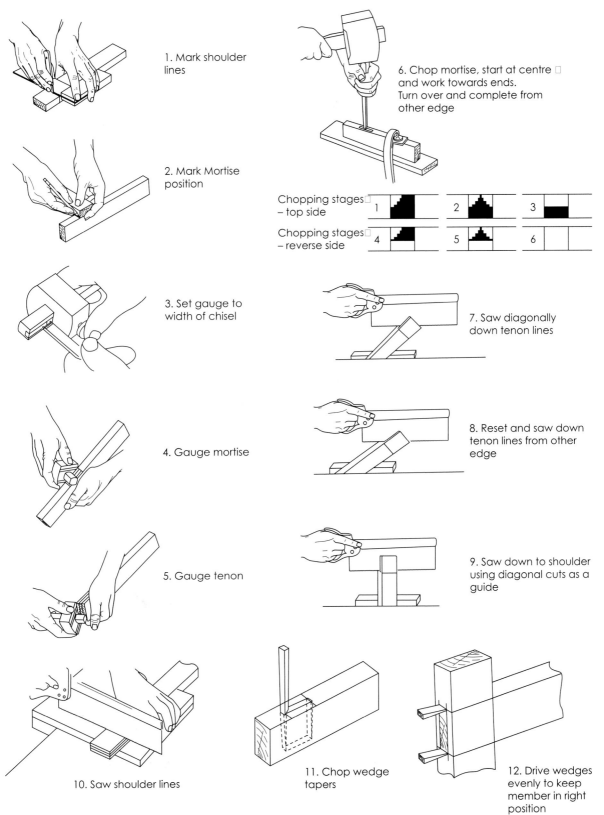

1. Mark shoulder lines

2. Mark Mortise position

3. Set gauge to width of chisel

4. Gauge mortise

5. Gauge tenon

6. Chop mortise, start at centre □ and work towards ends. Turn over and complete from other edge

| Chopping stages □ – top side | 1 | 2 | 3 |
| Chopping stages □ – reverse side | 4 | 5 | 6 |

7. Saw diagonally down tenon lines

8. Reset and saw down tenon lines from other edge

9. Saw down to shoulder using diagonal cuts as a guide

10. Saw shoulder lines

11. Chop wedge tapers

12. Drive wedges evenly to keep member in right position

▲ **Figure 3.47** Forming a through mortise and tenon joint

7. Drive the chisel with a mallet to the depth of 3 to 4 mm. Work back wards towards the end of the mortise in 3 to 4 mm steps.

8. Each time the chisel is driven it will cut progressively deeper so that when the mortise end is reached it will be about halfway through. Use a to-and-fro rocking action to release the chisel and break the waste.

9. Turn the chisel round and work from the centre back towards the other end.

10. Turn the member over; tap out any loose waste before securing to avoid bruising the work. Chop out the waste from the other edge using the same process, until the mortise is cut through.

11. Secure the tenon member at an angle in a vice. Saw diagonally down each tenon line and across the end grain, keeping just to the waste side. Reset in the vice and saw down the tenon lines from the other edge. Place the member upright in the vice and saw down level to the shoulder using the diagonal saw cuts as a guide.

12. Hold the member on a bench hook and use a tenon saw along the shoulder lines to remove the waste. Take care not to cut too deep and weaken the tenon.

13. Dry assemble the joint. The tenon should fit straight from the saw, but may be pared with a chisel if it's too tight. Working a tenon saw along the shoulder on both sides with the joint assembled, will ease joints that do not pull up. Again take care not to cut into the tenon.

14. Working from the outside edge make a tapered cut with the chisel about 6 mm wide at either end of the mortise. Cut two wedges from a piece of timber the same thickness as the tenon.

15. Clean up the inside edges before applying the glue to the tenon and the shoulders. Assemble the joint and cramp if required.

16. Apply the glue to the wedges and tap them in, striking each alternately to keep the members in the right location and the wedges level.

17. Once the glue has cured clean up the faces with a smoothing plane.

18. Saw off any protruding ends of the tenon and wedges and finish by planing the ends flush.

Wedging stub tenons

This is normally done using **fox wedges** as illustrated in Figure 3.48.

1. Chisel out a tapered undercut to both ends of the mortise. Cut two wedges the same thickness as the tenon, about two-thirds of the tenon length and 3 to 4 mm at the widest point of the taper.

2. Make two saw cuts in the tenon about 6 mm in from either edge and just longer than the wedges.

3. Drill a small hole at the end of each cut to help prevent splitting the tenon.

4. Clean up the inside edges and apply glue to the tenon, shoulders and wedges.

5. Start the wedges in the saw cuts and assemble the joint. When cramping, the wedges will be driven into the cuts by the bottom of the mortise, causing the tenon to spread and fit tightly against the undercut mortise tapers.

6. Clean up the faces once the glue has cured.

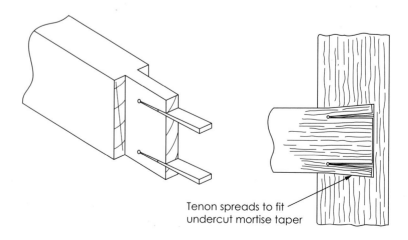

Tenon spreads to fit
undercut mortise taper

▶ **Figure 3.48** Fox wedges

Cutting haunches

These are cut out after the shoulders have been removed as illustrated in Figure 3.49.

1. Mark the tenon width from the previously cut mortise.
2. Mark the length of the haunch and cut along both lines with a tenon or panel saw.
3. Cut the hauncheon with a tenon saw and chisel out the waste.
4. A coping saw may be used to cut out the waste from central haunches between twin tenons.

The part of the tenon cut away to form the haunch may be used for cutting wedges.

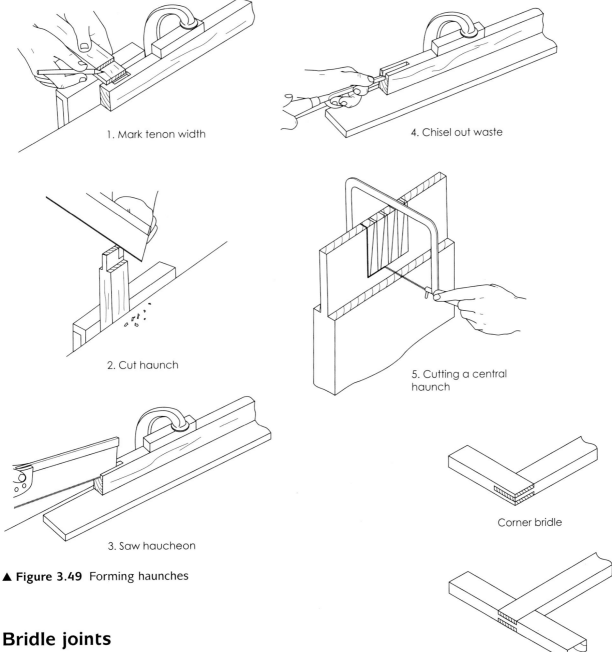

1. Mark tenon width

4. Chisel out waste

2. Cut haunch

5. Cutting a central haunch

3. Saw haucheon

Corner bridle

Tee bridle

▲ **Figure 3.49** Forming haunches

Bridle joints

These joints are a form of mortise and tenon, traditionally used for jointing heavy timber frames and roof trusses (Figure 3.50). Now used for light framing in joinery and cabinetwork. Also known as **forked mortise and tenon joints** when used in the middle of a member.

▲ **Figure 3.50** Bridle joints

Dowel joints

Dowels provide a simple method of joining solid timber framework and construction of cabinets, etc. They are in effect a butt joint that has been reinforced with small wood pegs, glued into holes drilled into both parts being jointed (Figure 3.51).

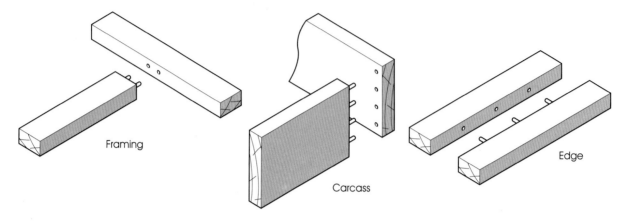

▲ Figure 3.51 Dowel joints

Dowels themselves are made from hardwood. They may be cut from a length of commercially produced dowel rod. Alternatively ready made chamfered-end fluted dowels are available (Figure 3.52).

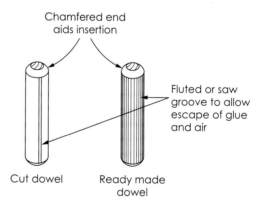

► Figure 3.52 Dowels

Use a dowel diameter that is between a third and a half the thickness of the pieces being joined and a minimum length of five times its diameter. Longer dowels give a greater gluing area and thus form a stronger joint.

When making dowels they should be prepared by:

- sawing to length;
- chamfering the ends to aid location and insertion;
- cutting a groove down the length of each dowel to allow excess glue and trapped air to escape when the joint is assembled.

Marking out dowel joints

Accurate positioning and drilling of dowel holes in each piece is essential. They may be marked out using a number of methods.

Edge-to-edge and framing joints may be marked out by placing the two pieces flush in the vice with the face sides out (Figure 3.53).

1. Square pencil lines across the edge at each dowel position.

156

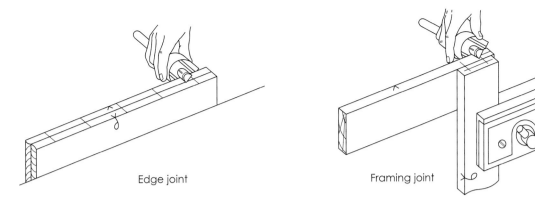

Edge joint   Framing joint

▲ **Figure 3.53** Marking out for dowels

2. Set a gauge to half the timber thickness and score the centre line on each, working from both face sides.

3. The number and spacing of dowels depends on size; at least three for edge joints 100 to 150 mm apart; at least two for framing joints, use three for rails over 100 mm wide.

4. Centre the drill where the lines cross and bore the holes. These should be a little deeper than half the dowel length, to ensure a tight fitting joint and create a glue reservoir.

Centre pins may be used to mark one part from the other. You can use panel pins or buy some proprietary dowel centre points (Figure 3.54);

■ *Using the panel pin method,* mark the dowel positions on either edge of an edge-to-edge joint, or on the end of the rail for framing joints. Drive panel pins into the centre marks and crop off the heads with a pair of pincers, leaving about 6 mm protruding. Lay both members flat on the bench and push together to transfer the centre points.

■ *Using proprietary centre pins,* mark and drill the holes in one edge or end of the rail. Insert the dowels pins into each hole and bring together as before.

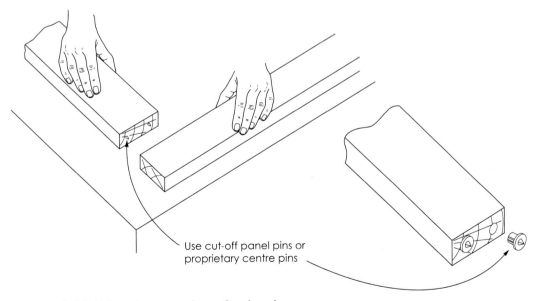

Use cut-off panel pins or
proprietary centre pins

▲ **Figure 3.54** Using pins to mark out for dowels

Templates may be made to aid marking out (Figure 3.55). Cut a thin piece of plywood or MDF the same size as the cross section of the timber or carcass panel. Mark the dowel centres on the template. Position over the joint and drive the panel pins to mark the centres in each piece.

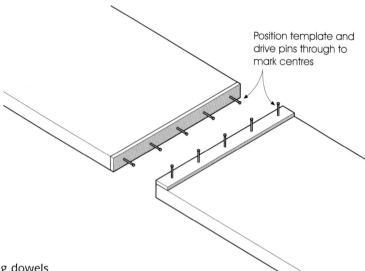

▶ **Figure 3.55** Use of template for marking dowels

Proprietary **dowelling jigs** are also available. These clamp onto the work piece, position the holes and guide the drill accurately (Figure 3.56). Most have interchangeable guide bushes for 6, 8 and 10 mm drills and dowels.

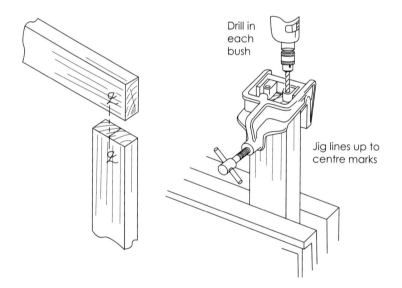

▶ **Figure 3.56** Proprietary dowel jig

Assembling frames

When making frames each joint should be cut separately and the entire frame dry assembled to check the fit of the joints, overall sizes, square and winding, before gluing and cramping up (Figure 3.57).

■ **Squaring up.** Frames are checked for square with a **squaring rod.** This consists of a length of rectangular section timber with a panel pin in its end. The end with a panel pin is placed in one corner of the frame. The length of the diagonal should then be marked in pencil on the rod. The other diagonal should then be checked. If the pencil marks occur in the same place, the frame must be square. If the frame is not square, the sash cramps should be angled to pull the frame into square.

■ **Winding.** Frames are checked for winding or twist with **winding strips**, which are two parallel pieces of timber. With the frame laying flat on a level bench, place a winding strip at either end of the job.

edge finish and whether a balancer is to be used. The following can be used as a general guide for a worktop with a balancer, decorative laminate top and a matching laminate front edge.

Cutting the laminate

Always take care to avoid running your hands and fingers along the edges of laminates, as they are very sharp.

The top and balancer should be cut slightly bigger than the worktop core board. Whenever possible it is best to cut laminate with the lined underside of the sheet following the longest dimension.

Support the entire sheet and either use a fine tooth tenon saw for cutting from the decorative face, or use a laminate cutter to score a line on the decorative face (Figure 3.62) Break the sheet along the scored line by lifting the waste piece upwards.

The edging strips are best cut using a marking or cutting gauge:
1. Set the gauge about 5 mm wider than the core board.
2. Gauge along the edge of the sheet to score a line.
3. Separate the strip by applying thumb pressure on the scored line starting at one end, whilst at the same time lifting up the edge of the strip.
4. Carefully continue to work along the line to separate the entire strip.

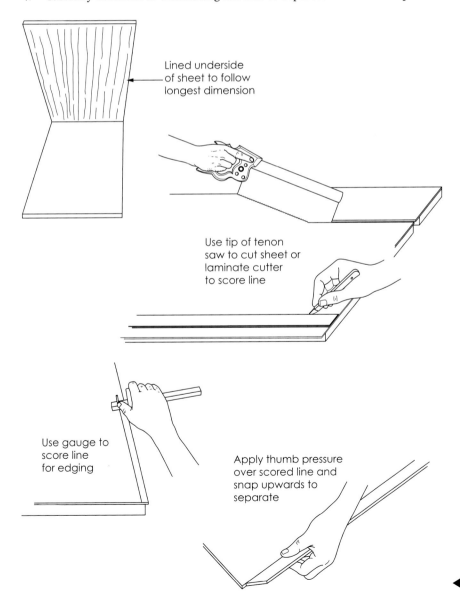

Lined underside of sheet to follow longest dimension

Use tip of tenon saw to cut sheet or laminate cutter to score line

Use gauge to score line for edging

Apply thumb pressure over scored line and snap upwards to separate

◀ **Figure 3.62** Cutting laminate

Applying the balancer

1. Dust off the balancer and the core.
2. Lay the balancer, face side down on the core.
3. Apply contact adhesive to the balancer; lay off in one direction using a serrated spreader (Figure 3.63). Always follow the adhesive manufacturer's instructions as many adhesives give off toxic and potentially explosive vapours.

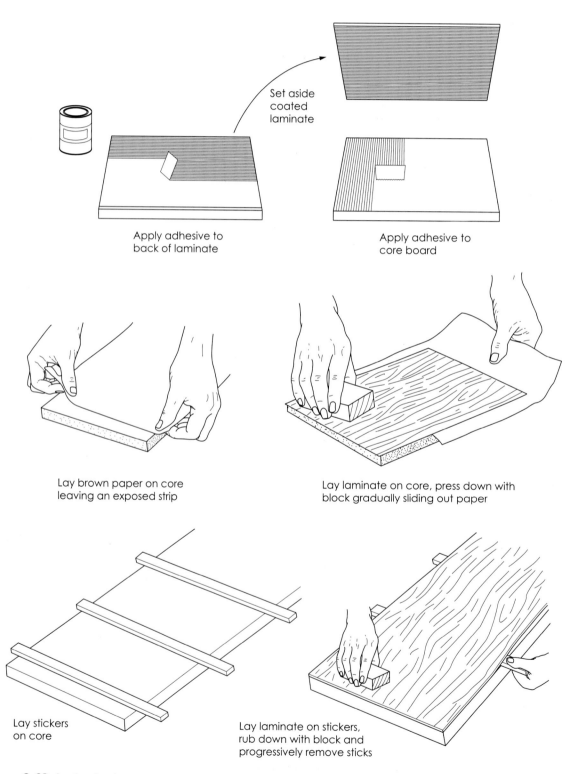

Set aside coated laminate

Apply adhesive to back of laminate

Apply adhesive to core board

Lay brown paper on core leaving an exposed strip

Lay laminate on core, press down with block gradually sliding out paper

Lay stickers on core

Lay laminate on stickers, rub down with block and progressively remove sticks

▲ **Figure 3.63** Laying laminate

4. Set aside the coated balancer and apply contact adhesive to the core board, using the same method as before except this time lay off at right angles to that of the balancer. This results in a better bond than would be achieved if both were laid off in the same direction.

5. After the manufacturer's recommended period of time (usually when the surface is touch dry) either lay a sheet of brown paper over the entire core leaving just a 50 mm band at one end exposed or alternatively lay small prepared strips of timber (stickers) across the core at about 150 mm intervals. The purpose of this is to keep the two surfaces separated until they are correctly positioned and eventually rubbed down.

6. Position the balancer on the brown paper or stickers. When correctly aligned either: press the balancer against the exposed strip, gradually slide the paper out pressing the balancer down with a block of wood to the core, working from the centre to the edges of the sheet each time to avoid air traps; or starting from one end remove the first sticker and start rubbing down as above, progressively removing the stickers as you progress. Finally starting at the centre and working outwards, rub over the whole surface to ensure good adhesion, using a block of wood or a rubber-faced roller. Pay particular attention to the edges.

7. Trim the edges preferably using a guided laminate trimming cutter in a portable powered router (Figure 3.64). Where a router is not available, a low angled block plane, file or cabinet scraper can be used.

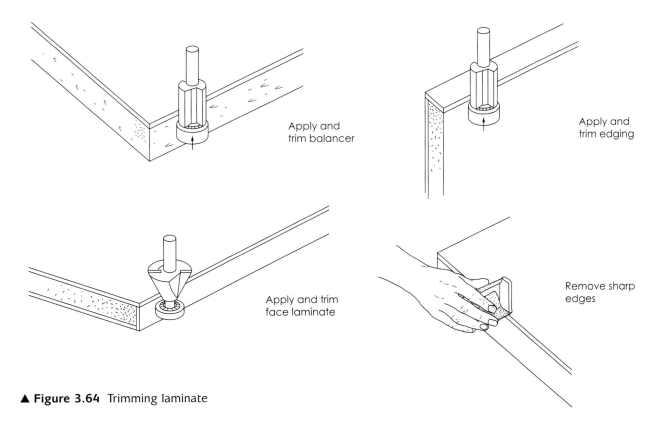

Apply and trim balancer

Apply and trim edging

Apply and trim face laminate

Remove sharp edges

▲ **Figure 3.64** Trimming laminate

Apply edging strip

1. Apply a coat of contact adhesive to the edge of the core and allow to dry. This acts as a primer to seal the absorbent surface.

2. Apply contact adhesive to the back of the edging strip and a second coat to the edge of the core.

3. When the adhesive is touch dry position the edging, working from one end, keeping an even overhang on both the balancer and core face. To ensure good contact use a block of wood and tap with a hammer along the entire edge.

4. Trim the edging flush with the core face and balancer, again preferably using a router.

Applying decorative face laminate. This stage is carried out in the same way as the balancer, except that extra care is required when hand trimming and cleaning up so as not to spoil the decorative face.

Finally the sharp arris should be removed using a fine abrasive paper and a hand block.

Chapter Four

Interpretation of drawings

This chapter covers the work of both the site carpenter and bench joiner. It is concerned with the types and interpretation of drawing used in the construction industry. It includes the following:

- ✏ scales and scale rules
- ✏ lines, symbols and abbreviations used on drawings
- ✏ orthographic and pictorial methods of projection
- ✏ types of working drawings
- ✏ bench and datum marks
- ✏ sketches.

Drawings are the major means used to communicate technical information between all parties involved in the building process. They must be clear, accurate and easily understood by everyone who uses them. In order to achieve this, it is essential that architects, designers and all others who produce drawings use standardised methods for scale, line-work, layout, symbols and abbreviations.

Scales and lines used on drawings

Scales use **ratios** that permit measurements on a drawing or model to relate to the real dimensions of the actual job. It is impractical to draw buildings, plots of land and most parts of a building to their full size, as they simply will not fit on a sheet of paper. Instead they are normally drawn to a smaller size, which has a known scale or ratio to the real thing. These are then called **scale drawings** as illustrated in Figure 4.1.

The main scales (ratios) used in the construction industry are:

> 1:1; 1:5; 1:10; 1:20; 1:50; 1:100; 1:200; 1:500; 1:1250; 1:2500

We say: '1:10' is a ratio or scale of 'one to ten' or sometimes 'one in ten'.

The ratio shows how much smaller the plan or model is to the original. Figure 4.2 illustrates 1 m length drawn to various scales. In a drawing to a scale of 1:10:

> 5 mm would stand for or relate to 50 mm on the actual job, or
>
> 50 mm would stand for or relate to 0.5 m (500 mm).

In a drawing to a scale of 1:20:

> 5 mm would stand for or relate to 100 mm on the actual job, or
>
> 50 mm would stand for or relate to 1 m (1000 mm).

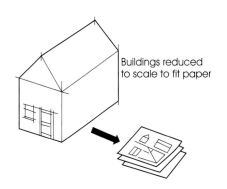

Buildings reduced to scale to fit paper

▲ **Figure 4.1** Scale drawings

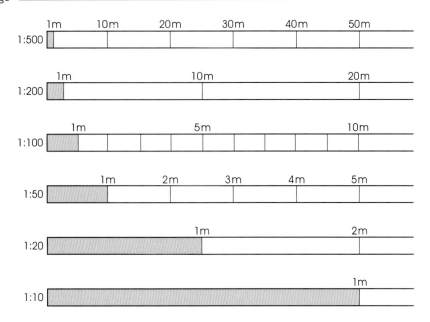

▶ **Figure 4.2** 1 m drawn to various scales

See illustrated examples in Figure 4.3. It is simply a matter of multiplying the scale measurement by the scale ratio.

Example

10 mm at a 1:50 scale ratio represents 500 mm in the building.

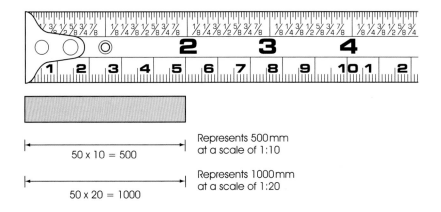

▶ **Figure 4.3** Actual and scaled dimensions

Scale rules

These are used to save having to calculate the actual size represented on drawings. These have a series of marks engraved on them so that dimensions can be taken directly from the rule, see Figure 4.4.

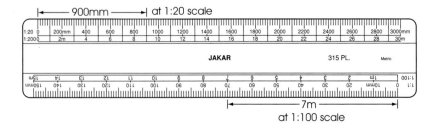

▲ **Figure 4.4** Scale rule

Although scale rules can be useful when reading drawings preference should always be given to written dimensions shown on a drawing. Mistakes can be made due to the dimensional instability of paper at differing moisture contents or a fold in the paper might make it impossible to get an accurate size.

The opening size shown in Figure 4.5 can be accurately determined by taking away the length of the stud partition from the overall width of the room, rather than trying to scale it with a rule, which may result in over or under measurement depending on the humidity and condition of the paper.

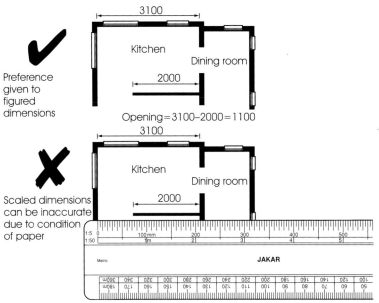

◀ **Figure 4.5** Taking off measurements

Dimensions shown on scale drawings normally include the symbol for the measurement units used. A lower case letter 'm' is used for metres and lower case letters 'mm' for millimetres. Units should not be mixed on drawings, $2\frac{1}{2}$ metres can be shown either as 2.5 m, 2.50 m 2.500 m or 2500 mm. Units are not always included with the dimension, in these cases whole numbers are taken as millimetres (e.g. 1500) and decimal numbers to one, two or three places are taken as metres (1.5 or 1.50 or 1.500). Note that 1500 mm = 1.5 m (or 1.50 or 1.500 m).

On some drawings, to avoid confusing or missing the position of the decimal point an oblique stroke is used to separate metres from millimetres e.g. 1/500 for 1 m 500 mm (1.500 m).

Where the dimension is less than a metre a nought is added before the stroke e.g. 0/750 for 750 mm.

Sequence of dimensioning. The recommended sequence for expressing dimensions is illustrated in Figure 4.6. Length is normally expressed first, followed by width and then thickness:

$$L \times W \times T = 2100 \times 225 \times 50 \text{ mm}$$

Where only the sectional size is quoted, width is normally stated before thickness e.g. 225 × 50 mm. Beware that the sectional size on drawings may be stated with the size seen in plan first e.g. a 100 × 50 mm section may be shown as 50 × 100 mm when used as a joist or 100 × 50 mm for a wall plate.

Lines used on drawings

A variety of different line types and thickness or 'weight' is used on drawings for specific purposes; these are illustrated in Figure 4.7.

■ *Solid lines* are used for the actual parts of an object that can be seen. Thick or dark lines for the outside boarder; medium or lighter lines for

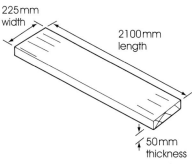

2100 x 225 x 50mm
Expressed as (length) (width) (thickness)

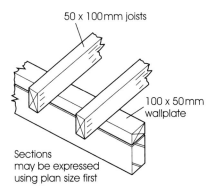

Sections may be expressed using plan size first

▲ **Figure 4.6** Sequencing of dimensions

Line type		Used for
———————————	Thick	Main outlines
———————————	Medium	General details and outlines
———————————	Thin	Construction and dimension lines
——————⋀——	Breakline	Breaks in the continuity of a drawing
▬▬·▬▬·▬▬·▬	Thick chain	Pipe lines, drains and services
————·————·—	Thin chain	Centre lines
▲——————————▲	Section line	Showing the position of a cut (the pointers indicate the direction of view)
- - - - - - - - - - - - - -	Broken line	Showing details which are not visible
⊢———————⊢	Dimension line	Showing the distance between two points

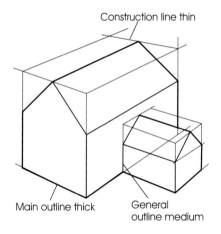

▶ **Figure 4.7** Line types and weights

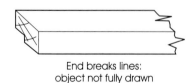

▲ **Figure 4.8** Use of line weights

Construction line thin

Main outline thick

General outline medium

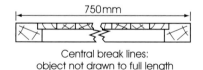

End breaks lines: object not fully drawn

750mm

Central break lines: object not drawn to full length

▲ **Figure 4.9** Use of break lines

internal boarders and general detail; thin or light lines for construction and dimensioning. Figure 4.8 shows a simple building outline that illustrates the use of these different line weights.

■ *Break-lines in a zig-zag pattern* are used to show a break in the continuity of a drawing as illustrated in Figure 4.9. End break-lines are used to indicate that an object has not been fully drawn. Central break-lines are used to indicate that the object has not been drawn to its full length.

■ *Chain lines in a long/short or dot/dash pattern* are used for centre lines and services (Figure 4.10).

■ *Section lines* are shown on a drawing to indicate an imaginary cutting plane, at a particular point through an object. Pointers or arrows on the line indicate the direction of view that will be seen on a separate section drawing. Where more than one section is shown these will be labelled up as A–A, B–B etc. according to the number of sections. The section drawings themselves may be included on the same drawing sheet or be cross-referenced to another drawing (Figure 4.11).

■ *Broken lines* indicate hidden details that cannot be seen on the object as drawn, or for work that is to be removed, as illustrated in Figure 4.12.

■ *Dimension lines* are used to indicate the distance between points on a drawing. These are lightly drawn solid lines with arrowheads terminating against short cross-lines, see Figure 4.13. Open (birdsfeet) arrowheads are often used for the basic/modular 'unfinished' distances in a drawing. Closed (solid) arrowheads are the preferred method for general use and 'finished' work sizes.

Separate or running dimensions

The actual figured dimensions may be shown as either separate or running dimensions as shown in Figure 4.14.

■ Separate dimensions are normally written above and centrally along the line, figures on vertical lines should be written parallel to the line, so that they are read from the right.

■ Running dimensions are shown cumulatively commencing with zero from a fixed point or datum. These are often written on the cross-line at right angles to the dimension line. Care must be taken not to confuse separate and running dimensions when reading them off a drawing.

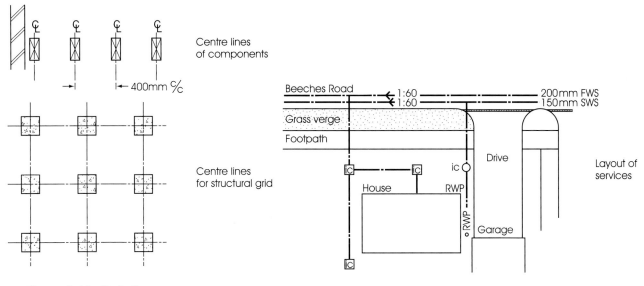

▲ **Figure 4.10** Chain lines

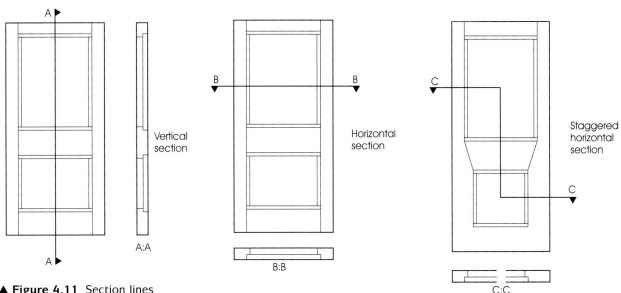

▲ **Figure 4.11** Section lines

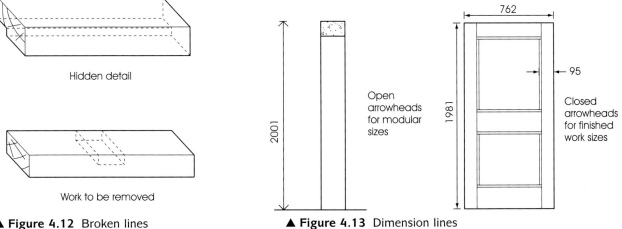

▲ **Figure 4.12** Broken lines

▲ **Figure 4.13** Dimension lines

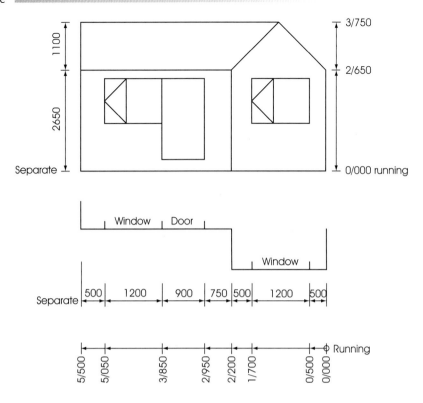

▶ **Figure 4.14** Separate and running dimensions

Symbols and abbreviations used on drawings

Symbols are graphical illustrations used to represent different materials and components in a building drawing (Figure 4.15).

Abbreviations. These are a short way of writing a word or group of words. They allow maximum information to be included in a concise way. Table 4.1. shows some of the abbreviations commonly used in the building industry.

Table 4.1

Aggregate	agg	BS tee	BST	Foundation	fdn	Polyvinyl acetate	PVA
Air brick	AB	Building	bldg	Fresh air inlet	FAI	Polyvinylchloride	PVC
Aluminium	al	Cast iron	CI	Glazed pipe	GP	Rainwater head	RWH
Asbestos	abs	Cement	ct	Granolithic	grano	Rainwater pipe	RWP
Asbestos cement	absct	Cleaning eye	CE	Hardcore	hc	Reinforced concrete	RC
Asphalt	asph	Column	col	Hardboard	hdbd	Rodding eye	RE
Bitumen	bit	Concrete	conc	Hardwood	hwd	Foul sewers	FS
Boarding	bdg	Copper	Copp cu	Inspection chamber	IC	Sewers surface water	SWS
Brickwork	bwk	Cupboard	cpd	Insulation	insul	Softwood	swd
BS* Beam	BSB	Damp proof course	DPC	Invert	inv	Tongue and groove	T & G
BS Universal beam	BSUB	Damp proof membrane	DPM	Joist	jst	Unglazed pipe	UGP
BS Channel	BSC	Discharge pipe	DP	Mild steel	MS	Vent pipe	VP
BS equal angle	BSEA	Drawing	dwg	Pitch fibre	PF	Wrought iron	WI
BS unequal angle	BSUA	Expanding metal lathing	EML	Plasterboard	pbd		

*BS = British Standard

172

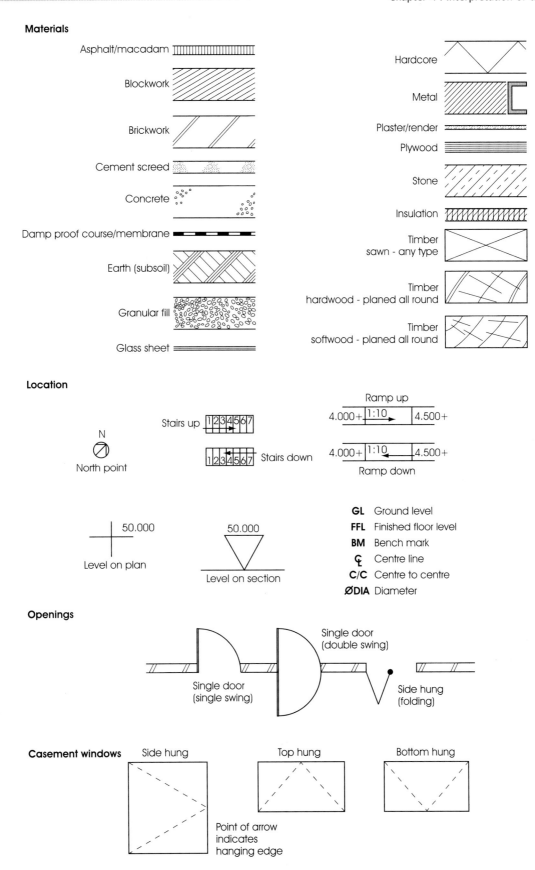

Figure 4.15 Symbols used on drawings

Methods of projection

Drawings of objects can be produced as either a series of plan or elevation views called *orthographic projection* or in a form that closely resembles the three-dimensional appearance called *pictorial projection*.

Orthographic projection

This method is used for working drawings, plans, elevations and sections. A separate drawing of each face of all the views of an object is produced in a systematic manner on the same drawing sheet, see Figure 4.16. Each view is at right angles to the face:

- *Plan view* details the surface of an object when looking down on it vertically. Floor plans in building drawings are normally drawn as a section taken just above windowsill height.
- *Elevation view* of an object details the surface from either side, front or rear.
- *Section view* details the cut surface produced when an object is imagined cut through with a saw.

First and third angle projection

The actual position on the drawing sheet of the plan, elevations and section will vary depending on the method of projection used, either first or third angle. The viewing position in relationship to an object is illustrated in Figure 4.17.

- **First angle projection** (Figure 4.18) is generally used for building drawings where in relation to the front view the other views are arranged as follows: the view from above is drawn below; the view from below is drawn above; the view from the left is drawn to the right; the view from the right is drawn to the left; the view from the rear is drawn to the extreme right. A sectional view may be drawn to the left or the right where space permits.
- **Third angle projection** (Figure 4.19) a form of orthographic projection used for engineering drawings. It is also termed **American projection**. In relation to the front elevation the other views are arranged as follows: the view from above is drawn above; the view from below is drawn below; the view from the left is drawn to the left; the view from the right is drawn to the right; the view from the rear is drawn to the extreme right. A sectional view may be drawn to the left or the right where space permits.

Pictorial projection

This is a method of drawing objects in a three-dimensional form. Often used for design and marketing purposes as the finished appearance of the object can be more readily appreciated by the general public. Differing views are achieved by varying the angles of the base lines and the scale of the side projections.

- **Isometric.** The most widely used form of pictorial drawing where all vertical lines are drawn vertical, and all horizontal lines are drawn at an angle of 30 degrees to the horizontal. The length, width and height are all drawn to the same scale, see Figure 4.20.
- **Planometric.** A form of pictorial drawing where all vertical lines are drawn vertical, horizontal lines on the front elevation are drawn at 30 degrees and those on the side elevation at 60 degrees to the horizontal, giving a true plan shape. The length, width and height are all drawn to the same scale, see Figure 4.21.
- **Axonometric.** A form of pictorial drawing similar to planometric, except that the horizontal lines in both the elevations are drawn at

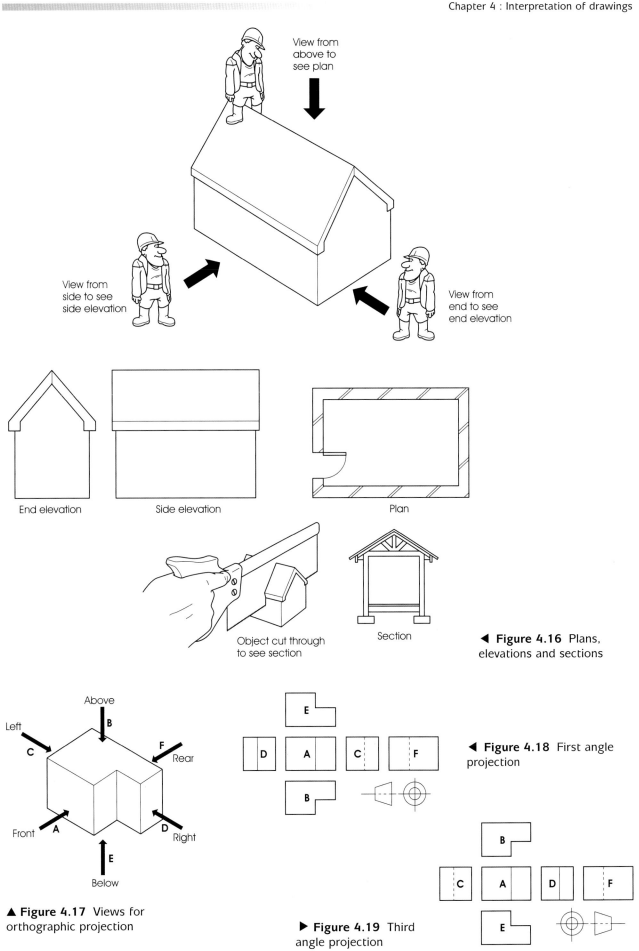

View from
above to
see plan

View from
side to see
side elevation

View from
end to see
end elevation

End elevation

Side elevation

Plan

Object cut through
to see section

Section

◀ **Figure 4.16** Plans,
elevations and sections

Above

Left

C

B

F

Rear

Front

A

D

Right

E

Below

▲ **Figure 4.17** Views for
orthographic projection

E

D A C F

B

◀ **Figure 4.18** First angle
projection

B

C A D F

E

▶ **Figure 4.19** Third
angle projection

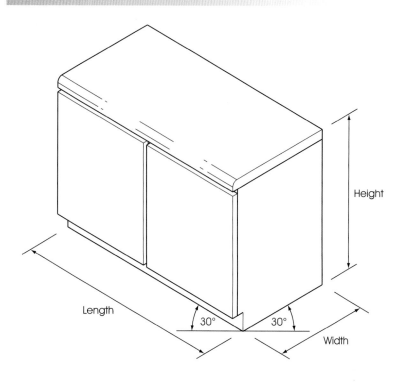

▶ **Figure 4.20** Isometric projection

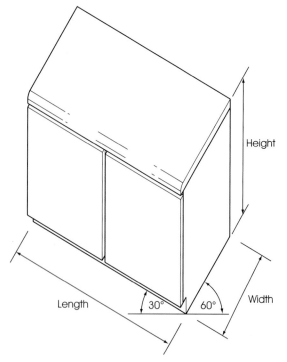

▲ **Figure 4.21** Planometric projection

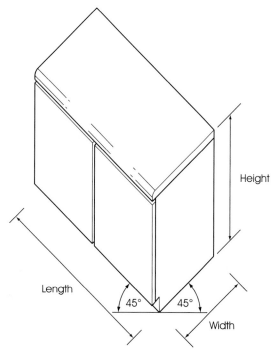

▲ **Figure 4.22** Axonometric projection

45 degrees to the horizontal, again giving a true plan shape. The length, width and height are all drawn to the same scale, see Figure 4.22.

■ **Oblique.** A form of pictorial projection, which uses a true front elevation. The side elevation can be either cabinet or cavalier. All vertical lines are drawn vertical and all horizontal lines in the front elevation are drawn horizontal, while all horizontal lines in the side elevations are drawn at 45° to the horizontal. In cavalier these 45° lines are drawn to their full or scale length, while in cabinet they are drawn to half their full or scale length to give a less distorted view, as illustrated in Figure 4.23.

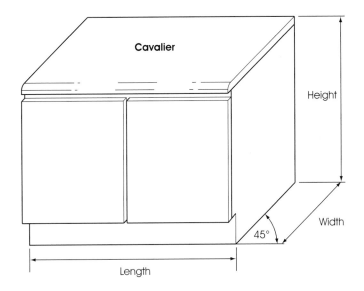

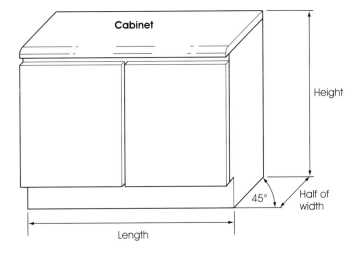

▲ **Figure 4.23** Oblique projection

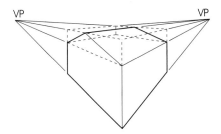

▲ **Figure 4.24** Parallel (one point) perspective projection

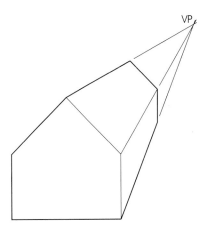

▲ **Figure 4.25** Angular (two-point) perspective projection

■ **Perspective.** A form of pictorial projection where the horizontal lines disappear to one or more points on the imaginary horizon. These disappearing points are termed as VPs or 'vanishing points' as illustrated in Figure 4.24. **Parallel perspective** has a true front elevation with the sides disappearing to one vanishing point (one-point perspective). **Angular perspective** is drawn with the elevations disappearing to two vanishing points (two-point perspective). All upright lines in both forms are drawn vertically. Sloping lines such as the pitch of the roof are best first drawn in a rectangular box, with the ridge being positioned mid-way across the box as shown in Figure 4.25.

Working drawings

These are scale drawings showing the plans, elevations, sections, details and locality of an existing or proposed building.

Layout of working drawings. Drawings can be produced either by hand drafting using a drawing board and instruments or on a computer using a

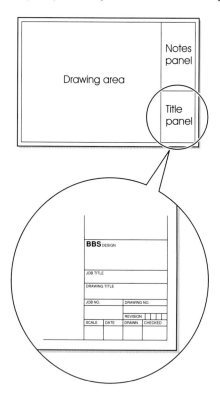

▲ **Figure 4.26** Layout of working drawing sheet

▶ **Figure 4.27** International paper sizes

CAD (computer-aided design) programme. In addition to the actual labelled drawings they will also include figured dimensions, printed notes to explain exactly what is required, and a title panel which identifies and provides information about the drawing (Figure 4.26).

Paper sizes. Drawings are normally produced on a range of international paper sizes as illustrated in Figure 4.27. 'A0' is the base size consisting of a rectangle having an area of one square metre and sides which are in the proportion of 1:√2. All the A series are of this proportion, enabling them to be doubled or halved in area and remain in the same proportion, which is useful for photographic reproduction. A1 is half A0; A2 is half A1; A3 is half A2; A4 is half A3. Where sizes larger than A0 are required the A is proceeded by a number, 2A being twice the size of A0 and 4A being four times the size of A0 etc.

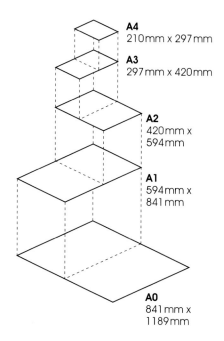

Types of working drawings

These drawings can be divided into a number of main types.

Location drawings

- **Block plans**, to scales of 1:2500 or 1:1250, identify the proposed site in relation to the surrounding area, see Figure 4.28.
- **Site plans,** to scales of 1:500 or 1:200, giving the position of the proposed building and the general layout of roads, services and drainage etc. on the site see Figure 4.29.
- **General location plans,** to scales of 1:200, 1:100 or 1:50, showing the general position occupied by the various items within the building and identify the location of the principal elements and components see Figure 4.30.

Component drawings

- **Range drawings** (Figure 4.31), to scales of 1:100, 1:50 or 1:20, show the basic sizes and reference system of a standard range of components.
- **Detail drawings** (Figure 4.32), to scales of 1:10, 1:5 or 1:1, show all the information that is required in order to manufacture a particular component.

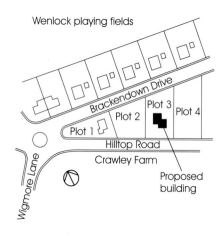

▲ **Figure 4.28** Block plan

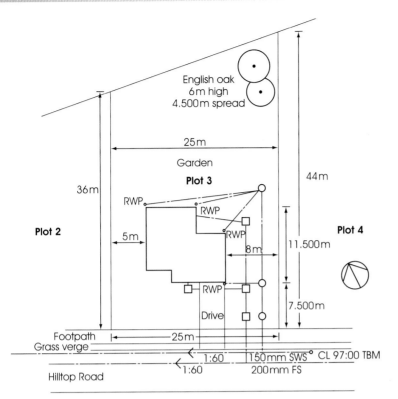

▲ **Figure 4.29** Site plan

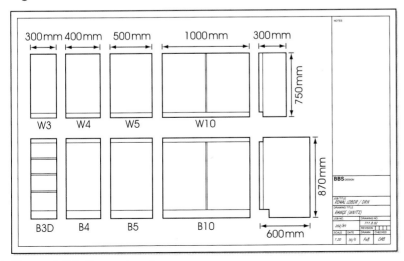

▲ **Figure 4.30** General location plan

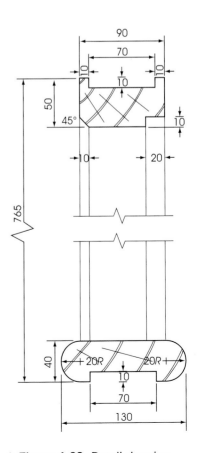

▲ **Figure 4.31** Range drawing

▲ **Figure 4.32** Detail drawing

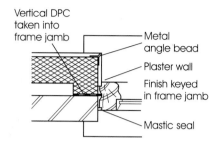

▲ Figure 4.33 Assembly drawing

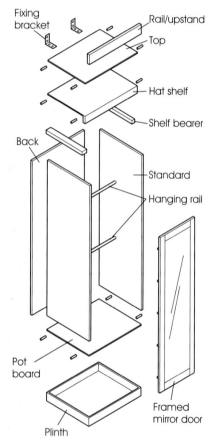

▲ Figure 4.34 Exploded view of single wardrobe unit

Assembly drawings:

- **Assembly details** (Figure 4.33), to scales of 1:20, 1:10 or 1:5, showing in detail the junctions in and between the various elements and components of a building.
- **Exploded views** may be used to pictorially illustrate how components are assembled, see Figure 4.34.

Drawing registers. These list the drawings relevant to a particular job and who they have been distributed to, as well as drawing numbers, revision numbers, formats, sizes and scales (Figure 4.35). Always check the drawing that you are using with the register to ensure it is the latest revision.

Schedules. These contain mainly written information in the form of a list or table (Figure 4.36). They are used to collect specialised information about repetitive parts of a building such as doors, windows and finishes, etc. Obtaining information from a schedule about any particular item is fairly straightforward and is known as 'extracting' or 'taking off'. Details relevant to a particular item are indicated in a schedule by a dot, cross or tick; a number may also be included where more than one item is required.

The information that schedules contain is essential when preparing estimates and tenders. In addition, schedules are also extremely useful when measuring quantities, locating work and checking deliveries of materials and components.

Figure 4.37 illustrates an extract from a typical door/ironmongery schedule and shows its relationship with various building drawings.

Design and production drawings

Designers often issue **design drawings** or sketches to manufacturers, showing their design intent for a particular item. The manufacturer will interpret these design details and prepare draft working or **production drawings** for the designer's approval prior to the commencement of manufacture. On receipt of the draft production drawings the designer will check them over, if all is okay they will sign them off and return them as approved. The manufacturer in turn will change the status of the drawings from draft to production drawings (Figure 4.38).

Redline drawings. Where draft drawings are not approved the designer will return them to the manufacturer for alteration. The alterations required will be marked in red pen, so that they stand out (hence the name 'redline drawings'). Details to be amended are often highlighted or circled in a 'cloud' to emphasise them. The manufacturer will have to amend the drawing and re-submit it to the designer for approval. Illustrated in Figure 4.39 is a design sketch of a desk along with the manufacturers draft that has been redlined by the designer with details inside 'clouds' like speech bubbles.

Internal drawings. These consist of a floor plan of a particular room or area, along with elevations of each wall. They are particularly useful to show the layout of furnishings and fitments manufactured to the production drawings (Figure 4.40).

Bench and datum marks

These are identified points from which all other positions are either taken or are related to.

- *Bench mark or Ordnance Survey mark* is a mark related to the Ordnance Datum (OD), which is the mean sea level at Newlyn in Cornwall, England. From this datum the Ordnance Survey has established Ordinance Bench Marks (OBMs) throughout the country. They are

DRAWING REGISTER

Project: **ALFA HOUSE REFURBISHMENTS**
Location: **NOTTINGHAM**
Client: **CREASTA HOTELS GROUP**
Project No: **14/035**

**CAB ARCHITECTS LTD
BRETT HOUSE
BEASTON
NOTTINGHAM
NG10 2XF
TEL: 01159 000010
FAX: 01159 000011
EMAIL: CAB@ARCH.COM**

CAB ARCHITECTS

Issue Codes:		Issue Date:	23.06.05	14.07.05	28.07.05								
1. Information	5. Tender Issue												
2. Preliminary	6. For Construction												
3. For Approval	7. Planning Application												
4. Billing	8. Building Control Application												
	9. Contract Issue												

Purpose of Issue and Revision:			5	5	6								
Description:	Dwg No:	Size:											
BEDROOM 45 : SURVEY	14/0351	A3	✓										
BEDROOM 45 : PLAN	14/0352	A3	✓	A	B								
BEDROOM 45 : ELEVATIONS	14/0353	A3	✓	A	B								
BEDROOM 45 : AXONOMETRIC	14/0354	A3	✓										
UTILITY ROOM : SURVEY	14/0355	A2	✓										
UTILITY ROOM : PLAN	14/0356	A2	✓	A	B								
UTILITY ROOM : ELEVATIONS	14/0357	A2	✓	A	B								
UTILITY ROOM : AXONOMETRIC	14/0358	A3	✓										

Distribution:		Number of Copies/Media Issue (E) e-mail (D) Disk/CD (H) Hard									
BBS DESIGN: JAMES PETERS		1D	1H	1E							
RE-INTERIORS: ALLEN GRANT		1D	1H	1E							
THE JOINERY COMPANY: JANE GAW			1H	1E							

File			1	1	1								
Issued by:			CAB	CAB	CAB								

Description	D1	D2	D3	D4	D5	D6	D7	D8	D9	10D	D11	D12	D13	D14	D15
DOOR TYPE															
EXT 1	✓														
EXT 2						✓									
EXT 3										✓					
INT 1		✓	✓		✓		✓		✓		✓			✓	✓
INT 2				✓								✓	✓		
INT 3								✓							
DOOR SIZE															
813 X 2032 X 44	✓														
835 X 1981 X 44						✓				✓					
762 X 1981 X 35		✓	✓		✓		✓		✓		✓			✓	✓
686 X 1981 X 35				✓								✓	✓		
610 X 1981 X 35								✓							
DOOR FINISH															
HARDWOOD/POLISH	✓														
SOFTWOOD/PAINT						✓				✓					
MOULDED H.B./PAINT		✓	✓	✓	✓		✓	✓	✓		✓	✓	✓	✓	✓
HANGING															
100 MM BRASS BUTTS	✓3														
100 MM STEEL BUTTS						✓3									
75 MM CHROME BUTTS		✓2	✓2	✓2	✓2		✓2	✓2	✓2		✓2	✓2	✓2	✓2	✓2
250 MM TEE HINGES										✓3					
FASTENING															
RIM NIGHT LATCH	✓														
MORTICE DEADLOCK	✓									✓					
MORTICE LOCK/LATCH		✓	✓		✓	✓					✓			✓	✓
MORTICE LATCH				✓			✓	✓	✓		✓	✓			
100 MM CHROME BOLT	✓2					✓2			✓						
MISCELLANEOUS															
LETTER PLATE	✓														
SECURITY CHAIN	✓														
HINGE BOLT	✓2					✓2									
LOCK/LATCH LEVER		✓	✓		✓	✓									
LATCH LEVER				✓			✓	✓	✓		✓			✓	✓
COAT HOOK				✓2					✓2			✓	✓		
KEYHOLE ESCUT.	✓2									✓2					

Notes:

Ex = External
INT = Internal

REFER TO FLOOR PLAN FOR DOOR LOCATION

BBS Design

Job Title
THORNEY WOOD REFURBISHMENT

Drawing Title
DOOR & IRONMONGERY SCHEDULE

Job No.	Drawing No.
031550	**CAB51**

Scale	Date	Drawn	Checked
N/A	**15.3.03**	**JPB**	**PSB**

▲ **Figure 4.36** Door and ironmongery schedule

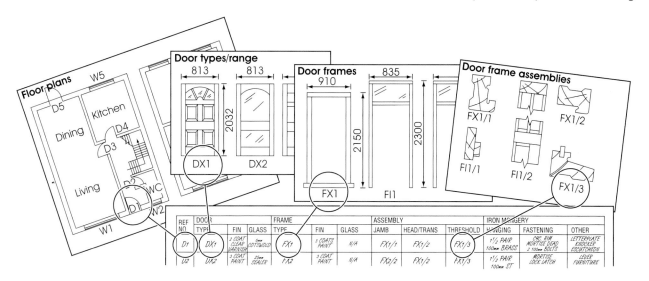

▲ **Figure 4.37** Relationship between drawings and schedules

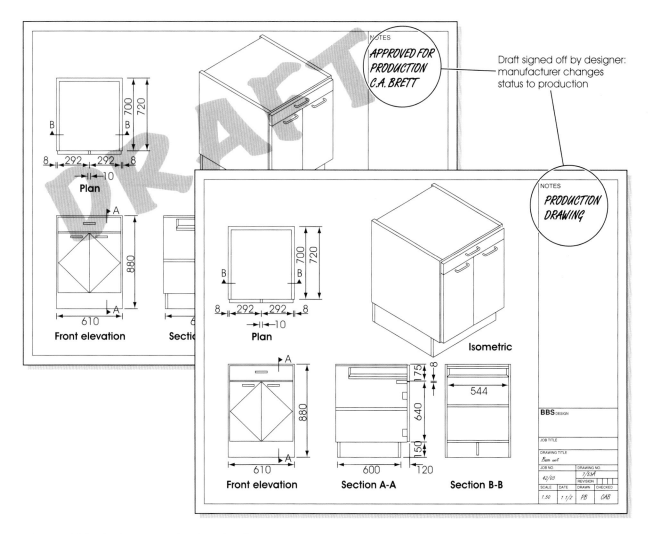

▲ **Figure 4.38** Draft and production drawings

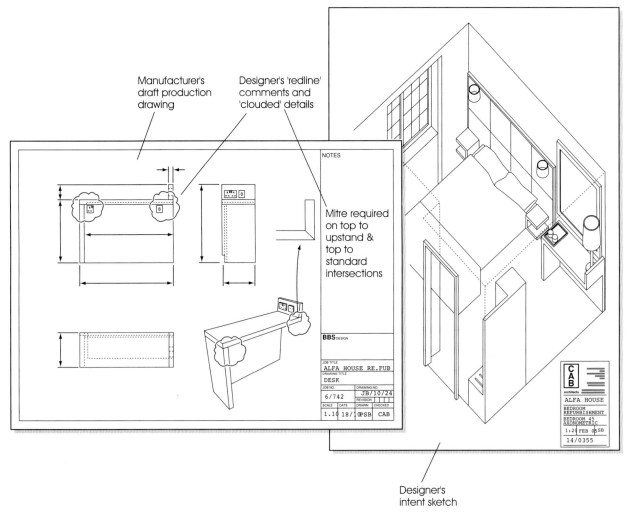

▲ **Figure 4.39** Redline drawing

normally lines cut into the vertical faces of stone or brick permanent structures. However, bolt-and-pivot bench marks are also to be found. A chiselled broad arrowhead points to the mark, bolt or pivot as illustrated in Figure 4.41. The level value above OD of an OBM can be obtained from the relevant Ordnance Survey map, or from the local authority's planning department.

■ *Temporary bench marks (TBMs).* These are marks that have been transferred from the nearest OBM back to the building site, from which all on-site levels are taken from. TBMs are normally steel or timber pegs driven into the ground to a convenient level value and set in concrete for protection (Figure 4.42). The relationship between the TBM and the levels of the building is illustrated in Figure 4.43.

■ *Datum marks and lines.* These are temporary marks or lines that are established on a site or in a room or area, from which other points nearby are measured. A broad arrowhead as illustrated in Figure 4.44 should be used to indicate the datum line or mark.

Sketches

These are rough outlines or initial drafts of an idea before full working drawings are made. Alternatively, sketches may be prepared to convey thoughts and ideas. It is often much easier to produce a sketch of your intentions rather than to describe in words or produce a long list of instructions. Sketches can be produced either freehand, that is without the use of any equipment, or be more accurately produced using a ruler

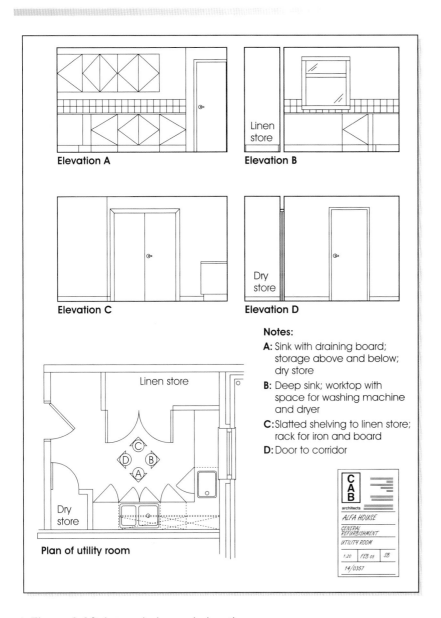

▲ **Figure 4.40** Internal plan and elevations

(Within figure 4.40:)

Elevation A

Elevation B

Linen store

Elevation C

Dry store

Elevation D

Notes:

A: Sink with draining board; storage above and below; dry store

B: Deep sink; worktop with space for washing machine and dryer

C: Slatted shelving to linen store; rack for iron and board

D: Door to corridor

Linen store

Dry store

Plan of utility room

CAB architects

ALFA HOUSE

GENERAL REFURBISHMENT

UTILITY ROOM

1:20 | FEB 05 | SB

14/0357

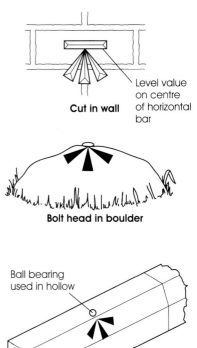

Level value on centre of horizontal bar

Cut in wall

Bolt head in boulder

Ball bearing used in hollow

Pivot in kerbstone

▲ **Figure 4.41** Ordinance Benchmarks (OBM)

and set square to give basic guidelines (Figure 4.45). Methods of projection in sketches follow those used for drawings, i.e. orthographic or pictorial.

Pictorial sketching is made easy if you imagine the object you wish to sketch with a three-dimensional box around it. Draw the box first, lightly, with a 2H pencil, then draw in the object using an HB grade pencil.

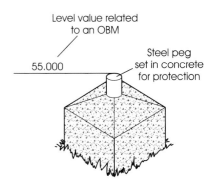

Level value related to an OBM

55.000

Steel peg set in concrete for protection

▲ **Figure 4.42** Temporary benchmark (TBM)

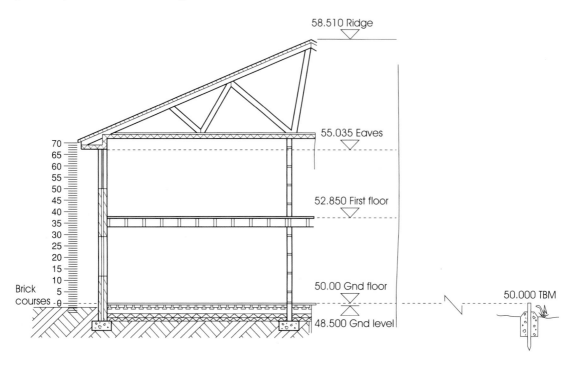

▲ **Figure 4.43** Part section through building showing relationship with TBM

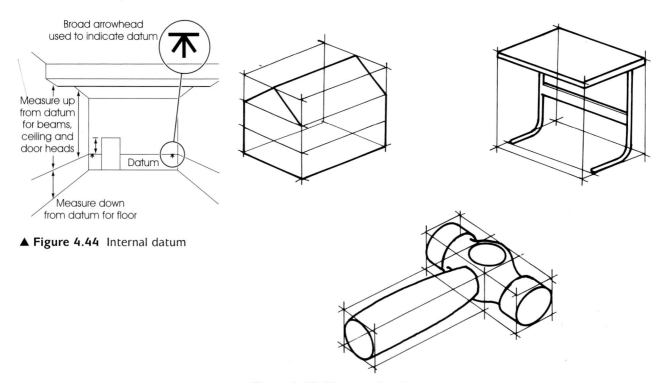

▲ **Figure 4.44** Internal datum

▲ **Figure 4.45** Pictorial sketching

Chapter Five

Portable powered hand tools

This chapter covers the work of site carpenters and bench joiners. It is concerned with the principles of setting up and using portable powered hand tools. It includes the following:

❡ use of a range of portable powered hand tools for ripping and cross-cutting, cutting curves, mitres and compound bevel cuts, planing, rebating, grooving and moulding, drilling and boring, finishing and the insertion of screws and nails

❡ observation of safe working practices and the following of instructions when maintaining, setting up, using and changing tooling for portable powered hand tools.

The carpenter and joiner have at their disposal a wide range of powered hand tools, enabling many operations to be carried out with increased speed, efficiency and accuracy.

Requirements for power tools and power supplies

Power tools have a label attached to the body that gives the following typical details, see Figure 5.1:

■ Makers name
■ The **CE mark** showing that the tool has been designed and manufactured to European standards.
■ The **double insulation mark** showing that the motor and other live parts are isolated from any section of the tool that the operator can touch (for mains powered tools).
■ Model and serial number, which must be quoted when purchasing replacement parts.
■ Electrical information (for mains powered tools):
  – V = voltage (electrical pressure for safe working);
  – Hz (Hertz) = frequency range of the power supply in cycles per second;
  – A = amperes or amps (rate at which the electric flows under load conditions);
  – W = watts (amount of electrical energy or power required to drive the equipment).
■ Electrical information (for battery power tools):
  – V = voltage (electrical pressure);
  – Ah = amp hours (capacity of battery).
■ Air supply usage information (for air-powered tools):
  – L/min = Litres per minute (average air consumption);
  – Bar = operating air pressure (1 bar is equal to atmospheric pressure).

Double
insulation
symbol

▲ **Figure 5.1** Typical power tool label

■ Capacity information:
  – RPM = Revolutions per minute of the motor.
  – Minimum and maximum chuck opening size, maximum depth of cut, etc., depending on the type of tool.

Power supplies and safety procedures

Electricity is the main source of power supply commonly used in workshops, building sites and installation on customer's premises. **Compressed air** is used to a lesser extent. Typical sources of power are illustrated in Figure 5.2.

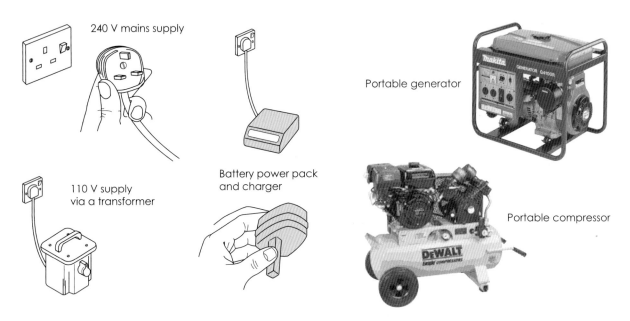

240 V mains supply

110 V supply via a transformer

Battery power pack and charger

Portable generator

Portable compressor

▲ **Figure 5.2** Power supplies

Electricity is supplied from either:

■ the mains at 240V or stepped down to 110-V via transformer;
■ a portable, petrol or diesel powered generator at 240V or 110V;
■ rechargeable batteries at various voltages for portable power tools.

Compressed air is supplied from a portable or fixed air compressor.

Electricity supplies

240-V mains

Mains-driven equipment is normally supplied via a 13-A 3-pin plug and a 2- or 3-core flexible cable that can be plugged directly into a standard outlet (Figure 5.3a). The live wire (brown sheathing) is protected by a fuse, which should be matched to the ampere rating of the tool. If a fault occurs the earth wire (green and yellow banded sheathing) carries the current safely to earth. As a result of a fault an increased flow of electricity passes through the fuse, causing the fuse to 'blow' (burn out) and cut off the electric supply to the tool.

Double insulation. All new power tools are double insulated and may be supplied without an earth wire. However, the live wire must still be fused to suit the ampere rating of the tool (Figure 5.3b). Although double insulted tools are safer than the older single insulated ones, both present a safety hazard to the operator if the cable carrying the electricity is cut during the work or a fault develops in the cable.

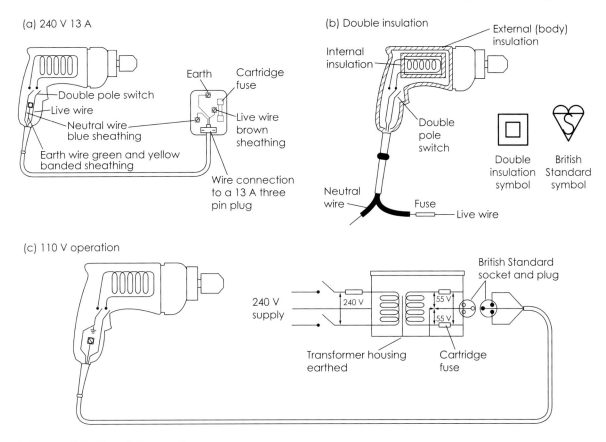

(a) 240 V 13 A

Double pole switch
Live wire
Neutral wire blue sheathing
Earth wire green and yellow banded sheathing
Earth
Cartridge fuse
Live wire brown sheathing
Wire connection to a 13 A three pin plug

(b) Double insulation

External (body) insulation
Internal insulation
Double pole switch
Neutral wire
Fuse
Live wire

Double insulation symbol
British Standard symbol

(c) 110 V operation

240 V supply
240 V
55 V
55 V
British Standard socket and plug
Transformer housing earthed
Cartridge fuse

▲ **Figure 5.3** Electricity supplies

110-V mains

110-V mains equipment normally operates from a 240-V supply via a **step-down transformer** with a centrally tapped earth. In the event of a fault, the maximum shock an operator would receive would be 55 volts (Figure 5.3c).

This voltage reduction provides safer operation by lowering the fault current if or when it passes through the body:

- Low levels of current may only cause an unpleasant tingling sensation, but the consequent momentary lack of concentration may be sufficient to cause involuntary reactions or injury (unintentional contact with moving parts, or falling, etc.).
- Medium levels of current passing through the body result in muscular tension and burning, in addition to the involuntary reactions or loss of concentration.
- High levels of current passing through the body again causes muscular tension or spasms and burning, but in addition affect the heart and can result in death or serious injury.

Since lower voltage means lower fault currents which means lower risk, reduced voltage schemes are the accepted procedure for safe working.

Reduced voltage equipment operating at 110 V is the specified voltage for using power tools on building sites.

Residual current device (RCD)

The use of RCDs provides increased protection to the operator from electric shock. RCDs work by monitoring the flow of electricity in the live and neutral wires of the circuit. In the event of a fault an imbalance in the flow occurs causing the device to trip, cutting off the electrical supply almost instantaneously.

RCDs are available as either (Figure 5.4):

- A permanently wired 13 A socket outlet, with RCD fitted to a ring main circuit.
- An RCD adaptor fitted between the tool plug and socket outlet.
- A combined RCD and plug permanently fitted to the tool power lead.

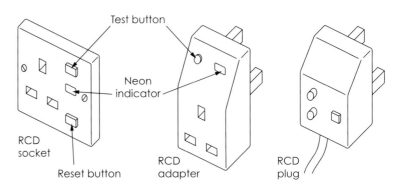

▶ **Figure 5.4** Residual current device (RCD)

The combined RCD and plug is recommended, as it is impossible to inadvertently leave it out of the power supply system, as could be the case with the other two options.

Before each use the 'test' button should be pressed to check the RCD mechanically and electrically. If everything is working correctly the device will 'trip' requiring the pressing of the 'reset' to restore the power supply.

Industrial power supplies

Building sites and some works may be wired using colour-coded industrial shielded plugs and sockets (Figure 5.5a):

> Blue for 240 V
> Yellow for 110 V.

The position of the key and pin layout are different, making it impossible to plug into the wrong power supply.

When 110-V power tools are used 'off-site', a step down transformer should be used from the 240 V mains supply (Figure 5.5b). In cases where an extension lead is required this should be used on the 110 V side between the transformer and the tool. If the extension lead is stored on a drum, it must be fully unwound before use, in order to prevent overheating within the drum when supplying power.

(a) Industrial type shielded plug and socket

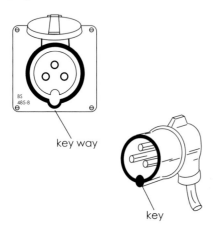

key way

key

(b) Arrangement for using 110 V power tools from 240 V mains supply

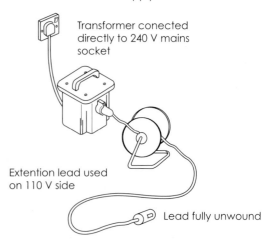

Transformer conected directly to 240 V mains socket

Extention lead used on 110 V side

Lead fully unwound

▲ **Figure 5.5** Options for connecting to the electricity supply

Portable electricity generators and tools

Where mains electricity is not available, power for equipment can be provided by a petrol, LPG or diesel portable generator. These normally have outlets for both 240 V and 110 V. Care must be taken when placing generators in order to minimise any nuisance from emission of fumes and noise.

Battery-operated equipment. These tools are termed '**cordless**'. These are powered by removable **rechargeable batteries** located in the tool, often the handle. When the battery has become discharged through use, a recently charged spare battery may be inserted to replace it.

To recharge a battery it is removed from the tool and placed in a compatible **battery charger**. Chargers are available to run off the mains 240 V supply, from a 12 V car cigarette lighter socket or from a 110 V supply.

The time taken to recharge and the charging cycle varies between manufacturers. A typical three-stage charging cycle is:

1. Fast charge – 20 minutes to 1 hour to provide the maximum charge.
2. Equalisation charge – 1 to 3 hours additional charge to top-up cells that have not been fully charged at the end of a fast charge cycle.
3. Maintenance charge – a trickle charge to maintain a full battery at peak efficiency, topping up any self-discharge.

Battery care. To ensure maximum battery life a complete three-stage charge is recommended every 20 fast charges.

The amount of work a fully charged battery can do before it requires recharging is dependent on the type and size of the battery, the power required to drive the tool, the nature of work being undertaken, and the age of the battery pack.

Take care when handling and charging batteries to ensure maximum life:

■ Fully charge batteries before storage; however, they will still start to self-discharge.
■ Store batteries in a cool, dry environment, between 4°C and 40°C. Never allow the battery to freeze.
■ Remove the battery during transportation, as the accidental continuous switch operation (when in a tool bag) can cause permanent damage to the battery and nearby equipment.
■ Allow a heavily used and rapidly discharged battery to cool down and stabilise before recharging.
■ Recharge the battery when the tool no longer performs its intended use.
■ Frequent overloading or stalling of the tool will reduce the battery life.

Compressed air (pneumatic) power supplies

Air above the atmospheric pressure of 1 bar is used as a power source for pneumatic tools. Air compressed to between 5 and 8 bar is used either to turn vanes fixed to a drive shaft to give rotary motion for drills, etc., or a piston to give the motive force for driving nails and staples. Compressors used to supply the compressed air may be a permanent fixture in a workshop or a portable unit for site work and work at customers' premises.

Air compressors consist of the following items as illustrated in Figure 5.6a:

■ Petrol, LPG, diesel or electrically powered motor.
■ Air pump, where a piston compresses the air and passes it into a receiver. The body of the pump is normally finned for strength and cooling and should be well ventilated as heat is generated during compression.
■ Air vessel in the form of a storage tank for the compressed air, which evens out the pulsating pump delivery, so that the air is available for use at a constant pressure. A drain-off point is included in the bottom of the tank to periodically drain off condensation.

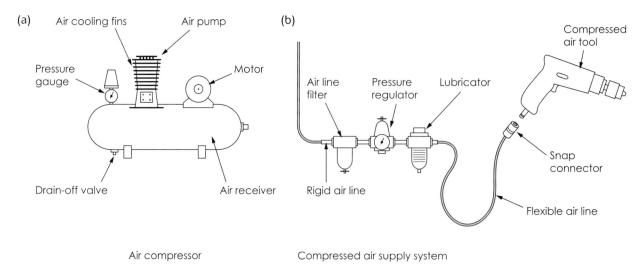

▲ Figure 5.6 Compressed air equipment: (a) compressor; (b) delivery system

Delivery of compressed air to a power tool is via the following as illustrated in Figure 5.6b:

- A system of ridged airlines to transport the compressed air to the fixed outlet.
- A valve and connector for a flexible hose to the tool. The size of the line and hose must be compatible with the tool to be used. Any variation in the bore of the line will cause a variation in the power supplied to the tool.
- Tools used from portable compressors will be supplied directly from the compressor via a flexible hose. Ideally the compressor should be positioned close to the work, as long hoses cause pressure drops through friction and also create potential tripping hazards.

For maximum efficiency the air supply system should incorporate as close to the tool as possible:

- a filter to remove any moisture from the air;
- an adjustable pressure regulator to control the air pressure reaching an individual tool;
- a lubricator to allow a controlled amount of lubricant to reach the tool being used.

Some tools do not require air lubrication and this should be turned off as it may perish the internal seals causing failure. Check manufacturer's instructions if in doubt.

Safety procedures for power supplies and tools

General power tool safety

- Never use a power tool unless a competent person has properly trained you in its use.
- Never use a power tool unless you have your supervisor's permission.
- Always select the correct tool for the work in hand. If in doubt consult the manufacturer's instructions.
- Ensure that the power tool and supply are compatible.
- Ensure that the cable or air hose is:
  - free from knots and damage;
  - firmly secured at all connections;
  - unable to come into contact with the cutting edge or become fouled during the tool's operation.

- Route all cables and hoses with care so as to avoid any tripping or trapping hazards.
- Before making any adjustments, always disconnect the tool from its power supply.
- Always use the tool's safety guards correctly and never remove or tie them back.
- Never put a tool down until all the rotating parts have stopped moving.
- Always wear the correct personal protective equipment (PPE) for the job. These may include (Figure 5.7):
  - ear defenders;
  - safety goggles or face screen;
  - dust mask or respirator;
  - safety helmet.
- Always use dust extraction equipment when operating power tools (Figure 5.8). This is in addition to and not in place of PPE.
- Avoid using power tools for extended periods of time, especially those that cause vibration, such as sanders, hammer drills and pneumatic tools, as this may result in hand or arm vibration syndrome HAVS. See Chapter 6 for further details.

Ear muffs

Dust Mask

Respirator

Goggles

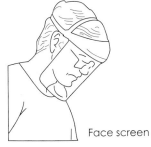

Face screen

▲ **Figure 5.7** Personal protective equipment (PPE)

▲ **Figure 5.8** Industrial vacuum cleaner/extractor

- Loose clothing and long hair should be secured so that they cannot be caught up in the tool. In the event of entanglement switch off the power supply immediately.

- All power tools should be properly maintained and serviced at regular intervals by a suitably trained person. Never attempt to service or repair a power tool yourself. If it is not working correctly or its safety is suspect, return it to the stores with a note stating why it has been returned. In any case, tools in constant use should be returned to the stores for inspection at least once every seven days.

- Ensure that the material or workpiece is firmly cramped or fixed in position so that it will not move during the tool's operation.

- In general, compressed air tools must be started and stopped under load, whereas electric tools must not.

- Never use an electric tool where combustible liquids or gases are present.

- Never carry, drag or suspend a tool by its cable or hose.

- Think before and during use. Tools cannot be careless, but their operators can be. Most accidents are caused by simple carelessness.

Compressed air equipment

Compressed air equipment, if correctly used, can be perfectly safe. However, if misused, they can cause severe personal injury. Compressed air entering the body causes painful swelling. If it is allowed to enter the bloodstream it can make its way to the brain, burst the blood vessels and cause death.

The following additional safety points must be observed whenever compressed air equipment is used:

- The compressor must be in the control of a fully trained competent person at all times.

- All equipment must be regularly inspected and maintained.

- Training must be given to all persons who will use compressed air equipment.

- Position portable compressors in a well ventilated area.

- Ensure all hose connections are properly clamped.

- Route all hose connections to prevent snaking, risk of persons tripping or traffic crossing them, as any squeezing of the hose causes excess pressure on the couplings.

- Always isolate the tool from the air supply before investigating any fault.

- Never disconnect any air hose unless protected by a safety valve.

- Never use an air hose to clean away waste material or anything else that may result in flying particles.

- Never use compressed air to clean down yourself or anyone else as this carries a great risk of injury to the eyes, ears, nostrils and rectum.

Power drills and drivers

As with all types of equipment, power tool manufacturers supply tools for various markets, ranging from the cheap, almost 'throw away' type for the bottom end of the do-it-yourself market suitable only for light/occasional use, through to the more sophisticated and powerful industrial models, which will stand up to constant heavy use.

The power drill is the most common type of portable power tool in use. Your choice of drill (Figure 5.9) will depend on the following factors:

- power supply available;
- amount of use;
- size of holes required;
- material being drilled.

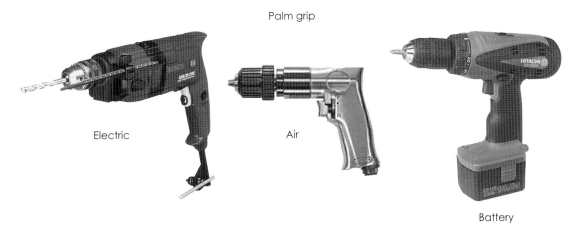

Palm grip

Electric

Air

Battery

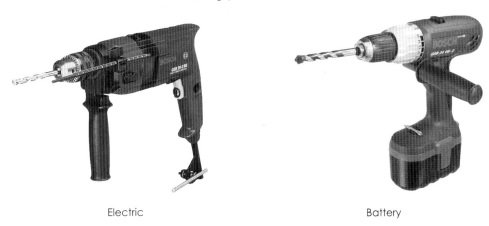

Palm grip with side handle

Electric

Battery

'D' shape back handle with side handle

Electric

Battery

▲ **Figure 5.9** Types of power drill

Drill features

Figures 5.10 and 5.11 refer.

Chucks

Drill and screwdriver bits are held securely in the chuck, see Figure 5.10, normally by three self-centring jaws that grip the shank of the bit. Traditionally chucks are operated with a toothed key to open and close the jaws. Other more recent chucks are 'keyless' being operated by griping and turning the cylinder collars. Some of the larger drills are fitted with a fast

'Keyless' three
jaw chuck

Key operated
three jaw chuck

Chuck key

▶ **Figure 5.10** Drill chucks

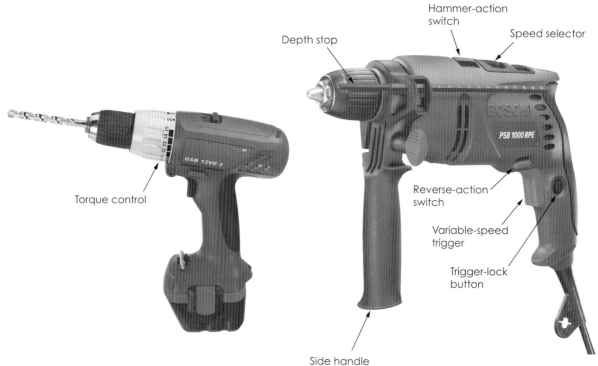

Depth stop

Hammer-action
switch

Speed selector

Torque control

Reverse-action
switch

Variable-speed
trigger

Trigger-lock
button

Side handle

▲ **Figure 5.11** Typical drill features

action chuck (SDS). When the chuck collar is pulled back, it automatically opens enabling a special bit with grooved shanks to be inserted.

Chuck capacity describes the range of drill bits that the chuck can accept. This also corresponds to the maximum size of hole that the tool is capable of boring in steel. Small drills have a chuck capacity of 1 to 10 mm in diameter and larger ones 1.5 to 13 mm. Both drills may be used to bore larger holes in wood when fitted with reduced shank drills or spade bits, etc.

Speed selection

Drills perform better when the speed of the bit can be adjusted to suit the size of hole and material being worked. In general use a fast speed for boring holes in wood and a slower speed for drilling holes in metal, masonry and for driving woodscrews.

■ *Basic drills* may only have one non-load speed (2400 rpm). Other basic drills may be equipped with gearing enabling two fixed non-load speeds (900 and 2400 rpm) via a lever or switch.

■ *Variable speed drills* vary from zero to maximum speed (0 to 2400 rpm) according to the pressure applied to the trigger. Some are equipped with a dial to pre-set optimum speeds by limiting the trigger movement. Variable speed drills may also be equipped with two mechanical gears; the slower speed provides more torque (rotating power).

A *soft start drill* is a useful feature as it minimises the initial jolt of a high-speed electric motor, which can cause drill bit skidding or damaged screw heads.

Other drill features

- **Reverse action.** Many air and mains electric power drills and all battery drills are fitted with a switch to change the direction of rotation for removing screws. This feature is also useful for releasing drill bits that have become jammed during drilling.

- **Torque control.** Most battery drills are equipped with adjustable torque control enabling screws of different sizes to be driven to flush or below the fixing surface. Adjustment is via a collar behind the chuck.
  Use the lower torque numbers for smaller screws and the higher numbers for large screws or when driving screw heads below the surface. When boring holes ensure the torque control is set to the drill position (maximum torque) to avoid slipping.

- **Hammer action.** Both mains electric and battery powered drills may be equipped with a hammer action, enabling easy boring in brick, stone and concrete. Operated by switch before or during operation the hammer action puts typically 4000 blows per minute (BPM) behind the rotating drill bit to break up the masonry as work progresses. It is essential that special percussion drill bits are used for this work.

- A **palm grip handle** is used for lighter drills, for single-handed use. The design ensures that the pressure is exerted directly in line with the drill or screwdriver bit. On heavy drills a secondary side handle is fitted near the chuck, for two-handed operation.

- A **D-shaped back handle** is used for large, heavy drills, along with an adjustable front handle for two-handed operation. This design is suited to percussion or hammer action drills for boring in masonry and concrete using special percussion drill bits.

- **Side handle with depth stop.** Most mains electric drills and some battery drills can be fitted with the additional side handle for two-handed control in use. Integral depth stops can be adjusted to come into contact with the workpiece when the drill has reached the required depth.

- **Trigger lock.** A button adjacent to the trigger can be pushed in with the drill running to lock the trigger for continuous drilling. Resqueezing the trigger releases the lock and stops rotation.

- **Angle drilling.** Purpose-made angle drills or attachments for standard drills as illustrated in Figure 5.12, enable holes to be bored or screws to be driven in confined spaces.

- **Drill stand.** For repetitive drilling work and light mortising a drill can be fitted into a drill stand as illustrated in Figure 5.13. In order to comply with safety requirements the stand is fitted with a retractable chuck and drill guard, which must always be in position when drilling. Various size chisels along with their auger bits are available for use in a drill stand for mortising.

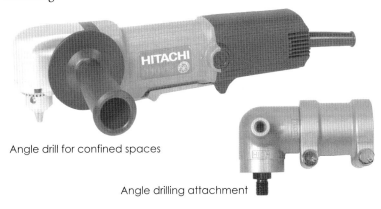

Angle drill for confined spaces

Angle drilling attachment

▲ **Figure 5.12** Angle drill and angle attachment

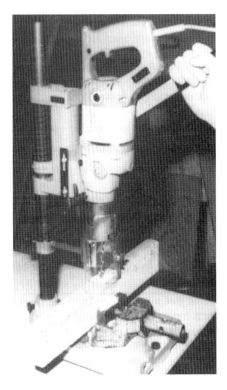

▲ **Figure 5.13** Drill stand

Operation of power drills

During use:

1. Hold the drill firmly.
2. Cramp the work piece.
3. Do not force the drill.
4. Remove the drill from the hole frequently to clear the dust and allow it to cool.
5. Drill a small pilot hole first to act as a guide when drilling larger holes.
6. Reduce the pressure applied on the drill as it is about to break through to avoid snatching or twisting.

Twist drills, bits and accessories

Twist drills and bits used in power drills should be of a high quality to withstand the heat generated when drilling at speed. A range of twist drill bits and accessories are illustrated in Figure 5.14.

- **Twist drills** are available in sets ranging in size from 1 mm increasing by 0.5 mm to 13 mm in diameter. **Carbon steel drills** are suitable for woodwork. But **high-speed steel drills (HSS)**, which stay sharper longer, are considered better quality and are suitable for drilling wood, metal and plastics. **Titanium-coated HSS drills** are also available for extended working life. Larger twist drills from 13 to 25 mm are made with reduced shanks to fit standard drill chucks. Standard length twist drills termed **jobber drills** range from 35 mm in length for the 1 mm diameter up to 150 mm in length for the 13-mm diameter. Extra length twist drills are available for drilling deeper holes. Twist drills are not easily centred, so it is worthwhile centre punching the centre of the hole first to avoid skidding off the mark.

- **Brad point bits** also known as **lip and spur bits** or **dowel bits**. These are wood bits that have a centre point to prevent them skidding off the mark when the power is applied. These are available in a range of sizes from 3 mm to 20 mm in diameter. Sizes over 13 mm have reduced diameter shanks enabling standard chuck use.

- **Spade bits** also known as **flat bits**. These are relatively cheap bits for power drilling larger diameter holes in wood. They are available from 6 mm to 50 mm in diameter. Make sure the point is fully inserted in the wood before starting to drill.

- **Masonry drills** are similar to twist drills but have a tip of tungsten carbide fitted into a slot on the cutting end, used for boring into masonry. When used in a hammer action/precision drill ensure the tipped drill is recommended by the manufacturer to withstand the vibration produced, otherwise the tip is likely to shatter.

- **Forstner bits** are superior quality bits which leave a clean, accurate and (if required) flat bottom or blind hole. They are available in sizes from 10 mm to 50 mm in diameter. Unlike other drills which are guided by their centre point or spiral, Forstner bits are guided by their rim, enabling them to bore holes that run over the edge of a piece and overlap other holes.

- **Countersink bits** make a countersunk or tapered recess to accommodate the head of a countersunk wood screw. A pilot and clearance hole should be drilled first with a twist drill. The countersunk bit is located in the clearance hole and run at high speed for a clean finish.

- **Combined drill and countersink bits** are available in a range of sizes to suit the most common wood screws. They drill a pilot hole, shank clearance hole and countersink in one operation. Versions are available to drill a counter-bore hole to accept a wooden plug that conceals the screw head.

- **Counter-bore bits** drill a pilot, clearance and counter-bore hole in one operation. The counter-bore hole can then be subsequently filled with a wooden pellet to conceal the screw head.

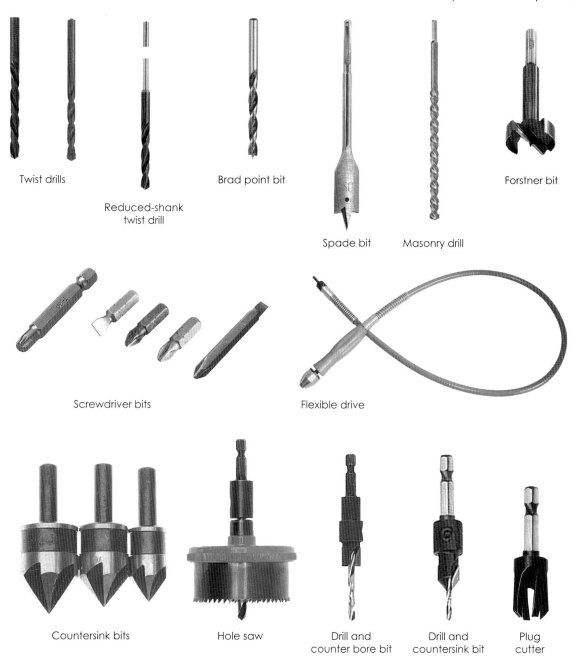

Twist drills

Reduced-shank
twist drill

Brad point bit

Spade bit   Masonry drill

Forstner bit

Screwdriver bits

Flexible drive

Countersink bits

Hole saw

Drill and
counter bore bit

Drill and
countersink bit

Plug
cutter

▲ **Figure 5.14** Twist drills, bits and accessories

- **Plug cutters** should match the counter bore bit. They cut a wooden plug from timber, which closely matches the grain and colour of the work. Plug cutters can only be effectively used in a pillar drill or drill stand attachment.

- **Concealed hinge boring bits** available in 25, 30 and 35 mm diameters, for boring concealed hinges. Similar to a Forstner bit, many have a tungsten carbide tipped cutting edge. For control it is best used in a pillar drill or with a drill stand.

- **Hole saws** consist of a cylindrical shaped saw blade, which is held in a backing plate secured to a twist drill passing through its centre. Hole-saws are typically available form 14 mm to 150 mm in diameter. As the saw blade peripheral speed is high for the larger diameters, select a slower speed and steadily feed the saw into the work.

- **Screwdriver bits.** A range of slot, cross and security head tips are available as screwdriver bits for power drills and drivers. When driving or removing screws, the slowest speed should be selected and pressure maintained during the entire operation to avoid the bit jumping out of the screw.

- **Flexible drives** allow use of a drill in awkward places. It consists of a 1-m drive cable sheathed in a flexible casing, with a 6-mm capacity chuck at one end and a spindle at the other for securing into a standard power drill chuck.

- **Hinge drill.** A twist drill secured within a spring-loaded housing having a chamfered end to suit the hinge countersink holes. The hinge is used as the template, with the chamfered end positioned in the screw hole; pressure is applied to bore perfectly positioned pilot holes.

- **Portable drill stand.** A telescopic device that attaches to the standard 43-mm drill collar which enables the precision drilling of holes at right angles; can also be adjusted for holes up to 45°. Ideal for use with plug cutters and concealed hinge boring bits.

- **Collar size.** Many drills have an international standard 43-mm collar directly behind the chuck. This enables the drill to be used with a range of attachments.

Power operated screwdrivers

In addition to the dual use of power drills as screwdrivers there is a range of dedicated power screwdrivers available as illustrated in Figure 5.15. The body and motor is similar to that of an electric power drill, although a reduction gear is fitted to give the correct speed for screwdriving. Where a two-speed tool is used, the slow speed should be used for wood screws and the high speed used for self-tapping screws into thin metalwork. Most tools are manufactured with a reverse gear to enable screws to be removed.

The front housing of the tool holds the screwdriver bits and contains a clutch assembly that operates in two stages:

1. The tool's motor will run but the screwdriver bit will not rotate until sufficient pressure is exerted on the tool to enable the clutch to operate and engage the main drive.

2. When the screws have been driven and are tight in the required position, the second stage of the clutch operates and stops the screwdriver bit rotating.

Dry wall screwdrivers have a single clutch, which allows screws to be driven through plasterboard panels into the studwork without pre-drilling Bugle head screws are normally used and the clutch will disengage when the head is just below the surface, but without breaking through the paper covering.

Power screwdrivers with adjustable clutch settings, enabling different size screws to be driven to a predetermined depth below the work surface before the clutch disengages.

Automatic feed power screwdrivers use screws that are attached to a plastic cartridge belt, which automatically feeds the screws into the driving position, one after the other. Collated screws are available in a range of sizes in both woodscrew and drywall patterns. Typically each cartridge contains 1000 screws.

Screwdriver bits and sockets

Various screwdriver bits and sockets are available to suit different types and sizes of screws. It is a simple operation to change the type of bit when required. The hexagonal shank of the bit is simply pushed into the front housing of the tool and retained in position by a spring-loaded steel ball which locates in a groove around the top of the shank. The bit is removed by simply pulling it out of the front housing. In addition hex shank twist drills are available for use as pilot drills in power screwdrivers.

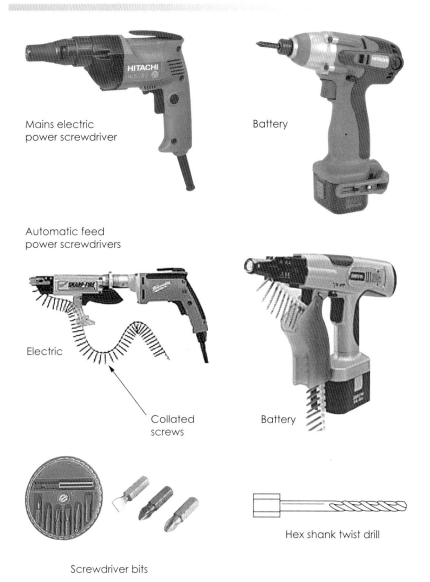

Mains electric
power screwdriver

Battery

Automatic feed
power screwdrivers

Electric

Collated
screws

Battery

Hex shank twist drill

Screwdriver bits

◀ **Figure 5.15** Power-operated
screwdrivers

Operation of power screwdriver

1. Select a screwdriver bit that is compatible with the tool being used.
2. Drill a pilot hole, clearance hole and countersink (where required) before screwing, to avoid overloading the motor and splitting the material.
3. Maintain a steady, firm pressure on the screwdriver so that the bit cannot jump out and damage the screw head or the workpiece.

Portable power saws

There are many types of power saw available to the woodworker each with its own specific range of functions.

Circular saws are often termed **skill saws** (Figure 5.16). They are used for a wide range of sawing operations including cross-cutting, rip sawing, bevel and compound bevel cutting, rebating, grooving, trenching and the cutting of sheet material. Blade diameters range from 130 mm to 240 mm giving a vertical maximum depth of cut ranging from 40 mm to 90 mm. Angle cuts up

Mains electric   Battery

▲ **Figure 5.16** Circular saws

to 45° are possible but at reduced depths. Various teeth profiles are available to suit the work in hand. However the use of tungsten carbide tipped saw blades is preferable in all situations, but particularly when cutting plywood, MDF, chipboard, fibreboard, plastic laminates and abrasive timbers.

Operation of a circular saw

Follow these steps referring to Figure 5.17:

1. Select and fit the *correct blade* for the work in hand (rip, cross-cut, combination or tungsten-tipped, etc.).
2. Adjust the *depth of the cut* so that the teeth project about half their length through the material to be cut.
3. Check the *blade guard* is working properly. It should spring back and cover the blade when the saw is removed from the timber.
4. Check the *riving knife* is rigid and set at the recommended distance from the blade.
5. Set the saw to the required *cutting angle*. This is indicated by a pointer on the pivot slide.
6. Insert the *rip fence* (if required) and set to the width required. When cutting sheet material or timber where the rip fence will not adjust to the required width, a straight batten can be temporarily fixed along the board to act as a guide for the sole plate of the saw to run against.
7. Check to ensure that all adjustment levers and thumbscrews are tight.
8. Ensure that the material to be cut is properly supported and securely fixed down. As the saw cuts from the bottom upwards, the face side of the material should be placed downwards. This ensures that any breaking out which may occur does not spoil the face of the material.
9. Rest the front of the saw on the material to be cut. Depress the safety lock with your thumb and pull the trigger to start the saw. Circular saws should not be fitted with a trigger lock that enables continuous running.
10. Allow the blade to reach its full speed before starting to cut. Feed the saw into the work smoothly and without using excess pressure. The blade guard will automatically retract as the saw is fed into the work.
11. If the saw binds in the work, ease it back until the blade runs free.
12. When the end of the cut is reached, remove the saw from the work, allowing the blade guard to spring back into place and then release the trigger. Do not release the trigger before the end of the cut has been reached.

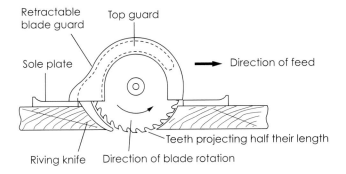

Retractable blade guard

Top guard

Sole plate

Direction of feed

Riving knife

Direction of blade rotation

Teeth projecting half their length

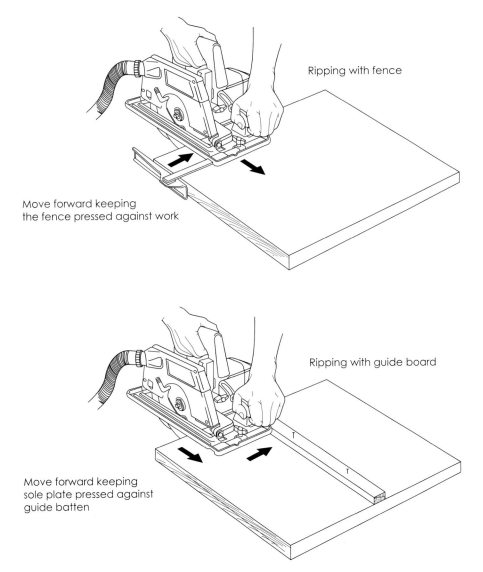

Ripping with fence

Move forward keeping the fence pressed against work

Ripping with guide board

Move forward keeping sole plate pressed against guide batten

▲ **Figure 5.17** Operation of the circular saw

Cross-cutting

Square and bevelled cross-cutting can be undertaken by cramping or pinning a guide batten to the workpiece with the blade set in the vertical position as illustrated in Figure 5.18.

The blade is set at an angle to the baseplate for **compound bevel cutting**.

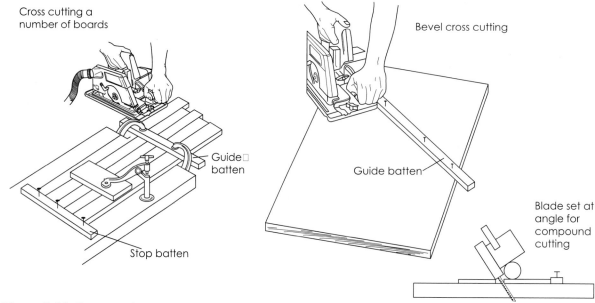

Cross cutting a
number of boards

Guide
batten

Stop batten

Bevel cross cutting

Guide batten

Blade set at
angle for
compound
cutting

▲ **Figure 5.18** Cross-cutting

Rebating, grooving and trenching

Rebating is done using the rip fence as a guide (Figure 5.19). Make two saw cuts along the grain, one from the edge and the second from the face. The blade must be set to the required depth.

Plough grooves are again formed using the rip fence as a guide. Set the fence to cut the two sides of the groove. Reset the fence and gradually remove the waste by making successive passes with the saw.

Trenching is carried out by making a series of parallel saw cuts across the grain, with the blade set to the required depth. Use a guide batten on either side of the trench and make successive passes of the saw to remove the waste.

Pocket cutting

This is a cut that starts and finished within the length or width of a board or floor, etc. It is this cut that can be used when cutting traps for access to services in completed floors. When cutting pockets the following stages should be followed, see Figure 5.20. This is in addition to the normal operating stages.

1. Mark the position of the pocket to be cut. If on a floor, check that the lines to be cut are free of nails, etc.

2. Adjust the depth of the cut so that the saw blade will penetrate the floorboards by less than 1 mm. If the blade is allowed to penetrate deeper there is a danger that the saw might cut into any services, which may be notched into the tops of the joists, e.g. electric cables, water and

Rebate cut in two stages

Plough groove
formed by
series of cuts

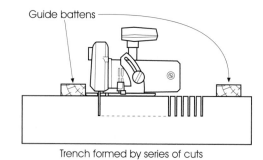

Guide battens

Trench formed by series of cuts

▲ **Figure 5.19** Rebating, grooving and trenching

gas pipes, etc. Always check with a metal-detecting sounding device to be on the safe side.

3. Tack a batten at the end of the cut to act as a temporary stop.

4. Place the leading edge of the saw's sole plate on the work surface against the temporary stop.

5. Partially retract the blade guard using the lever and start the saw.

6. Allow the blade to obtain its full speed and gently lower the saw until its sole plate is flat on the work surface.

7. Release the trigger, allow the blade to stop, remove the saw. Turn it around and complete the cut in the opposite direction to the corner.

8. Repeat the previous stages on the other three sides to complete the pocket or access trap.

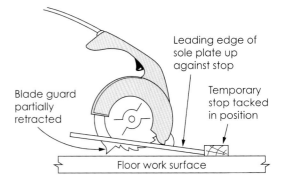

▲ **Figure 5.20** Circular saw for pocket cutting

Jig saws

These are also known as **reciprocating saws** (Figure 5.21). Although they may be used with a fence for straight cutting they are particularly useful for circular, shaped and pierced work. In addition many models have adjustable sole plates that allow bevel cutting to be carried out.

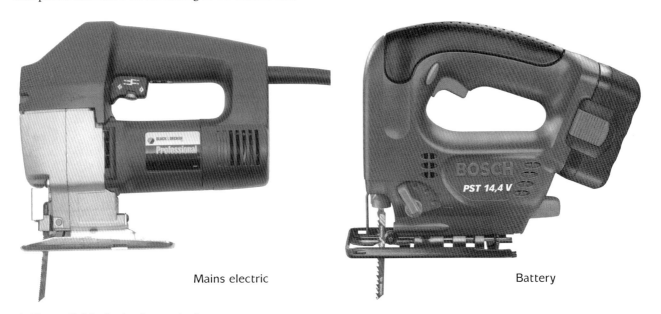

Mains electric   Battery

▲ **Figure 5.21** Jig (reciprocating) saws

Basic jig saws have a simple up and down action with the blade cutting on the upward stroke. Other better quality saws are equipped with a pendulum action that advances the blade into the work on the upward cutting stroke and moves it away on the down stroke. This cutting action prolongs the life of the blade by minimising tooth wear and friction on the down stroke. In addition it also helps to clear the kerf of waste.

As the blade cuts on the upward stroke, it tends to chip or splinter on both sides of the kerf on the upper surface of the workpiece. Either apply a length of masking tape over the line to be cut, or cut with the 'good' or finished face to the underside.

Some models are equipped with replaceable inserts, which fit into the sole to fill the space around the blade. This supports the material either side of the kerf, thus minimising splintering.

A range of different blades are available, suitable for cutting a wide variety of materials.

Operation of jig saw

Follow these steps, referring to Figure 5.22:

1. Select the correct blade for the work in hand.
2. Select the correct speed: slow speed for curved work and metal cutting, high speed for straight cutting.

▶ **Figure 5.22**
Operation of jig saw

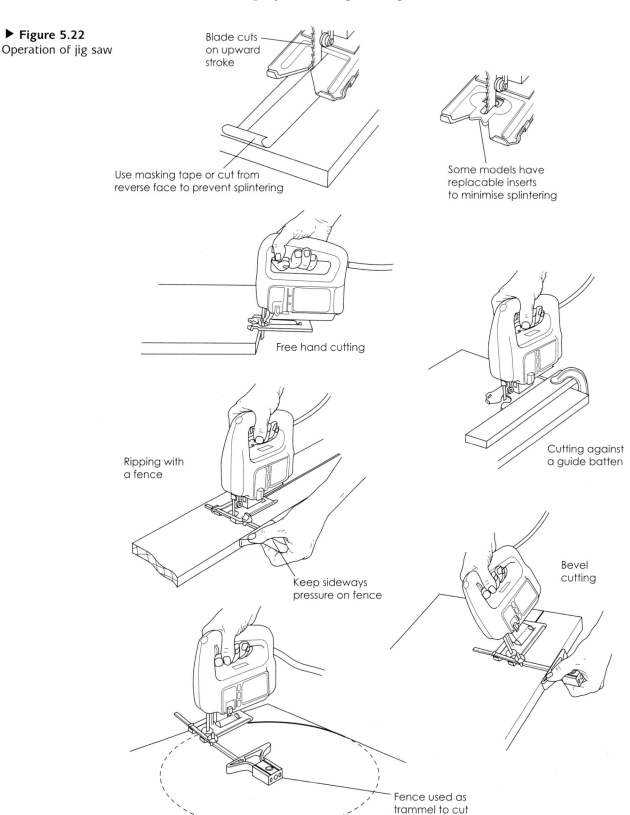

Blade cuts on upward stroke

Use masking tape or cut from reverse face to prevent splintering

Some models have replaceable inserts to minimise splintering

Free hand cutting

Cutting against a guide batten

Ripping with a fence

Keep sideways pressure on fence

Bevel cutting

Fence used as trammel to cut circles

3. Ensure that the material to be cut is properly supported and securely fixed down.

4. Rest the front of the saw on the material to be cut and pull the trigger to start the saw.

5. When the blade has reached its full speed, steadily feed the saw into the work, but do not force it as this may cause it to wander off line.

6. When the end of the cut is reached release the trigger, keeping the sole plate of the saw against the workpiece, but making sure the blade is not in contact.

Pocket cutting

To cut a pocket or aperture in a board, a starter hole for the blade is first bored in the waste (Figure 5.23). Insert the blade in the hole, switch on and cut along the first line into the corner. Back-up the saw about 25 mm and make a curved cut on to the second line of the aperture. Repeat this operation at each corner until the aperture is complete. Finally go back to each corner and remove the triangular waste piece by sawing back in the opposite direction.

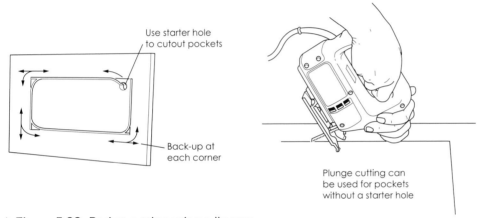

Use starter hole to cutout pockets

Back-up at each corner

Plunge cutting can be used for pockets without a starter hole

▲ **Figure 5.23** Pocket cutting using a jig saw

Alternatively a plunge cutting technique can be used to cut apertures without the need for a starter hole. After setting the saw up and following the safety precautions as before:

1. Hold the saw over the line to be cut and tip forward until the front edge of the sole plate rests firmly on the workpiece.

2. With the blade clear of the work surface and keeping a firm grip on the saw, pull the trigger. Pivot the saw from the front edge so that the blade cuts into the surface until full penetration is made.

3. Proceed with normal cutting to complete the shape. If the pocket being cut is square or has sharp corners, it will be necessary to repeat the plunge cut along its other sides.

Mitre saw

Also known as a **chop saw** or **down-stroking saw** they consist of a circular saw which pivots from a pillar above a baseplate, as illustrated in Figure 5.24. The baseplate and saw can be rotated in either direction from the square by up to 45° in relation to the fence. This enables square and mitred cross-cuts to be made. In addition the motor and blade assembly can be adjusted so that the blade can be set at a bevel angle to the baseplate. Some models can be bevelled in both directions, others to the left only due to the motor projection. Standard adjustment is between 0° and 45° from the vertical. This adjustment allows bevelled and compound mitred crosscuts to be made.

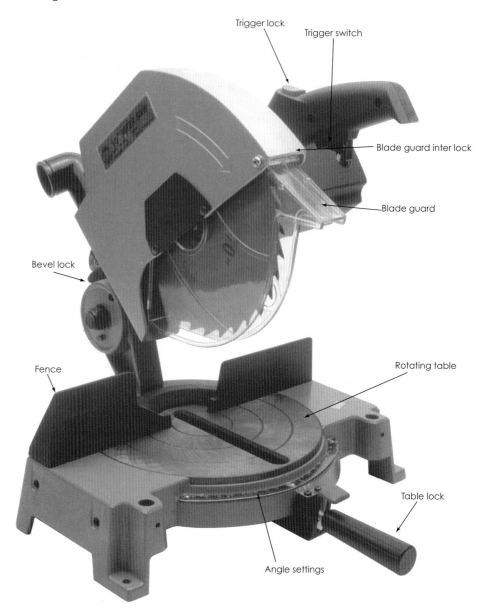

Trigger lock

Trigger switch

Blade guard inter lock

Blade guard

Bevel lock

Rotating table

Fence

Table lock

Angle settings

▲ **Figure 5.24** Mitre saw

The operator starts the saw using the handle trigger and blade guard interlock device. When the saw is running at full speed, a downward movement on the handle retracts the blade guard and brings the blade into contact with the workpiece to make the cut.

The size of material that can be cut is limited by the size of the saw blade. Typical cross-sectional capacities for a 190-mm blade are:

> 105 mm × 50 mm for square cross-cutting.
>
> 65 mm × 50 mm for 45° mitring
>
> 35 mm × 35 mm for 45°/45° compound mitres.

Models with larger diameter blades will have a bigger cutting capacity. Some more advanced saws are available with a pull-out function in addition to the plunge or chop action. This allows the cross-cutting of wider boards.

Some models project a laser line onto the workpiece which indicates the position of the saw cut.

Most mitre saws are equipped with dust extraction facilities. This should be used at all times. In addition you should always wear ear defenders, eye protection goggles and a dust mask when using a mitre saw. Any assistant or others working near to you will also need to wear the same personal protective equipment.

Operation of a mitre saw

See Figure 5.25.

Cross-cuts at 90° to the edge of the workpiece are made with the mitre table set above 0°.

Mitred cross-cuts are made with the table set above 0° in either direction.

Bevel cuts are made with the mitre table set to 0° and the blade set at an angle between 0° and 45°.

Compound mitred cross-cuts are a combination of a mitre and a bevel cut.

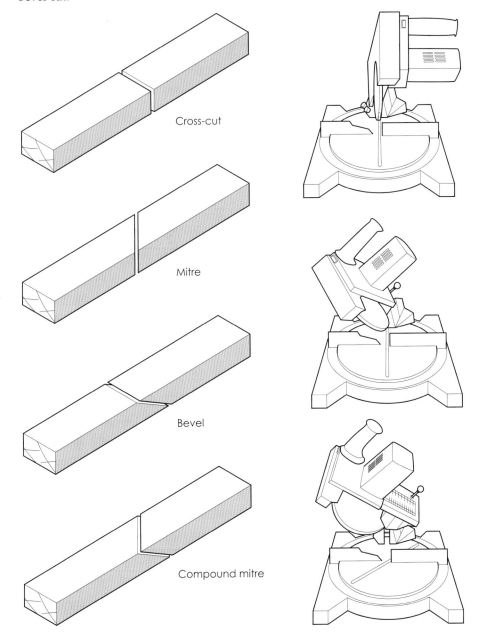

Cross-cut

Mitre

Bevel

Compound mitre

▲ **Figure 5.25** Types of cut using a mitre saw

1. Before use read the manufacturers instruction as adjustments and specific procedures can vary.
2. Rotate the mitre table, so that the indicator aligns with the required angle. Lock the table in position.
3. Check all adjustments are tight.
4. Perform a dry run of the cutting action to check for problems before plugging into the power supply.
5. Position the workpiece on the base table, with one edge against the fence. Where the timber is warped place the convex edge to the fence. Putting the concave edge to the fence is dangerous as it can result in snatching of the workpiece and jamming the blade. Long lengths will require an end support level with the saw table.
6. Preferably cramp the workpiece to the fence. Smaller sections may be held in position with the left hand. However ensure it is well clear of the saw's cutting area.
7. Grip the handle firmly, squeeze the trigger and allow the blade to reach its maximum speed.
8. Operate the blade guard interlock lever and slowly lower the saw blade into and through the workpiece.
9. Release the trigger and allow the saw blade to stop rotating before raising it from the work. Short off-cuts may be dangerously projected across the workshop if the blade is lifted whilst still running.
10. Always disconnect from the power supply before making any further adjustments or changing the saw blade.

Biscuit jointer

▲ **Figure 5.26** Biscuit jointer

This is a miniature plunging circular saw, developed principally to make a form of tongued-and-grooved joint for furniture and cabinetwork (Figure 5.26).

The joint works like a dowel joint, except that compressed beech oval biscuits fit into matching saw slots (Figure 5.27). When glue is applied to the joint, the biscuit expands to fill the slot, producing an easy fitting butt joint. Biscuit jointers can also be used to cut grooves and trim sheet material.

Biscuit jointing is ideal for both framed and cabinet carcass construction, using solid timber and board material. Joints can be butted, mitred or edge to edge.

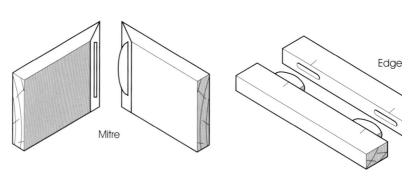

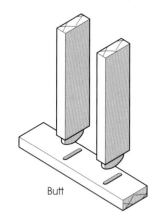

▲ **Figure 5.27** Use of biscuits for jointing

Operation of a biscuit jointer

Follow these steps referring to Figure 5.28:
1. Mark the centre line of the joint on both pieces.
2. Mark the biscuit slot centres along the centre line, typically spaced about 100 mm to 150 mm apart.
3. Set the depth of the cut to suit the size of the biscuit and adjust the fence to align the blade with the centre line.

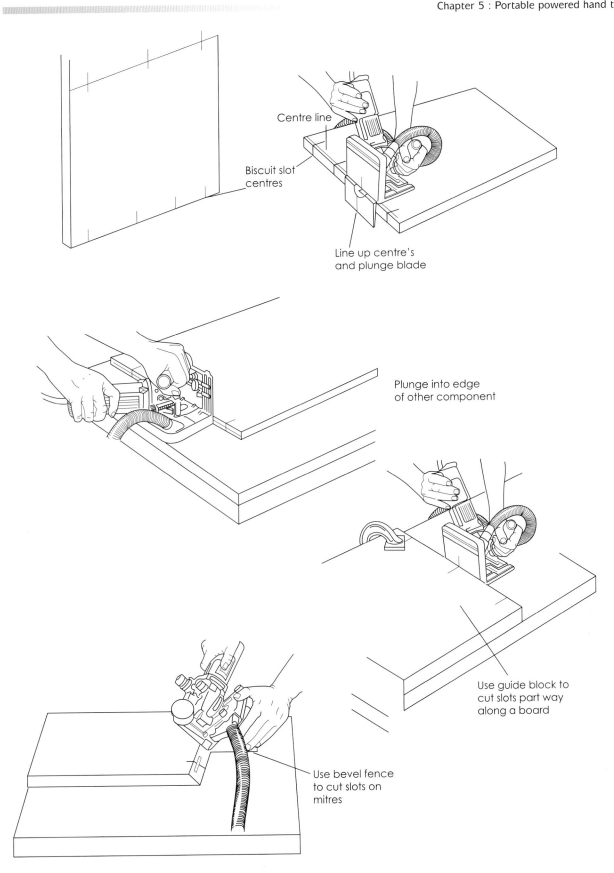

Centre line

Biscuit slot centres

Line up centre's and plunge blade

Plunge into edge of other component

Use guide block to cut slots part way along a board

Use bevel fence to cut slots on mitres

▲ **Figure 5.28** Biscuit joint preparation

4. Line up the centre cutting mark on the tool with the central mark of the slot. Press the fence against the work. Start the motor and plunge the blade to make the slot. Retract the blade and allow the motor to stop before moving on.

5. Repeat the above stages to cut the remaining slots in both pieces to be jointed.

Power planers and routers

Planers

The portable planer (Figure 5.29) is mainly used for edging work although it is capable of both chamfering and rebating. On site it is extremely useful for door hanging and truing up the edges of sheet material. Surfacing and cleaning up of timber can be carried out when required but it tends to leave ridges on surfaces that are wider than the width of the tool.

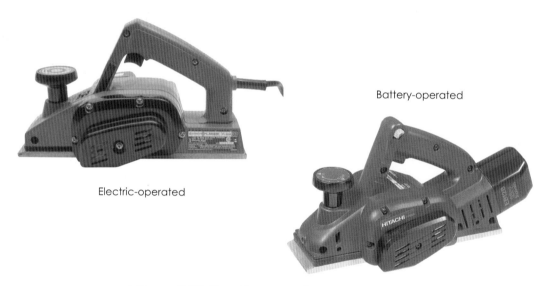

Battery-operated

Electric-operated

▲ **Figure 5.29** Portable power planers

Operation of a planer

Follow these steps, referring to Figure 5.30:

1. Check that the cutters are sharp and set correctly.
2. Adjust the fence to run along the edge of the work as a guide.
3. Rest the front of the plane on the workpiece, ensuring that the cutters are not in contact with the timber.
4. Pull the trigger and allow the cutters to gain speed.
5. Move the plane forward keeping pressure on the front knob. The depth of the cut can be altered by rotating this knob.
6. Continue planing, keeping pressure both down and up against the fence.
7. When completing the cut, ease the pressure off the front knob and increase the pressure on the back. This prevents the plane tipping forward and causing the cutter to dig in when the end of the cut is reached.
8. Allow the cutters to stop before putting the plane down; otherwise the plane could take off the revolving cutters.

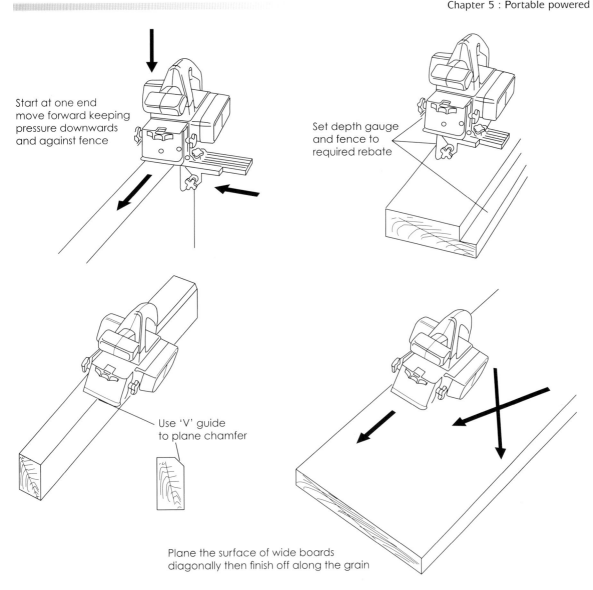

Start at one end move forward keeping pressure downwards and against fence

Set depth gauge and fence to required rebate

Use 'V' guide to plane chamfer

Plane the surface of wide boards diagonally then finish off along the grain

▲ **Figure 5.30** Using a planer

Rebating

■ Adjust the side fence and depth gauge to the required rebate dimensions.

■ Use repeated passes to plane to the required depth, keeping the fence pressed up against the workpiece at all times.

Chamfering

■ Use the 'V' guide on the in-feed sole plate to locate the planer on the 90° corner of the workpiece.

■ Use a number of passes to plane the required 45° chamfer.

■ The side fence may be adjusted to run along the edge of the workpiece to guide the plane as the chamfer widens, or when angles other than 45° are required.

Surfacing a wide board

Plane diagonally across the board, in two directions, overlapping each pass.

Plane parallel to the edge of the board, using a number of overlapping passes to cover the entire board width.

Powered routers

The portable power router has largely taken the place of a wide range of hand moulding, grooving and rebating planes (Figure 5.31). In addition they are also used for trimming, recessing, housing, slot mortising, dovetailing and on many types drilling and plunge cutting. Mechanically, the majority of routers are similar, with a router cutter (bit) mounted directly below a motor housing that is fitted with a handle on either side. The motor moves up and down on two spring-loaded columns attached to the baseplate. The router works by spinning the cutter at a very high speeds up to 30,000 rpm.

Laminate trimmer

Plunging router

▲ **Figure 5.31** Powered routers

Router operation

Refer to Figure 5.32:

■ The cutter can be plunged (lowered) into the workpiece by pressing down on the motor against the spring-loaded columns. This action also allows the retraction of the cutter safely above the baseplate before lifting the router from the work.

■ The cutter can be set at a predetermined depth, which is then entered into the workpiece sideways.

■ Alternatively, the router may be mounted upside down in a table. With the cutter protruding above the table the workpiece is fed past the cutter.

Collet capacity. The shank of router cutters fit into a tapered collet and are secured by a locking nut. Collet sizes are usually either: 6 mm, 8 mm or 12 mm in diameter to suit the cutter shank. Larger routers have interchangeable collets, enabling all three sizes of cutter shanks to be used.

Spindle speed. Maximum spindle speed varies between 20,000 and 30,000 rpm, depending on the motor's power output. The maximum speed can be used for most operations. Slower speeds should be selected when working some hardwoods or those with interlocking or troublesome grain.

Large diameter moulding cutters have a very high peripheral speed and should be run at lower spindle speeds for safety. Lower speeds should also be used when routing plastics and soft metals.

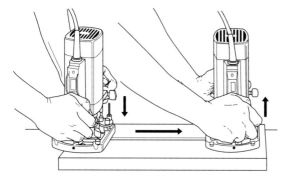

Plunge, rout and retract

Router inverted under table,
feed workpiece past the cutter

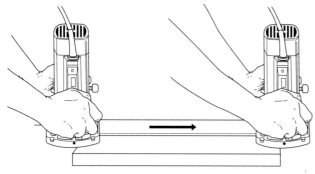

Preset cutter projection and rout

▲ **Figure 5.32** Router operation

Router cutters

Router cutters are available in a wide variety of sizes and shapes as illustrated in Figure 5.33. They fall into a number of main categories:

- **Groove-forming cutters** are a range of basic cutters in various widths, used to make recesses both with and across the grain. They normally require guiding by a fence or guide batten.

- **Edge-forming cutters** have a guide tip which runs against the edge of the workpiece, making the use of a fence or guide batten unnecessary. Fixed guide tips can cause the wood to burn due to friction. This can normally be removed with a finely set plane. However, cutters with a ball-bearing guide tip are available. These run against the work edge without causing the burn blemish.

- **Trimming cutters.** There are two types available for trimming plastic laminate. Both have a ball-bearing guide wheel that runs on the laminated board. The parallel profile is used to trim the edging strip flush with the board and the chamfered profile for trimming the top laminate to a slight angle.

High speed steel (HSS) cutters are suitable for most general woodworking operations, but **tungsten carbide-tipped cutters,** although more expensive, hold their edge longer, especially when working abrasive timbers, hard glue lines, sheet material and plastics.
Solid tungsten carbide cutters are also available; they can be used with the same materials as tipped cutters, but are considerably more expensive.

The edges of HSS cutters can be honed by the woodworker using an oiled slip-stone to extend their usage. But tungsten tipped and solid tungsten cutters can only be resharpened using specialist grinding facilities.

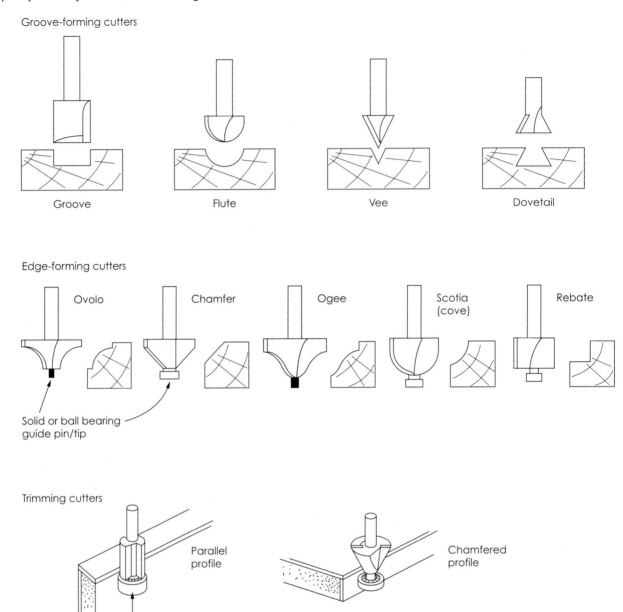

Groove-forming cutters

Groove

Flute

Vee

Dovetail

Edge-forming cutters

Ovolo

Chamfer

Ogee

Scotia (cove)

Rebate

Solid or ball bearing guide pin/tip

Trimming cutters

Parallel profile

Chamfered profile

Ball-bearing guide

▲ **Figure 5.33** Router cutters

Health and safety

When using routers fine dust particles are produced especially when working with man-made board materials. Even a short session can cover both the workshop and the operator with a cloud of lung-clogging dust. Most routers can be equipped with a clear plastic hood that fits over the base to enclose the cutter. The hood is connected by a flexible hose to a vacuum extraction unit that collects the dust and larger waste material as it leaves the cutter. It is strongly recommended that this facility is always used with the router even for short periods of working. This is in addition to the wearing of standard personal protective equipment, earmuffs, dust masks and goggles for yourself and anyone working near you.

Operation of a router

For most operations, once having set up the router, cramped the workpiece in position and connected it to the power supply, two alternatives may be used depending on the type of router and the work in hand.

Plunging technique:

1. Place the router on the workpiece ensuring that the cutter is clear. Taking a firm grip of the router, start the motor and allow it to obtain maximum speed. Plunge the router to the pre-set depth and lock. For deep cuts, several passes will have to be made to achieve the required depth, see Figure 5.34.

2. Applying a firm downward pressure move the router steadily forwards. Routing too slowly will cause the cutter to overheat resulting in a poor finish and leaving burn marks on the work.

3. At the end of the cut release the lock, which will cause the cutter to spring up clear of the work. Switch off the motor and allow the cutter to stop rotating before moving the router. It is most important that the router is fed in the correct direction with respect to the rotation of the cutter. The cutter rotates clockwise when viewed from the top of the router, see Figure 5.35.

Pre-set depth technique:

1. Set the router cutter to the required projection below the baseplate. Deep cuts will have to be done in several stages.

2. Rest part of the router baseplate on the workpiece. Ensure the projecting cutter is clear of the work. Start the motor and allow it to obtain maximum speed.

3. Applying a firm downward pressure and side pressure against a guide batten or fence, move the router steadily forward to make the cut.

4. When the cutter emerges at the other end, switch off and allow the cutter to stop rotating before lifting the baseplate off the workpiece.

Housing or trenching using guide battens

Follow these steps, referring to Figure 5.36:

1. Cramp a batten across the workpiece. Ensure that the batten projects beyond both edges. When positioning the batten, an allowance must be made equal to the distance between the cutting edge of the bit and the baseplate. For housings wider than the cutter two battens are used.

2. Lower and lock the router's plunge mechanism to the required cutter projection.

3. Rest the router baseplate on the workpiece with the cutter clear of the work.

4. Switch on and allow the motor to reach its working speed.

5. Keeping the router baseplate pressed against the batten, feed the router into and through the workpiece at a steady rate until the cut is complete. For wide housings the first cut is made against the left-hand batten. Then move the router across to the right-hand batten to complete the cut. Using this method the rotation of the cutter helps to hold the router base against the guide battens.

6. Switch off with the cutter just clear of the work. Allow the cutter to stop before lifting the baseplate off the workpiece.

Grooving using the fence

Most grooves are required to be parallel with and close to the edge of the workpiece. The bolt-on side fence is ideal for this operation (Figure 5.37):

1. With the router unplugged, rest the router on the workpiece with the cutting edge of the cutter aligned with the edge of the groove marked on the surface.

2. Adjust the side fence up to the edge of the workpiece and tighten its thumbscrews.

3. Cut the groove using either the pre-set depth or plunge technique. In either case keep side pressure on the fence throughout the cut.

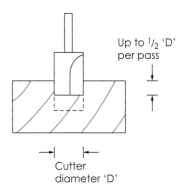

Up to ¹/₂ 'D' per pass

Cutter diameter 'D'

▲ **Figure 5.34** Deep cuts are made in several passes

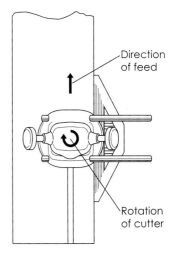

Direction of feed

Rotation of cutter

▲ **Figure 5.35** Direction of feed and cutter rotation

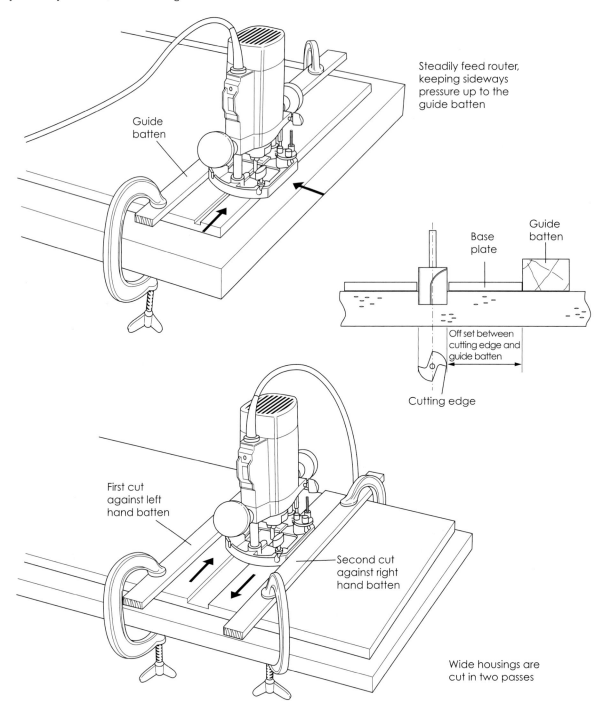

Steadily feed router, keeping sideways pressure up to the guide batten

Guide batten

Base plate

Guide batten

Off set between cutting edge and guide batten

Cutting edge

First cut against left hand batten

Second cut against right hand batten

Wide housings are cut in two passes

▲ **Figure 5.36** Housing or trenching with guide battens

Rebating with a straight cutter

Rebates are best cut using an edge-forming cutter with a guide pin or ball-bearing guide (Figure 5.38). However, they can be cut with a straight cutter and side fence using a similar procedure as that used for cutting grooves.

Moulding with an edge-forming cutter

Moulding or rebating the edge of a workpiece is achieved using an edge-forming cutter with a guide pin or ball-bearing guide:

■ *Moulding all four edges of a panel.* Work the outer edge of the workpiece in an anti-clockwise direction, as this ensures the cutter rotation pulls it

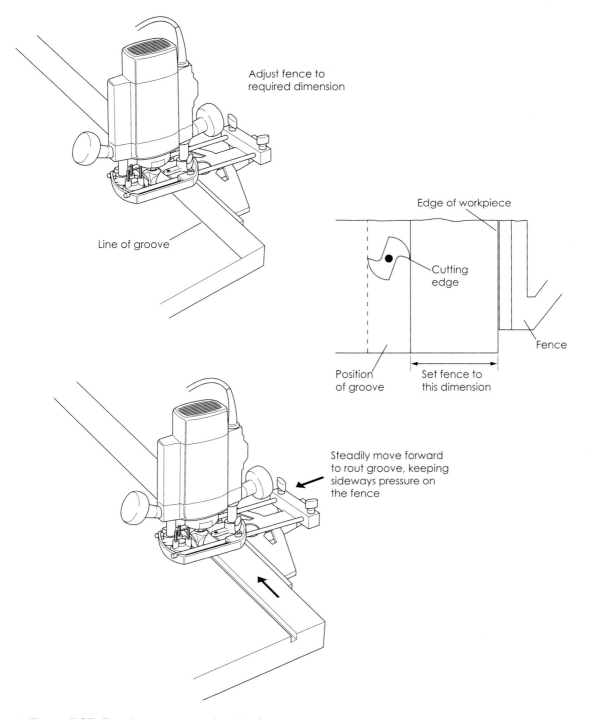

Adjust fence to
required dimension

Line of groove

Edge of workpiece

Cutting
edge

Fence

Position
of groove

Set fence to
this dimension

Steadily move forward
to rout groove, keeping
sideways pressure on
the fence

▲ **Figure 5.37** Forming a groove using the fence

into the work rather than pushing it off (Figure 5.39). Where the panel is
of solid timber, mould the end grain before the sides. This ensures that
any breakout at the far edge of the cut is removed when the sides are
moulded.

■ *Moulding the inside of an assembled frame.* This is achieved using
  the same technique as moulding the outer edge of a panel,
  except this time move the router in a clockwise direction to keep
  the cutter in contact with the workpiece, Figure 5.40. The cutter will
  leave rounded corners; for mouldings these can be an additional
  feature. However, rebates for glazing will require squaring up with
  a chisel.

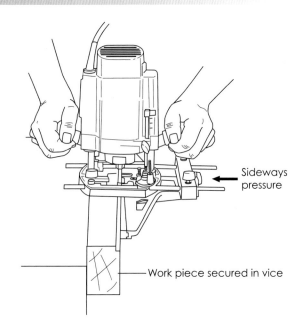

▶ **Figure 5.38** Rebating with a straight cutter.

Sideways pressure

Work piece secured in vice

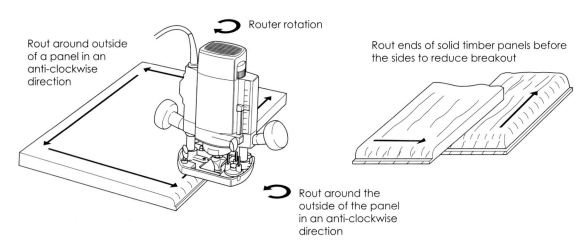

Router rotation

Rout around outside of a panel in an anti-clockwise direction

Rout around the outside of the panel in an anti-clockwise direction

Rout ends of solid timber panels before the sides to reduce breakout

▲ **Figure 5.39** Moulding with an edge-forming cutter

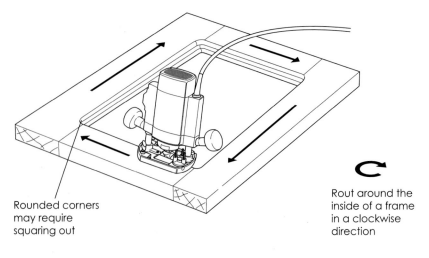

Rounded corners may require squaring out

Rout around the inside of a frame in a clockwise direction

▲ **Figure 5.40** Moulding the inside of an assembled frame

Recessing with a template

This operation requires the fitting of a template guide to the baseplate of the router and a template of the recess required (Figure 5.41). When making the template an allowance must be made all round equal to the distance between the cutting edge of the bit and the outside edge of the template guide.

1. Fix the template in the required position.
2. Place the router base on the template, taking a firm grip on the router. Start the motor and allow it to attain its working speed. Plunge the router to the pre-set depth and lock.
3. Applying a firm downward pressure, move the router around the edge of the template before working the centre. It is most important to feed the router in the opposite direction to the rotation of the cutter. (Study Figures 5.39 to 5.43.)
4. On completion, retract the cutter, switch off the motor and allow the cutter to stop rotating before putting the router down. Rounded corners will be left by the cutter that can easily be squared up if required with a chisel (hinges and lock faceplate are available with rounded corners for recessing with a router).

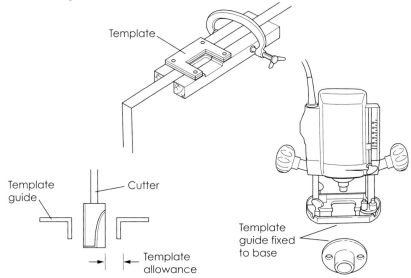

◀ **Figure 5.41** Recessing with a template

Cutting circles

This is achieved with the aid of a trammel bar, which is fixed to the baseplate at one end and has a centre pin at the other (Figure 5.42).

The router is rotated in an anti-clockwise direction about the centre pin to create either a circular hole in the workpiece or a circular disk.

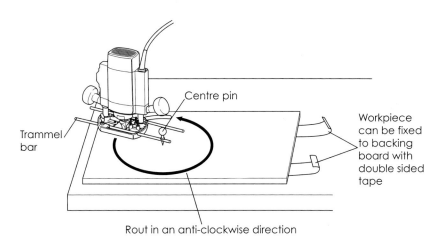

◀ **Figure 5.42** Routing circles

221

Cutting with a template

Routing shapes against a template is the ideal way of cutting shapes or making a number of identical components (Figure 5.43).

Templates are best cut from MDF: accuracy is essential as any minor defect will be re-produced in the finished components.

A template guide is fixed to the baseplate to run against the edge of the template. Therefore an allowance has to be made in the template equal to the distance between the cutting edge and the outer edge of the template guide.

Templates can either be pinned to the workpiece or fixed to it using double-sided tape, depending on the intended finish.

On outside edges rout in an anti-clockwise direction and on the inside edges rout in a clockwise direction to keep the template guide in contact.

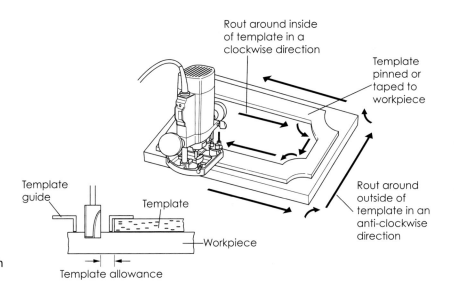

▶ **Figure 5.43** Routing shapes with a template

Laminate trimming

Small routers for single-hand operation are manufactured specifically for laminate trimming, although it is possible to fit a ball-bearing guided laminate trimmer to a standard plunging router.

Cutting joints with a router

The router can be used to cut a wide range of joints.

Rebated and grooved joints

Lapped, bare-faced tongues and halving joints can all be cut using a similar technique (Figure 5.44):
1. Cramp the components to be jointed side by side to the bench top.
2. Use a straight cutter in the router, with the baseplate against a guide batten to make the cut.

Halving joints will require several passes to cut the required width.

Tongue-and-groove joints

Follow these steps, referring to Figure 5.45:
1. Set up the router with a straight cutter and guide fence.

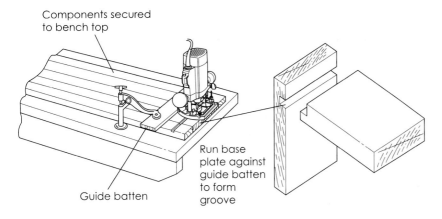

Components secured
to bench top

Run base
plate against
guide batten
to form
groove

Guide batten

◀ **Figure 5.44** Routing grooved joints

2. Secure the piece to receive the tongue between two scrap pieces of timber in the vice. This gives a wider flat surface to give better support to the baseplate.
3. Pass the router along both sides of the piece to form the tongue.
4. Secure in the vice the piece to be grooved, again using two scrap pieces for support.
5. Reset the guide fence to position the cutter in the centre and pass the router along the piece to form a matching groove.

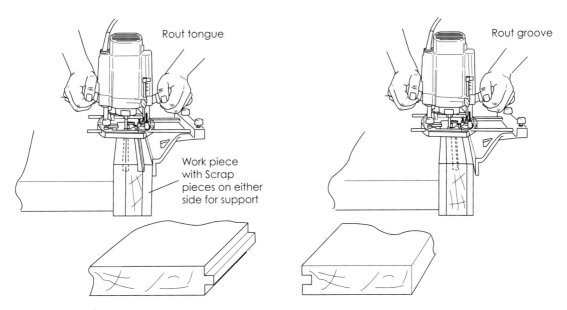

Rout tongue

Work piece
with Scrap
pieces on either
side for support

Rout groove

▲ **Figure 5.45** Routing tongue-and-groove joints

Mortise and tenon joints

Follow these steps, referring to Figure 5.46:
1. Set up the router with the required mortise width straight cuter and guide fence.
2. Secure the piece in the vice using scrap timber on either side to increase the supporting width.
3. Cut the mortise in stages as a short stopped groove, gradually increasing the depth of cut with each pass. Stub mortises can be routed from one edge; though mortises may require routing from both edges.
4. Pencil lines can be used as a guide to the positions for plunging and retracting the router, but for greater accuracy pin stop blocks to one of the scrap pieces.

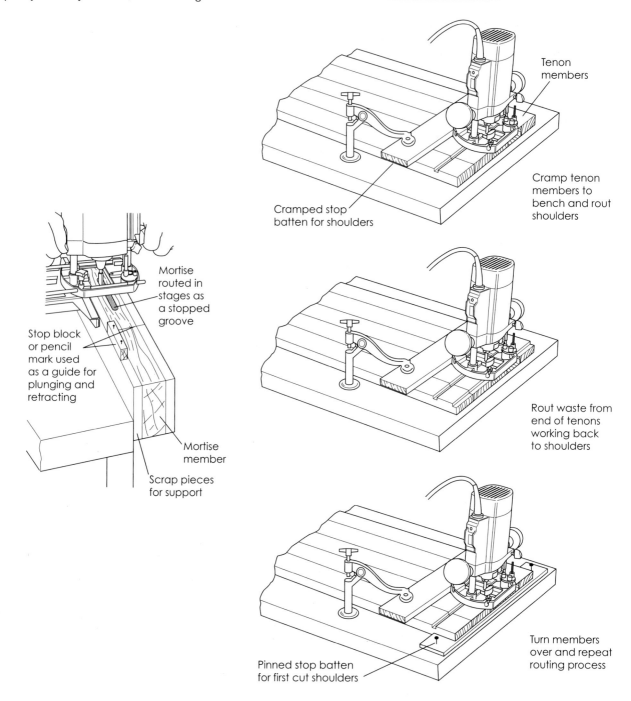

Tenon members

Cramp tenon members to bench and rout shoulders

Cramped stop batten for shoulders

Mortise routed in stages as a stopped groove

Stop block or pencil mark used as a guide for plunging and retracting

Mortise member

Scrap pieces for support

Rout waste from end of tenons working back to shoulders

Pinned stop batten for first cut shoulders

Turn members over and repeat routing process

▲ **Figure 5.46** Routing a mortise and tenon joint

5. Use a chisel to square up the rounded ends of the mortise.
6. Lay the pieces to be tenoned side by side on the bench. Secure using a cramped guide batten parallel with but offset from the shoulder.
7. Set the cutter projection to the required depth and cut all the shoulders in one operation.
8. Remove the remaining waste freehand starting at the ends of the tenon for support gradually working back towards the shoulder.
9. Turn pieces over, butt the cut shoulders against a stop batten pinned to the bench.
10. Secure with a cramp and batten and repeat the previous stage to complete the tenons.

Dovetail joints can be cut with a router, set up with a dovetail cutter and a manufacturer's dovetail jig, as illustrated in Figure 5.47. The set-up and use of these vary; detailed instructions are supplied by the jig's manufacturer and must be followed:

1. Cramp, the two pieces to be joined in the jig; these should be inside out and slightly offset.

2. Set the dovetail cutter to the required depth and cut the joint, passing the router in and out of each finger of the jig template.

3. Remove both pieces from the jig and turn them around ready for assembly.

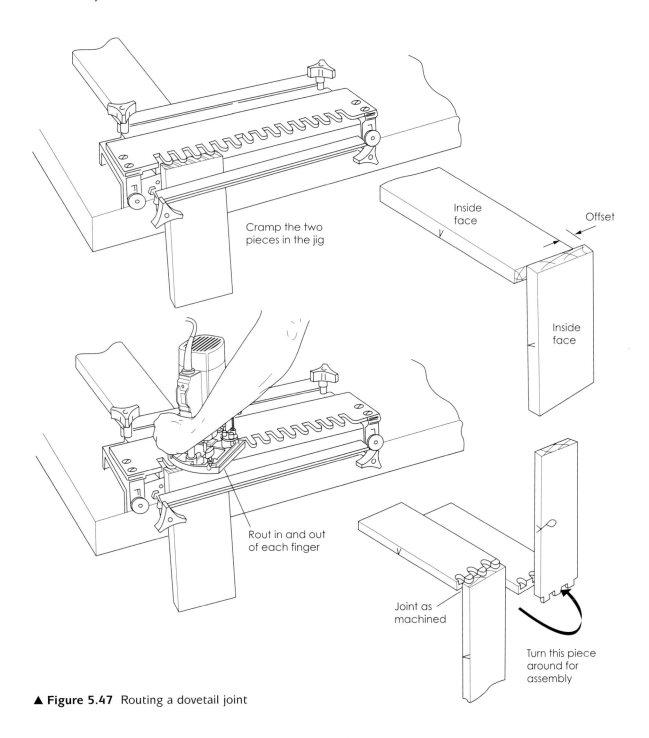

Cramp the two pieces in the jig

Inside face

Offset

Inside face

Rout in and out of each finger

Joint as machined

Turn this piece around for assembly

▲ **Figure 5.47** Routing a dovetail joint

Sanders

The use of power sanders has taken most of the effort out of the process of shaping and finishing work. However, even the finest of finishing sanders still leave minute surface scratches that may require a final period of hand sanding to remove, especially when the surface is to receive a polished or varnished finish. A typical range of sanders is illustrated in Figure 5.48.

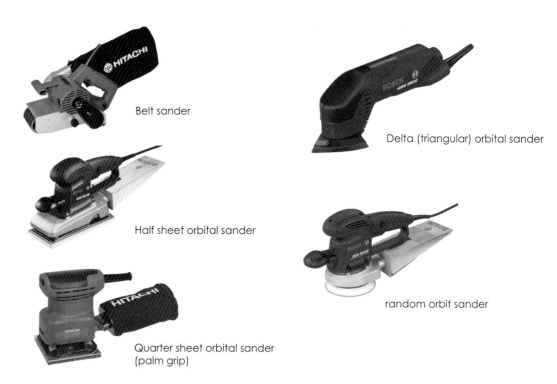

Belt sander

Delta (triangular) orbital sander

Half sheet orbital sander

random orbit sander

Quarter sheet orbital sander (palm grip)

▲ **Figure 5.48** Types of power sander

Belt sanders are used for jobs requiring rapid stock removal. When fitted with the correct grade of abrasive belt they can be used for a wide range of operations, such as smoothing and finishing joinery items, block flooring and even the removal of old paint and varnish finishes.

The sanding or abrasive belt is fitted over two rollers. The front roller is spring-loaded and can be moved backwards and forwards by the belt-tensioning lever. This movement allows the belt to be changed easily and it also applies the correct tension to the belt. When changing the belt it is necessary to ensure that it will rotate in the correct direction. This is indicated on the inside of the belt by an arrow. If the belt is inadvertently put on the wrong way round, the lap joint running diagonally across the belt will tend to peel. This could result in the belt breaking, with possible damage to yourself and the work surface.

To keep the belt running central on the rollers, there is a tracking control knob on the sander that adjusts the front roller by tilting it either to the left or right as required. The tracking is adjusted by turning the sander bottom upwards with the belt running and rotating the tracking knob until the belt runs evenly in the centre without deviating to either side.

Orbital sanders are also known as the **finishing sander** as it is mainly used for fine finishing work. The sanders base has a 3 mm orbit which operates at 12,000 rpm.

Various grades of **abrasive paper** may be clipped to the sander's base. It is best to start off with a coarse grade to remove any high spots or roughness and follow on with finer grades until the required finish is obtained, although where the surface of the timber has machine marks or there is a definite difference in the levels of adjacent material, the surface should be levelled by planing before any sanding is commenced.

The baseplate size is based on a proportion of the standard size sheet used for hand work. Larger machines are termed half- and third-sheet sanders. Smaller, single-handed palm-grip machines use a quarter-sheet. Delta sanders have a small triangular baseplate for finishing in tight corners.

Random orbit sanders work with an off-centre (eccentric) orbital action plus a simultaneous rotation, to give a surface that it virtually scratch free. The circular rubber backing pad is flexible to enable convex and concave surfaces to be finished. Abrasive discs are self-adhesive, with either a peel-off backing or Velcro backing, for efficient fitting and replacement.

Most sanders are fitted with a **dust collection bag**, or provision for connecting to a vacuum extract system. Some types require the use of perforated abrasive sheets to allow the dust and fine abrasive particles produced when working to be sucked up.

Always use the dust collection/extraction system, because as well as creating a health hazard, the dust and abrasive particles will clog the sheet and cause further scratches to the work.

Operation of sanders

Refer to Figure 5.49:

1. Always start the sander before bringing it into contact with the work surface and remove it from the work surface before switching off. This is because slow-moving abrasive particles will deeply scratch the work surface.
2. Do not press down on a sander in use: the weight of the machine itself is sufficient pressure. Excessive pressure causes clogging of the abrasive material and overheating of the motor.

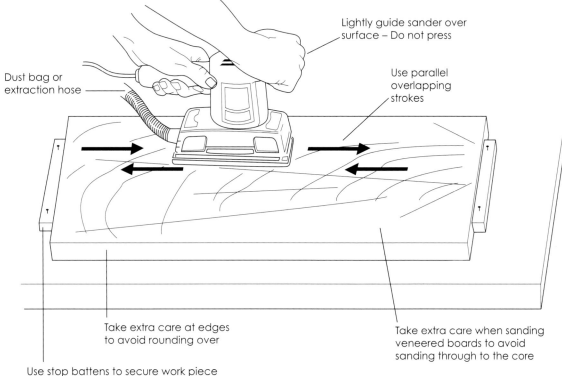

Lightly guide sander over surface – Do not press

Dust bag or extraction hose

Use parallel overlapping strokes

Take extra care at edges to avoid rounding over

Take extra care when sanding veneered boards to avoid sanding through to the core

Use stop battens to secure work piece to bench, especially when using a belt sander

▲ **Figure 5.49** Operation of sanders

3. For best results lightly guide the sander over the surface with parallel overlapping strokes in line with the grain.

4. Always use the abrasive belts and sheets specifically recommended by the manufacturer for the particular model as makeshift belts and sheets are inefficient and can be dangerous.

5. Always use the dust collecting bag or extraction system where fitted. In any case always wear a dust mask as inhaling the dust from many species of wood causes coughing, sneezing and runny eyes and nose.

Automatic pin and staple drivers

Various pin gun nailers and staplers are available to suit a wide range of fixings. The smaller drivers are usually operated by spring power, the larger ones by compressed air, although a limited range of gas-operated nailers and electric tackers are also available. A typical selection is illustrated in Figure 5.50.

Operation of drivers

Operation of automatic drivers will vary depending on the type being used but the following are a number of general points:

1. Do not operate the trigger until the baseplate is in contact with the fixing surface.

Hand-held spring stapler

Compressed air
pin gun nailer/stapler

Hand hammer-action stapler

Cordless gas-operated
nail gun

▲ **Figure 5.50** Automatic drivers

2. Keep fingers clear of the baseplate.
3. Maintain a firm pressure with the fixing surface during use. Failure to do so can result in kickback of the tool and ricochet of the nail or staple.
4. On tools with a variable power adjuster, carry out trial fixings at a low setting and gradually increase until the required penetration is achieved.

Cartridge-operated fixing tools

The cartridge-operated fixing tool is an invaluable aid to making fast and reliable fixings to a variety of materials including concrete and steel (Figure 5.51) that normally requires drilling and plugging or threading before a fixing can be made.

▲ **Figure 5.51** A general-purpose cartridge tool

Most tools operate using what is known as an indirect-acting system. This means that they have a piston in their barrel. When the cartridge is detonated, the released gas acts on the piston, accelerating it from rest to drive the fastener into the base material at a maximum velocity of up to 100 m/s. The piston is held captive in the tool and once the piston stops so will the fastener. This virtually eliminates the risk of through-penetration (fastener passing through soft base material) or ricochets (deflection of fastener within base material), even if the base material is inconsistent (Figure 5.52).

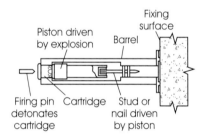

▲ **Figure 5.52** Indirect acting system

Cartridges

Cartridges are available in a range of explosive strengths to suit the insertion of various fasteners into materials of different hardness. In order to identify the cartridge strength, each individual cartridge is colour-coded by a touch of colour on its crimped end and a power rating on its base, see Table 5.1 for details.

Table 5.1 Colour-coding of cartridges

Cartridge Strength	Code Number	Identification Colour
Weakest	1	Grey
	2	Brown
	3	Green
	4	Yellow
	5	Blue
	6	Red
Strongest	7	Black

Note: Manufacturers may only produce a limited range of cartridge strengths, with variation between strengths being obtained by varying the volume of the combustion chamber

Cartridges are supplied either singly or in plastic strips for repeated firing. They should always be carried and stored in the manufacturer's packaging until required (Figure 5.53). This packaging should include the following information:

■ The manufacturers name.

■ A colour coded label indicating the cartridges strength.

■ The cartridge size.

■ The number of cartridges in the box.

■ The tool or range of tools in which they can be used.

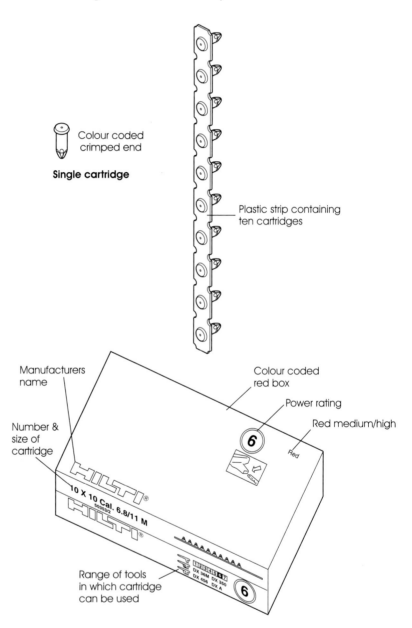

▶ **Figure 5.53** Packaging of cartridges

Some cartridge tools use only one or a limited range of cartridge strengths. The required power level is obtained by adjusting a thumb-wheel on the side of the tool that varies the volume of the combustion chamber.

Fasteners

Fastenings made by cartridge tools are of two types:

■ Those where the object is fixed permanently, e.g. nailed.

■ Those where the fastening must allow the object to be removed, in which case a threaded stud should be used.

The basic types of permanent and detachable fastener available are illustrated in Figure 5.54.

- Type 1: A nail for fixings into concrete/brickwork and steel.
- Type 2: A nail for fixings into steel.
- Types 3 and 4: Studs with various threads for fixings to concrete/brickwork. Suitable types are also available for fixing into steel.
- Type 5: Eyelet pin for fixings into concrete and brickwork.

Base material

Suitable base materials are those that allow the fastener to travel in a straight line, to the correct depth and with the maximum holding power (Figure 5.55):

- Materials that are too hard (marble, hard bricks and cast iron etc.) may cause recoil of the tool resulting in deflection or ricochet of the fastener.
- Materials that are too brittle (glass, slates and glazed tiles, etc.) will crack or shatter.
- Materials that are too soft (lightweight building blocks and hollow constructions, etc.) may allow through penetration of the fastener.

Hand hammer test. When cartridge fixing into a suspect material, the best procedure to follow is to carry out the hand hammer test, which involves attempting to hand hammer a fastener into the material (Figure 5.56). Wear eye protectors for this test.

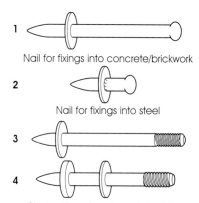
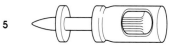

1   Nail for fixings into concrete/brickwork

2   Nail for fixings into steel

3

4

Studs with various threads for fixings into concrete/brickwork, versions are also available for fixings into steel

5   Eyelet pin for fixings into concrete/brickwork

▲ **Figure 5.54** Types of fasteners

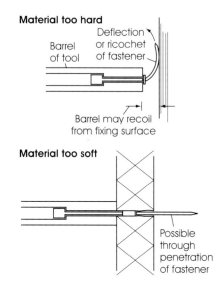

▲ **Figure 5.55** Suitability of base system

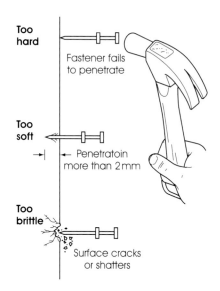

▲ **Figure 5.56** Hand hammer test to determine suitability of base material

A cartridge fixing is not suitable if any of the following occur:

- The point of the fastener is blunted, or fails to penetrate up to 2 mm (too hard).
- The surface of the material cracks, crazes or is damaged (too brittle).
- The surface is easily penetrated, in excess of 2 mm (too soft).

Where the strength of the material into which a fixing is being made is not known, a **test fixing** should be carried out in order to establish the required cartridge strength.

Always make test fixings using the lowest strength cartridge first, increasing by one strength each time until the required fixing is achieved (Figure 5.57).

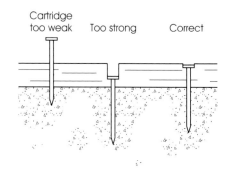

▲ **Figure 5.57** Making test fixings to check correct cartridge strength

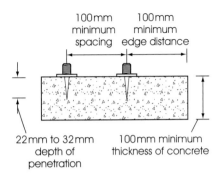

100mm minimum spacing

100mm minimum edge distance

22mm to 32mm depth of penetration

100mm minimum thickness of concrete

▲ **Figure 5.58** Fixings in concrete

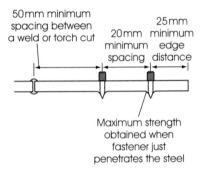

50mm minimum spacing between a weld or torch cut

20mm minimum spacing

25mm minimum edge distance

Maximum strength obtained when fastener just penetrates the steel

▲ **Figure 5.59** Fixings in steel

Fixing to concrete. When a fastener is fired into concrete, the concrete is compressed against the fastener shank creating a friction hold assisted by the heat generated during the driving process causing sintering (fusing together) of the concrete to the fastener's tip. Maximum strength is achieved when that factors illustrated in Figure 5.58 are observed.

Fixing to steel. When a fastener is fired into steel, the steel is displaced and flows back against the fastener shank creating a keying (knurled shank) or friction hold (smooth shank). In addition the heat generated during the driving process causes a partial fusion fastener to the steel. Maximum strength is achieved when the factors illustrated in Figure 5.59 are observed.

Safe operation of cartridge-operated fixing tools

Safe working procedures are most important when using cartridge tools, and these can never be overemphasised. All persons using and maintaining cartridge-operated tools must be trained in the particular tool they are using. Manufacturers provide training and issue certificates of competence for users of their products. It is important to undertake the course for each tool that you use, as each tool can have different operation, maintenance, storage and misfire procedures.

Before operating a tool, always run through the following points:
1. Ensure you have had the correct training.
2. Ensure the tool is in good repair (cleaned and serviced daily).
3. Ensure you understand the misfire procedure.
4. Always wear the recommended PPE (eye protection, safety helmet and ear protectors).
5. Ensure the base material is suitable (try the hand hammer test).
6. Ensure the correct pin, piston and cartridge combination is used.
7. Never lay down a loaded tool.
8. Ensure you understand the handling and storage of the tool and cartridges: stored in a metal box, in a secure store, to which only authorised persons have access.

Misfire procedure

Misfires are when a cartridge fails to detonate and can be caused by either a faulty cartridge or a fault in the firing mechanism. The general procedure to follow in the event of a misfire is:
1. Keep the tool in position and attempt to re-fire.
2. If the tool fails to fire the second time, wait 15 seconds, remove tool from the workface, keeping the tool pointing away from people and down to the ground. Remove the cartridge following the manufacturers instructions.
3. Replace tool in its storage case and return to the manufacturer for investigation.

Security

Cartridge tools and their accessories must always be kept with their manufacturer's case and be under the control of an authorised person. After use it should be unloaded and replaced in its lockable case and placed in a secure store. Spare cartridges should be replaced in their correct colour coded packaging and also placed in a secure storage place.

Chapter Six

Machine utilisation

This chapter covers the work of site carpenters and bench joiners. It is concerned with the safety requirements and principles of setting up and using traditional woodworking machines. It includes the following:

- ❡ accident statistics for woodworking machines
- ❡ woodworking machinery legislation
- ❡ safe use of woodworking machines for sawing, planing, moulding and jointing
- ❡ health hazards associated with noise, wood dust and vibration
- ❡ maintenance of woodworking machines.

The extent to which carpenters and joiners are expected to use woodworking machines vary widely from firm to firm. There are many where the operation of any machine is exclusively the province of the trained machinist, others where carpenters and joiners carry out all the machining. Also there are the firms that fall between these two extremes, where machinists carry out bulk work but the joiners have at their disposal a limited range of machines for the occasional one-off job or replacement member.

Accident statistics and legislation

When compared to other industries woodworking accounts for a disproportionately high number of machine accidents. In an HSE survey of 1000 woodworking machine accidents 73% of the total were attributable to four machine types as illustrated by the pie chart in Figure 6.1.

- ■ **Circular saw benches** accounted for 35% of the total, with many resulting in the loss of fingers. 83% of this total occurred whilst undertaking ripping or cross-cutting operations. In most cases the saw guard was either incorrectly adjusted or missing altogether. The Health and Safety Executive (HSE) states that many of the accidents could have been avoided if the saw guard was correctly adjusted and a push stick used.

- ■ **Planing machines** accounted for 20% of the total, with the result being the slicing of fingers. With a cutter block making 10,000 cuts a minute, if your fingers are in contact for only a split second many slices will be removed, e.g. 16 slices in 0.1 s. 80% of the total planing accidents occurred during normal flatting and edging operations rather than the more specialised rebating, moulding or chamfering operations. The HSE stated that although in most cases a bridge guard was provided, accidents were the result of failure to adjust it correctly.

- ■ **Vertical spindle moulders** accounted for 14% of the total, frequently involving the loss of several fingers. 42% occurred on straight-through

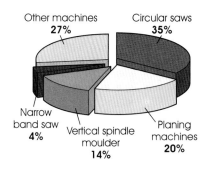

▲ **Figure 6.1** HSE survey of woodworking machine accidents work, 34% on stopped work and 15% on curved work. The HSE stated that the straight work accidents were due mainly to the lack of use of false fences and Shaw guards to enclose the cutters. Stopped and curved work accidents mainly resulted from the failure to use backstops, jigs and workpiece holders.

■ **Narrow band saws** accounted for 4% of the total. Most were due to contact with the blade whilst presenting material to the moving blade or removing it from the table after cutting. In addition, accidents also occurred when undertaking, setting, adjusting, cleaning and performing maintenance operations whilst the blade was still moving.

The HSE statistics go on to show that accidents are disproportionately high in small premises as illustrated in Figure 6.2. Over 50% of those injured had only received 'on the job training'; 24% had not received training or instruction on the machine they were using, and of these only 5% were under supervision; finally 25% of accidents involved formally trained operators, which indicates that safe working methods were being bypassed.

To help prevent accidents at woodworking machines all concerned should:

■ Assess all the risks in the workplace and put precautions in place to eliminate them.
■ Ensure that all necessary guards are in position and used at all times.
■ Ensure all operators are suitably trained on the machine they are using and in addition that they are properly supervised.
■ Check that machinists are following safe working methods at all times.

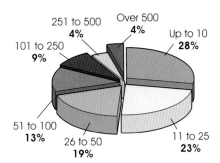

▲ **Figure 6.2** HSE statistics showing accidents and size of firm

Legislation

The main legal requirements covering the supply and use of woodworking machines are contained in the following:

■ *Provision and Use of Work Equipment Regulations* (PUWER).
■ *Management of Health and Safety at Work Regulations* (MHSWR).
■ And a supporting Approved Code of Practice (ACOP) *Safe use of Woodworking Machinery*.

PUWER: Provision and Use of Work Equipment Regulations

PUWER requires risks to people's health and safety from equipment that they use at work to be prevented or controlled. **Employers** and the **self-employed** who provide equipment for use at work or *persons who control or supervise the use of equipment* have a duty under the regulations to ensure that it is:

■ suitable to use for the purpose and in the conditions which they are used;
■ maintained in a safe condition;
■ inspected by a competent person to ensure that it is and remains in a safe condition.

The regulations also require that risks created by the use of equipment are eliminated wherever possible or controlled by taking a combination of the following appropriate measures:

■ Hardware measures such as the provision of guards, protection devices, system controls, warning devices, safety markings and personal protective equipment (PPE);
■ Software measures such as following safe systems of work and providing machinists with adequate operating information, instruction and training.

Employees do not have duties under PUWER but have duties under other safety legislation such as MHSWR and *Health and Safety at Work etc Act* (HSW Act) to comply with the following:

- to take care at all times and ensure that their actions do not put 'at risk' themselves, their workmates or any other person;
- to co-operate with their employers to enable them to fulfil the employers' health and safety duties;
- to use the equipment and safeguards provided by the employers;
- to never misuse or interfere with anything provided for health and safety.

ACOP: Safe Use of Woodworking Machinery

PUWER relates directly to work equipment in general rather than specifically to woodworking machines. The ACOP gives advice on the precautions to take to ensure the safe use of woodworking machines. It is accepted that the guidance it contains reflects the precautions that are widely accepted and used in industry to reduce the risks from woodworking machinery, and in addition mirror many of the requirements contained in the revoked *Woodworking Machines Regulations*.

It is essential that every user of woodworking machines, including students and trainees, has a thorough knowledge of the relevant requirements and that they fully implement them for their own safety and that of others.

It is not possible to gain skills in the use of woodworking machines and powered hand tools by reading alone. Therefore all this section sets out to do is to identify the various machines, their safe working procedures and cutting principles.

> *It is essential to receive the relevant practical training by a competent person before using any machine.*

General operator safety requirements

The safety requirements of the ACOP, *Safe Use of Woodworking Machinery,* which are relevant wherever woodworking machines are in use, may be summarised as follows.

Young people

Those who have not yet reached the age of 18 years should not be allowed to operate high-risk woodworking machinery (circular saws, band saws, surface planers and spindle moulders) unless they have the necessary maturity, competence and have satisfactorily completed appropriate training. They must be adequately supervised during training and this must continue after training until they are sufficiently mature.

Suitability of work equipment

This includes the suitability of the actual machine, the place it is located and the actual work process it is being used for. It is recognised that some operations can be carried out on more than one machine. The most suitable machine that presents a lower risk should always be selected for the task. For example cutting a rebate on a properly guarded spindle moulder presents a lower risk than using the cutter block of a surface planer. Likewise removing a thin sliver on a circular saw is a higher risk than removing it on a surface planer.

In some cases it is possible to undertake the work safely provided additional safety measures are taken, for example rebating on a surface planer can be carried out safely provided the workpiece is properly supported, a Shaw guard is used and the table gap is guarded on both sides of the fence.

- *Limited cutter projection tooling* also known as 'chip thickness limitation tooling' should be used for hand-fed machines. Evidence shows that the use of this tooling reduces the severity of injury if the machine operator's fingers come into contact with the rotating tool, see Figure 6.3.
- *Tool speeds.* No tool should be run at a greater working speed than that marked on the tool or specified in the manufacturer's information.

Round profile tool

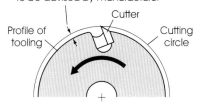

Separate limiter

Built-in limiter

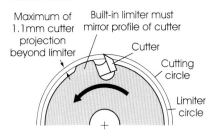

▲ **Figure 6.3** Limited cutter projection tooling

- *Cylindrical cutter blocks* only must be used on hand-fed planing machines.
- *Detachable cutters* must be the correct thickness for the cutter block and the manufacturer's recommendations for balancing and mounting them should be followed.
- *Choosing tooling.* Cutters must be of the correct type for the work in hand. They must be in good condition and kept sharp. Damaged or cracked cutters increase the risk of injury through disintegration (breakup), they must not be used and should be thrown away.

Maintenance

Every machine must be maintained in an efficient state, efficient working order and in good repair. Efficient in this context refers to how the condition of the machine may affect health and safety; it is not concerned with any productivity issues. Where a machine has a maintenance log, the log must be kept up to date.

The responsibility for the maintenance of hired machinery should be agreed between the hire company and the person responsible for hiring the equipment. Both parties should agree exactly what and when each is responsible for.

Specific risks

Wherever possible risks must be controlled by:

- *eliminating the risk*; or if that is not possible,
- taking 'hardware' measures such as the *provision of guards to control the risks;* but if the risks cannot be adequately controlled,
- taking 'software' measures such as *following safe systems of work* and the provision of information, instruction and training to deal with the remaining risk.

The use of machines should be restricted to people who are properly trained and have enough information and instruction, particularly where the machine is hand-fed. Machine safety requires the combination of the use of guards, protection devices and appliances, the selection of competent people to use the equipment and the following of safe working practices and systems of work. In addition appropriate precautions are required to avoid the risks associated with setting, adjusting, removing off-cuts, maintenance and the general cleaning of machines and their tooling.

Information and instructions should be made available to workers in an easily understandable form. This should cover the following points:

- all health and safety information relating to the use of the woodworking machinery;
- any limitations on the use of specific machines;
- any foreseeable difficulties that could arise;
- the methods to deal with these difficulties;
- any practical tips gained from experience of using the machinery.

Training should be provided for supervisors, machine operators, those who assist in machining processes and those who set, clean or maintain woodworking machines. All training schemes should include the following elements:

- *General.* Instruction in the knowledge and safety skills common to woodworking processes, including good housekeeping and an awareness of the dangers such as 'taking off', 'dropping on' and 'kickback'.
- *Machine specific.* Practical instruction in the safe use of a particular machine, to include:
  - the potential dangers of the machine and any limitations to its use;
  - the main causes of accidents and relevant safe working practices, including the correct use of guards, protection devices and appliances and the operation (where fitted) of a manual brake.
- *Familiarisation.* On-the-job training under close supervision.

Demonstrating competence. Employers need to be satisfied that their workers are adequately trained and can demonstrate competence in the work that they are required to undertake. Assessment of all workers should be carried out, including existing staff, people changing jobs, new employees and those new to wood machining.

Assessment of competence is best carried out in two stages before and after training: an initial assessment to identify the training needs of the individual followed by an assessment after training to measure its success, see Figure 6.4. Employers are recommended to keep a record of all training and assessments and periodically review them for each person. These records can be used as the basis for compiling a list of authorised machine operators. Figure 6.5 illustrates a typical format.

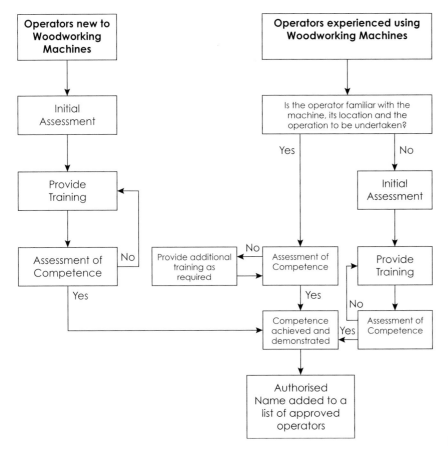

◀ Figure 6.4 Assessment of competence

A person demonstrates competence when the required knowledge and safe working practices are used consistently when working at the machine. A competent worker will be able to demonstrate:

1. Selection of the correct machine and tooling for the job, e.g. the worker has the competence to say 'This is the wrong machine for the job, it can be done more safely on a (name of alternative machine).
2. The purpose and adjustment of guards, protection devices and appliances.
3. A knowledge of safe methods of working including appropriate selection of jigs, holders, push-sticks and similar protection appliances.
4. A practical understanding of legal requirements, for example the need to provide and use guards, as well as their correct adjustment and positioning, etc.
5. Knowledge of the nature of wood and the hazards produced including dust, kickback, snatching and ejection.

237

List of Authorised Machine Operators:

The authorised woodworking machine trainer of BBS Joinery Co. is   **JOHN WILSON**

I certify that:

- I have carried out the training shown in the individual training records, for machines and individuals listed below.

- I am satisfied that these individuals have demonstrated competence in the operation of the machines as indicated and have achieved all the training objectives, including:
  1. Selection of the most suitable machine for the type of work to be undertaken;
  2. Purpose and adjustment of guards and other safety devises;
  3. Selection and use of safety devises, including push sticks, jigs and work holders;
  4. Practical understanding and application of legal requirements;
  5. Demonstration of safe working practices, including feeding, setting, cleaning and taking off

Signed: _J D WILSON_ (Authorised Trainer)

Operator's Name:	Type of Machine									
	Circular ripsaw	Crosscut saw	Dimension saw	Narrow bandsaw	Surface planing machine	Thickness planing machine	Single ended tenononer	Spindle moulder	Other:	Other:
DAVID SIMS	✓	✓	✓							
RAY SHUTER	✓	✓	✓	✓	✓	✓	✓	✓		
PAUL CHAMBERS	✓	✓	✓	✓	✓					

▲ **Figure 6.5** List of authorised machine operators

Refresher training. Training needs are likely to be greatest for new recruits, however, as skills decline if they are not used regularly, periodic refresher training should be provided for all workers. Refresher training is also particularly relevant where:

- a worker has not operated a particular machine for some time;
- the method of control of the machine has changed;

- new equipment or technology has been introduced;
- the system of work has been changed.

Risk assessment. Employers are required to take effective measures, in order to *prevent access to dangerous parts of machinery* or stop their movement before any part of a person enters a danger zone.

Risk assessment should be undertaken to identify the hazards presented by woodworking machinery. If a hazard presents a reasonably foreseeable risk of injury to a person, then the part of the machine that generates the risk is a 'dangerous part'. The purpose of risk assessment is to identify measures that can be taken to overcome the risks presented by hazards.

The following measures should be taken to prevent access to dangerous parts of a woodworking machine:

1. Use of fixed enclosing guards.
2. Use of other guards and/or protection devices such as interlocked guards and pressure mats.
3. Use of protection appliances such as jigs, holders and push-sticks, etc.
4. Provision of information, instruction, (re)training and supervision.

Protection against specified hazards. Employers are required to take adequate measures to control the risks from specific hazards such as 'kickback' that can result in the workpiece being ejected from the machine or disintegration of the tooling.

- **Kickback** is common and dangerous on hand-fed machines. It occurs when the tooling bites into the wood, causing the wood to be forcibly ejected from the machine often into the path of the machine operator. The use of power feeds, workpiece holders or jigs, spring loaded guards, sectional feed rollers, anti-kickback fingers and riving knifes, all reduce the risk of kickback.
- **Disintegration** of the tooling is minimised by the use of proprietary tools maintained in a good and sound condition. In addition the machine guard should be such as to provide sufficient protection against a disintegrating tool or cutter being ejected from the machine.

Other preventative measures include:

- **Stop controls.** Braking devices should be fitted to all woodworking machines that have a rundown time after switching off of more than 10 s. Existing machines should have braking devices fitted retrospectively and new machines built to European Standards should be equipped with an automatic brake to bring the machine to rest in less than 10 s.
- **Stability.** All machines other than those of a portable nature should be secured down to the floor, bench or similar fixture. This is in order to prevent the machine moving unintentionally during use and also to minimise the risk from noise and vibration.
- **Markings.** The safe working speed should be displayed or marked on the machine and tooling. Where it is not possible to mark the tooling, a table should be available to those who select or use the tool showing the tool speed range. The ACOP specifically mentions that circular sawing machines should be marked with the diameter of the smallest blade that should be used.
- **Warnings** should be displayed on work equipment where appropriate for reasons of health and safety. The ACOP specifically mentions that a warning notice must be displayed, stating that *only one workpiece at a time shall be fed into the machine* on a combined surface planing and thicknessing machine that is not equipped with sectional feed rollers or another anti-kickback device.

General requirements for woodworking machines

Woodworking machines often have high-speed cutters, which, owing to the nature of work they do, cannot be totally enclosed. It is therefore particularly important to maintain a clean and safe working environment that takes into account the capabilities and weaknesses of those working in it.

Use of safety appliances:

- The use of safety appliances such as push sticks and jigs, keep the operators hands in a safe positions, whilst maintaining full control of the workpiece during cutting operations.
- Power feeds reduce the need for hands to approach the cutters and should be used whenever reasonably practicable.

Machine controls:

- All machines should be fitted with a means of isolation from the electric supply, which should be located close to the machine.
- Lockable isolators can be used to prevent unauthorised use of a machine and give increased protection during maintenance.

Working space:

- Machines should be located in such a position that the operator cannot be pushed, bumped or easily distracted.
- There should be sufficient space around machines for items to be machined, for finished workpieces and for bins for waste materials without obstructing the operator.
- Wherever possible machine shop areas should be separated from assembly or packaging areas, also areas used by forklift trucks or other means of transportation.
- All access and escape routes must be kept clear.
- Waste bins should be emptied at regular intervals and waste sacks containing wood dust should be stored outside the workroom.

Floors:

- The floor surface of the work area must be level, non-slip and maintained in good condition.
- The working area around a machine must be kept free from obstruction, off-cuts, shavings, etc.
- Supply cables and pipes should be routed at high level or set below floor level, in order to prevent a tripping or trapping hazard.
- Polished floors should be avoided as they present a risk of slipping.
- All spillages should be promptly cleared up to avoid the risk of slipping.

Lighting:

- Machinists require good lighting (natural or artificial) in order to operate safely.
- Lighting should be positioned or shaded to prevent glare and not shine in the operator's eyes.
- Adequate lighting must also be provided for gangways and passages.
- Windows should be shaded when necessary to avoid reflections from worktables and other shiny surfaces.

Heating:

- Low temperatures can result in a loss of concentration and cold hands can reduce the operator's ability to safely control the workpiece.
- A temperature of 16°C is suitable for a machine shop.
- In a sawmill where heavier work is undertaken a temperature between 10°C and 16°C is considered suitable.

■ Where it is not possible to heat the entire area, radiant heaters can be provided near or adjacent to the work area, to enable operators to warm themselves periodically.

Dust collection:

■ Wood dust is harmful to health.

■ Woodworking machines should be fitted with an efficient means of collecting wood dust and chippings.

■ Local exhaust ventilation systems should be regularly maintained to prevent their efficiency deteriorating.

Training

■ No person should use any woodworking machine unless they have been properly trained for the work being carried out.

■ People under 18 years of age are prohibited from operating certain machines, unless they have successfully completed an approved training course.

Specific machine use and safety requirements

The safety requirements of the ACOP, *Safe Use of Woodworking Machinery*, are supported by a series of *Wood Information Sheets* published by the Health and Safety Executive. These contain practical guidance on the safe use of woodworking machines, which although not compulsory is considered to be good practice. This general guidance and that applicable to specific types of machines is illustrated and briefly described in the following section together with each machine's use and cutting principles.

Circular saw benches

The three main types of circular saw are:

■ **Cross-cut saw** used for cutting to length.

■ **Rip saw** used for cutting to width and thickness.

■ **Dimension saw** used for the precision cutting of prepared timber and sheet material.

There are two basic types of operation as seen in Figure 6.6. In the first, used for cross cutting only, the material which is being cut remains stationary on the table while the revolving saw is drawn across it. The second type, where the material is fed past the revolving saw, is suitable for both rip and cross-cutting.

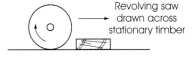

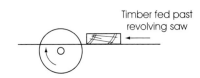

▲ **Figure 6.6** Circular saw (basic methods)

Cross-cut saw

The saw unit is drawn across the material cutting it to length (Figure 6.7). Adjustable length stops may be fitted where repetitive cutting to the same length is required. With most models it is also possible to carry out the following operations (see Figure 6.8):

■ cross-cutting

■ compound cutting

■ cutting birdsmouths

■ cutting housings

■ cutting notches

■ cutting halving joints

▲ **Figure 6.7** Cross-cut saw

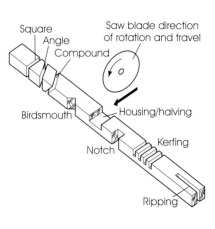

▲ **Figure 6.8** Cross-cut saw operations

- kerfing
- ripping (with riving knife fitted)
- trenching, tenoning and ploughing with special cutters.

*S*tarting, stopping and isolation controls. In common with most woodworking machines, the motor is controlled by a recessed start button and a mushroom-head stop button. When connected up, all machines should also be fitted with an isolating switch, so that the machine can be completely isolated (disconnected) from the power supply when setting, adjusting, or carrying out maintenance work on the machine (see Figure 6.9). The purpose of recessing the start button is to prevent accidental switching on. The stop button is mushroomed to aid positive switching off and should be suitably located to enable the operator to switch the machine off with their knee in an emergency.

In addition, to reduce the risk of contact with the operator during the rundown, machines should be fitted with a braking device that brings blades and cutters safely to rest within 10 s.

Rip saw

Two operations are involved in cutting timber to the required section (see Figures 6.10 and 6.11) as follows:

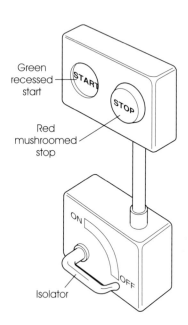

▲ **Figure 6.9** Start, stop and isolation controls

▶ **Figure 6.10** Rip saw

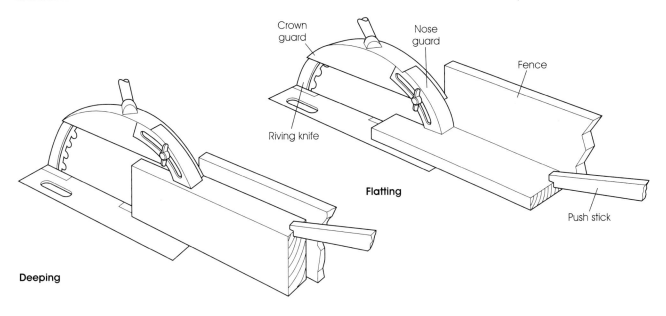

Flatting

Deeping

▲ **Figure 6.11** Deeping and flatting operations

■ Cutting the timber to the required width, which is known as **flatting**.
■ Cutting the timber to the required thickness, which is known as **deeping**.

Machine operators often make up their own bed pieces and saddles to enable them to carry out **bevel and angle ripping** as shown in Figure 6.12.

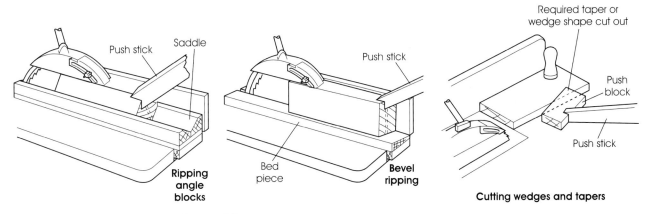

▲ **Figure 6.12** Bed pieces, saddles and jigs

Some saws have a recess on each side of the blade where it enters the table. These recesses are intended to receive felt packings and a hardwood mouthpiece. The packing helps to keep the saw cutting in a true line. The mouthpiece helps to prevent the underside of the timber breaking out or **'spelching'**.

When setting the machine up for any ripping operation, the fence should be adjusted so that the arc on its end is in line with the gullets of the saw teeth at table level. Binding will occur if it is too far forward and, if it is too far back, the material will jump at the end of the cut, leaving a small projection.

Dimension saw

This is used for cutting material to precise dimensions (see Figure 6.13). Most sawing operations are possible although on a lighter scale than the previous two machines.

▶ **Figure 6.13** Dimension saw

The cross-cut fence adjusts for angles and the blade may be tilted for bevels/compound cutting and can be moved up and down for trenching. The large sliding side table, used for cross-cutting also serves to give support when cutting sheet material.

Saw blades

A range of circular saw blades is available for various types of work. Figure 6.14 illustrates the section of two in common use:

- **Plate**, also known as **parallel plate** for straightforward rip and cross-cutting work.
- **Hollow ground** for dimension sawing and fine finished work.

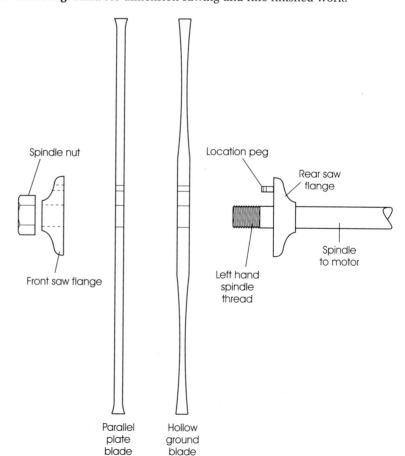

▶ **Figure 6.14** Circular saw blade section

Other section blades with a thin rim are available but have limited uses, for example, swage, ground-off and taper. They are used for rip sawing thin sections. Each has its own particular application, although the purpose of each is the same: to save timber by reducing the width of the saw kerf.

Saw teeth require setting so that the **kerf** produced (width of the saw cut) is wider than the thickness of the blade. Otherwise it will bind on the timber and overheat as a result of the friction, causing the blade to wobble and produce a wavy or 'drunken' cut.

The teeth can be set in two main ways as shown in Figure 6.15:

■ **Spring set teeth**, where adjacent teeth are sprung to the opposite side of the blade. This is the same method as that used for hand saws.
■ **Swage set teeth**, mainly used for setting thin rim rip saws. The point of each tooth is spread out evenly on both sides to give it a dovetail-shaped look.

Hollow ground and tungsten tipped saws do not require setting as the hollow grinding provides the necessary clearance or the tip side overhang respectively.

For efficient cutting the shape of the saw teeth must be suitable for the work being carried out (see Figure 6.16).

■ Rip saws require **chisel-edge teeth** which incline towards the wood (they have positive hook). Teeth for ripping hardwood require less hook than those for ripping softwood.
■ Cross cut saws require **needle-point teeth** which incline away from the wood (they have negative hook). The needle-point teeth for hardwood cross-cutting must be strongly backed up.
■ Dimension sawing ideally requires a **combination blade** of both rip and cross-cut teeth, although, as dimension saw benches are rarely used for ripping, a **fine cross-cut blade** is often fitted.

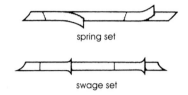

▲ **Figure 6.15** Types of saw blade setting

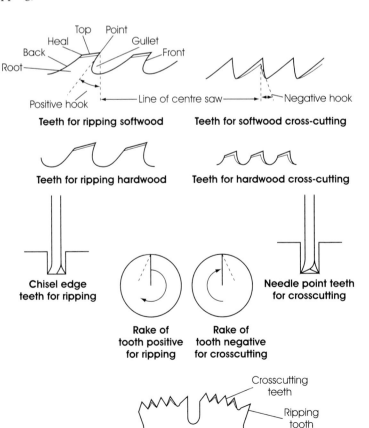

Combination teeth for dimension saw

◀ **Figure 6.16** Circular saw teeth

The use of wear-resistant tungsten carbide tipped teeth saws (Figure 6.17) is to be recommended when cutting abrasive hardwoods, plywood and chipboard. This reduces excessive blunting, extending the period before resharpening is required.

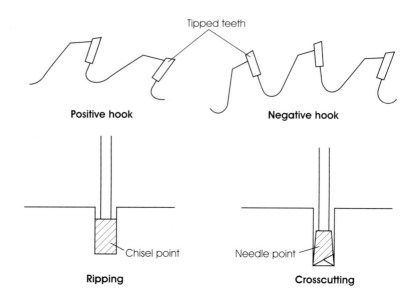

▶ **Figure 6.17** Tungsten carbide-tipped teeth circular saw

Saw blade maintenance

After a period of use, saw blades will start to dull (lose their cutting edge). This will progressively cause a poor finish to the saw cut including burning of both the timber and the blade and possibly cause blade wobble due to overheating. In addition, it will require excessive pressure by the operator to force the timber through the saw.

■ The sharpening of circular saw blades is normally carried out on a saw sharpening machine or by hand filing.

■ To ensure true running of a saw blade, it should be fitted in the same position on the saw spindle each time it is used. This can be achieved by always mounting the blades on the spindle with the location/driving peg uppermost and, before tightening, pulling the saw blade back onto the peg.

■ Resin deposits on saw blades should be cleaned off periodically. They can be softened by brushing with an oil/paraffin mixture and scraped off. A wood scraper is preferable, as it will avoid scratching the saw blade.

Safe use of manually operated cross-cut saws

The guidance applicable to cross-cut saws is illustrated in Figure 6.18 and summarised in the following points:

1. The non-cutting part of the blade must be totally enclosed with a fixed guard, which should extend down to at least the spindle.
2. Guards or a saw housing should be provided so that there is no access to the saw blade when in its rest position.
3. A nose guard should be fitted to prevent contact with the front edge of the blade during cutting and when the saw is at rest.
4. The maximum extension or stroke of the saw should be set so the nose guard cannot extend beyond the front of the saw table.
5. A braking device should be fitted to the machine that brings the blade to rest within 10 s, unless there is no risk of contact with the blade during rundown.
6. A fence is required on either side of the cutting line and should be high enough to support the timber being cut. The gap in the fence should be just sufficient to allow the passage of the nose guard. When straight

cutting on a machine that is capable of angled cuts, any excessive gap in the fence should be closed by the use of renewable fence inserts or false fence.

7. It is recommended that 'no hands' areas be marked in yellow hatching on the table 300 mm either side of the blade. Operators should be trained not to hold timber in these areas during cutting operations.

8. Workpiece holders or jigs should be used when cutting small workpieces or narrow sections.

9. Offcuts and woodchips should only be removed when the saw has stopped and is in the rest position; even then it is good practice to use a pushstick rather than the hands.

10. In order to reduce the likelihood of distorted timber binding on the saw causing kickback, any bow should be placed against the bed and any spring against the fence, with packers being used to prevent rocking.

11. Although some machines have the facility to turn the cutting head through 90º to allow rip sawing, a circular saw bench is considered a safer more suitable option.

12. Jigs and workpiece cramps should be used when undertaking operations such as trenching and the pointing of stakes or pales in order to provide workpiece stability and prevent kickback.

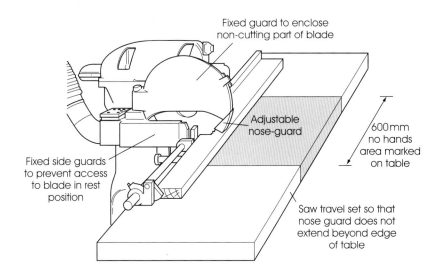

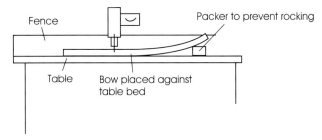

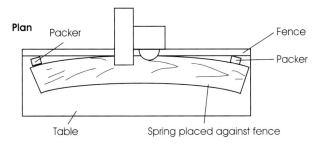

◀ **Figure 6.18** Safe use of manually opperated cross-cut saws

Safe working practices for circular saw benches

The guidance applicable to circular saw benches is illustrated in Figure 6.19 and summarised in the following points:

1. The part of the saw below the saw table must be fully enclosed.

2. In order to reduce the risk of contact with the moving saw blade during rundown a braking device must be fitted to the machine that brings the blade to rest within 10 s.

3. A riving knife must be fitted directly behind the saw blade. Its purpose is to part the timber as it proceeds through the saw and thus prevent it jamming on the blade and being thrown back towards the operator. Whenever the saw blade is changed the riving knife must be adjusted so that it is as close as practically possible to the saw blade and, in any case, the distance between the riving knife and the teeth of the saw blade should not exceed 8 mm at table level. The distance between the top of the riving knife and the top of the blade should be no more than 25 mm, except for blades over 600 mm diameter where the riving knife should extend at least 225 mm above the table. Riving knifes should have a chamfered leading edge and be thicker than the saw blade but slightly thinner than the width (kerf) of the saw cut.

4. The upper part of the saw blade must be fitted with a strong adjustable saw guard (crown guard), which has flanges on either side that cover as much of the blade as possible and must be adjusted as close as possible to the workpiece during use. An extension piece may be fitted to the leading end of the crown guard. This guards the blade between the workpiece and crown guard. If when cutting narrow workpieces the guard cannot be lowered sufficiently because it fouls a fixed fence, a false fence should be fitted. In all circumstances the extension piece must be adjusted as close as possible to the workpiece during use.

5. The diameter of the smallest saw blade that can safely be used should be marked on the machine. A smaller blade less than 60% of the largest saw blade for which the machine is designed, will not cut efficiently due to its lower peripheral (tip of teeth) speed.

6. Saw benches should be fitted with local exhaust ventilation above and below the table, which effectively controls wood dust during the machine's operation.

7. Where an assistant is employed at the outfeed (delivery) end of the machine to remove the cut pieces, an extension table must be fitted so that the distance between the saw blade spindle and the end of the table is at least 1200 mm. The assistant should be instructed to remain at the outfeed end of the extension table and not to reach forward towards the saw blade.

8. The operator's hands should never be in line with the saw blade or be closer than necessary to the front of the saw. A suitable push stick should be used in the following circumstances:
   - feeding material where the cut is 300 mm or less;
   - feeding material over the last 300 mm of the cut;
   - removing cut pieces from between the saw blade and fence unless the width of the cut piece exceeds 150 mm.

In order to reduce the risk of contact with the saw blade, it is recommended that a demountable power feed is used whenever possible. This is not a substitute for the riving knife, which must be kept in position at all times.

A fence should be used to give support to the workpiece during cutting. For shallow or angled cutting the normal fence may need replacing with a low fence to enable the use of a push stick or prevent the canted blade touching the fence.

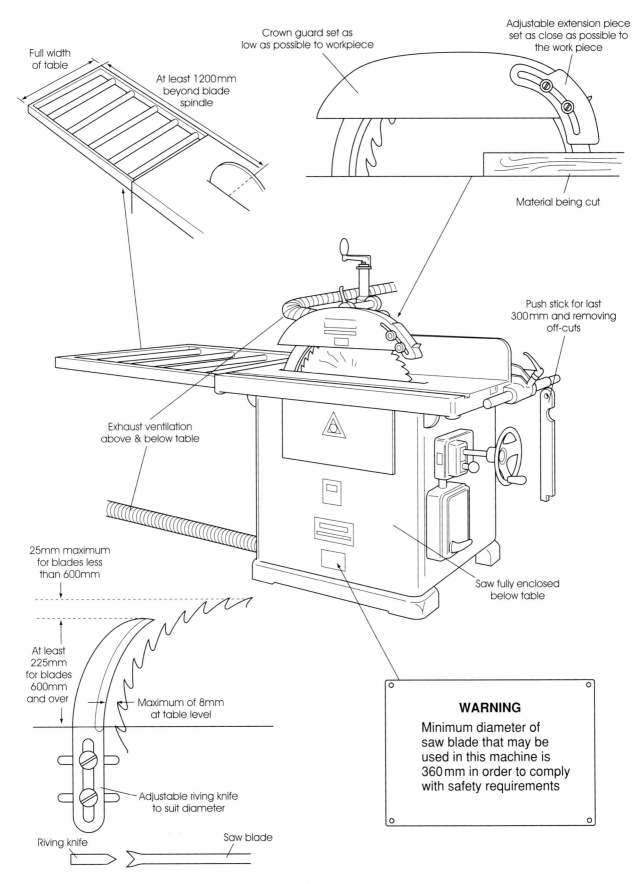

Full width of table

At least 1200mm beyond blade spindle

Crown guard set as low as possible to workpiece

Adjustable extension piece set as close as possible to the work piece

Material being cut

Push stick for last 300mm and removing off-cuts

Exhaust ventilation above & below table

Saw fully enclosed below table

25mm maximum for blades less than 600mm

At least 225mm for blades 600mm and over

Maximum of 8mm at table level

Adjustable riving knife to suit diameter

Riving knife

Saw blade

WARNING

Minimum diameter of saw blade that may be used in this machine is 360mm in order to comply with safety requirements

▲ **Figure 6.19** Safe working practices of circular saw benches

Circular saws must not be use for the following operations:

- Cutting tenons, grooves, rebates or mouldings unless effectively guarded. These normally take the form of Shaw 'tunnel-type' guards which, in addition to enclosing the blade, apply pressure to the workpiece, keeping it in place.
- Ripping unless the saw teeth project above the timber, e.g. deeping large sectioned material in two cuts is not permissible.
- Cutting round section timber unless the workpiece is adequately supported and held by a gripping or cramping device.

Angle cuts and bevels can be made on tilting arbour saws by inclining the blade; the fence should be set in the low position or a false fence used to prevent the rotating blade coming into contact with it. On non-tilting fixed position saws jigs can be used to provide workpiece support during cutting.

Narrow band saw

This machine, as its name implies, has a fairly long, endless narrow blade up to 50 mm wide (Figure 6.20). Its main function is for cutting curves and general shaping work, although it is also capable of ripping and bevel cutting.

Figure 6.21 shows a close-up view of the thrust wheel and guides, which are fitted above the table. There is also a similar arrangement below the table. The purpose of the thrust wheel is to support the back of the blade and stop it from being pushed back during the cutting operation. These should be set up approximately 1 mm away from the back of the blade when it is stationary. The guides are set up to just clear the blade. Their purpose is to stop any sideways movement of the blade and keep it running true on its intended path.

▲ **Figure 6.20** Narrow band saw

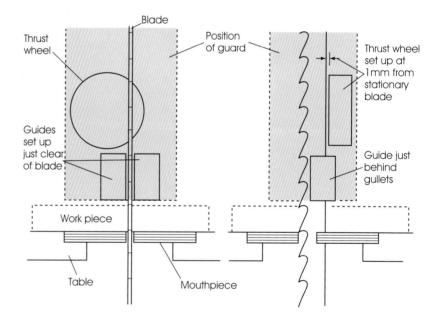

Front section   **Side section**

▶ **Figure 6.21** Sectional views showing thrust wheel and guides at table level

Also shown is the hardwood mouth where the saw passes through the table.

Various widths of band saw blades are available and in general the narrower the blade the tighter the curve that can be cut, i.e. widest blades for straight cuts and large sweeping curves and the narrowest blades for small radius curves as illustrated in Figure 6.22.

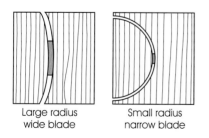

Large radius wide blade   Small radius narrow blade

▲ **Figure 6.22** Band saw widths

Band saw blade storage

The blades are supplied, and should be stored, in a folded coil consisting of three loops. After they have been replaced they should be refolded for safety reasons. The procedure to do this is illustrated in Figure 6.23:

- Wearing strong gloves, grip the blade in your outstretched hands with palms uppermost.
- The blade will twist when you simultaneously turn your hands inwards and over.
- As you lower the blade down to floor level it will fold into three coils.
- Store blade back in the box it was supplied in, or a purpose-made one.

Safety in the use of narrow band saws

The guidance applicable to narrow band saws is illustrated in Figure 6.24 and summarised in the following points:

- All moving parts must be totally enclosed with substantial guards with the exception of the cutting section from the top pulley down to the table. On new machines manufactured to European Standards, these guards should be interlocked with the motor to prevent it starting if any guard is not fully in place.
- An adjustable guard should be provided to cover the part of the blade between the top pulley wheel enclosure and the table. This guard should be attached to and move with the top blade guide.
- In use the guard and top blade guide should be adjusted as close as possible to the workpiece and firmly secured in position.
- The blade below the table must be guarded at all angles of table tilt.
- For efficient cutting the blade must be of the correct type, width and thickness; sharp and properly set; correctly tensioned and tracked.
- The blade will keep its condition longer if relaxed at the end of each work shift. A suitable notice e.g. 'Slack blade, re-tension before switching on' should be placed on the machine to remind the next user.
- The use of a demountable power feed and a fence is recommended when straight cutting, as it removes the risk to the operator.
- A fence should always be used when hand-feeding straight cuts in order to guide the workpiece. A wooden guide block should be used to hold the workpiece against the fence, whilst a pushstick is used for feeding close to the blade and the safe removal of off-cuts.
- When cutting without a fence (curved work) the hands should be kept as far away from the blade as possible. When hands approach the blade towards the end of a cut they should be placed on either side of the cutting line.
- For repetitive curved work the use of a guide pin running on a template fixed to the workpiece is recommended for increased safety.
- Bevel cutting can be undertaken by tilting the table and using the fence to support the workpiece. On fixed table machines a jig will be required for workpiece support. Push sticks should always be used towards the end of the cut.
- Simple tenons can be cut using a fence and backstop. For repetitive work jigs can provide a safer system of work.
- The use of a workpiece holder is recommended for the safe cutting of small wedges.
- The use of a jig is recommended when cutting circular discs.
- Round stock must be held in a suitable cramped jig when cross-cutting to prevent it from rotating with the cutting action of the blade.

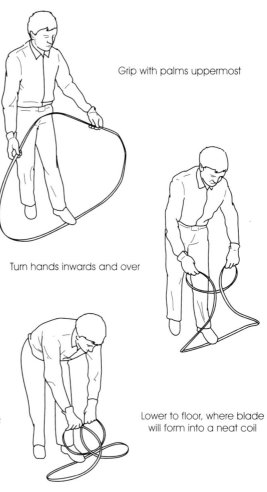

Grip with palms uppermost

Turn hands inwards and over

Lower to floor, where blade will form into a neat coil

▲ **Figure 6.23** Folding a band saw blade

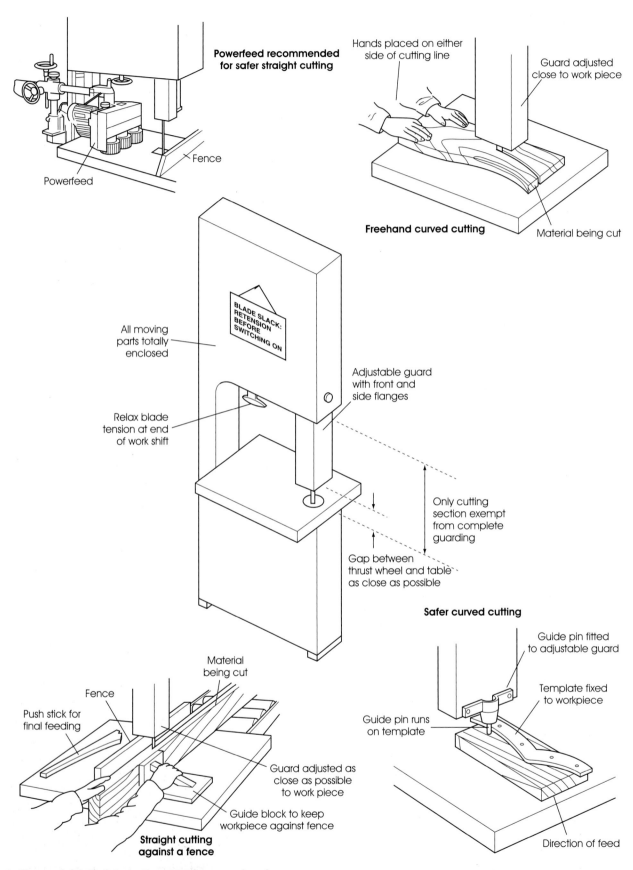

Powerfeed recommended for safer straight cutting

Powerfeed

Fence

Hands placed on either side of cutting line

Guard adjusted close to work piece

Freehand curved cutting

Material being cut

BLADE SLACK: RETENSION BEFORE SWITCHING ON

All moving parts totally enclosed

Relax blade tension at end of work shift

Adjustable guard with front and side flanges

Only cutting section exempt from complete guarding

Gap between thrust wheel and table as close as possible

Safer curved cutting

Guide pin fitted to adjustable guard

Template fixed to workpiece

Guide pin runs on template

Direction of feed

Material being cut

Fence

Push stick for final feeding

Guard adjusted as close as possible to work piece

Guide block to keep workpiece against fence

Straight cutting against a fence

▲ **Figure 6.24** Safety in the use of narrow band saws

Planing machines

The three main types of planing machine are:

- hand feed surface planer;
- power feed panel planer or thicknesser;
- combination hand and power feed planer.

With a block containing two cutters revolving at between 4000 and 6000 rpm, a series of cutter marks or ripples, two per revolution, are produced on the timber surface. The distance between the marks is called the pitch. True flatness of a surface is approached as the pitch is reduced to a minimum.

A slow feed speed produces a small pitch (larger number of short cutter marks) and, as a result, a smoother finish. A fast feed speed, on the other hand, produces a long pitch (small number of long cutter marks) which results in an irregular surface. This principle is illustrated in Figure 6.25. Acceptable cutter mark pitches for items of joinery work are:

> External and general – 2.5 mm and less
>
> High quality and cabinet – 1.25 mm and less

Much time in sanding and hand finishing can be saved if timber is planed at the outset to a high standard.

Hand feed surface planer

This machine has two main uses (Figure 6.26):

- **Surfacing.** To produce a smooth, flat and straight face side on a piece of timber.
- **Edging.** To produce a smooth, flat and straight face edge which is at right angles to the face side.

In addition this machine can also be used for bevel edging, by fitting a pressure (Shaw) guard and tilting the fence.

A section through a surface planer is shown in Figure 6.27. The outfeed table should be level with the top of the cutting circle of the block, while the infeed table is adjusted below the cutting circle to determine the depth of the cut.

The timber should be fed so that its grain runs with the cutter, otherwise the grain may be 'picked up' leaving a poor irregular finish. Distorted timber should be surfaced convex side up so that its ends, where possible, make contact with the table. Badly distorted timber should be cross-cut into shorter lengths before surfacing.

Power feed panel planer

This machine is also known as a **thicknesser** (Figure 6.28). Its purpose is to plane timber that has previously been surfaced and edged to the required width and thickness.

Combination hand and power feed planer

Combination machines combine the functions of both the surface and panel planer (Figure 6.29).

When planing timber to width and thickness on a panel or combination planer, the timber should be planed to width before being planed to thickness. This is so that the tendency for the timber to tip over in the machine and distort is reduced to a minimum.

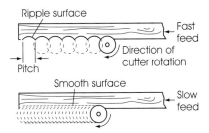

▲ **Figure 6.25** Effect of feed speed on surface finish

▲ **Figure 6.26** Hand feed surface planer

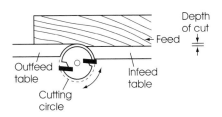

▲ **Figure 6.27** Section through surface planer

▶ **Figure 6.28** Power feed panel planer (thicknesser)

▲ **Figure 6.29** Hand and power feed combination planer

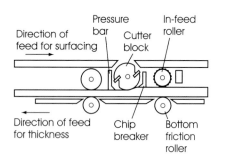

▲ **Figure 6.30** Section through a combination planer

A section through a combination planer is shown in Figure 6.30. A panel planer has the same design although without the top table. The timber is mechanically fed through the machine at a constant rate by the power-driven feed rollers, the first of which is serrated to grip the sawn surface. Either side of the cutter block is a spring-loaded pressure bar/chipbreaker to keep the timber in contact with the bottom friction rollers and provide a breaking edge for the wood chips.

Safe use of hand-fed planing machines

The guidance applicable to hand-fed planing machines is illustrated in Figure 6.31 and summarised in the following points:

■ There must be no access to the planing block or drive mechanism below the table.

■ The cutter block must be cylindrical and on machines post-1995 the maximum cutter projection should be no more than 1.1 mm.

■ Knives should be sharpened at regular intervals to reduce the risk of snatching, kickback and injury.

■ The gap between the cutting circle and the lips of the infeed and outfeed tables should be set as small as practically possible, generally 3 mm ± 2 mm. The gap between the two tables must also be kept as close as possible.

■ Every machine must be equipped with a bridge guard fitted centrally over the cutter block. This guard must be:
  – strong and rigid and made from a material that in the event of accidental contact with the cutter, neither the guard nor cutter will disintegrate;
  – constructed so that it cannot be easily deflected and expose the cutter block;
  – long enough to cover the cutter block with the fence at maximum adjustment – telescopic guards are available for large machines;
  – at least as wide as the diameter of the cutter block;
  – easily adjustable both horizontally and vertically without the use of tools.

■ In order to reduce the risk of contact with the cutter block during run down a braking device should be fitted to the machine that brings the blade to rest within 10 s.

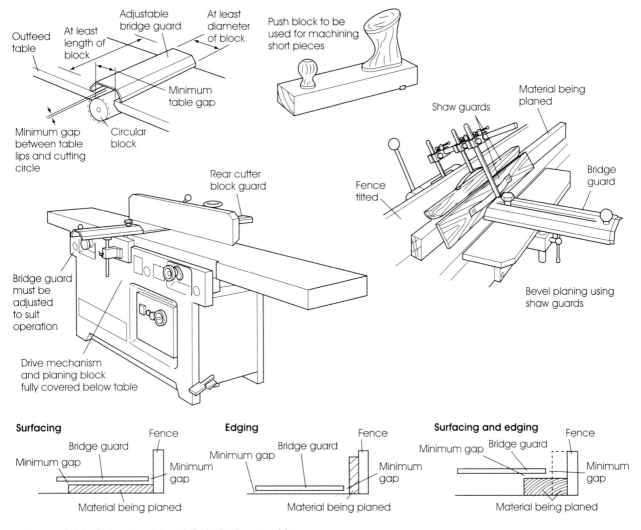

▲ **Figure 6.31** Safe use of hand-fed planing machines

- In use the bridge guard must be adjusted so that in all planing operations the gap between the guard, table, fence and workpiece is as small as possible.
- A guard must be fitted to the fence to cover the full width and length of the cutter block on the non-working side of the fence.
- A push block should be used when machining short lengths of timber. This is in order to reduce the risk of the workpiece dipping as it leaves the lip of the infeed table and snatching on the cutters causing kickback.
- Rebating or moulding etc. should not be carried out on a planing machine. New machines manufactured to European Standards are designed so that it is not possible to carry out rebating. Rebating may be permissible on pre-1995 machines if they are effectively guarded, normally by a Shaw guard, but the use of a properly guarded vertical spindle moulder presents a far lower safety risk.
- Bevel or angle planing can be carried out provided that:
  - the fence is tilted to support the workpiece;
  - the table gap on both sides of the fence is covered;
  - a tunnel (Shaw) guard is formed that prevents the operator's hands from coming into contact with the cutter block.
- It is recommended that the use of a demountable power feed be considered, as it removes the need for the operator's hands to approach the cutters, as well as overcoming the risk of kickback. However the use of a power feed does not remove the need for the guarding measures stated above.

Mortisers

There are two main ways of machining mortises (see Figure 6.32):

■ **Hollow chisel mortisers.** The mortise is cut by a hollow chisel, inside which an auger bit rotates. The auger bit drills a round hole, thereby removing most of the waste, leaving the chisel to square up the hole.

■ **Chain mortisers** use a cutting chain that runs around a guide bar. A chipbreaker is required on the upward running (exit) side of the chain to prevent the edge of the mortise breaking out.

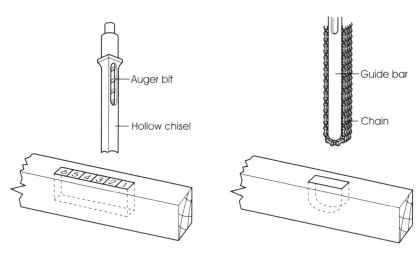

▲ **Figure 6.32** Machining mortises

▲ **Figure 6.33** Hollow chisel mortiser

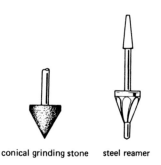

conical grinding stone   steel reamer

▲ **Figure 6.34** Means of sharpening hollow chisels

The chisel mortiser is to be preferred where neatly cut mortises are required. But in production situations, the chain mortiser is often used because of its ability to cut mortises at a much greater speed.

Combined machines are also available on which either cutting action can be selected as required.

Hollow-chisel mortiser

See Figure 6.33. Various sized chisels and bits are available, from 6 mm square up to 25 mm square. In order to accommodate this range of chisels and bits, different size collets must be fitted to both the chisel and bit, so that they can be tightened correctly.

When correctly set up, the bit should project 1 mm to 2 mm below the chisel. The machine edges of these chisels are sharpened by either a conical grinding stone, which is fitted to some machines, or a steel reamer used in a carpenter's ratchet brace. Figure 6.34 shows both the conical grinding stone and steel reamer for sharpening hollow mortise chisels.

Chain mortiser

See Figure 6.35. Various sized chains and guide bars are available to cut mortises in one penetration, from 4.5 mm wide and 18 mm long to 32 mm wide and 75 mm long. Most chain machines are fitted with a grinding attachment to facilitate the semiautomatic sharpening of the chain cutting edges.

The start and stop control of both mortising machines normally works in conjunction with the hand lever which starts the cutting action as it is pulled downwards and stops the cutting action when it is raised.

As with the hand method of mortising, through mortises are better cut from both sides in two operations, reversing the timber between them. When carrying out this operation, it is essential to keep the same side of the timber against the fence to avoid stepped mortises and twisted frames. For these reasons it is common practice to position the face side of the timber against the fence of the machine.

Note: Mortises are cut before tenons as chisels and chains are fixed sizes. A tenoning machine can be adjusted to fit a mortise, but not the other way round.

Tenoning machines

The cutting of tenons on a typical machine can utilise two cutting blocks, two scribing blocks and a cut-off saw as illustrated in Figure 6.36. The two main cutter blocks or tenoning heads are mounted one above the other on horizontal motors. Once these are set and revolving the timber is passed between them to form the tenon. The tenoning heads have vertical adjustment to vary the tenon thickness and horizontal adjustment to enable the cutting of tenons with unequal shoulders (long and short shoulders for rebated framing).

▲ **Figure 6.35** Chain mortiser

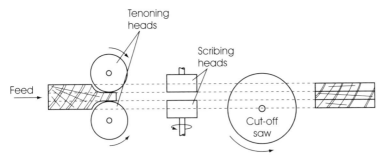

▲ **Figure 6.36** Section through tenoning machine

The tenoning heads are fitted with small shoulder cutters or spurs to cross-cut the shoulders just before the tenon cutters. These spurs give a clean shoulder without any breakout.

Scribing heads are vertically mounted behind the tenoning heads to form where required single or double through scribed shoulders.

Finally a cut-off saw is positioned at the rear of the machine to allow prepared tenons to be cut to the required length.

Various types of tenons that tenoners may produce are shown in Figure 6.37.

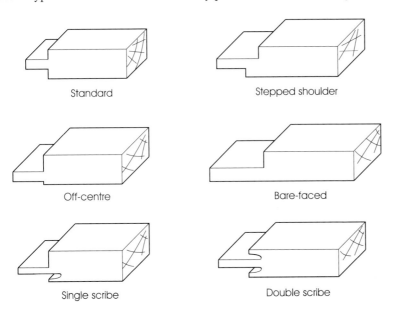

◀ **Figure 6.37** Various types of tenon

Single-ended tenoner

See Figure 6.38. Once the cutters have all been set up correctly the member can be placed face side down on the sliding table and secured after moving it into contact with the guide fence and shoulder length stop. The table is then moved forward by the operator so that the timber passes between the rotating cutter heads.

▶ **Figure 6.38** Traditional single-ended tenoner

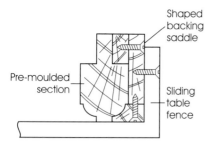

▲ **Figure 6.39** Use of backing saddle to prevent spelching

Further forward movement of the table is required for the scribing heads. Where the cut-off saw is required the table is pushed forward to the limit of its travel. The member must then be reversed and passed again through the machine in order to tenon its other end. After reversing the member ensure that its face side is still in contact with the table.

Double-ended tenoners. These machines process both ends of the member at one pass through the machine.

It is normal practice to mortise and tenon timber before any moulding operations have been carried out, although it is sometimes necessary to tenon premoulded sections. In these circumstances it is necessary to prepare and fix to the fence a shaped backing saddle or breakout piece. This must be the reverse profile of the moulding and prevents the member spelching out as the cutters leave it (see Figure 6.39).

Safe use of single-end tenoning machines

The guidance applicable to single-end tenoning machines is illustrated in Figure 6.40 and summarised in the following points. This information deals mainly with the upgrading of traditional hand-fed machines. New machines built to European Standards should be suitably equipped by the manufacturer.

- All machines should be fitted with a suitable braking device that will bring the machine to a safe stop.
- The tenoning and scribing heads should be enclosed to the greatest extent practicable. A combination of fixed and adjustable guards is

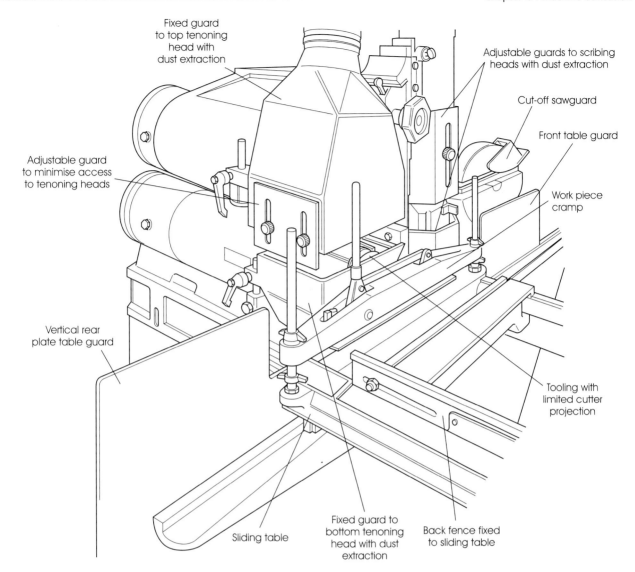

Fixed guard
to top tenoning
head with
dust extraction

Adjustable guards to scribing
heads with dust extraction

Cut-off sawguard

Front table guard

Adjustable guard
to minimise access
to tenoning heads

Work piece
cramp

Vertical rear
plate table guard

Tooling with
limited cutter
projection

Sliding table

Fixed guard to
bottom tenoning
head with dust
extraction

Back fence fixed
to sliding table

▲ **Figure 6.40** Safe use of 'traditional' single-ended tenoning machine

recommended. Exhaust ventilation hoods should be incorporated in the
fixed guards, to enable their positioning as close as possible to the
cutting operation for optimum performance. Openings through which
the workpiece has to pass should be kept as small as possible by the use
of adjustable guards.

■ Cut-off saws should be fully enclosed, ideally by spring-loaded or gravity-
fall guards that open only when the workpiece is being cut or
withdrawn.

■ Vertical plates should be fitted to the machining side of the sliding table
to prevent the operator's hands coming into contact with the cutters as
the workpiece is moved through the machine.

■ Fixed guards should enclose all belts, pulleys, chains, sprockets, gears
and revolving shafts, etc. Any access doors in them should be interlocked
to prevent the machine starting unless they are closed.

■ Third-party access around the sides and rear of the machine should be
deterred by an enclosure or barrier, unless this is already prevented by
walls, other fixed structures or other machines. Care should be taken
not to create a risk of being trapped between the barrier and the moving
table.

■ Only limited cutter projection tooling should be used.

▶ **Figure 6.41** Spindle moulder

■ All workpieces must be securely cramped to prevent movement during the machining process. Where more than one workpiece is machined at the same time additional cramps should be used.

■ The risk of the operator's fingers becoming trapped between the cramping device and the workpiece can be reduced by: using two-stage cramping; reducing the gap between the workpiece and cramp to less than 6 mm; guarding the cramping area.

■ The fence and breakout strip should be securely fixed to the sliding table. Where it is possible for them to come into contact with the tooling they should be made of a material that will not damage the tooling or cause additional risk.

■ Exhaust ventilation should be provided for the safe removal of wood waste and dust, and the area around the machine should be kept clear of loose chippings at all times. Where possible the extraction system should include a flexible hose for general machine cleaning.

Spindle moulder

This is a most versatile machine (Figure 6.41) capable of edging, rebating, grooving and moulding either straight or curved members. In addition the fitting of attachments to facilitate dovetailing, corner locking joints, stair string trenching and tenoning may increase this range of operations.

Basically the machine consists of a vertical spindle driven at its lower end by a belt from the motor (see Figure 6.42). Set spindle speeds from 3000 to 15,000 rpm are achieved by motor switching and moving the drive belt onto different diameter pulleys.

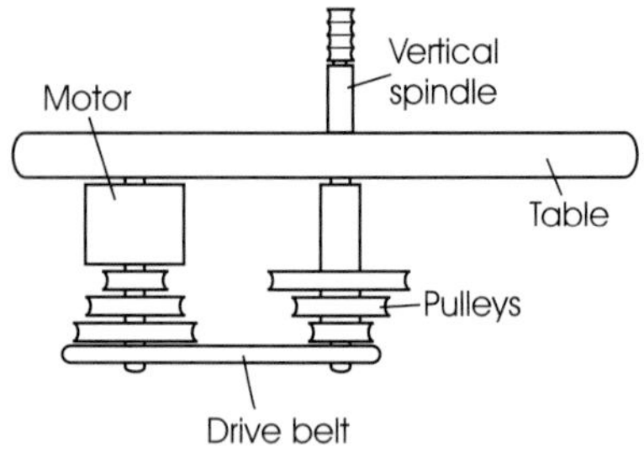

Cutter heads

The upper end of the spindle, which projects through the work table, is designed to accept various different cutter heads. These should be of the limited cutter projection type. Some older machines may have a slot in the spindle to accommodate cutters known as a 'French head'; these should no longer be used, due to the excessive cutter projection.

▲ **Figure 6.42** Section through vertical spindle moulder

With the exception of solid profile cutters, all other cutter set-ups require balancing before use. The greater the spindle speed the more crucial balancing becomes.

The correct speed to operate a spindle moulder is dependent on the type of cutter head and the work in hand, but in general, the larger the cutter head the slower the speed. Higher speeds are required for curved work than for straight work.

Excessive speed is potentially dangerous. Always consult the machine handbook and cutter manufacturer for specific information.

Fences

The standard straight fences and Shaw type pressure guards are suitable for use when running straight work. For curved work a ring fence and template is normally required. The template (usually 9-mm plywood or MDF) cut to the required shape is fixed to the member and kept in contact with a ring fence or other suitable guide while the member is worked. The template itself should have a 'lead in' on either end to enable safe working without the cutters snatching.

Safe use of vertical spindle moulding machines

The guidance applicable to vertical spindle moulding machines is illustrated in Figure 6.43 and summarised in the following points.

General safeguarding:

- In order to reduce the risk of contact with the cutter block during rundown a braking device should be fitted to the machine that brings the blade to rest within 10 s.

- Table rings should be used to reduce the gap between the spindle and table to a minimum. The use of rings reduces the risk of the workpiece dipping and snagging on the edge as it passes over the gap.

- The cutters must be enclosed to the maximum possible extent. A false fence should be used to minimise the exposure of moving parts. Jigs with suitable handholds must be used when the work in hand makes it impracticable to fully enclose the cutters.

- Stopped work where the material is fed in the same direction as the cutters (backfeeding), is not recommended, due to the risk of the cutter snatching. Even with the aid of a jig with suitable handholds it should be discouraged. Wherever possible the workpiece should be fed to the tool against the direction of spindle rotation.

Tooling:

- Only limited cutter projection tooling should be used. Detachable cutters and limiters must be of the correct thickness for the block in which they are used (thin cutters may become detached from their block in use with disastrous results).

- Only use tooling marked 'MAN' meaning manual or hand-feed, even if a demountable power feed unit is being used.

- The selected rotational speed of the machine must be appropriate for the tooling being used. The designed speed range should be marked on the cutter block.

Straight work:

- The cutters, cutter block and the spindle behind the fence should by fully enclosed by a suitable guard that includes provision for the connection of a dust exhaust outlet.

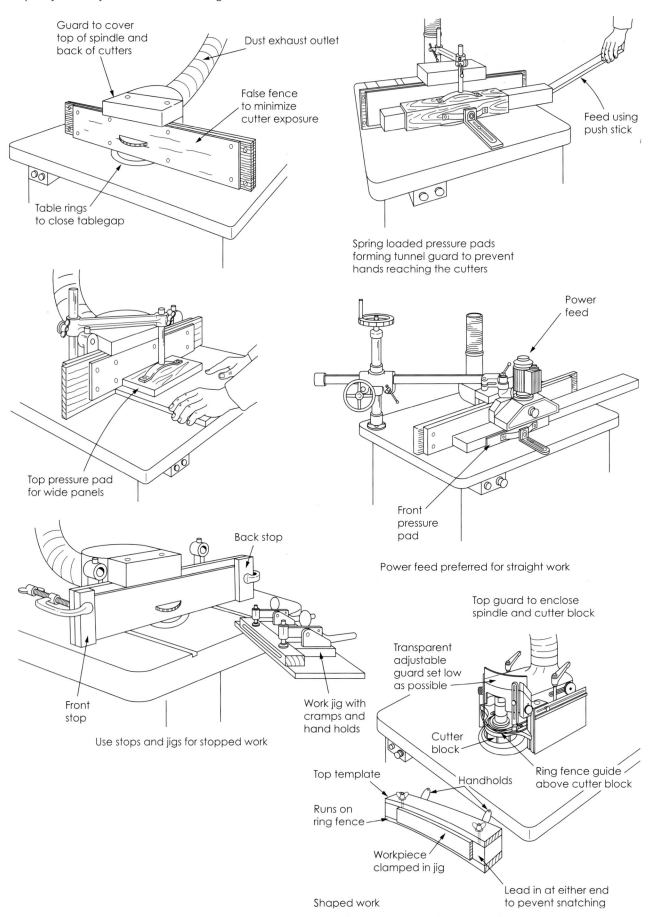

Guard to cover top of spindle and back of cutters

Dust exhaust outlet

False fence to minimize cutter exposure

Table rings to close tablegap

Feed using push stick

Spring loaded pressure pads forming tunnel guard to prevent hands reaching the cutters

Top pressure pad for wide panels

Power feed

Front pressure pad

Power feed preferred for straight work

Back stop

Front stop

Work jig with cramps and hand holds

Use stops and jigs for stopped work

Top guard to enclose spindle and cutter block

Transparent adjustable guard set low as possible

Cutter block

Ring fence guide above cutter block

Top template

Handholds

Runs on ring fence

Workpiece clamped in jig

Lead in at either end to pevent snatching

Shaped work

▲ **Figure 6.43** Safe use of vertical spindle moulding machines

- A false fence must be used to close the gap between the outfeed and infeed fences, so that only the part of the cutter that is cutting is exposed. This provides good workpiece support and prevents the workpiece dipping and snagging between the fences.

- Before the cutter is broken through a false fence the area should be protected by top and side pressure pads of sufficient size to prevent cutter access. Breaking through is achieved by either pushing/adjusting the fence assembly back onto the cutter or by raising the cutter up into the false fence, depending on the work in hand.

- Extension tables or roller trestles should be used to provide support for long lengths of timber.

Full-length straight cuts:

- Wherever possible a demountable power feed unit in conjunction with a side pressure pad should be used for straight cuts.

- When a power feed cannot be used, the cutting area should be enclosed by vertical and horizontal spring-loaded pressure pads that form a tunnel through which the workpiece can be safely fed with the aid of a pushstick.

- The pressure pads, normally made from hardwood, should be the same width and thickness as the workpiece and long enough to prevent the operators' hands reaching the cutters.

- The machining of wide thin panels may only require the use of a top pressure pad.

- To provide adequate support to heavily moulded thin workpieces a packer may be required on the out-feed side of the fence. This should be a mirror image of the moulding.

Straight stopped work:

- Where the cutters have to break into the solid face of the workpiece or break out before reaching the end, a suitable cramped jig with handholds should be used.

- Front and backstops fixed to the false fence allow greater control of the jig, give stability and prevent kick-back when 'dropping on'. The jig containing the workpiece is typically positioned at an angle to the fence with its end against the backstop, fed slowly onto the fence and into the cutters to break in, then fed forward along the fence up to the front stop to complete the cut.

Shaped or curved work:

- The straight fence is removed when setting up for shaped or curved work and a ring fence or guide, together with an adjustable guard, is fitted to enclose as much of the spindle and cutter block as possible.

- The use of a jig incorporating a template that runs on the ring fence/guide is required for all curved work. Where possible the ring fence/guide should be positioned above the cutters.

Machining faults: causes and remedies

The more common faults that occur when machining timber are listed in Table 6.1, along with their probable causes and suggested remedies.

Table 6.1 Machining faults: causes and remedies

Operations	Fault	Probable causes	Remedies
Ripping	The saw blade starts to wobble	Blade overheating because of: (a) too tight packing (b) dull teeth (c) insufficient set (d) abrasive timber (e) incorrect blade tension	(a) Reduce thickness of packing (b) Sharpen blade (c) Set teeth (d) Use tungsten-tipped blade (e) Replace blade
Ripping	The timber being sawn moves away from the fence or binds against the fence	(a) Fence not parallel to blade (b) Arc on fence not set in line with gullets of teeth	(a) Re-align fence (b) Adjust fence
Ripping	Rough sawn finish	Uneven setting or sharpening of teeth	Resharpen and reset correctly.
Ripping	The blade binds in the saw kerf	(a) Dull teeth (b) Insufficient set (c) Case hardening or twisted timber	(a) Sharpen blade (b) Set teeth (c) Avoid if possible. If not, use tungsten-tipped blade and feed slowly forward, easing back when binding occurs
Surface planning and edging	Tail end of board drops at end of surfacing or edging operation, leaving a dip in the end of the timber	Out-feed table set too low	Raise out-feed table

Dip in end

Operations	Fault	Probable causes	Remedies
Surface planning and edging	Timber will not feed onto out-feed table	Out-feed table set too high	Lower out-feed table

Timber will not feed

Operations	Fault	Probable causes	Remedies
Edging	Edge of timber is not planed square	Fence is not square to table	Adjust fence

Edge not square

Operations	Fault	Probable causes	Remedies
Surface planing, edging and thicknessing	Large number of cutter marks visible	(a) Timber fed too fast (b) Unequal projection of cutters, therefore only one cutting	(a) Feed more slowly (b) Set cutters.

Cutter marks visible

Operations	Fault	Probable causes	Remedies
Surface planing, edging and thicknessing	Poor irregular finish (grain picked up)	Timber fed against direction of grain	Reverse timber and plane with the grain

Poor irregular finish

Table 6.1 *continued*

Operations	Fault	Probable causes	Remedies
Thicknessing	Timber will not feed, or has difficulty in feeding	(a) Too much timber being planed off in one go (b) Resin build-up on rollers (c) Incorrectly adjusted feed or bottom rollers	(a) Reduce amount to be planed off (b) Clean off with solvent (c) Adjust rollers
Thicknessing	Bruise marks or scores on the face of the timber	(a) Resin build-up on rollers (b) Knots or splinters of wood wedged in gap between table and rollers	(a) Clean off with solvent (b) Remove knot or splinter
Thicknessing	Chatter marks or ripples appear at the end of the board being thicknessed	(a) Incorrectly adjusted rollers or pressure bar (b) Long board not being supported as it nears the end of its cut	(a) Adjust. (b) Support the end of the board

Chatter marks at end

Operations	Fault	Probable causes	Remedies
Mortising	Chisel overheats	(a) Not enough clearance between chisel and auger (b) Too much clearance between chisel and auger (c) Dull chisel and auger (d) Resinous timber clogging chisel and auger (e) Bent auger	(a) Reset clearance (1–2mm) (b) Reset clearance (c) Sharpen chisel and auger (d) Clean with solvent (e) Replace auger
Mortising	Mortise has zigzag edges	Chisel not set square to fence	Square chisel to fence

Zig zag edges

Operations	Fault	Probable causes	Remedies
Mortising	Stopped mortises and haunches have uneven bottoms	Too much clearance between chisel and auger	Reset clearance
Mortising	Chisel or auger will not fit into mortiser or will not tighten correctly	Incorrect size collets being used	Select correct size collets for chisel and auger
Chain mortising	Irregularly shaped mortise	(a) Chain too slack (b) Worn guide bar and rollers	(a) Adjust chain (b) Replace
Chain mortising	One edge of mortise breaks out	Incorrectly set chip breaker	Reset chip breaker
Chisel or chain mortising	Mortise not vertical	(a) Timber not seating correctly up against fence (b) Badly sharpened chisel with uneven points, causing it to pull to one side (c) Worn guide bar and rollers	(a) Ensure timber is correctly seated (b) Square up points and resharpen correctly (c) Replace

Mortise not vertical

Operations	Fault	Probable causes	Remedies
Band sawing	Rough sawn finish	(a) Uneven setting or sharpening of teeth (b) Timber being fed too slowly causing teeth to become dull	(a) Resharpen or reset correctly (b) Sharpen blade and feed timber faster

Table 6.1 *continued*

Operations	Fault	Probable causes	Remedies
Band sawing	Blade twists and wanders from vertical	(a) Incorrectly aligned or adjusted guides (b) Incorrect blade tension	(a) Reposition guides correctly (b) Adjust tension
Band sawing	Band saw breaks	(a) A too wide or narrow blade being used for work in hand (b) Offcuts wedged in mouthpiece (c) Incorrectly aligned or adjusted guides and thrust wheels (d) Cracked or badly jointed blade	(a) Use correct blade, wide for heavy work, narrow for small radius work (b) Renew mouthpiece (c) Reposition guides and thrust wheels correctly (d) Blade requires rejointing
Tenoning	Tenon shoulder profile reversed	Faced side reversed when turning member end to end	Always keep face side of the member in contact with the table
Mortising	Mortise stepped	Faced side reversed when turning member to end	Always keep face side of the member in contact with the fence

Workshop organisation and safety procedures

Small to medium joinery shops will normally contain the majority if not all of the following machines:

- Cross-cut saw
- rip saw
- dimension saw
- band saw
- surface planer ⎫
- panel planer ⎬ or a combined machine
- mortiser
- tenoner
- spindle moulder
- sanding machine(s).

The layout of a woodworking machine shop is most important to its efficient running. A shop must be planned to keep the timber moving, as far as possible, in a continuous flow from the timber store with the minimum of back tracking right through all the machine operations to the finishing, assembly, painting and dispatch areas.

Figure 6.44 illustrates a typical layout and workflow for a woodworking machine shop.

The movement of component parts from one machine or area to another is normally done with the help of trolleys. Component parts are taken off one trolley, passed through the machine, stacked on another trolley and then moved on to the next stage.

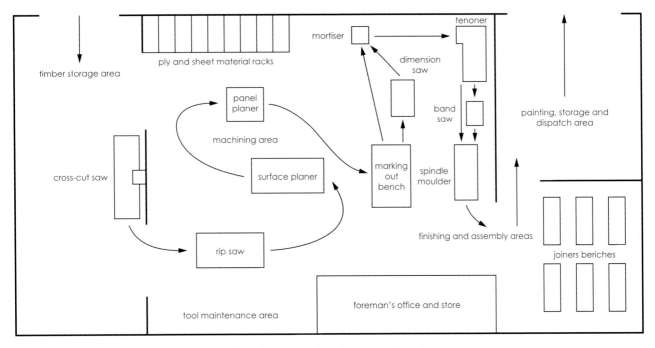

▲ **Figure 6.44** Typical layout and workflow for a woodworking machine shop

The workflow or stages for producing a typical item of joinery would be as follows:

1. Cross cut all timber to the required or manageable length in the timber store. This does away with the need to bring long lengths of timber into the shop.
2. Rip saw all the timber to the approximate section.
3. Machine face side and edge on the surface planer.
4. Plane to the required width and thickness on the panel planer (thicknesser).
5. Mark out the timber for joints and mouldings, etc., on the marking-out bench.
6. Cut sheet material to width and follow by fine dimensioning to length, on the dimension saw.
7. Cut joints on the mortiser and the tenoning machine.
8. Cut any curved work to the required shape on the band saw.
9. Run the required rebates, grooves and mouldings, etc., on the spindle moulder.
10. Pass all components to the finishing and assembly shop for joiners to finish, assemble and clean up the joinery items.

Hazards associated with wood dust

The wood dust created when machining timber can lead to the following health problems:

- Skin disorders: dermatitis, reddening of the skin, itching, swelling, dry flaking skin, blisters and weeping sores.
- Obstructions in the nose.
- Inflamed and watery eyes.
- Respiratory irritation: coughing, sneezing, runny nose, nosebleeds and asthma (severe breathing difficulties).
- Nasal cancer.

The most harmful effects are created by the fine airborne dust produced when sanding or sawing in confined, poorly ventilated areas, although the

Table 6.2 Toxic woods and their effects

	Asthma	Bronchial effects	Central nervous system effects	Coughing	Decrease in lung function	Dermatitis	Eye irritation	Headache	Mucous membrane irritation	Nettle rash	Nose bleeds	Reaction to light	Respiratory disorders	Runny nose	Skin irritation	Sneezing	Splinters go septic	Throat irritation	Vomiting
Abura																			*
Afrormosia			*												*			*	
Afzelia						*										*			
Agba														*					
Alder		*				*								*					
Ash				*															
Beech				*	*	*													
Birch						*													
Bubinga						*													
Cedar	*		*			*			*				*	*					
Chestnut						*													
Douglas Fir		*				*								*				*	
Ebony						*			*										
Gaboon	*			*		*	*												
Greenheart																	*	*	
Gum						*													
Hemlock		*											*						
Idigbo														*					
Iroko	*					*			*										
Larch						*			*										
Limba				*					*									*	
Mahogany						*			*				*						
Makore		*				*			*				*						
Mansonia							*			*			*		*			*	
Maple				*															
Meranti/Lauan															*				
Oak	*						*									*			
Obeche						*			*				*		*	*			
Opepe			*			*						*							
Padauk						*									*			*	*
Pine				*								*			*				
Poplar						*										*			
Ramin						*							*						
Rosewood						*							*						
Sapele															*				
Spruce												*	*						
Teak						*			*				*						
Utile															*				
Walnut						*								*		*			
Wenge			*			*												*	
Whitewood						*													
Yew						*		*											

severity of the effect will depend on the sensitivity of the individual concerned and the species of timber being worked. The common timbers that are known to have irritant properties are listed in Table 6.2, along with the type and degree of effect.

In addition to the health hazards associated with wood dust there are the following:

■ Wood dust on the floor results in slipping and tripping risks.

■ Wood dust in the air can impair the machine operators' vision.

■ Concentrations of fine dust particles in the air form a mixture that can explode if ignited. Explosions can occur in dust extraction equipment unless special precautions are taken. Such explosions can also dislodge any accumulated dust deposits on walls, floors, pipes and ledges, causing a secondary explosion.

■ Wood dust will also burn easily if ignited. Fires can start from a variety of heat sources, including overheating motors, pilot lights, sparking power tools, heating appliances and cigarettes.

Under the *Control of Substances Hazardous to Health (COSHH) Regulations* employers are required to:

■ Carry out an assessment of the health risks associated with wood dust and the actions required to prevent or control the risks

■ Prevent the exposure to wood dust, or where this is not reasonably practicable take measures to adequately control it.

Dust from both hardwoods and softwoods have been assigned maximum exposure limits (MELs) of $5\,mg/m^3$ over an average 8-hour period.

Since it is the dust that causes the problems, the most effective precaution is the use of local exhaust ventilation at each machine to reduce the dust level below the MEL. However, this is only part of the answer, as a certain amount of airborne dust cannot be avoided. Therefore the following precautions should be taken:

■ Maintain good levels of housekeeping; keep floors clear and free from dust and wood chips. Use vacuum cleaning equipment to clean walls, ceilings, pipes and ledges regularly to prevent dust accumulation.

■ Always use the appropriate PPE, dust mask or respirator, eye protection, overalls and barrier cream or disposable plastic gloves.

■ Use vacuum cleaning equipment to remove dust from clothing.

■ Thoroughly wash or shower as soon as possible after exposure to remove all traces of dust.

Note: Compressed airlines should not be used for general cleaning or removal of wood dust as it will create potentially explosive clouds and redistribute the dust.

Hazards associated with noise

Exposure to loud noise can permanently damage the hearing resulting in deafness or tinnitus (ringing in the ear).

The level of noise can vary from machine to machine depending on the timber being machined, the tooling being used, the machine set-up and the extraction system. Typical examples at machines where no noise reduction measures have been taken are:

97 dB(A) at sanding machines

100 dB(A) at spindle moulders

101 dB(A) when using portable power tools

102 dB(A) at circular saw benches

104 dB(A) at thicknessers.

Under the *Noise at Work Regulations* certain duties are placed on manufacturers, suppliers, employers and employees to reduce the risk of hearing damage to the lowest reasonably practicable level.

Manufacturers/suppliers. It is a requirement to provide low-noise machinery and to provide information to purchasers concerning the level of noise likely to be generated and its potential hazard.

Employers. It is a requirement to reduce the risk of hearing damage. Specific actions are triggered at daily personal noise exposure levels of 85 and 90 dB(A). At 85 dB(A) employers must:

- make a noise assessment to identify workers exposed and the actions to be taken;
- provide suitable ear protectors on request;
- provide information to employees about the risks to hearing and the legislation.

At 90 dB(A) employers must:

- reduce noise exposure as far as is reasonably practicable by means other than hearing protectors, e.g. by erecting acoustic enclosures around machines, changing to quieter tooling, modifying dust extraction equipment or changing patterns of work to ensure combinations of the most noisier machines are not used at the same time;
- designate the work area as an ear protection zone, by suitable signs indicating 'Ear protection must be warn';
- provide suitable ear protectors and ensure that all who enter the area wears them.

Employees have the duty to use the noise control equipment and hearing protection that is provided and to report any defects.

Hazards associated with vibration

Exposure to vibration can cause permanent damage to a machine operator or power tool operator. Effects include impaired blood circulation, damage to nerves and muscles and damage to bones in the hands and arms.

Hand-arm vibration syndrome (HAVS)

This is likely in any process where hands are exposed to vibrations from vibrating tools or workpieces. The effect of the vibration dose received by an operator over a day depends on:

- the vibration frequency;
- the duration of exposure;
- the exposure pattern;
- the grip and force required guiding the tool or workpiece.

Precautions should be taken where:

- Any tingling or numbness is felt after 5–10 minutes of use.
- High-risk operations are carried out, such as the use of hand-held sanders, hand-fed or hand-held circular saws, pneumatic nailing or stapling tools.

Employers should take preventive measures to reduce vibration, such as limiting the duration of exposure, the selection of suitable work methods, use of reduced vibration tools and appropriate PPE. In addition, health surveillance of people who are exposed to vibration should also be undertaken, especially where high levels or long durations are concerned.

Manufacturers have a duty to reduce vibration levels and supply vibration data for their equipment. Figure 6.45 illustrates a typical labelling system for use particularly on power tools, which give the operator guidance on vibration levels and recommended daily maximum duration of use.

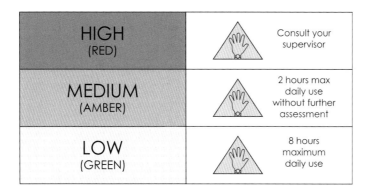

◀ **Figure 6.45** Manufacturer's labelling system to warn users of vibration levels

Maintenance requirements

In order to function efficiently and with safety, woodworking machines and portable power tools require adequate maintenance. If this is not carried out, the result will be premature wear, breakdowns and damage. This could be very costly in terms of repairs, lost production time and missed delivery dates in addition to the potential source of danger it creates for all workshop personnel. There are a number of maintenance techniques that can be used. However a combination of planned and condition-based maintenance is recommended as the most efficient.

Planned maintenance is where all of the equipment is checked and serviced at, say, weekly intervals. This might consist of general cleaning, oiling, greasing and adjustment, the checking and replacing as required of worn, strained or distorted parts. The main advantages of planned maintenance, also known as **preventative maintenance**, are:

- the reduction in repair costs, as minor faults can often be identified and remedied before they develop into major items;
- equipment lasts longer;
- increased production rates because there is less likelihood of equipment breakdowns.

Condition-based maintenance. During the planned maintenance inspection a note should be made of items which are worn due to normal wear and tear but which do not require immediate replacement. This is so that replacement parts may be obtained in advance and their fitting planned to be carried out at a convenient time. This may be the shutdown period, after normal work has finished or during a meal break, depending on the nature of the fault and its priority.

Breakdown, unplanned or emergency maintenance is where worn or damaged parts are only replaced when the equipment breaks down. This method is costly, dangerous and totally inefficient: the workshop simply muddles through, often in an unsafe way from one breakdown to another.

Undertaking machine maintenance

The manufacturer's maintenance schedule supplied with each machine gives the operator information regarding routine maintenance procedures. The schedule will detail the parts to be lubricated, the location of grease nipples and the type, frequency and amount of grease.

A typical procedure might be:

- Remove all rust spots with fine wire wool.
- Clean off resin deposits and other dirt, using an oil/paraffin mixture and wooden scraper.
- Wipe over the entire machine using clean rags.
- Apply a coat of light grade oil to all screws and slides. Excess should be wiped off using a clean rag.

- Clean off grease nipples and apply correct grade and amount of grease using the correct gun.
- Check freeness of all moving parts.
- Check worktables are smooth, free of obstruction and damage.
- Check all guards are freely adjustable over the full range and that they continue to fulfil their safety function.
- Check any mechanical feed systems for track and smooth operation.
- Check the operation of switches and other protection devices to ensure they are in effective working order.
- Check that all tooling is sharp and undamaged.
- Check that all tool holders and cramping systems function freely and safely.
- Check that all jigs, workpiece holders and pushsticks, etc., are fit for safe use and are stored to minimise the risk of damage.

Chapter Seven

Setting out, marking out and levelling

This chapter covers the work of site carpenters and bench joiners. It is concerned with the principles of setting and marking out in the workshop and setting out and levelling on site. It includes the following:

- design of joinery, including function, production and materials, aesthetics and anthropometrics
- internal and external building surveys
- setting and marking out for joinery
- setting out and levelling on-site.

Design of joinery

A large proportion of the joinery used in the building industry today is mass-produced by large firms who specialise in the manufacture of a range of items (doors, windows, stairs, and units) to **standard designs** and specifications. However, there is still a great need for independent joinery works to produce joinery for high quality work, for repair and replacement and one-off items to individual or **specialised designs.**

The design of this joinery is normally the responsibility of either an architect or designer, who should supply the joinery works with a brief consisting of scaled working drawings, full-size details and a written specification of their client's requirements, see Figure 7.1. However, these joinery details are often little more than a brief outline, leaving the interpretation of the construction details to the joinery works.

Often the best joinery is produced when the architect or designer discusses their design at an early stage with the joinery manufacturer, so that each can appreciate the requirements and difficulties of the other, and amend the design accordingly. This communication between the two parties enables the work to be carried out efficiently and therefore have a noticeable effect on the finished joinery item. In addition the joinery manufacturer may also be involved in the joinery design for small works directly with the customer when an architect or designer has not been employed.

Design considerations:

When designing and detailing joinery four main aspects must be taken into account:

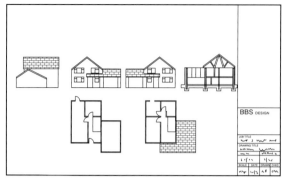

Scaled drawings

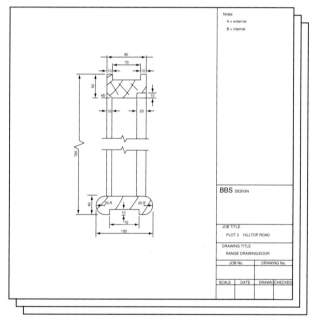

Joinery design details

Specification

▲ **Figure 7.1** The architect's or designer's brief to the joinery works

- function
- production and materials
- aesthetics.
- anthropometrics.

All of these design aspects are important although, depending on the nature of the work in hand, more or less priority may be given to any one aspect in order to achieve a satisfactory design.

Function

This aspect is the first to be considered and concerns the general efficiency of an item. The designer will consider this by asking himself/ herself a series of questions, such as:

- What are the main functions of the item, e.g. access, security, ventilation, seating, etc?
- Who will mainly use the item, e.g. adults, teenagers or small children, etc. Each will have a different size and user requirement.
- In what environment will it be used, e.g. temperature, humidity, weather, likelihood of vandalism, harsh treatment, etc?

■ What statutory regulations might affect the design, e.g. stairs, fire doors, etc?

An analysis of the answers to these questions will point to suitable materials and construction details, such as sizes and finishes, resulting in a satisfactory functional design. However this functional design may require further amendment after considering the production, aesthetic aspects and anthropometrics.

This analysis can be applied to even the most simple or common joinery requirements. Although this can be a mental process, Table 7.1 shows a typical checklist for use as the design is committed to paper.

Table 7.1 Functional design considerations

Joinery requirement	Kitchen/rear access door and frame to new house
Main functions	• Access from kitchen to garden for a disabled person (wheelchair access required) • Through vision • Daylight admission • Weather exclusion • Security
Special functions	• Wider than standard door width (to accommodate wheelchair) • Inward opening off ramp • Standard door height • Low-level glazing with clear glass (toughened) at line-of-sight when seated • Threshold flush with internal floor and outside ramp • Solid rebate frame • 5-lever security mortise lock fixed lower than standard height
Environment	• Domestic usage (may take additional abrasion at low level) • Adequate space both internally and externally • Aspect of door exposed to weather
Special features	• Hardwood door and frame for increased resistance to abrasion • Kick plates fitted to both sides • Water bar at threshold with the minimum projection • Additional weather strips fitted to frame • Canopy roof desirable over ramp
Statutory regulations	• See Building Regulations
Design	• This is mainly fixed by the above specification • Materials requirements • Production requirements • Aesthetic requirements

Production

Construction details should be designed not only to avoid unnecessary handwork, enabling the maximum possible use of machinery and power tools, but also to utilise the minimum amount of material to the best possible effect. This consideration is vital to the economic production of joinery.

Standard sections. The size and profile of a section will be determined by the functional considerations and the desired finished appearance. Figure 7.2 illustrates a range of standard joinery profiles, which can be economically produced by machine.

Figure 7.3 illustrates the typical application of various standard profiles to produce one section. This may be described in a specification as 'a three-times grooved, once chamfered, rebated, throated, weathered and pencil-rounded sill section'.

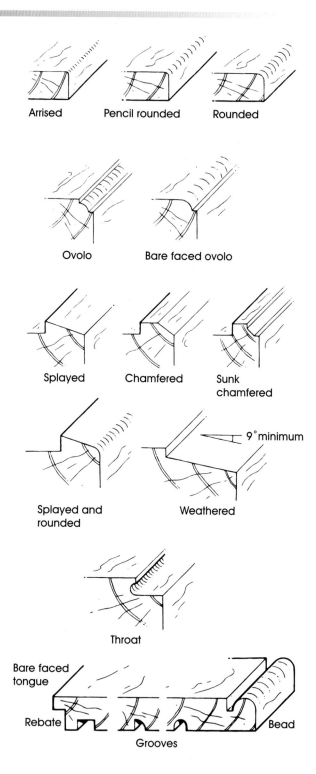

▶ **Figure 7.2** Standard joinery profiles

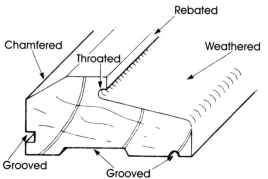

▶ **Figure 7.3** Application of standard profiles to sill section

The design and proportions of the profile should also take into account the type of joint to be used at intersections, as additional handwork can be involved at joints. Figure 7.4 illustrates a number of profiles and their suitability for machine scribed joints. Pencil-rounded and steeply chamfered profiles are best hand mitred as the razor edge produced by scribing them is difficult to machine cleanly and is easily damaged during assembly. These problems are avoided by the use of a sunk chamfer or ovolo profile. It is impossible to scribe bead or other undercut profiles, so hand mitring is the only option.

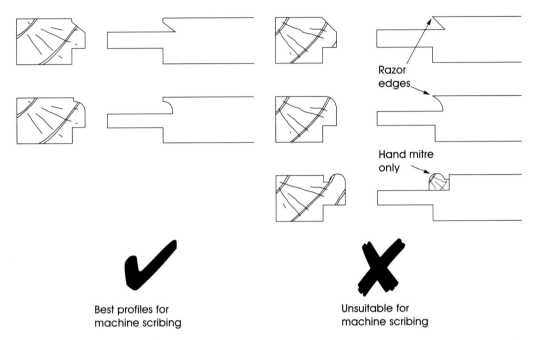

Best profiles for machine scribing

Unsuitable for machine scribing

▲ **Figure 7.4** Machine scribing of joints

Routing. An alternative to either scribing or hand mitring is the technique of routed profiles, see Figure 7.5, where rectangular sections are framed up and assembled, prior to being worked with a router.

Stopped rebates (see Figure 7.6) should be avoided whenever possible, as they are expensive to produce. This is because they require a separate machine operation, and also the curve left by the cutter on exit has to be squared by hand.

Materials. When considering both the functional design and production of joinery items an understanding of the nature of the material is essential. After a study of Chapter 1 you will realise that a careful consideration at the design and setting-out stages is required. Otherwise an item of joinery may quickly deteriorate, become unsightly or even become unfit for its purpose.

The **key factors** which must always be borne in mind are:
1. Select an appropriate species of timber.
2. Specify an appropriate method of conversion.
3. Specify an appropriate moisture content (this must be maintained during production and delivery).
4. Use construction details that both minimise effects of movement yet allow movement to take place without damage.

Aesthetics

This is concerned with the appearance or 'beauty' of an item and can thus be down to individual opinion. What is in good taste or acceptable to one person may be the complete opposite to another.

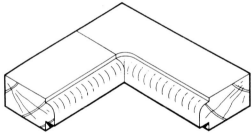

▲ **Figure 7.5** Routed profile applied after framing

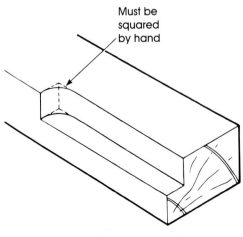

Must be squared by hand

▲ **Figure 7.6** Stopped rebate

The aesthetics of joinery is the province of the architect or designer who has a sensitive trained eye and can consider the complexities of proportion, shape, harmony, finish and compatibility to produce a design that will have the desired effect.

In addition the aesthetic effect can be considerably enhanced or marred in the joinery works by the degree of enthusiasm and craftsmanship exercised by the machinist and joiner during each stage of manufacture and installation.

Quality control

Many joinery works operate an internal quality control procedure, whereby the joiners are responsible for checking their own work on completion and attaching a quality label to it. Illustrated in Figure 7.7 is a typical quality label that identifies the job and the person undertaking the checks. In addition to the quality checks undertaken by the joiner, random sample quality checks are normally carried out by a foreman or manager on a daily or regular basis. This provides a further quality control measure.

A typical company's quality check sheet for joinery assembly work is shown in Figure 7.8. It contains guidance for making the quality check and includes a grading system, which can be used as statistical quality evidence.

BBS Quality Assured Joinery

Another Quality Joinery Product
Supplied by BBS. Tel. 01159434343

Part No. _____ Date. _____
Description. _____
Order No. _____
Checked by. _____

▲ **Figure 7.7** Quality check label

BBS Quality Assured Joinery

JOINERY STANDARDS

- ❑ All joinery, units and sub-assemblies to be clean and free of shavings etc.
- ❑ All items to be free of surplus glue
- ❑ All seen surfaces and edges to be free from scratches and blemishes
- ❑ All seen surfaces and edges to be free from machine marks and excessive sanding marks. No cross grain sanding marks on clear finished work
- ❑ All joints and mitres to be tight fitting
- ❑ All sharp arrises to be removed
- ❑ All glass, mirrors and other brightwork to be cleaned and smear-free
- ❑ All doors, drawers and other moving parts should operate smoothly
- ❑ All ironmongery should operate correctly
- ❑ All dimensions to be within + or – 1 mm
- ❑ All items to be square where applicable
- ❑ All finished items should conform to all details supplied
- ❑ All finished items should carry a completed quality label

Quality control check % rating

ELEMENT	CONFORMS		Non-conforming
	YES	NO	Reduction %
Overall dimensions and squareness			−10%
Joints and mitres			−10%
Door and drawer operation			−10%
Ironmongery operation			−5%
Surface and edge finish			−10%
Cleanliness			−3%
Correct to detail			−10%
Quality label			−3%
	Total Rating		%

Comments:

Inspected by: _____ Date: _____

▶ **Figure 7.8** Typical quality check sheet

Anthropometrics

This is concerned with the dimensions and needs of the human body or frame. If an item of joinery is to satisfactorily fulfil its function, it is essential that its ergonomics (shape and size) be related to the bodily characteristics of the intended users.

In normal circumstances it is rarely practical to produce an item of joinery to one specific individual's bodily characteristics. Instead standard anthropometrical data is used to suit as wide a range of people as possible. Figure 7.9 illustrates the more common, average body characteristics of adult males and females.

Although the average data can be used for many circumstances, in certain situations allowances must be made to cater for people who fall either side of the average. For example, where headroom is concerned the design should accommodate tall people (above average measurement), since short people, in these circumstances, are not at a disadvantage. On the other hand, where reach is concerned, short-armed people should be accommodated (below average measurement) since average and long-armed people are not at a disadvantage, as they will achieve the smaller reach with ease.

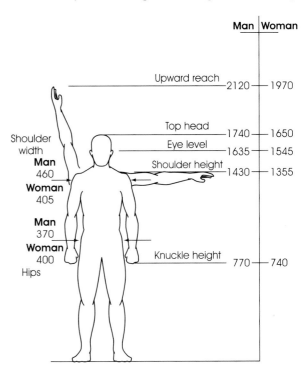

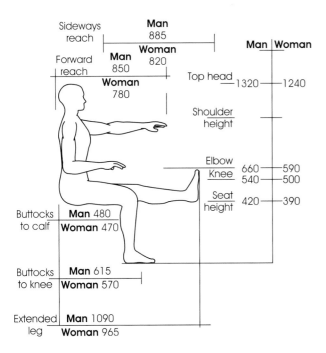

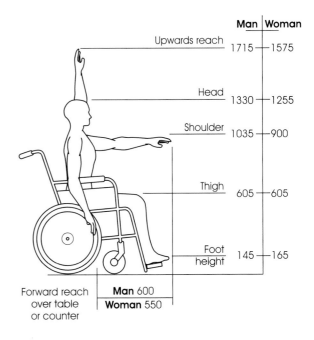

▲ **Figure 7.9** Average anthropometrical data (measurements in mm)

Apart from the obvious need for doors and stairs the main items of joinery that require the interpretation and application of anthropometrical data are: worktops, tables, seating, shelving and counters.

In addition, other fixed dimensions must be considered, particularly for storage compartments. For example, the space between the shelves of a bar back fitting must allow for the upright storage and easy removal of bottles, etc. Examples of **typical standard joinery items** along with suitable dimensions are illustrated in Figure 7.10.

Offices

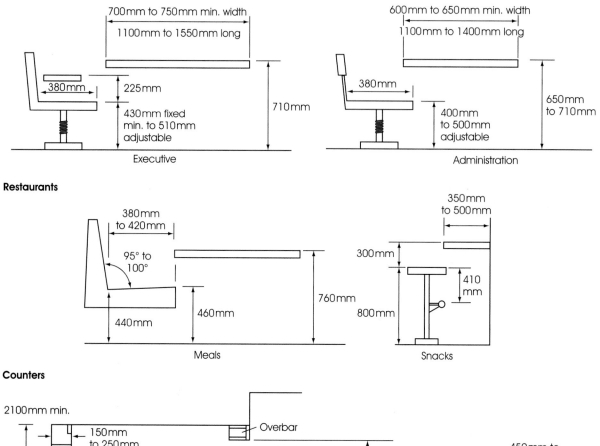

Executive

Administration

Restaurants

Meals

Snacks

Counters

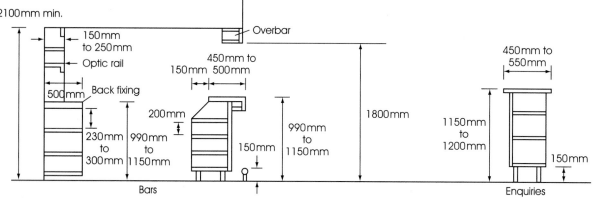

Bars

Enquiries

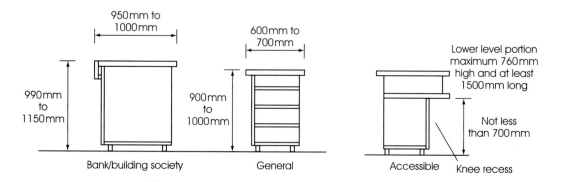

Bank/building society

General

Accessible   Knee recess

▲ **Figure 7.10** Typical joinery dimensions

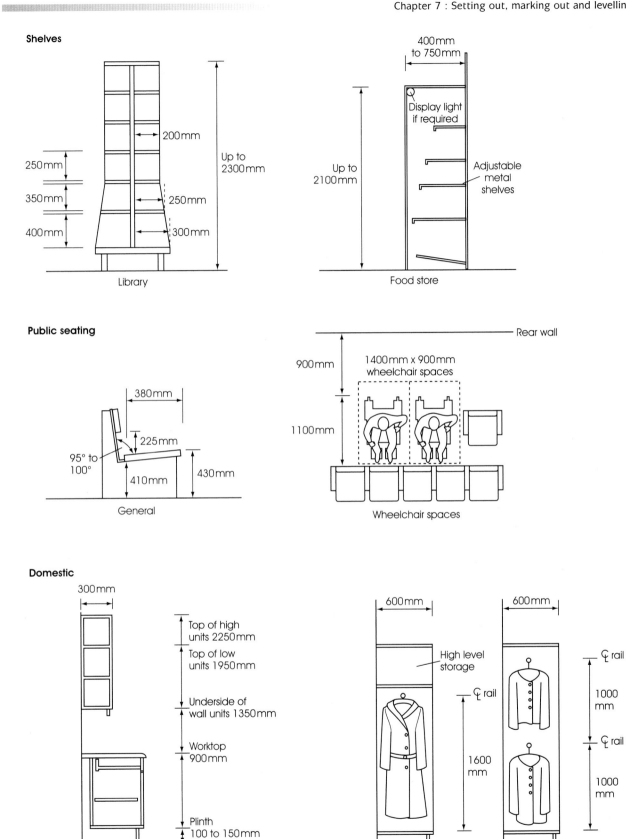

Shelves

Library

Food store

Public seating

General

Wheelchair spaces

Domestic

Kitchens

Bedrooms

▲ **Figure 7.10** *continued*

Building surveys

Before details for manufacturing can be drawn up, it is often necessary to make a site survey to check the actual measurements. It is preferable for this to be carried out by the joinery manufacturer's setter-out, since they will later use the information when setting out and deciding the allowances to be made for fitting and fixing.

Joinery items

Joinery items for existing buildings and rehabilitation work will always require a site survey, whereas the need for a site survey for joinery in new buildings will depend on the specification.

The two main methods of specifying joinery items are built-in joinery and fixed-in joinery.

- **Built-in joinery.** Where the joinery item is specified as 'built-in' or positioned during the main construction process, the work can normally be carried out directly from the architect's drawings and specifications without any need to take site dimensions. In many cases these may not even exist at the time.
- **Fixed-in joinery.** In cases where the joinery item is specified as 'fixed-in' or inserted in position after the main construction process, it is the joinery manufacturer's responsibility to take all measurements required for the item from the building and not the architect's drawings.

Extent of survey

The extent of the measurements and details taken during the site survey will depend on the nature of the work in hand. It can clearly be seen that the requirements of a survey for a small reception desk in a new building will be completely different to those of a survey for the complete refurbishment of an existing office block. The details taken may range from a single dimensioned sketch to a full external and internal survey of the whole building.

Survey procedure

Each survey is considered separately, sufficient measurements and details being taken in order to fulfil the survey's specific requirements. But a methodical approach is always required to avoid later confusion. The following survey procedures can be used to advantage in most circumstances. Refer to Chapter 8 for examples of measurement.

Drawings and equipment

Make enquiries to the building's owner and the local authority before surveying, to determine whether there are any **existing drawings** concerning the property. If so these can simplify the task by forming the basis of the survey sketches.

Equipment requirements will vary considerably depending on the survey requirements, but a list of basic equipment suitable for most tasks is as follows:

- A4 or A3 sketch pad, pens and pencils
- 30-m tape
- 2-m tape
- 2-m sectional measuring rod
- 1-m folding rule
- spirit level
- 2-m straightedge
- plumb bob.

In addition certain of the following items may be required for more detailed or specialist survey:

- stepladder
- extension ladder
- high-power torch
- moisture meter
- penknife
- camera
- binoculars
- moulding template
- hammer, bolster and floorboard saw.

Reconnaissance. Before the actual survey the building should be looked over, both internally and externally, to determine its general layout and any likely difficulties.

External survey

Sketch an outline plan and elevations of the building and then add the measurements. Wherever possible running dimensions are preferred to separate dimensions for plans (see Figure 7.11), since an error made in recording one separate dimension will throw all succeeding dimensions out of place and also make the total length incorrect.

- **Running dimensions** are recorded at right angles to the line; an arrowhead indicates each cumulative point. To avoid confusing the position of the decimal point, an oblique stroke is often used to separate metres and millimetres.

- **Separate dimensions** are recorded on the line; *arrowheads at both ends* indicate their extent. It is important that this distinction between the two methods is observed, as in certain situations it may be necessary to use both on the same sketch.

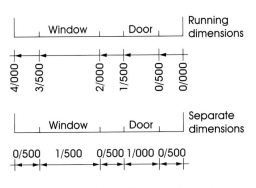

▲ **Figure 7.11** Outline dimensions

Running dimensions are taken in a clockwise direction around the building. Vertical dimensions on the elevations are taken from a level datum, often a damp-proof course. Typical external survey sketches are shown in Figure 7.12.

Estimated dimensions. Where measurements cannot be taken because they are inaccessible (gable end or chimney, etc.) they can be estimated by counting the brick courses and relating this to brickwork lower down that can be accurately measured.

Materials and finishes. All external details of materials and finishes, etc., should be recorded on the elevation sketches. Photographs of the elevations are often taken as a back-up to the sketches, especially where fine or intricate details have to be recorded.

Measuring openings. Where the survey is for a shop front, doors or windows, etc., accurate dimensions of the opening will be required (Figure 7.13). Vertical measurements should be taken at either end and at a number of intermediate positions. Horizontal measurements at the top and bottom are required. The diagonals should be taken to check the squareness and accuracy of the opening. The reveals should also be checked for plumb and straightness. In addition the head of the opening should also be checked for level and the slope of the pavement or exterior surface measured. The slope can be determined by means of a long straightedge, spirit level and rule.

Internal survey

Dimensioned sketches are made of each floor room, starting at ground-floor level (Figure 7.14). All rooms should be named or numbered and corridors lettered. These sketches can be traced from the external outline plan of the building. Measure through door or window openings to determine the thickness of the walls. Each floor plan should show a horizontal section

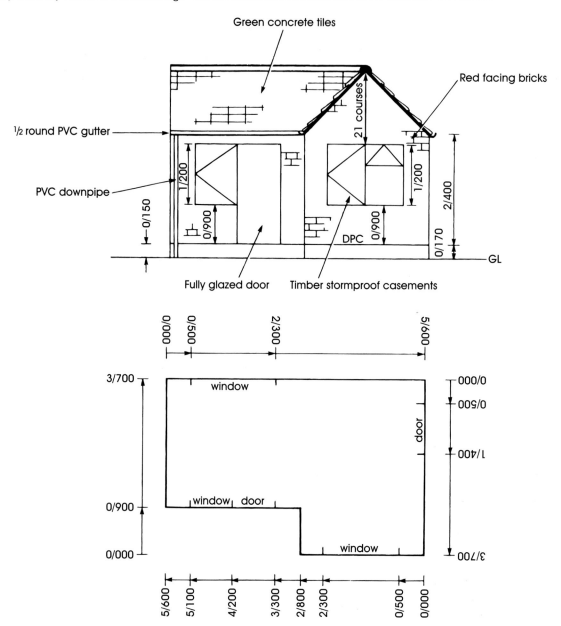

▲ **Figure 7.12** Typical external survey sketches

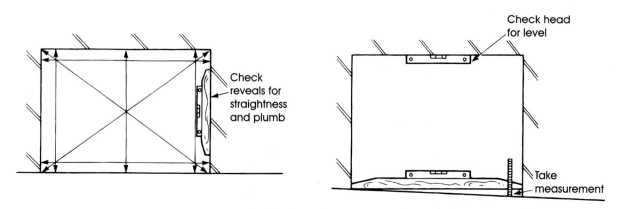

▲ **Figure 7.13** Measuring an external opening and checking slope of ground

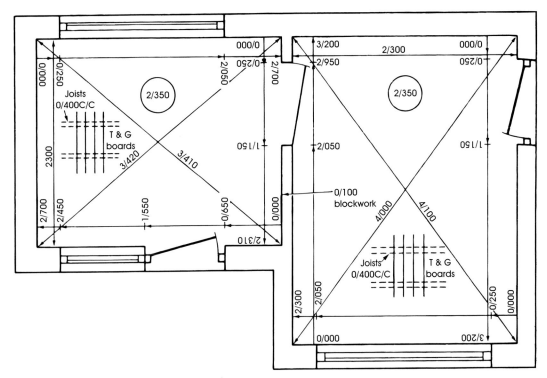

▲ **Figure 7.14** Internal survey floor plans

through the building about one metre above floor level. Measurements should be taken and recorded on the sketches in a clockwise direction around each room. Diagonal measurements from corner to corner check the shape of the room and enable it to be redrawn later.

Floor to ceiling heights are circled in the centre of each room. Floor construction and partition wall details are also shown on the floor plans. The floorboards run at right angles to the span of floor joists. The lines of nails indicate joist spacings. Pattern staining on walls and ceiling indicates positions of grounds or battening and ceiling joists.

The construction of walls can be identified by the sound made when tapped with the fist: brick walls sound solid, thin blockwork walls tend to vibrate, and stud walls sound solid over the studs and hollow between them.

Full size details

Where joinery items are to be repaired or replaced, full-size details of the sections and mouldings must be made to enable them to be matched later at the workshop. This task can be eased considerably by the use of a moulding template. The pins of the template are pressed into the contours of the moulding. It is then placed on the sketchpad and drawn around (see Figure 7.15). The exact location of where a moulding is taken should be noted, as these may vary from room to room.

Sketches of internal elevations or photographs may be required, especially where intricate details are concerned.

Vertical sections

Sketches of the vertical sections taken at right angles to the building's external walls complete the main sketches. A typical section is shown in Figure 7.16. Sections should include door and window heights, as well as the internal height of the roof. The thickness of upper floors and ceilings can be measured at the stairwell or loft trap door opening. Only details that can be seen and measured are sketched. No attempt should be made to guess details; therefore foundations, floor construction and lintels, etc., are not shown.

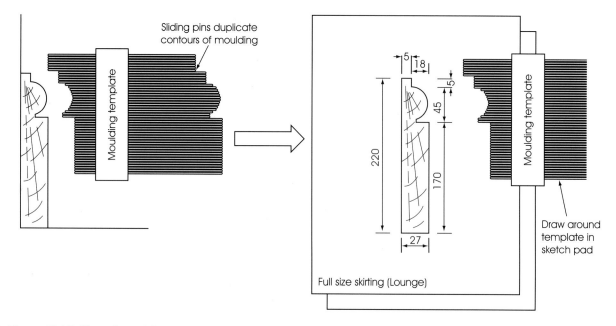

▲ Figure 7.15 Use of moulding template

▶ Figure 7.16 Vertical section
sketch

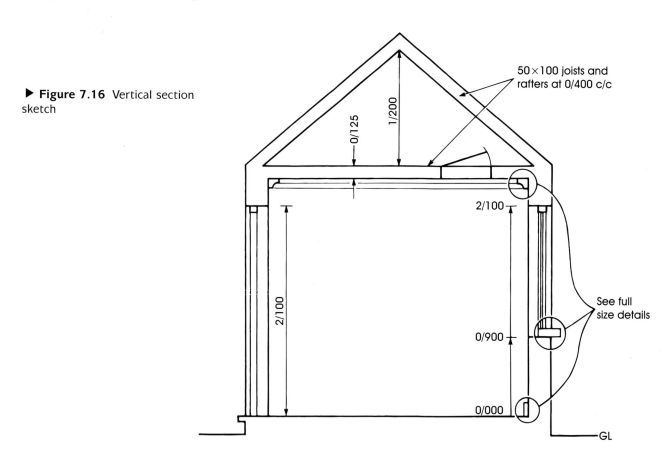

Stairs. Where a new flight of stairs is required the total rise and total
going should be measured. The total **rise** is the vertical distance from the
finished floor level at the bottom of the flight to the finished floor level at
the top. The total **going** is the overall horizontal distance of travel from the
nosing of the bottom step to the nosing of the upper floor or landing. Other
items to check are the length and width of the opening in the floor, the
position of the doorways at either end of the stairs and finally the floor level
(Figure 7.17).

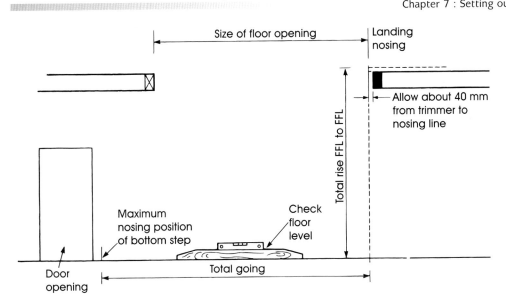

▲ Figure 7.17 Site measured for stairs

Internal openings, and areas or recesses for screens, partitions or fitments are measured in the same way as external openings.

Other details

Depending on the nature of the survey, service details such as outlets, sockets and switches/valves for gas, water, electric, television, telephone, etc., may be shown, although these are often recorded on separate **service plans** to avoid overloading and confusing the main floor plans.

In addition brief notes should be taken, recording details of structural defects and signs of decay and deterioration. This may entail lifting several floorboards and the partial removal of panelling or casings, etc.

Scale drawings

On returning to the workshop or office the sketches can be drawn up to produce a set of scale drawings and the brief notes can be used to form the basis of the survey report. It is at this stage that the necessity of taking all the dimensions and details is realised. One vital missing dimension can be costly as it will result in a further visit to the building at a later stage to take the dimension.

Joinery manufacturing operations and preparation for site work

The basic operations that are undertaken during the small to medium scale production of joinery items follow the traditional sequence of working by hand. However, machinists undertake most of the work, with the joiner only being involved at assembly. These operations are briefly described in Table 7.2. They are listed in workshop sequence and apply to traditional solid timber, framed joinery manufacture. Typical machines for small to medium joinery works are listed. Others including larger works will have additional or alternative machines, such as: a multi-head planer and moulder; a double-ended tenoning machine; or even CNC (computer numerical control) machines.

Table 7.2 Typical sequence of operations for a small joinery works

Operation	Description	Typical machine used
1. Setting out	The translation of design drawings and specification into production drawings, rods and cutting lists.	
2. Cross-cutting	The selection and cutting to length of timer shown on the cutting list.	Pull over cross-cut saw
3. Ripping	The cutting of listed timber to its sawn or 'nominal' width and thickness.	Circular rip saw bench
4. Surface planning	The accurate preparation of the face side and edge.	Surface planer or combination planer
5. Thicknessing	The preparation and reduction in size of the material to the required width and thickness.	Panel planer or combination planer
6. Marking out	The selection and marking out of timbers, to show the exact position of joints, mouldings, sections and shapes.	
7. Mortising	The cutting of mortises and haunches.	Chisel or chain mortiser
8. Tenoning	The cutting of tenons to suit mortises and the production of scribed shoulders.	Single-ended tenoner
9. Moulding	The running of mouldings and other sections.	Spindle moulder
10. Assembly	The fitting of joints, gluing, cramping, squaring and final cleaning up of an item of joinery.	

Setting out: workshop rods

Before making anything but the most simple, one-off item of joinery, it is normal practice to set out a workshop rod. **Setting out** is done by the setter-out, who translates the architect's scale details, specification and their own survey details into full-size vertical and horizontal sections of the item. In addition, particularly where shaped work is concerned, elevations may also be required.

Rods are usually drawn on thin board, thin plywood, white painted hardboard or rolls of decorator's lining paper.

When the job with which they are concerned is complete and they are no longer required for reference, boards may be planed or sanded off and used again. Plywood and hardboard rods may be painted over with white emulsion.

Although paper rods are often considered more convenient, because of their ease in handling and storage, they are less accurate in their use. This is because paper is more susceptible to dimension changes as a result of humidity and also changes due to the inevitable creasing and folding of the paper. In order to avoid mistakes, the critical dimensions shown in Figure 7.18 should be included where paper rods are used.

■ Sight size is the dimension between the innermost edges of the components; also known as daylight size as this is the height and width of a glazed opening, which admits light.

■ Shoulder size is the length of the member (rail or muntin) between shoulders of tenons.

■ Overall size (O/A) is the extreme length or width of an item.

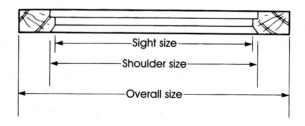

▶ **Figure 7.18** Critical dimensions

Where figured dimensions are different from the rod, always work to the stated size.

Rods for windows and doors: A typical rod for a casement window is shown in Figure 7.19. The drawings on the rod show the sections and positions of the various window components on a height and width rod. All of the component parts of the window can then be marked accurately from the rod. The rod should also contain the following information:

- rod number;
- date drawn;
- contract number and location;
- the scale drawing from which the rod was produced;
- the number of items required.

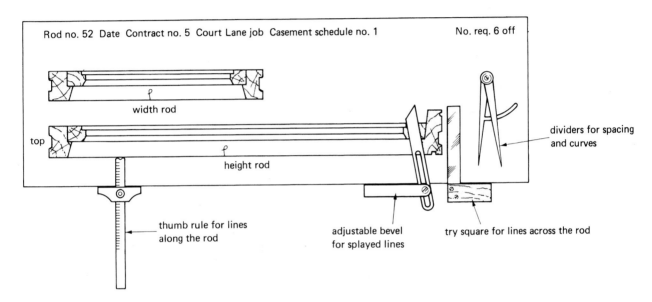

▲ **Figure 7.19** Casement window rod

The drawing equipment the setter-out will use to produce the rod is also shown in Figure 7.19:

- a thumb rule for lines along the rod;
- an adjustable bevel for splayed lines;
- a try square for lines across the rod;
- dividers for spacing and curves.

It is standard practice to set out the height rod first, keeping the head or top of the item on the left of the rod and the face of the item nearest the setter-out. The width rod is drawn above the height rod, keeping identical sections in line, see Figure 7.20.

Framed, ledged and braced door. A workshop rod for a framed, ledged and braced door is illustrated in Figure 7.21. In this rod the position of the mortises have been indicated by crosses, as is the practice in many workshops. The rear elevation has also been included for the joiner's information when fitting the braces.

Preparation stages. Figure 7.22 shows the three easy stages can be used to build up a detailed section. This method can be used when producing both workshop rods and scale drawings:

- Stage 1. The components are drawn in their rectangular sections.
- Stage 2. The square and rectangular sections are added.
- Stage 3. All other details are then added including hatching.

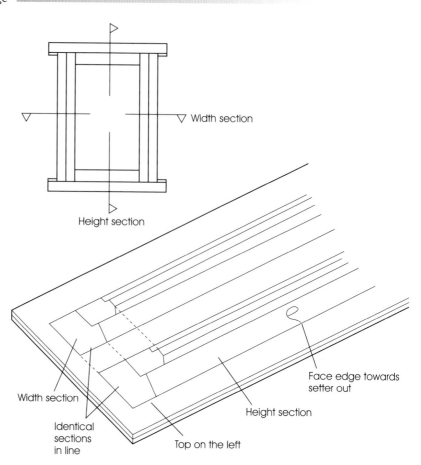

▶ **Figure 7.20** Layout of a workshop rod for a casement window

Width section

Height section

Face edge towards setter out

Height section

Width section

Identical sections in line

Top on the left

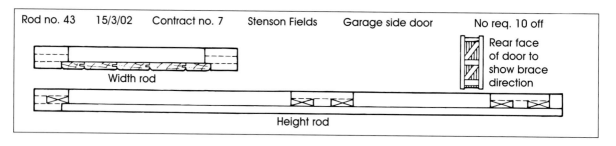

▲ **Figure 7.21** Rod for framed, legged and braced door

Rod no. 43   15/3/02   Contract no. 7   Stenson Fields   Garage side door   No req. 10 off

Width rod

Height rod

Rear face of door to show brace direction

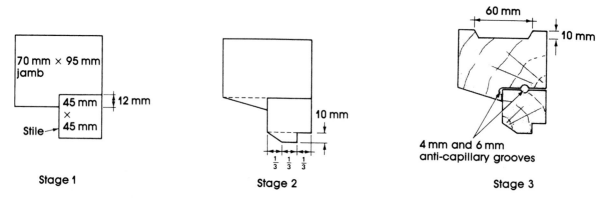

▲ **Figure 7.22** Building up a drawing

70 mm × 95 mm jamb

45 mm × 45 mm

12 mm

Stile

Stage 1

10 mm

⅓ ⅓ ⅓

Stage 2

60 mm

10 mm

4 mm and 6 mm anti-capillary grooves

Stage 3

It is good practice to keep the square and moulded sections the same depth. This eases the fitting of the joint, as the shoulders will be in the same position.

Sometimes it is not practical to set out the full height or width of very large items. In such cases the section may be reduced by broken lines and inserting an add-on dimension between them for use when marking out as shown in Figure 7.23.

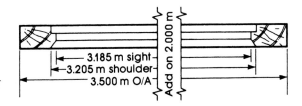

▲ **Figure 7.23** Add-on dimensioning

Detailing at openings When determining details for doors and windows the setter-out must take into account their opening radius. A number of applicable door and window sections are illustrated in Figure 7.24.

- Detail A shows that the closing edge must have a leading edge (bevelled off) to prevent jamming on the frame when opened.
- Detail B applies to the use of parliament and easy-clean hinges which both have extended pivot points. Here both the opening edge and frame jamb have been bevelled off at 90° to a line drawn between the pivot point and the opposite inner closing edge.
- Detail C shows how splayed rebates are determined for narrow double doors, bar doors and wicket gates, etc.

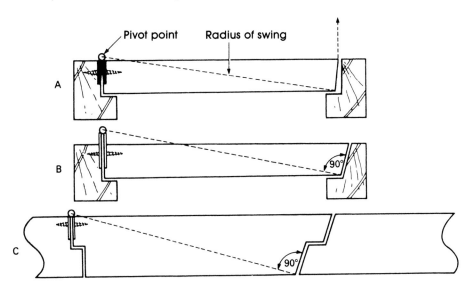

▲ **Figure 7.24** Opening details (doors and windows)

Glazed door with diminished stiles. Illustrated in Figure 7.25 is a workshop rod for a glazed door with diminished stiles and a shaped top rail. As the stiles section is different above and below the middle rail, two width rods are required. Also included in the rod is a half elevation of the curved top rail as its shape is not apparent from the sections.

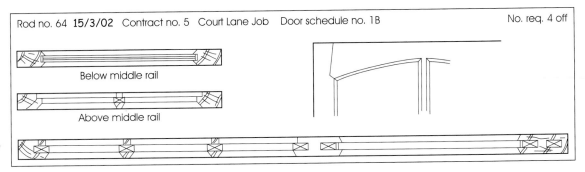

▲ **Figure 7.25** Rod for half-glazed door

Rods for tables and cabinets

Rods for three-dimensional joinery items such as tables and cabinets are drawn using broken details and add-on dimensions for one or more of the sections. This is in order to represent the three, often considerable, framed dimensions on a narrow rod.

Figure 7.26 illustrates a rod for a dining table. The length and height sections are shown in full along the rod, with the full width section being positioned under the length.

Figure 7.27 illustrates a rod for a cabinet (floor unit), to be constructed from melamine-faced chipboard (MFC) with hardwood trims. A circled detail (not part of the rod) shows that the carcass is rebated out and screwed together. The hardwood trims are glued and pinned in position after assembly to conceal the screw fixings. Also included on this rod are notes on assembly for use by the marker-out and joiners.

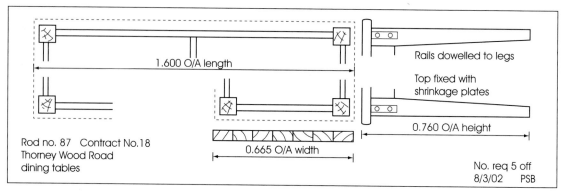

▲ **Figure 7.26** Rod for three-dimensional item (dining table)

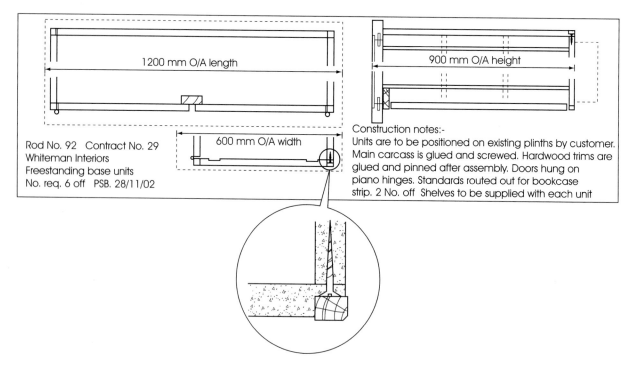

▲ **Figure 7.27** Rod for cabinet (floor unit)

Rods for stairs

These are often drawn showing the top and bottom details with the width section positioned between them (Figure 7.28).

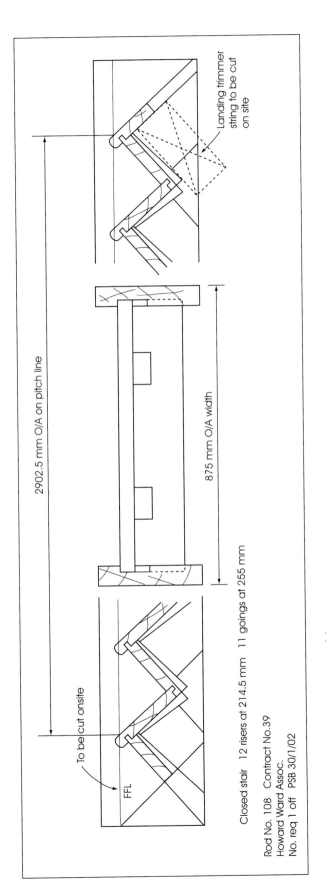

▲ Figure 7.28 Rod for a closed stair (no newels)

Closed stair 12 risers at 214.5 mm 11 goings at 255 mm

2902.5 mm O/A on pitch line

875 mm O/A width

To be cut onsite

FFL

Landing trimmer string to be cut on site

Rod No. 108 Contract No.39
Howard Ward Assoc.
No. req 1 off PSB 30/1/02

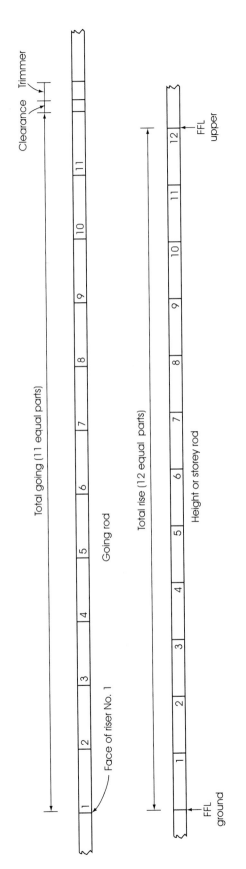

▲ Figure 7.29 Storey and going rods

Total going (11 equal parts)

Going rod

Face of riser No. 1

Total rise (12 equal parts)

Height or storey rod

Clearance Trimmer

FFL upper

FFL ground

Where newels form part of the construction these should be drawn to show the housings for the treads and risers, and mortises for the string. To accompany these details the setter-out should supply storey and going rods on timber battens, as shown in Figure 7.29. These are set out using the total rise and going of the stair taken during the site survey. Further information is not required, as the marker-out will make a set of templates from these details in order to complete the task.

Before setting out storey and going rods for a flight of stairs, it is necessary to determine the individual rise and going for each step.

- The **rise** of each step is determined by dividing the total rise (vertical measurement from finished floor to finished floor) by the number of risers required.
- The **going** of each step is determined by dividing the total going (horizontal measurement from bottom step nosing to landing nosing) by the number of treads required (one less than the number or risers).

These dimensions are controlled by the Building Regulations. Chapter 5, Stairs, in *Carpentry and Joinery Book 2*, contains details for typical situations.

Example

Internal flight for a dwelling house with a total rise of 2574 mm and a restricted going of 2805 mm, assuming that the minimum number of steps are required:

Minimum number of risers = total rise ÷ maximum permitted rise for location of stair (220 mm in this case) = 2574 mm ÷ 220 mm = 11.7, say 12 (each measuring less than the maximum permitted).

Individual rise = total rise ÷ number of risers = 2574 mm ÷ 12 = 214.5 mm.

Individual going = total going ÷ number of treads (always one less than the number of risers) = 2805 mm ÷ 11 = 255 mm.

Maximum pitch (in this case) = 42°. Draw rise and going full size and check the angle with a protractor, see Figure 7.30. In this case it measures 40°, which is permissible. Alternatively, trigonometry can be used to check the pitch (see Chapter 8, Applied calculations).

Where the pitch measures greater than 42° it can be reduced by introducing an extra rise and going, thereby slackening the pitch. However, this will increase the total going, so that stairs with restricted goings (such as those with a doorway at the bottom of the stairs) will require a total redesign in order to comply, possibly by the introduction of a landing to change the direction on plan.

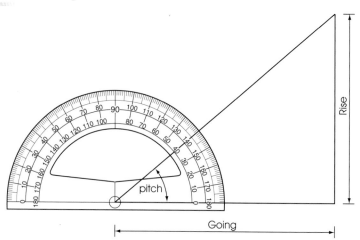

▶ **Figure 7.30** Use of protractor to check pitch of stairs

Cutting lists

When the rod has been completed the setter-out will prepare a cutting list of all material required for the job. The list will accompany the rod throughout the manufacturing operations. It is used by the machinists to select and prepare the required materials with the minimum amount of waste. The cutting list, or a duplicate copy, will finally be passed on to the office for job costing purposes.

There is no standard layout for cutting lists; their format varies between firms. However it is important that they contain the following information as a minimum:

- Details of the job, job title or description, date, rod and contract number.
- Description of each item.
- Quantity of each item required (No. off).
- Finished size of each item.

A typical cutting list for six casement windows is shown in Figure 7.31. The length of each item shown on the cutting list should be the precise length to be cut. It must include an allowance over the lengths shown on the rod for manufacturing purposes.

- Between 50 mm and 75 mm is the normal allowance for each horn on heads and sills, to take the thrust of wedging up and to allow for 'building in'.
- A horn of at least 25 mm is required at each end of stiles for both wedging and protection purposes.
- An allowance of at least 10 mm in length for rails that are to be wedged.

An alternative more detailed cutting list for the same six casement windows is illustrated in Figure 7.32. In addition to the previous cutting list, it also includes the following:

- An item number that can be crayoned on each piece to allow its easy identification during all stages of manufacture.
- The sawn sectional sizes of the items to simplify the timber selection, machining and final costing of the job.
- The type of material to be used for each item, e.g. softwood, hard wood or sheet material, etc.

Cutting list		
Rod no. 52 Date		Contact no. 5
Job title Casement window		
Item	No. off	Finished size (mm)
Frame:		
Jambs	12	70×95×1000
Head	6	70×95×700
Sill	6	70×120×700
Casement:		
Stiles	12	45×45×900
Top rail	6	45×45×500
Bottom rail	6	45×70×500

▲ **Figure 7.31** Cutting list

Cutting list					
Rod no. 52		Date		Contract no. 5	
Job title		Casement window			
Item no.	Item	No. off	Finished size (mm)	Sawn size	Material
	Frame:				
1	Jambs	12	70×95×1000	75×100×1000	Redwood
2	Head	6	70×95×700	75×100×700	Redwood
3	Sill	6	70×120×700	75×125×700	Oak
	Casement:				
4	Stiles	12	45×45×500	50×50×500	Redwood
5	Top rail	6	45×45×900	50×50×900	Redwood
6	Bottom rail	6	45×70×500	50×75×500	Redwood

▲ **Figure 7.32** Detailed cutting list

Computer-aided setting out

Many medium to large joinery firms now utilise computers as an aid to producing workshop rods. Setters-out input the job details into a CAD (computer-aided design) software programme, to produce both scale and full-size drawings/rods. These can be printed out on a plotter for workshop use. The use of CAD in this way has the added advantage for specifying future jobs. Standard details can be stored in the computer for a range of commonly produced joinery items requiring only dimensions to be added/adjusted in order to customise the rod before printing out for each new job.

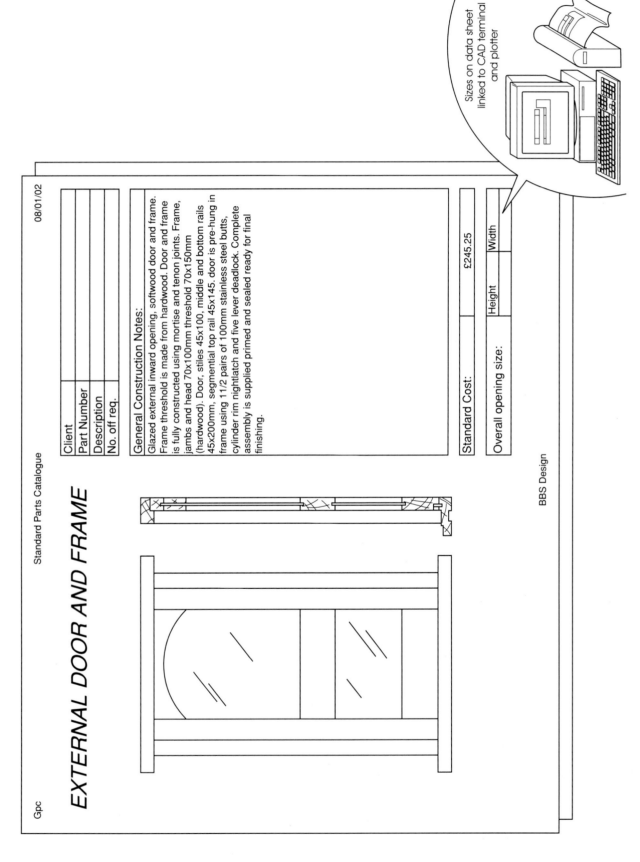

Gpc

Standard Parts Catalogue

08/01/02

EXTERNAL DOOR AND FRAME

| Client |
| Part Number |
| Description |
| No. off req. |

General Construction Notes:

Glazed external inward opening, softwood door and frame. Frame threshold is made from hardwood. Door and frame is fully constructed using mortise and tenon joints. Frame, jambs and head 70x100mm threshold 70x150mm (hardwood). Door, stiles 45x100, middle and bottom rails 45x200mm, segmential top rail 45x145. door is pre-hung in frame using 11/2 pairs of 100mm stainless steel butts, cylinder rim nightlatch and five lever deadlock. Complete assembly is supplied primed and sealed ready for final finishing.

| Standard Cost: | £245.25 |

| Overall opening size: | | |
| Height | Width |

Sizes on data sheet linked to CAD terminal and plotter

BBS Design

▲ **Figure 7.33** Standard data sheets

Data sheets. Illustrated in Figure 7.33 is a typical joinery shop standard data sheet for an external glazed door. The computer software has been developed so that on inputting the required sizes, the rod will automatically be adjusted to suit and can then be printed off on the plotter. In addition the computer can also be programmed to produce cutting lists at the same time.

Timber selection

Radial sawn sections are normally preferred for joinery, as these remain fairly stable after conversion, with little tendency to shrink or distort. However, for clear finished work this factor may take second place, as the important consideration will then be which face of a particular timber is the most decorative.

Grain and defects

The timber's grain and defects must be considered when marking out and machining. Careful positioning of a face mark (the leg of the mark points towards the face edge) may allow defects such as knots, pith and wane, etc. to be machined out later by a rebate or a moulding as shown in Figure 7.34. Knots and short graining should be avoided, especially near joints or on mouldings. All defects should be positioned so that they will be on the unseen face when the item has been installed.

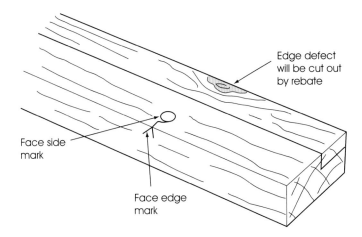

Edge defect will be cut out by rebate

Face side mark

Face edge mark

◀ **Figure 7.34** Positioning of face mark

Grain direction must be considered also when marking up faces, in anticipation of later machining operations. For the best finish, diagonal grain should slope into the cutter rotation as shown in Figure 7.35. The grain will tend to break out if marked the other way.

The visual effect of grain direction and colour shading for painted work is of little importance, but careful consideration is required for hardwood and clear finished joinery.

Decorative matching

The aim is to produce a decorative and well-balanced effect. Figure 7.36 illustrates a pair of well-matched panel doors. The meeting stiles have been cut from one board so that their grain matches. Likewise, adjacent rails have been cut from one continuous board to provide a continuity of grain. Any heavily grained, or darker shaded timber is best kept to the bottom of an item for an impression of balance and stability. If these were placed at the top, the item would appear to be 'top-heavy'. Panels should be matched and have their arched top grain features pointing upwards. See Figure 7.37.

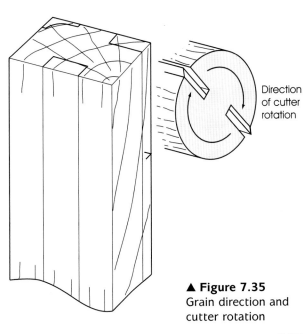

Direction of cutter rotation

▲ **Figure 7.35**
Grain direction and cutter rotation

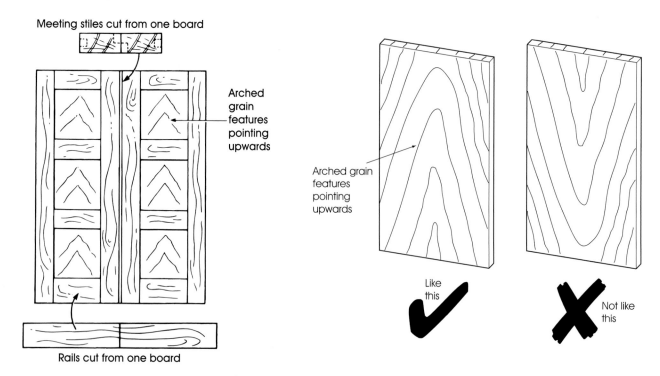

Meeting stiles cut from one board

Arched grain features pointing upwards

Rails cut from one board

▲ **Figure 7.36** Grain matching

Arched grain features pointing upwards

Like this ✔

Not like this ✘

▲ **Figure 7.37** Marking out panels

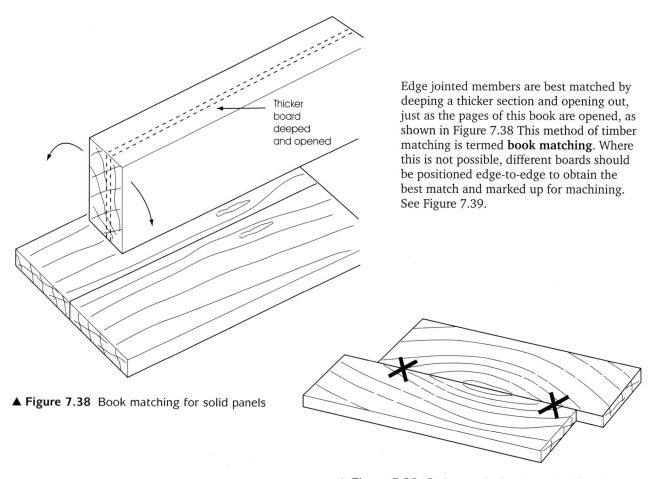

Thicker board deeped and opened

▲ **Figure 7.38** Book matching for solid panels

Edge jointed members are best matched by deeping a thicker section and opening out, just as the pages of this book are opened, as shown in Figure 7.38 This method of timber matching is termed **book matching**. Where this is not possible, different boards should be positioned edge-to-edge to obtain the best match and marked up for machining. See Figure 7.39.

▲ **Figure 7.39** Grain matched and marked for jointing

Veneers

Where veneered sheet materials are used for clear finished work their surface should be veneered with a radially (quarter cut) or tangentially (crown cut) sliced veneer, depending on the required finished appearance.

When veneering panels for doors and drawer fronts, general carcass work and wall panelling, etc., there are a number of different ways in which the veneers may be matched in order to create different decorative effects.

Marking out

Marking out involves referring to design drawings, workshop rods and cutting lists produced during the setting-out process and the selection and marking out of timbers to show the exact position of joints, mouldings, sections and shapes. In addition it may also include the making of **jigs** for later manufacturing or assembly operations.

After the timber has been prepared and faces marked, the actual marking out of the item can be undertaken. Depending on the size of joinery works and the volume of work it handles, setting out and marking out may either be undertaken by one person or treated as separate roles to be undertaken by different people.

Framed joinery

A workshop rod for a glazed door is illustrated in Figure 7.40. It shows how a stile and rail are laid on the rod, the sight, shoulder and mortise position lines are squared up with the aid of a set square. The mortises, tenons and sections, etc., are marked out as shown in the completed stile, see Figure 7.41. Most details are simply pencilled on; however mortises should be marked with gauge lines and the shoulders of tenons marked with a marking knife.

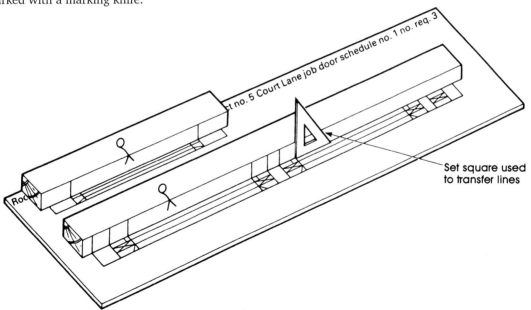

▲ **Figure 7.40** Marking out from rod

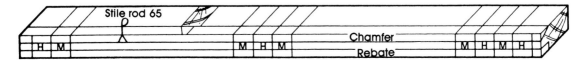

▲ **Figure 7.41** Marked out stile

In many joiners' shops it is standard practice to sketch the section on a member to enable all who handle it to instantly see how it should look when finished. The marked-out member is also marked up with the rod or job number so that it may be identified with the rod/job throughout manufacturing operations.

Where a paired or handed member is required (stiles and jambs) the two pieces can be placed together on a bench with their face sides apart, as all squaring should be done from the face side or edge for accuracy. The lines can then be squared over onto the second piece as shown in Figure 7.42. When pre-sectioned timber has to be marked out, a box square, as shown in Figure 7.43, can be used to transfer the lines around the section.

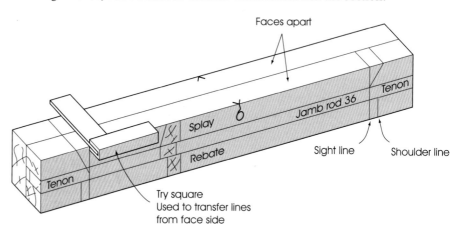

▶ **Figure 7.42**
Transferring marks to the second of a pair

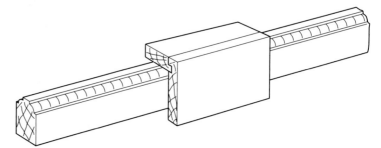

▶ **Figure 7.43** Use of a box square

Using a pattern

Whenever more than one joinery item of a particular design is required, the first to be marked out becomes a pattern for the rest of the job. After checking the patterns against the rod for accuracy, they can be used to mark out all other pieces and for setting up the machines. The positions of the mortises are normally marked out on every member as it is not economical to spend time setting up chisel or chain mortising machines to work to stops, except where very long runs are concerned. Shoulder lines for tenons are required on the pattern only, as tenoning machines are easily set up to stops enabling all similar members with tenons in a batch to be accurately machined to one setting.

Information for the machinist should be included on the pattern as to how many or how many pairs are required. Illustrated in Figure 7.44 are the pattern head and jamb for a batch of 10 doorframes.

Figure 7.45 shows a pattern being used to mark out a batch of paired stiles. As any distortion of the timber could result in inaccuracies they must be firmly cramped together.

The use of this method ensures greater accuracy than if each piece were to be individually marked from the rod. Alternatively, a batch of stiles can be cramped between two patterns and the positions marked across with the aid of a short straightedge. At the end of a run the patterns can be machined, fitted and assembled to produce the final item.

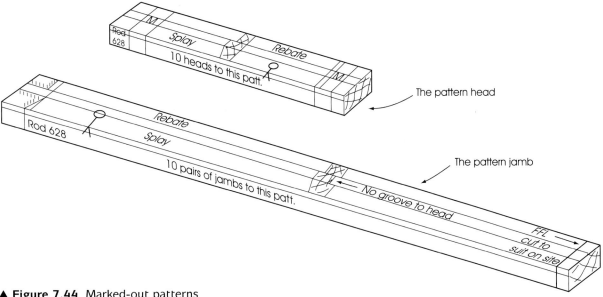

▲ **Figure 7.44** Marked-out patterns

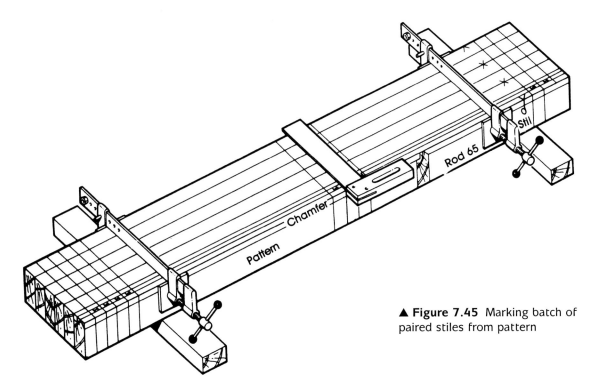

▲ **Figure 7.45** Marking batch of paired stiles from pattern

Three-dimensional joinery

Marking out for three-dimensional items follows the methods used for framed joinery, each component part having its own individual pattern containing information for the machinist. Where sheet material is concerned this may be marked out on the actual component part or a thin plywood or MDF template/jig may be produced.

Illustrated in Figure 7.46 is the design drawing for a base unit, a data sheet for one of its standards and a plywood template marked up and drilled out, to be used as a machining jig.

Joinery works with computer-controlled boring or routing facilities will programme the machines directly from the data sheet without any need to mark out panels or make templates.

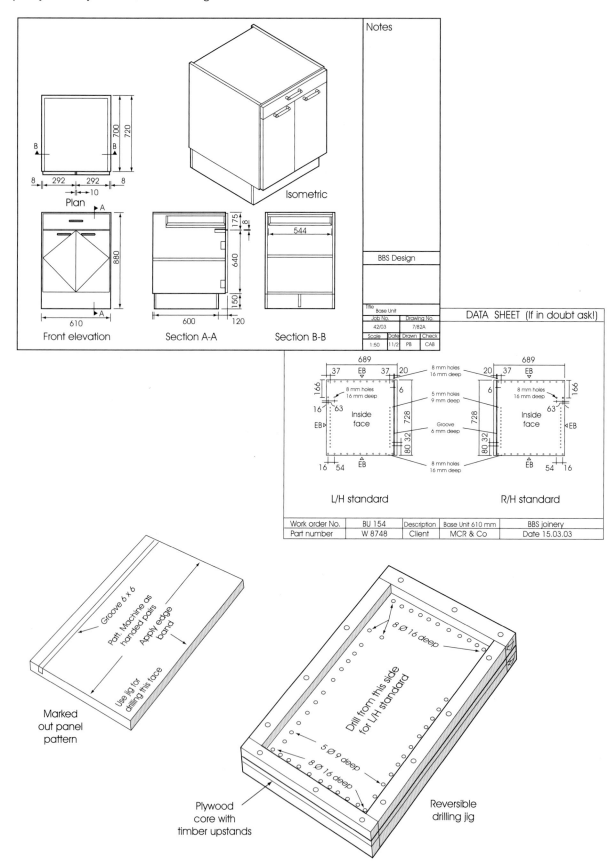

▲ **Figure 7.46** Base unit details

Stairs

The marker-out will make a number of templates out of thin plywood or MDF to assist in the marking out of stairs, see Figure 7.47.

- *Pitchboard* is an accurately prepared right-angled triangle with its two shorter sides conforming precisely to the individual rise and going dimensions of those on the rods.
- *Margin template* is made as a tee piece with its projection equal to the distance above the tread and riser intersection to the upper edge of the string.
- *Combined pitch and margin template* may be made as an alternative and is often preferred for ease of use.
- *Tread and riser templates* are equal to the shape of the tread and riser plus the allowance for wedging. It is good practice to make the slope for the wedge the same on both templates, so that only one type of wedge needs to be used when the stairs are assembled.

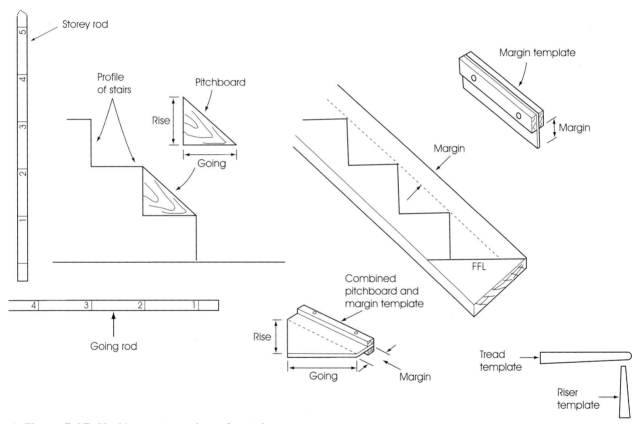

▲ **Figure 7.47** Marking out templates for stairs

Marking out strings. These are marked out as a left- and right-handed pair for closed stairs, or as a handed wall and open string for stairs open on one side, see Figure 7.48.

1. Mark face side and edge on machined timber. Place paired strings on the bench face sides up and face edges apart.
2. Pencil on pitch or margin line on strings, using either margin template or adjustable square.
3. Set pair of dividers to hypotenuse (longest side) of pitchboard and step out along the margin line. This establishes the tread and riser intersection points. It is good practice to step out the intersections rather than just rely on the pitchboard as each step could progressively grow by the thickness of the pencil line.

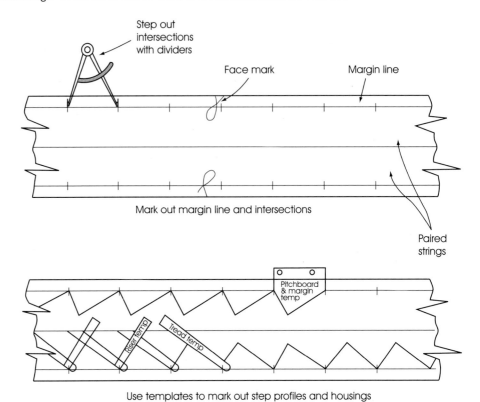

▲ **Figure 7.48** Marking out string for housing

4. Line up pitchboard and margin template on each stepped-out position in turn and pencil on the rise and going of each step.

5. Using the tread and riser templates, pencil their housing positions on to the strings. It is normal practice to leave the ends of strings long and square for cutting to suit on site. The outer string for an open side stair will require marking out at both ends for tenons, see Figure 7.49.

6. Mark out tenon at bottom of stair. A bull-nosed step is being used in this case. The face of the second riser in this case is taken to be the centre line of the newel. Measure at a right angles to the riser, half the newel thickness. Mark a plumb line using the riser edge of the pitchboard. This will give the shoulder line of the tenon. On occasions the shoulder is recessed into the newel face by about 5 mm to conceal shrinkage; if this is the case the shoulder will require extending forwards to allow for it. Alternatively a barefaced tenon may be specified to conceal shrinkage on one face.

7. Repeat the previous stage at the top end of the string except this time the centre line of the newel will be the face of the top riser.

8. From the shoulder line measure at a right angle again, three-quarters of the newel thickness, to give the tenon length. Divide the tenon as shown to give a twin tenon with a central haunch. The bottom end of the string will have an additional haunch at its lower edge whilst the upper part of the tenon at the top end of the string will be cut off level. This arrangement of tenons is to avoid undercutting the mortises or weakness caused by short grain if other arrangements are used.

Marking out newel. The marking out is illustrated in Figure 7.50. The four faces of the newel are drawn out to show the housing and mortise positions. The upper edges of both newels will also require mortising to receive the handrail.

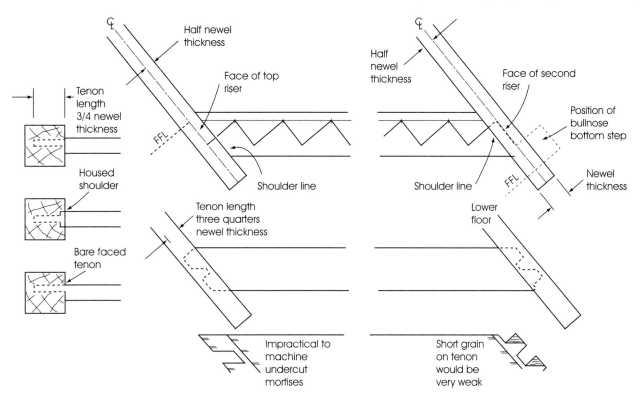

▲ **Figure 7.49** Marking out string for tenons

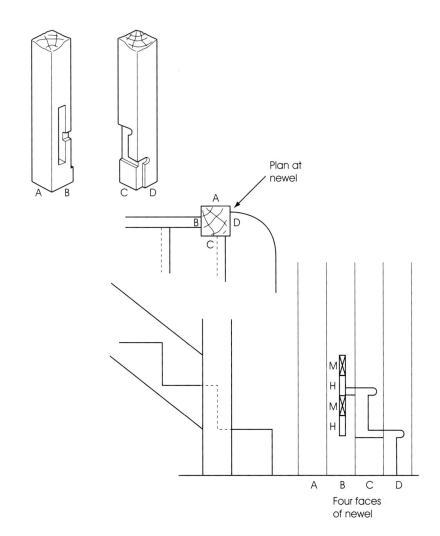

◀ **Figure 7.50** Marking out newel

Marking out handrail and balustrade. The overall height of the handrail on the stairs is 900 mm above the pitch line. On upper landings this is either 900 mm or 1000 mm above the floor level, depending on the use of the building (Figure 7.51).

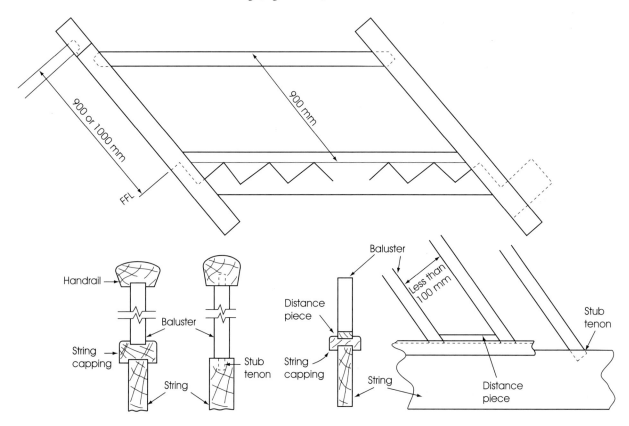

▲ **Figure 7.51** Marking out for handrail and balustrade

Handrails are normally tenoned into the mortised newels. The shoulder angle and line of the handrail tenons (and capping if used) will be the same as the string. Balusters may be either stub tenoned into the edge of the string or fitted in a groove run into a string capping, which itself is grooved over the string. At their upper end they are normally pinned into the groove run on the underside of the handrail, or again they may be stub tenoned. Distance pieces may be cut between balusters and pinned into the grooves to maintain the baluster spacings.

■ Determine the number of balusters required. Allow two per tread and one where there is a newel. For an 11-going flight with a bottom bull-nose step, there will be eight full treads and two part treads between newels, thus 18 balusters will be required.

■ Determine the horizontal distance between balusters. Divide the horizontal distance between the newels (less the space taken up by total width of all balusters) by the number of balusters plus one. (There will always be one more space than the number of balusters.)

Example

Distance between newels = 10 treads × going − newel width
= (10 × 255) − 100 = 2450 mm

Space taken by balusters = number of balusters × width
= 18 × 40 = 720 mm

Distance between balusters = (Distance between newels − space taken by balusters) ÷ (Number of balusters + 1)
= (2450 − 720) ÷ 19
= 91 mm

The maximum distance between balusters should be such that a 100-mm sphere (less than a small child's head) is not able to pass between them. At 91 mm the above example is acceptable for straight balusters. However, if turned balusters were to be used the space between the turned sections may exceed 100 mm. If this were the case an additional baluster should be added and the distance between recalculated.

Methods, briefings and costings

At some stage in the setting and marking out process, a **method statement** is often produced. This sets out the main operations that have to be undertaken in order to complete a specific task. Illustrated in Figure 7.52 is a typical joinery method statement for a batch of floor units. This would be passed onto the joiners for guidance prior to them starting the task.

It is standard practice in many joinery companies for a **workshop briefing** to take place prior to starting a task. This is particularly the case where prototypes for new designs, intricate assembly or specific technical issues are concerned.

The briefing meeting normally held by the setter/marker-out will often bring together the machinist and joiner to familiarise themselves with the task and its specific requirements. A further briefing may take place once the materials have been machined prior to assembly in order to consider methods and give assembly-specific instructions.

Determining costs

The setter-out/marker-out is in many organisations responsible for determining shop floor costs. This may involve both the calculation of material costs and the cost of labour.

Material costs are determined by taking off the volume (m^3) of solid timber and area (m^2) of sheet materials required for a job from the cutting list, plus the cost of any ironmongery. (An allowance is not normally made for consumables such as glue, nails and screws, etc., as these will be included in the overheads.) A percentage is added to these quantities for cutting and waste before the purchasing price of the materials is applied.

Labour costs are determined from the estimated length of time required to complete a task. These may be derived from:
- company standard times;
- records of past performance;
- built-up values using a task procedure costing sheet.

Illustrated in Figure 7.53 is a **task procedure sheet** developed by a joinery company for assembly work. It is used as a computer spreadsheet. The task is analysed and broken down into its individual elements and entered on the sheet for a one-off. The times for each task are calculated by the software and multiplied by the batch size (number off required), giving the total assembly time required.

Determining labour costs. The actual labour cost per hour charged by a company will be far in excess of the hourly rate paid to its employees. It will be the full cost of employing a person and is made up of the following:
- **Direct labour costs** of employing the hourly paid operatives, mainly wood machinists and joiners, including the cost of insurance and holidays, overtime rates and bonus payments, etc.
- **Indirect labour costs** of employing both hourly paid operatives such as general operatives and packing/dispatch operatives, etc., also salaried staff including workshop foremen, managers, the setter/marker-out and other technical staff, again including the cost of insurance and holidays, etc.

JOINERY METHOD STATEMENT

Works Order Number	15350	Del. Date	30-1-03	Part Number	BU14 MEW

Description	BASE UNIT 610 mm			Client	MEAL-E-WAY

Unit Time hrs	1·95	Quantity	10 off	Total Time hrs	19·5

No.	Operations
1	Assemble unit carcass to assembly details, using cascamite glue and dowels. Fix back of unit to base and sub top using 25 mm ring shank nails.
2	Assemble plinth using cascamite glue and plated screws. Fix to unit using 8 off white plastic modesty blocks.
3	Assemble drawer and install in unit, using bottom mounted steel runners. Fix slab front and ironmongery as detailed.
4	Hang doors and fit ironmongery as detailed.
5	Insert shelf studs and shelf.
6	Remove all arrises and wipe down all surfaces.
7	Undertake final quality check, label and transfer to pallet.
8	
9	
10	

(not definitive – provided as a guide only)

▲ Figure 7.52 Joinery method statement

ENTER TIMES FOR A 1 OFF ASSEMBLED, SHEET WILL ADJUST FOR BATCH QUANTITY

| DESCRIPTION |
| PART NUMBER |
| QUANTITY |
| |

JOINERY ASSEMBLY

ACTIVITY	QTY	DESCRIPTION	MEASURE	Mins	TOTAL Mins
GENERAL ASSEMBLY					
Pick up and position		LARGE - ABOVE 1 m	PER PIECE	1.00	
component part		MEDIUM - ABOVE 0.5 m	PER PIECE	0.70	
		SMALL - BELOW 0.5 m	PER PIECE	0.35	
Framed joints		FIT SINGLE MORTISE AND TENON JOINT	PER JOINT	2.25	
		HAND SCRIBE OR MITRE MOULDING AT JOINT	PER JOINT	3.00	
FIXINGS					
Nails/Pins/Dowels		GLUE AND POSITION WOODEN DOWEL	PER OCCASION	0.50	
		INSERT PIN OR NAIL	PER OCCASION	0.25	
		FILL PIN OR NAIL HOLE USING FILLER, INC SAND OFF	PER OCCASION	0.30	
Screws		POSITION AND SECURE SCREW	PER SCREW	0.50	
		FILL SCREW HOLE USING PELLET, INC SAND OFF	PER SCREW HOLE	1.25	
HAND & POWER TOOLS					
		USE CHOP SAW	PER CUT	0.75	
		USING POWER ROUTER	PER LIN M	1.25	
		USE HAND PLANE	PER LIN M	1.00	
		CUT GROOVE, APPLY GLUE, FIT BISCUIT	PER BISCUIT	1.25	
GLUE & CRAMPS					
		APPLY GLUE - PER BEAD	PER LIN M	0.75	
		SECURE AND REMOVE CRAMP	PER OCCASION	1.25	
Drawer		ASSEMBLE AND FIT DRAWER	PER ASSEMBLY	22.00	
Unit doors		ASSEMBLE AND FIT SWING DOOR TO UNIT	PER DOOR	12.50	
		ASSEMBLE AND FIT SLIDING DOORS TO UNIT	PER PAIR	14.75	
Building door		HANG C/W 2 No HINGES, LEVER FURN, AND LOCK/LATCH	PER DOOR	85.00	
IRONMONGERY					
Unit handles		FIT "D" TYPE HANDLE	EACH	2.10	
Unit catch		FIT BALES OR MAG. CATCH	EACH	2.50	
Surface bolt		FIT TOWER BOLT	EACH	2.75	
Additional hinges		FLUSH HINGE	PER HINGE	1.25	
		BUTT HINGE	PER HINGE	2.50	
		PIANO HINGE	PER LIN M	17.25	
Glass & metalwork		UNWRAP AND POSITION	PER ITEM	1.50	
SANDING & FINISHING					
		SAND - FACE	PER SQ. MTR	9.25	
		SAND - EDGE OR DE-ARRIS EDGE	PER LIN. MTR.	0.50	
LAMINATE & EDGING					
		APPLY FACE LAMINATE	PER SQ. MTR	30.00	
		APPLY LAMINATE EDGING	PER LIN. MTR.	12.50	
		APPLY TAPE EDGING (brush glue or iron-on)	PER LIN. MTR.	7.50	
		APPLY TIMBER EDGING	PER LIN. MTR.	17.50	
ESTIMATED ITEMS		DESCRIPTION	UNIT	TIME	
Estimated time for					
work required, but not					
included in sheet					
EXAMINE & LABEL					
		EXAMINE, AFFIX QUALITY LABEL TRANSFER TO PALLET	PER OCCASION	2.50	
OCCASIONAL ELEMENTS					

Recieve parts, ironmongery, check measurements, cutting list, plan method and sequence of work, consult on details, rectify faults, sort tools on bench or at workplace, turn work over or around on bench or workplace, examine assembly, machine items - grind off hardware to fit, find tools, pallets etc and clock on and off.

Percentage already included in rates shown above

TOTAL MINS. PER ITEM

TOTAL MINS. FOR COMPLETE BATCH QUANTITY

TOTAL HOURS FOR COMPLETE BATCH QUANTITY

▲ **Figure 7.53** Typical computer-based joinery task procedure costing spreadsheet

The typical cost of employing a person is two to three times the hourly rate paid and this does not include allowance for other overheads and profit.

Each firm will have their own system for recording the amount of time spent by hourly paid operatives on a particular task in order to calculate **standard timing** for tasks. Typical methods of **time collection** include (see Figure 7.54):

- **Job records** showing who has worked on a particular job. The foreman completes these.
- **Timesheets** kept by individual operatives, showing on a daily basis the type and duration of tasks undertaken.
- **Electronic smartcards,** which are being increasingly used even by smaller organisations. The card is 'swiped' into the system by an operative who enters a job code each time they undertake a different task.

In addition to determining the labour cost of a job, the information collected may also be used to:

- compare the actual cost with the estimated cost;
- compile standard times for tasks;
- compare individual performance against a standard time;
- calculate financial incentive scheme payments;
- compile statistical information on productivity and downtime.

Overall costs

Overheads and profits are added to the actual material and labour costs (direct and indirect) to determine the total price charged to the customer for a particular job. Typical costs recovered in overheads are:

- salaries of senior support staff and directors;
- building maintenance and running costs, including rents and rates;
- plant and machinery costs;
- consumable material costs.

Task planning

Before starting a job it is necessary to plan and organise the work. You should ask yourself the following questions:

- What is to be done?
- How is it to be done?
- When is it to be done?
- Where is it to be done?

In answering these questions as part of the planning process, you will refer to drawings, specifications, setting-out rods, method statements and the actual machined components, etc. Any discrepancies must be resolved before proceeding with the work, as any mistakes are more costly to rectify the further on they go unnoticed.

Depending on your employer's line of communication, queries may be directed via your workshop foreman/manager or direct with the setter-out/ marker-out.

The means of communication may involve direct face-to-face contact, a telephone call, e-mail or memo, or use of a standard form. The spoken word is often used as a quick and informal first point of contact. However, a written confirmation should be sought as evidence and as a means of updating information for the future.

Resolving irregularities

On checking the machined components with the drawings and details, if you notice that the data sheet for a veneered standard shows a vertical grain

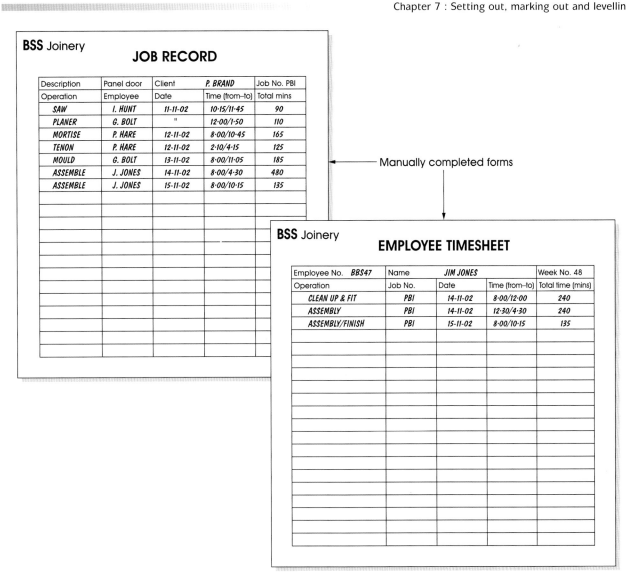

BSS Joinery
JOB RECORD

Description	Panel door	Client	*P. BRAND*	Job No. PBI	
Operation	Employee	Date	Time (from–to)	Total mins	
SAW	I. HUNT	11-11-02	10·15/11·45	90	
PLANER	G. BOLT	"	12·00/1·50	110	
MORTISE	P. HARE	12-11-02	8·00/10·45	165	
TENON	P. HARE	12-11-02	2·10/4·15	125	
MOULD	G. BOLT	13-11-02	8·00/11·05	185	
ASSEMBLE	J. JONES	14-11-02	8·00/4·30	480	
ASSEMBLE	J. JONES	15-11-02	8·00/10·15	135	

Manually completed forms

BSS Joinery
EMPLOYEE TIMESHEET

Employee No. *BBS47*	Name	*JIM JONES*		Week No. 48
Operation	Job No.	Date	Time (from–to)	Total time (mins)
CLEAN UP & FIT	PBI	14-11-02	8·00/12·00	240
ASSEMBLY	PBI	14-11-02	12·30/4·30	240
ASSEMBLY/FINISH	PBI	15-11-02	8·00/10·15	135

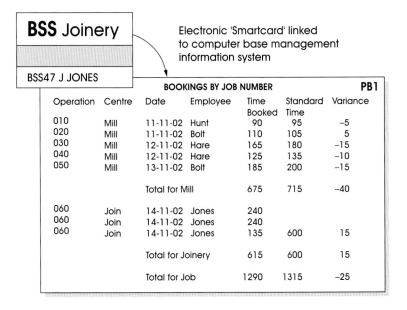

BSS Joinery

BSS47 J JONES

Electronic 'Smartcard' linked to computer base management information system

		BOOKINGS BY JOB NUMBER				PB1
Operation	Centre	Date	Employee	Time Booked	Standard Time	Variance
010	Mill	11-11-02	Hunt	90	95	–5
020	Mill	11-11-02	Bolt	110	105	5
030	Mill	12-11-02	Hare	165	180	–15
040	Mill	12-11-02	Hare	125	135	–10
050	Mill	13-11-02	Bolt	185	200	–15
		Total for Mill		675	715	–40
060	Join	14-11-02	Jones	240		
060	Join	14-11-02	Jones	240		
060	Join	14-11-02	Jones	135	600	15
		Total for Joinery		615	600	15
		Total for Job		1290	1315	–25

▲ **Figure 7.54** Time collection

direction, whereas the actual standard has been machined with a horizontal grain direction, see Figure 7.55, then:

- Your first action should be to speak directly or via the telephone with either your foreman/manager or setter-out/marker-out, to find whether there has been a last minute change or the standard has been incorrectly machined.
- If the data sheet has been revised and the grain direction as machined is correct, a new data sheet should be issued as confirmation.
- If the data sheet is correct you may be required to complete a corrective action report form. This will have the effect of re-ordering the 'incorrectly' machined part (Figure 7.56).

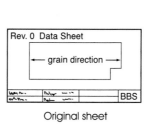

Original sheet

Machined standard

▶ **Figure 7.55** Irregularities between data sheet and cut panel

---

BSS Joinery

CORRECTIVE ACTION REPORT

Raised by:	JIM. JONES		Job number:	WC12		Date raised:	8-11-02
Description:	BASE UNIT					Client:	WHITES & CO.

Defect or Improvement				
Manufacturing		Design		Details
Saw:		Drawings:		General:
Plane:		Specification:		DATA SHEET FOR STANDARD,
Joint:		Data sheet:		SHOWS VERTICAL GRAIN, BUT HAS
Mould:		Method statement:		BEEN CUT HORIZONTALLY. CHECKED
Bore:		Setting out:		WITH DESIGN, DATA SHEET IS
Assembly:		Marking out:		CORRECT.
Handling damage:		Improvement:		
Other:	✔	Other:		

Suggested Corrective Action: (originator to complete) RE-MACHINE 10 NEW STANDARDS TO DATA SHEET. *J.J.*	Issued to: MACHINE SHOP
	Data received:
Corrective Action Taken: (recipient to complete)	Action taken by:
	Data completed:

☑ Tick appropriate box

▶ **Figure 7.56** Corrective action report form

Checking components

All materials and ironmongery should be inspected before use:

- Check you have the correct number of components shown on the cutting list.
- Inspect all components for damage. Any scratches on sheet material or bowing, shakes and other defects in timber, should be brought to the attention of your foreman/manager, who will decide whether thy are acceptable for use or require replacement.
- Ensure all ironmongery and other fittings are as specified and working correctly.
- Return defective items to the store for replacement.

Work tasks

Look at the method statement if provided, or write down a list of main tasks to be done in the order that they will be undertaken.

> **Example**
>
> Task: Apply hardwood edge to batch of worktops and overlay with laminate.
>
> Mitre edging at returns.
> Glue and loose tongue, edge to top.
> Clean edge flush with face.
> Overlay with laminate.
> Trim off laminate and profile edge
> Sand up edging ready for finishing.

Look at the method statement or your work tasks list. Ensure you have all the correct specialist tools and equipment ready to undertake the task. Make a list. Items, which you do not have, may require requisition from the stores, or even hiring from an outside supplier for the duration of the job. A typical **tools and equipment list** for the above task would be:

> Chop saw
> 900 mm sash cramps (10 off)
> 'Jay' roller
> Hand-powered router and profile cutter
> Orbital sander

Work timing

The time you take to complete a task will depend on your skill level and how familiar you are with it. You may be given either a target time to do the work or a required completion date.

You will have to consider whether you can complete the task on your own within the deadlines, or if not request assistance.

Review of task. On completion of the task, you should undertake a review. Think of any problems that occurred and steps you took to make things easier. Make a note of these and incorporate in future tasks. Figure 7.57 illustrates the planning, task and review procedure.

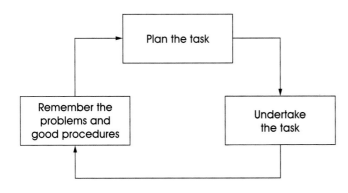

◀ **Figure 7.57** Planning, task and review procedure

Site setting out requirements and procedures

The setting out of a building can be divided into two distinct operations:

- Establishing the position of the building and setting up profiles.
- Establishing a datum peg and transferring required levels to various positions.

Establishing the position of the building

The basic requirements for establishing the position of a building are linear measurement (length), the setting out of angles and the setting out of curves.

Linear measurement

A 30-m steel tape is most often used for setting out; linen or plastic tapes should be avoided as they are liable to stretch, resulting in serious errors. Inaccuracies can also occur when using steel tapes if they are not fully stretched out (tensioned). A constant tension handle is available for use with steel tapes, which reduces errors to less than 3 mm in 30 m.

Linear measurements can also be made with surveyor's chains and other specialist equipment, such as electronic distance measurement devices (EDM).

Wherever linear measurements are made the tape must be kept horizontal if they are to be accurate. Where measurements are made over sloping ground the inaccuracies are potentially far greater. Figure 7.58 illustrates the correct methods to adopt. Where the slope is a slight one, pegs of different length can be used. On steeper slopes the tape must be held horizontal and the measurement plumbed down on to a peg. For longer measurements on steep slopes the distance should be divided and marked out in a number of stages. Alternatively the sloping measurement between two points may be taken and the horizontal distance calculated using Pythagoras rule. The difference in height between the two points being found by levelling (see use of optical level).

Setting out angles

Right angles can be simply set out using the following basic methods:

- **Builders' square** (see Figure 7.59). A large builders' square is set up on packers so that one leg is against the building line. The sideline should be positioned so that it runs parallel to the other leg of the square.

- **The 3:4:5 rule** (see Figure 7.60). A triangle is marked out having sides of 3 units, 4 units and 5 units. A measurement of 3 units and 4 units can be marked on the front and side lines. The side line can then be positioned so that the distance between the two marked points is five units.

- **Optical site square** (see Figure 7.61). This is an optical instrument consisting of a cylindrical head, which contains two telescopes permanently mounted at right angles (90°) to each other. Both of the telescopes can be individually adjusted vertically both up and down. This gives the instrument a range of between 2 m and 90 m. In use the head is mounted on a tripod, with adjustable sliding legs, and a datum rod pointed at one end to set up over a given position or mark and hollowed out at the other end to fit over the end of a nail.

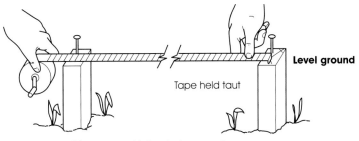

Level ground

Tape held taut

Measurement taken between nails
in setting out pegs

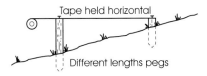

Tape held horizontal

Different lengths pegs

Sloping ground

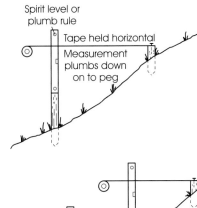

Spirit level or
plumb rule

Tape held horizontal

Measurement
plumbs down
on to peg

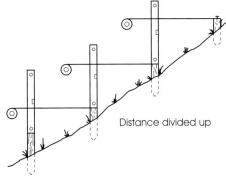

Distance divided up

◀ **Figure 7.58** Use of tape measure

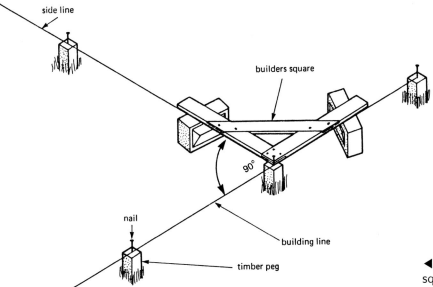

side line

builders square

90°

nail

building line

timber peg

◀ **Figure 7.59** Using a builder's square

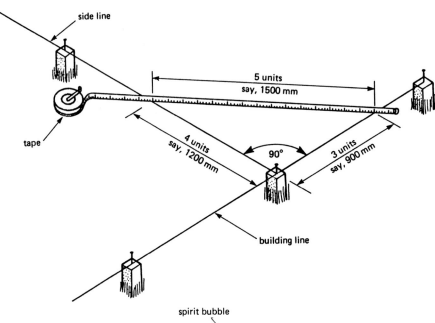

▶ **Figure 7.60** Setting out a right angle with a tape measure

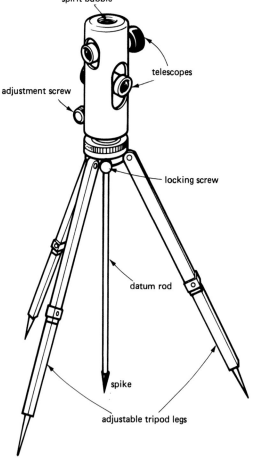

▶ **Figure 7.61** Optical site square

Setting up and using an optical square

1. Set up tripod; ensure all bolts and adjustable nuts are tight.
2. Place tripod over the corner peg of the building and extend the spike to fit over the nail in the top of the peg.
3. Carefully screw the instrument onto tripod and line: lower telescope along the front building line.
4. Adjust tripod legs so that the spirit bubble on top of the cylindrical head is in a central position. Tighten all tripod-adjusting nuts and recheck the spirit bubble. The instrument is now set up and ready for use.

Using the instrument (Figure 7.62):

1. Set up the instrument over corner peg A.
2. Sight through the lower telescope towards peg B. This is the furthest front corner of the building.
3. Adjust the fine setting screw and tilt the telescope until the spot on view is seen through it (see Figure 7.63a). The views through the telescope shown in Figures (b) and (c) are off the mark, and the telescope requires further adjustment to obtain the spot on view.
4. Sight peg C through the top telescope taking care not to move the instrument. Direct your assistant to move peg C sideways until the spot on view is seen through the telescope.

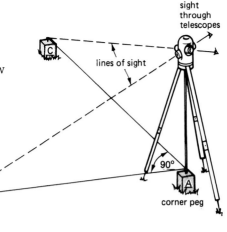

▲ **Figure 7.62** Using a site square

Whichever method has been used for setting out the first right angle, the fourth corner peg of a building can be set out by repeating the process over another corner or alternatively by making parallel measurements along the sides of the building. As a check for accuracy the diagonals should be measured. If these are the same and the sides are parallel the setting out must be square (Figure 7.64). When the diagonals are not the same, a check must be made through the previous stages to discover and rectify the inaccuracy before proceeding further.

Note: Surveyors will often use theodolites or optical levels with a graduated base ring to set-out or measure angles.

Projections from the main building for bay windows etc. are often set out using a timber template made up to the actual shape of the brickwork (see Figure 7.65). The template is positioned on packings up against the building line; the position of the foundation trench can be measured out from it. After excavation and concreting the template is used again to set out the position of the brickwork.

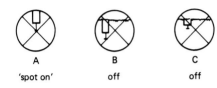

▲ **Figure 7.63** View through site square

▼ **Figure 7.64** Checking setting out for square

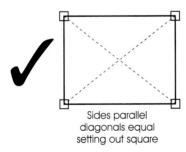

Sides parallel
diagonals equal
setting out square

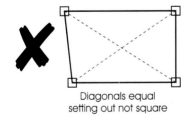

Diagonals equal
setting out not square

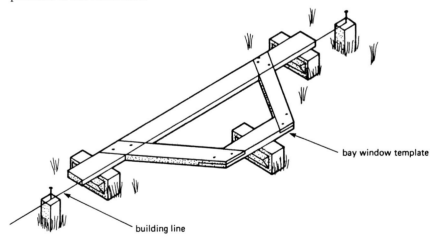

bay window template

building line

▲ **Figure 7.65** Bay window template

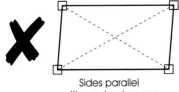

Sides parallel
setting out not square

Setting out curves

The methods used to set out a curve will depend on its size and whether its centre point is accessible or not. These methods include the following:

■ *Timber template.* Small repetitive curves for bay windows, etc., can set out using a timber template. Figure 7.66 shows a segmental bay window template in use.
■ *Radius rods* can be used to accurately set out all curved work of up to about 4 m radius (see Figure 7.67).
■ *Triangular frame:* Where the radius length is excessive, or in situations where the centre point is inaccessible, a timber triangular frame can be used. Two pegs are positioned on the building line indicating the ends of

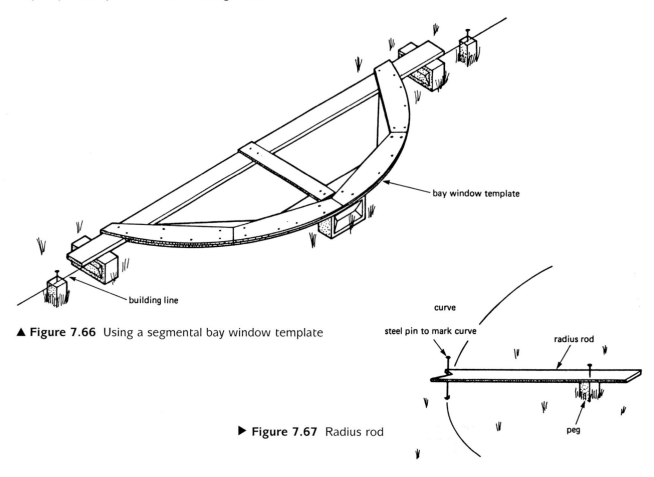

▲ **Figure 7.66** Using a segmental bay window template

▶ **Figure 7.67** Radius rod

the curve. The third peg is positioned at right angles to the building line indicating the maximum rise of the curve. For accuracy, nails are driven into the tops of the pegs to indicate the exact dimensions. A lightweight timber frame is then made over the pegs as shown in Figure 7.68. When the centre nail is taken out the frame can be moved across while still keeping it in contact with the other two nails, using a steel pin at the apex to mark the required curve.

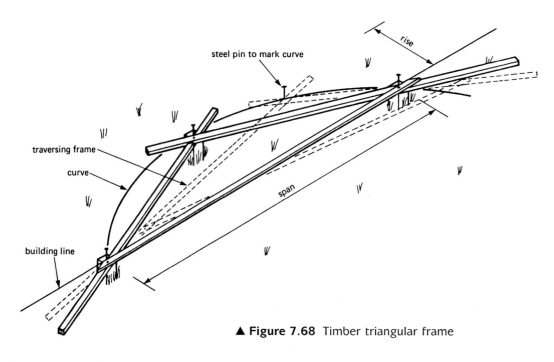

▲ **Figure 7.68** Timber triangular frame

■ *Calculated ordinates.* The previous methods are inappropriate for setting out curves of very large radius, as the size of the equipment makes it awkward to handle. One method that may be used is to calculate a number of ordinate lengths and peg these out from the base line. A shaped template is used between the pegs to mark the smooth line of the curve, as shown in Figure 7.69.

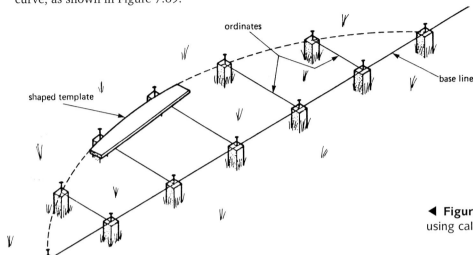

▲ **Figure 7.69** Setting out a curve using calculated ordinates

Pythagoras' rule can be used to calculate the ordinate lengths.

Example

To calculate the lengths of the ordinate required to set out a curve having a chord length of 20 m and rise of 2 m, first find the radius using the intersecting chord rule (see Figure 7.70):

$A \times B = C \times D$
So, $D = (A \times B) \div C = (10 \times 10) \div 2 = 50$

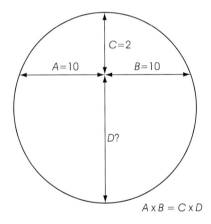

▲ **Figure 7.70** Intersecting chord rule

Therefore:

Diameter = 52 m (D + C)

Radius = 26 m (half diameter)

Divide the chord length into a number of equal parts, say at 2 m centres, giving ordinates 1 to 9 as shown in Figure 7.71.

Only ordinates 1–4 need to be calculated since 1 = 9, 2 = 8, 3 = 7, 4 = 6 and ordinate 5 is the rise.

Use Pythagoras' rule to calculate each individual ordinate length in turn. Figure 7.72 shows the method used to calculate the length of ordinate 1. Line Z is the distance between the chord and the centre point (radius minus rise) in the right-angled triangle ABC.

AC = radius

BC = distance of ordinate from centre line.

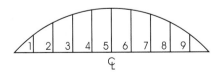

▲ **Figure 7.71** Chord divided into equal parts

Using Pythagoras's theorem:

$a^2 + b^2 = c^2$

So, $a^2 = c^2 - b^2 = 26^2 - 8^2 = 676 - 64 = 612$

Thence, $a = \sqrt{612} = 24.739$

Therefore:

Ordinate 1 = 24.739 – z = 24.739 – 24 = 739 mm

The length of ordinates 2 to 4 can be found using the same procedure.

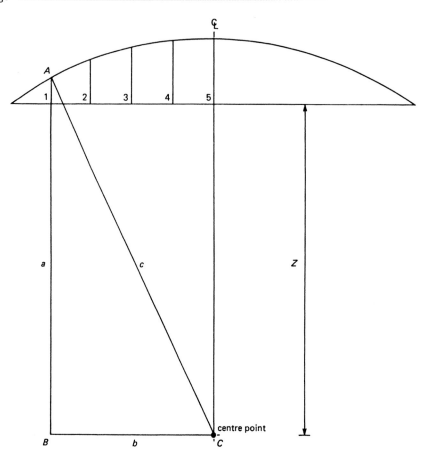

▶ **Figure 7.72** Calculating length of ordinate 1

Setting out small buildings

To establish the position of a small building the procedures illustrated in Figure 7.73 can be used, as follows.

1. All setting out is done from the building line. This will be indicated on the block plan and the local authority decides its position. The line is established by driving in 50 mm × 50 mm softwood pegs A and B on the side boundaries at the correct distance from, and parallel to, the centre line of the road. A nail in the top of the peg indicates the exact position of the line. Strain a line between these two nails.

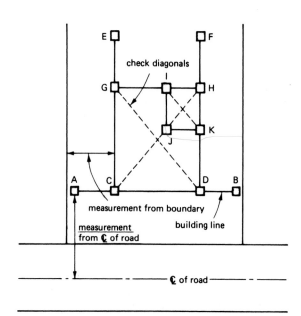

▶ **Figure 7.73** Setting out a small building

2. Drive two pegs C and D along the building line to indicate the front corners of the building. The position of the building in relation to the side boundaries will be indicated on the architect's drawings. Drive nails into the tops of the pegs to indicate the exact position of the corners on the building line.

3. Set out lines at right angles to pegs C and D and establish pegs E and F. Drive nails into the tops of pegs E and F to indicate the exact positions and tension lines between the four pegs.

4. Measure along lines CE and DF to establish pegs G and H in the far corners of the building. It is advisable at this stage to check the diagonals CH and DG. If these diagonals measure the same, the building must be square and the setting out can continue. Peg out the positions of offsets I, J and K. Check the smaller diagonals, which have been formed, to ensure accuracy.

5. Set up profile boards just clear of the trench runs at all of the corners and wall intersections of the building as shown in Figure 7.74.

6. Four nails are driven into the top of each profile board to indicate the edges of the foundation trench and the edges of the brickwork. An alternative, which is sometimes used instead of the nails, is to mark the positions on top of the profile boards with four saw cuts. Transfer positions of setting-out lines to the profile boards (see Figure 7.75).

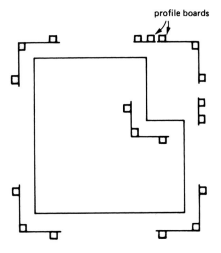

▲ **Figure 7.74** Layout of profile boards

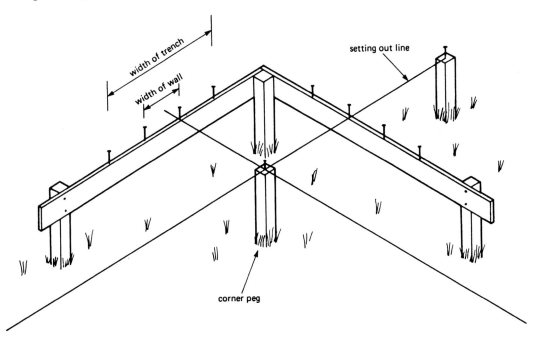

▲ **Figure 7.75** Transferring setting-out lines to profile boards

7. Lines can be strained between the nails on the profiles to indicate the exact positions for the excavators and bricklayers. During excavation these lines will be removed, and therefore the position of the trench should be marked by sprinkling sand on the ground directly under the lines to indicate the sides of the trenches.

Once the foundations have been excavated and concreted, one line is restrained around the nails on the profile boards to indicate the face of the brickwork. From this line the position of the brickwork can be marked on the foundation concrete as illustrated in Figure 7.76.

Setting out steel or concrete frame buildings using a structural grid. The centre lines of the columns, walls and beams, etc., form the basis of this grid, which can also be used as a referencing system to identify each

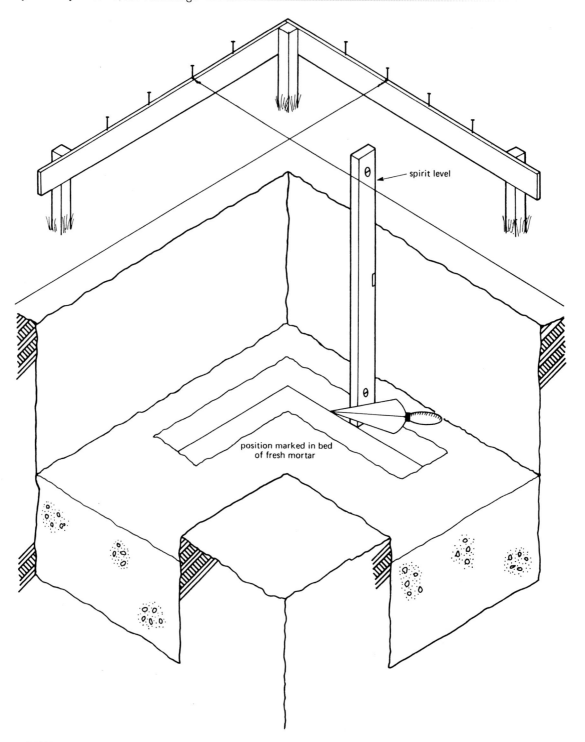

▲ **Figure 7.76** Marking out from a profile board

structural element (see Figure 7.77). The following procedure can be used to establish the grid:

1. Establish the four corner pegs of the building using the same method as before.

2. Set up continuous profile boards along each side of the building (see Figure 7.78). These should be well clear of the work so as not to obstruct the excavators.

3. Transfer corner points on to profiles. From these points mark out and drive nails along the profiles to indicate the centre lines of the columns and beams.

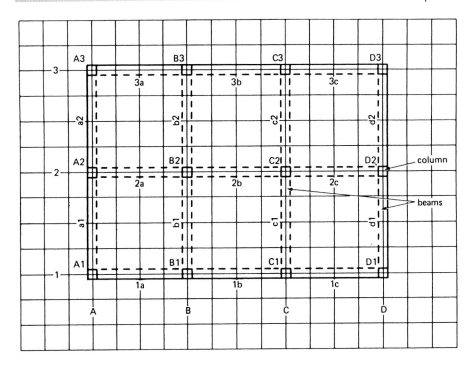

◀ **Figure 7.77** Using a structural reference grid

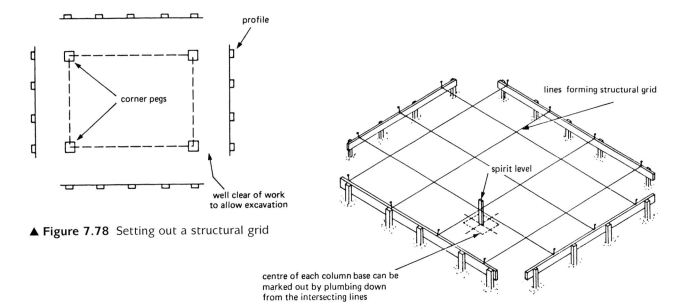

▲ **Figure 7.78** Setting out a structural grid

▲ **Figure 7.79** Marking out from a structural grid

4. Strain lines between the nails to indicate the centre line positions of the column bases, etc., which can be marked out by plumbing down from the line intersections (see Figure 7.79).

Establishing a datum peg and transferring levels

A site datum peg or temporary benchmark (TBM) is established on site as a reference point to which all other level positions are related to. This is normally related to an Ordnance Benchmark (OBM) (refer to Chapter 4). However, on some sites, particularly those remote from an OBM, a notional site datum value is assumed of say 50.000 or 100.000 m to represent the ground floor level.

Transferring levels

Levels on site can be simply transferred to their required positions using basic instruments such as a Cowley automatic level, a tilting level or an automatic level. Occasionally on small works certain more basic methods may be used, including a straightedge and spirit level, boning rods or a water level.

Cowley automatic level. The rectangular head contains a dual system of mirrors (Figure 7.80). This is mounted on the tripod or stand, the centre pin of which must point upwards in use. When the instrument's head is inserted onto the pin, a catch is released, setting the level ready for use.

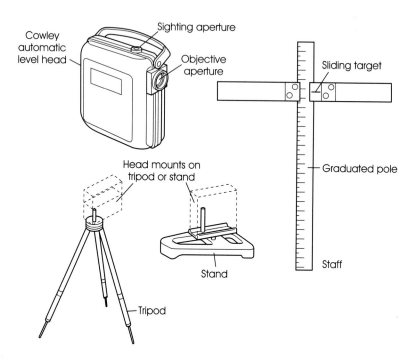

▶ **Figure 7.80** Cowley automatic level

The instrument is used in conjunction with a graduated staff, which consists of a pole with a moving target or crosspiece that can be adjusted up or down the pole. There is an arrow on the target or crosspiece, which indicates the exact measurement on the graduated pole.

The correct procedure for transferring a level from one point to another is described below and illustrated in Figure 7.81.

1.   Set up the tripod and insert the level on the pin of the tripod, checking that the pin is fully inserted.

2.   Sight through the sighting aperture in the top of the level and adjust the tripod so that the two mirrors are seen to form an approximate circle.

3.   Have an assistant hold the staff on the datum peg or temporary benchmark (TBM). **Note:** The assistant should ensure that the staff is held upright.

4.   Sight through the sighting aperture at the staff and have the assistant slide the target up or down until the target is seen level, as in Figure 7.82 view A or B. If view C or D is seen, the target requires further adjustment up or down the staff.

5.   Lock the slide in position on the staff. Ask the assistant to drive a peg into the ground until the target is once again sighted level. Peg A is now horizontally level to the datum peg or TBM.

Reduced or increased levels in relation to the TBM can be found by moving the target up or down the staff by the required distance and resighting until the target is seen level. Figure 7.83 shows how differences in levels can be measured by reading the graduated staff.

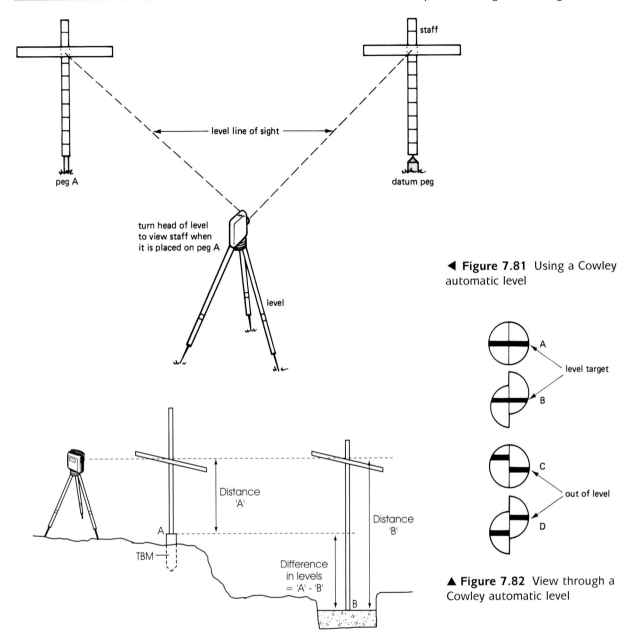

◀ **Figure 7.81** Using a Cowley automatic level

▲ **Figure 7.82** View through a Cowley automatic level

▲ **Figure 7.83** Measuring differences in levels

The **quickset tilting level** is basically a telescope with a spirit level on the side; it is mounted on a tripod and used in conjunction with a staff.

■ **Level,** Figure 7.84. The instrument is easily set up (hence the name quickset). It is attached to the tripod by a ball-and-socket mounting secured from underneath using the screw thread once the circular bubble has been centralised. Earlier types of tilting level (*dumpy level*) were more time consuming to set up, as the mounting consisted of a horizontal plate and three levelling footscrews, each requiring adjustment.

■ **Staff.** The most common is the telescopic staff, available in either wood or aluminium. It consists of box sections that slide within each other. When fully extended it has a total length of 4 to 5 m. The staff is graduated to give readings in metres, 100-mm and 10-mm divisions. Millimetres are read by estimation (see Figure 7.85). Older levels give an inverted image, causing the staff to be seen upside-down as shown in Figure 7.86. Ensure that the staff is held upright when readings are

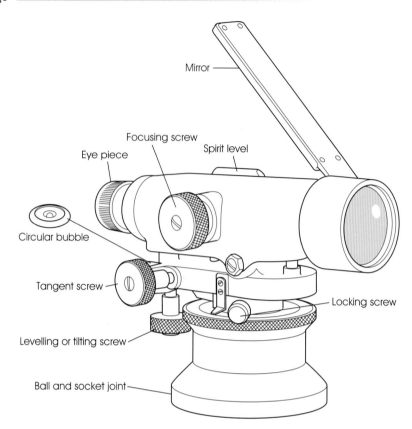

▲ **Figure 7.84** Quickset tilting level

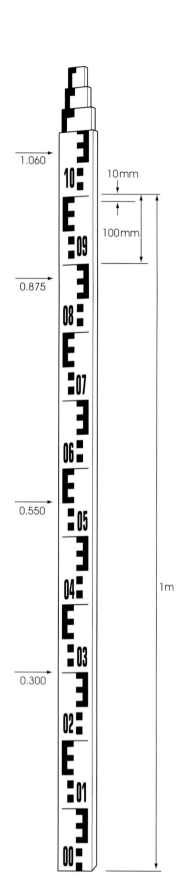

▲ **Figure 7.85** Reading a telescopic staff

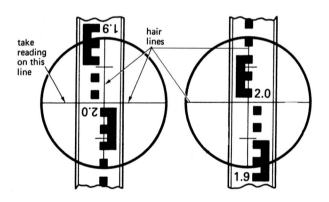

▲ **Figure 7.86** View of staff through level, left older style inverted image; right image correct way up

taken, otherwise the reading will be high (see Figure 7.87). Care should also be taken when using an extended staff to ensure that it is fully extended and the catch located in its correct position.

Setting up procedure:

1. Open up the tripod; extend the legs so that the instrument will be approximately at the user's eye level.

2. Firm the tripod into soft ground by treading on the steps at the bottom of each leg or use a baseboard on firm surfaces to prevent the legs slipping.

3. Ensure all nuts, bolts and screws are tight.

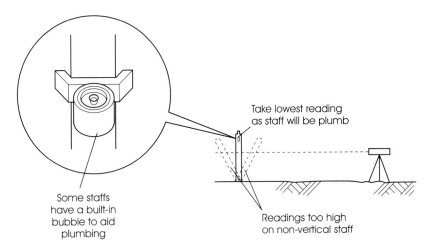

Some staffs have a built-in bubble to aid plumbing

Take lowest reading as staff will be plumb

Readings too high on non-vertical staff

◀ **Figure 7.87** Possible reading error

4. Place the instrument on the tripod, locate the screw thread and tighten when the circular 'bulls eye' bubble is central.

5. Adjust the eyepiece to focus the cross-hairs. This is done by placing the palm of one hand just in front of the telescope to direct a uniform amount of lighting to the instrument. Then, sighting through the eyepiece, rotate it until the cross-hairs appear as black and as sharp as possible.

6. Adjust the focusing screw to focus the telescope. This is done by sighting the staff through the telescope and then slowly rotating the focusing screw until the staff is seen clear and sharp. Rotate the tangent screw so that the staff is central. Care must be taken when adjusting the cross-hairs and focus as, if this is done incorrectly, errors in reading can occur owing to parallax. Parallax can be defined as the apparent separation between the cross-hairs and the staff image. You will notice this as a change in the staff reading when moving your eyes up and down. This should be used as a check each time the telescope is focused on a new staff position.

7. Adjust the spirit level by rotating the tilting screw so that the bubble is central when viewed in the mirror. It is most important that the bubble is centralised each time a reading is taken. Failure to do so results in inaccuracies as the line of sight is not truly horizontal.

Method of use. To find the differential in level between two points (Figure 7.88), set up the instrument midway between the points concerned using the previous sequence of operations.

■ Have the assistant hold the staff on the first position, sight the staff and take the reading.

■ Transfer the staff to the other position. Sight the new staff position, focus, adjust level, and take the reading.

■ The difference between the two readings will be the difference in level between the two points. High reading = lower level and Low reading = higher level

Example

Position 1 = 2.400; Position 2 = 1.250; Difference = 1.150
Therefore position A is 1.150 m lower than position B.

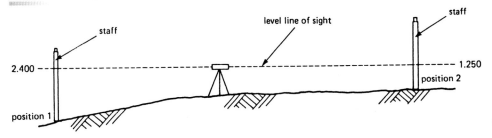

staff

level line of sight

staff

2.400

1.250

position 2

position 1

◀ **Figure 7.88** Using a quickset tilting level

327

Automatic levels are also available, being a simpler version of the tilting level. Providing the instrument is approximately level shown by the 'bulls eye' bubble, it will automatically adjust itself to the horizontal.

Digital levels are similar in use to automatic levels except that they include a device that scans a bar coded staff and displays the level reading and the horizontal distance on a screen.

The following levelling terms are used when transferring levels:

- *Backsight.* The first reading taken once the instrument has been set up in position.
- *Foresight.* The last reading taken before moving the instrument.
- *Intermediate sight.* All the readings taken between the backsight and foresight.
- *Height of collimation.* The height of the line of sight above the Ordnance datum.
- *Reduced level.* The height of any position above the Ordnance datum.

Recording readings. Staff readings should be recorded in a levelling book as they are taken, using either the collimation method or the rise and fall method. There is little to choose between the two methods, but the collimation method has been used in the following examples.

To establish a site datum peg (TBM)

Refer to Figure 7.89:

1. Set up the instrument midway between the nearest OBM and the intended position of the TBM. This should be in an accessible position just inside the site boundary but out of the way of site works and traffic, etc. When the distance between the two positions is greater than about 100 m a series of levels will have to be taken.

2. Take a reading with staff placed on the OBM: 1.115. this is a backsight and is recorded in the backsight column. See Figure 7.90 for a typical page of a levelling book. Since the reading is taken on the OBM its reduced level of 50.450 is recorded in the reduced-level column. The backsight and reduced level are added together to give the height of collimation, 51.565, which is recorded in the appropriate column.

3. To fix the value of the TBM, re-sight the staff with it placed on the TBM. It is often convenient to set this at the same reduced level as the ground floor slab of the new building, in this case a reduced level of say 50.000.

4. This reduced level must be subtracted from the height of collimation to find the required staff reading:
   - Height of collimation = 51.565
   - Reduced-level TBM = 50.000
   - Staff reading = 1.565

5. Therefore the datum peg must be driven until a reading of 1.565 is achieved. This will be the foresight and is recorded in the foresight column. The reduced level of the foresight, 50.000, has been recorded in the reduced-level column.

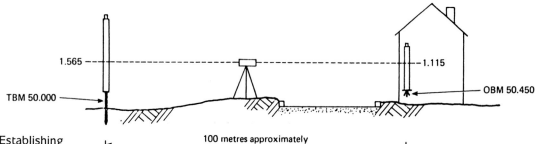

▶ **Figure 7.89** Establishing a site datum peg

Table 29 **Page from a collimation level book**

Back sight	Intermediate sight	Fore sight	Height of collimation	Reduced level	Notes
1.115			51.565	50.450	From OBM
		1.565		50.000	To TBM

Check   FS 1.565   FRL 50.450
         BS 1.115   LRL 50.000
         0.450   0.450

Differences equal OK ✓

▲ **Figure 7.90** Page from a collimation level book

6. As a check for calculation or recording errors, the difference between the sum of the backsights and the sum of the foresights should equal the difference between the first and last reduced level.

Setting a peg in the bottom of a foundation trench as a guide for concreting

Referring to Figure 7.91, take the reduced level of the top of the foundation concrete to be 48.500 and the reduced level of the TBM to be 50.000.

1. Set up level.
2. Sight staff on TBM and record as backsight 1.200.
3. Add backsight to reduced level of TBM: 1.200 + 50.000.
4. Record height of collimation 51.200.
5. Position peg in foundation trench.
6. Sight staff on peg and drive until reading is 2.700. This is found by taking the reduced level of the foundation 48.500 away from the height of collimation 51.200.
7. Record the foresight 2.700 and the reduced level 48.500. Then check.

From the previous examples it is clear that the staff reading required to set any reduced level can be found by subtracting the reduced level from the height of collimation.

Example

Formwork to the sides of a ground beam are to be set at a reduced level of 98.450. What staff reading is required if the reduced level of the TBM is 100.000 and the staff reading on the TBM is 1.225?

Height of collimation = TBM + staff reading on TBM = 100.000 + 1.225
                      = 101.225

Staff reading required = height of collimation − reduced level
                       = 101.225 − 98.450 = 2.775

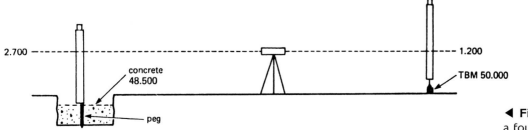

◀ **Figure 7.91** Setting a foundation peg

Basic levelling and lining

Straightedge and level

The straightedge and level is a very simple method, it can be used for most groundwork levelling operations, Figure 7.92. For transferring levels over distances shorter than the straightedge (up to 3 m), drive a 50 mm × 50 mm softwood peg in the required position. Rest the straightedge and level on top of the TBM and peg. Drive the peg into the ground until a horizontal level is achieved. Check for accuracy by first reversing the level on the straightedge and then reversing the straightedge on the TBM and peg. If the bubble in the level alters during these checks either the level is faulty or the straightedge is not parallel. In order to prevent any damage to the level it should be removed while driving in the pegs. When transferring levels over greater distances the procedure used is similar, except that additional pegs are driven into the ground at up to 3-m intervals and levelled from the previous peg.

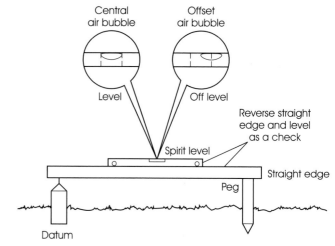

Use a straight edge and a spirit level for short distances

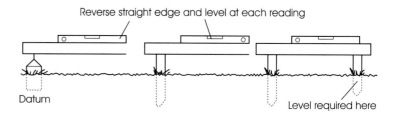

Transferring levels over longer distances

▶ **Figure 7.92** Use of a straightedge and level

Boning rods

These are used to check, transfer or 'sight' straight lines (horizontal or sloping) over fairly large distances. Boning is carried out using either three tee-shaped boning rods, or two fixed site rails and movable boning rod (traveller). Figure 7.93 shows how pegs driven into the bottom of a foundation trench to establish a horizontal level for the concreter to work to can be levelled with a boning rod held upright on the pegs at either end of the trench. The third boning rod is placed on each intermediate peg in turn. The intermediate peg is driven in until, when sighting from one end, the tops of all three boning rods appear in one line. It can be seen that peg B is too high and requires driving in further, but peg C is too low, so this must be removed and re-driven to the correct level. The two end pegs must be established at the correct level before the intermediate pegs can be 'boned'.

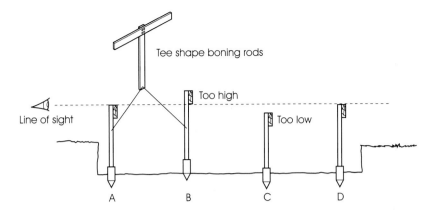

▲ **Figure 7.93** Use of boning rods

As well as levelling, boning rods can be used to establish an even gradient for the bottom of a drainage trench, etc. The method used is the same as before except that the end pegs are first driven to their respective levels. Figure 7.94 illustrates the use of site rails and a traveller to check the level of an excavation. The site rails are positioned at a height that is convenient for an operative to sight over and the traveller cut to the required length. By sighting all three points it can be seen where the surface is correct, too high or too low.

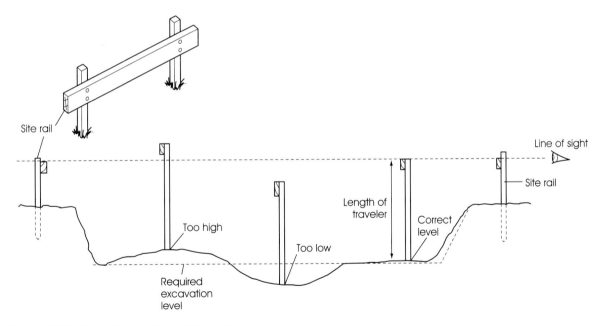

▲ **Figure 7.94** Use of site rails and traveller

Vertical setting out

The verticality (plumb) of a building can be controlled using a spirit level or a suspended **plumb bob** (Figure 7.95). More specialist techniques may be employed on larger structures, such as the use of a theodolite, optical auto-plumb or laser beam.

Vertical positioning of windows, floors and wall plates, etc., can be achieved with the use of a **storey rod** (Figure 7.96), which is marked out before work commences and fixed in one corner of the building.

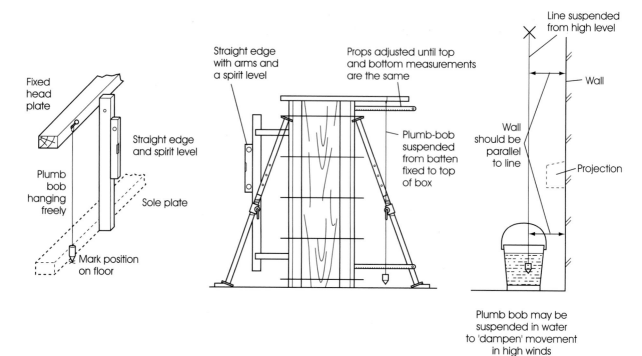

▲ **Figure 7.95** Use of spirit level and plumb bob

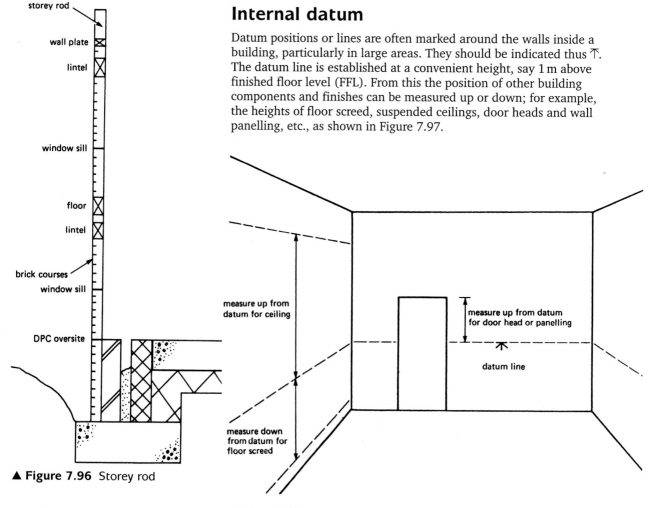

▲ **Figure 7.96** Storey rod

Internal datum

Datum positions or lines are often marked around the walls inside a building, particularly in large areas. They should be indicated thus ⊼. The datum line is established at a convenient height, say 1 m above finished floor level (FFL). From this the position of other building components and finishes can be measured up or down; for example, the heights of floor screed, suspended ceilings, door heads and wall panelling, etc., as shown in Figure 7.97.

▲ **Fiure 7.97** Measuring from datum line to establish internal levels

To establish the datum line, transfer a level position to each corner of the room using either a water level as shown in Figure 7.98 or an optical instrument as shown in Figure 7.99.

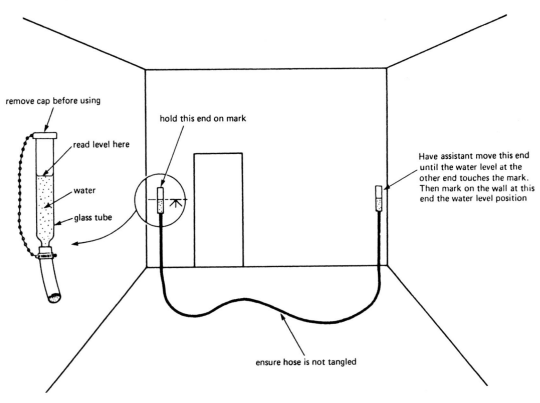

remove cap before using

read level here

water

glass tube

hold this end on mark

Have assistant move this end until the water level at the other end touches the mark. Then mark on the wall at this end the water level position

ensure hose is not tangled

▲ **Figure 7.98** Using a water level to establish datum line

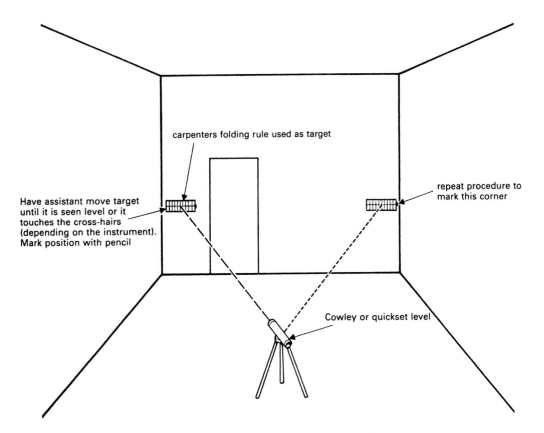

carpenters folding rule used as target

Have assistant move target until it is seen level or it touches the cross-hairs (depending on the instrument). Mark position with pencil

repeat procedure to mark this corner

Cowley or quickset level

▲ **Figure 7.99** Using an optical instrument to establish datum line

Having established the corner positions, stretch a chalk line between each two marks in turn and spring it in the middle, leaving a horizontal chalk dust line on the wall.

Setting up a water level

Before using a water level it must be prepared by filling it from one end with water, taking care not to trap air bubbles (Figure 7.100). Check by holding up the two glass tubes side by side: the levels of the water should settle to the same height. Electronic versions of the water level are also available. These have an electronic box at one end and a sighting tube at the other. The box is fixed to a datum and when the water in the sighting tube is level a bleeping signal is emitted.

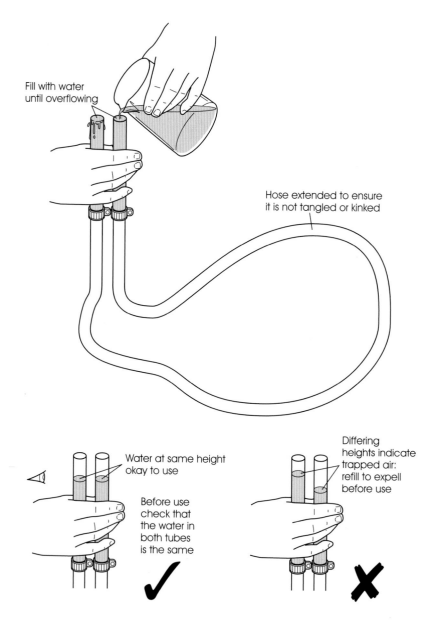

Fill with water until overflowing

Hose extended to ensure it is not tangled or kinked

Water at same height okay to use

Before use check that the water in both tubes is the same

Differing heights indicate trapped air: refill to expel before use

▶ **Fiure 7.100** Preparing a water level before use

Chapter Eight

Applied calculations

This chapter covers the work of both the site carpenter and bench joiner; it is concerned with the types of calculations that you may be required to undertake. It includes the following:

- ✍ basic rules of numbers
- ✍ units of measurement
- ✍ ratios, proportions and percentages
- ✍ statistics
- ✍ powers and roots
- ✍ angles, area and volume
- ✍ Pythagoras' rule and trigonometry
- ✍ measurement and costing of materials.

Carpenters and joiners use calculations as a means of communicating information on a daily basis. Typical questions encountered are:

- ■ How many do we need?
- ■ How long is it?
- ■ What is the area or volume?
- ■ How long will it take?
- ■ How much will it cost?
- ■ How much will I earn?

The answers to these questions will involve you in the manipulation of numbers in the form of calculations. Simple calculations may be carried out in your head; this is called **mental arithmetic**. Other calculations should be written down and worked out longhand or with the use of a battery-powered calculator. Readers are expected to have prior knowledge of the main basic rules of numbers and numerical processes; other skills are briefly covered in this chapter to provide a source of reference and revision as required. All examples have been written out longhand with the working out normally being undertaken with the aid of a calculator.

Basic rules of numbers

The four basic rules of number calculations are **addition, subtraction, multiplication** and **division**.

Mathematical operators are the signs or symbols used when operating with numbers:

- Addition is shown as + (plus sign) which means 'add' or 'sum'.
- Subtraction is shown as a – (minus sign) which means 'subtract' or 'take away'.
- Multiplication is shown as × (multiplication sign) which means 'multiply' or 'times'.
- Division is show as ÷ (division sign) which means 'divide' or 'share'.
- The result of addition, subtraction, multiplication or division is shown after the = (equals sign).

These are the most common symbols used in mathematical calculations.

Combined mathematical operations

You will need to understand the following combinations of mathematical statements:

$$12 - 6 + 4$$
$$3 \times (4 - 1) \text{ which can be written as } 3(4 - 1)$$
$$3 + 12 \div 4 \text{ which can be written as } 3 + 12/4$$
$$(40 \div 2) + (4 \times 6)$$
$$2(5 - 3) = 4$$

Rules for combined mathematical operations:

- You must work out the operation contained in the brackets ( ) first before proceeding with the remaining calculation.
- You must then do multiplication and division before addition and subtraction.

> ### Examples
> $$3(12 \div 3) = 3(4) = 3 \times 4 = 12$$
> $$40 \div 2 + (4 \times 6) = 40 \div 2 + (24) = 20 + 24 = 44$$

Fractions

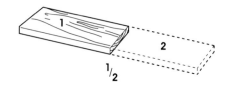

Fractions are parts of a whole number (Figure 8.1). A half is written as $\frac{1}{2}$ – it means one part of two parts. Two thirds is written as $\frac{2}{3}$ – it means two parts of three parts. Three quarters is as $\frac{3}{4}$ – it means three parts of four parts.

- The top part of a fraction is called the **numerator** and the bottom part is called the **denominator**.
- Fractions like $\frac{1}{2}$ and $\frac{2}{3}$ are called **proper fractions**.
- Fractions like $\frac{3}{2}$ and $\frac{5}{3}$ are called **improper fractions**.
- Fractions like $1\frac{1}{2}$ and $1\frac{2}{3}$ are called **mixed numbers**.

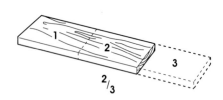

▲ **Figure 8.1** Fractions

Decimals

It is preferable to convert fractions into decimals before proceeding with addition, subtraction, multiplication or division operations.

Decimals are used for parts of a whole number (Figure 8.2). The word 'decimal' is short for 'decimal fraction'. Dividing the bottom number into the top number does the conversion.

$\frac{1}{4}$ becomes 0.25
$\frac{1}{2}$ becomes 0.5
$\frac{3}{4}$ becomes 0.75
$\frac{2}{3}$ becomes 0.667
$1\frac{1}{2}$ becomes 1.5 (as in Figure 8.2).

Note that:

- A **decimal point** (.) is used to separate the whole number from the decimal fraction.

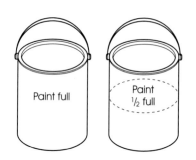

Paint full

Paint ½ full

▲ **Figure 8.2** 1.5 tins of paint

■ Decimal fractions are based on tenths and hundreds.
■ 1.5 means $1\frac{5}{10}$, which is the same as $1\frac{1}{2}$ (see above).
■ 4.25 means $4\frac{25}{100}$, which is the same as $4\frac{1}{4}$ (see above).

Using approximate answers

Common causes of incorrect answers to calculation problems are incomplete working out and incorrectly placed decimal points. Even when using a calculator, wrong answers are often the result of miskeying; even a slight hesitation on a key can cause a number to be entered twice.

Rough checks of the expected size of an answer and the position of the decimal point would overcome these problems. These rough checks can be carried out quickly using approximate numbers.

Example

4.65 × 2.05 ÷ 3.85
Rough check, say 5 × 2 ÷ 4 = 2.5
The calculator answer is 2.476 which is in close agreement

The rough check confirms that the answer is 2.476 and not 0.2476 or 24.76 etc.

Rounding numbers

Number of decimal places. For most purposes, calculations that show three decimal figures are considered sufficiently accurate. These can therefore be 'rounded off' to three decimal places. This, however, first entails looking at the value of the fourth decimal figure: if it is a five or above we **round up** the value by adding one to the third decimal figure. Where it is below five we make no change to the third decimal figure. This process maintains a good level of accuracy for the number.

Example

33.012357 becomes 33.012 for 3 decimal places, ignoring '357'
2.747642 is rounded up to 2.748 for 3 decimal places, adding 0.001 in place of '642'

Number of significant figures. On occasions a number may have far too many figures before and after the decimal place for practical purposes. This is overcome by expressing it to 2, 3 or 4 significant figures.

The term 'significant figures' is normally abbreviated to S.F. or sig. fig.

In the following examples, we have applied the same rule as for rounding numbers to a number of decimal places. If the next figure after the last S.F. is 5 or more then round up the last S.F. by 1.

After rounding up to a number of significant figures, you must take care not to change the place value of the decimal point.

■ 4582 is 4600 to 2 S.F., so 'trailing zeros' are added to the end in order to maintain the place value of the imaginary decimal point at the end. In this example '5' has been rounded up to '6'.
■ 4.582 = 4.6 to 2 S.F. Here there is no need to add 'trailing zeros' to maintain the place value of the decimal point.
■ 68.936102 is 68.94 to 4 S.F. using the rounding up rule for the 4th figure since the 5th figure is 5 or above – in this case 6.

Units of measurement

Metrication of measurements in the construction industry is almost total. Previously we used imperial units of measurement. However, you may still come across them occasionally, so knowledge of both is required. To avoid the possibility of mistakes all imperial measurements should be converted to metric ones. All calculations and further work can then be undertaken in metric units of measurement.

The **metric units** we use are also known as **SI units** (Systeme International d'Unites). For most quantities of measurement there is a *base unit*, a *multiple unit* 1000 times larger and a *sub-multiple unit* 1000 times smaller.

Example

- The base unit of length is the metre (m).
- Its multiple unit is the kilometre (km) which is a thousand times larger: $1\,m \times 1000 = 1\,km$.
- The sub-multiple unit of the metre is the millimetre (mm) which is a thousand times smaller: $1\,m \div 1000 = 1\,mm$.
- Note that $1\,mm \times 1000 = 1\,m$; $1\,km \div 1000 = 1\,m$.

Examples

$6500\,m = 6.5\,km$
$0.55\,m = 550\,mm$

It is an easy process to convert from one unit to another in the metric system. The decimal point is moved 3 places to the left when changing to a larger unit of length and three places to the right when changing to a smaller unit.

Units in common use

Time and angular measurements are the only units in everyday use that are not based on multiples of ten.

Measuring length

Length is a measure of how long something is from end to end, see Figure 8.3:

- *Long lengths* such as the distance between two places on a map are measured in **kilometres** (km).
- *Intermediate lengths* such as the length of a piece of timber are measured in **metres** (m).
- Small lengths such as the length of a screw or width of a board are measured in **millimetres** (mm).

Although the metre and the millimetre are the main units of length used by the carpenter and joiner, sometimes centimetres (cm) are encountered. Conversion into other metric units is as follows

$$1\,m = 100\,cm; \quad 1\,cm = 10\,mm = 0.01\,m$$

However, as the centimetre is not normally used in the construction industry all measurements should be stated in either metres or millimetres, e.g. 55 cm should be written as either 0.550 m or 550 mm.

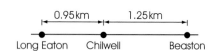

Long lengths in kilometres (km)

Intermediate lengths in metres (m)

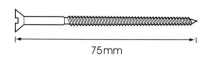

Small lengths in millimetres (mm)

▲ **Figure 8.3** Length

Measuring area

Area is concerned with the extent or measure of a surface. The floor illustrated in Figure 8.4 is 3 m wide and 4 m long. We say it has an area of 12 square metres (m^2).

- Two linear measurements multiplied together give the square area (square measure) created by the two lengths.
- Square metres and square millimetres are used for the units of area.
- These are written as m^2 or mm^2.

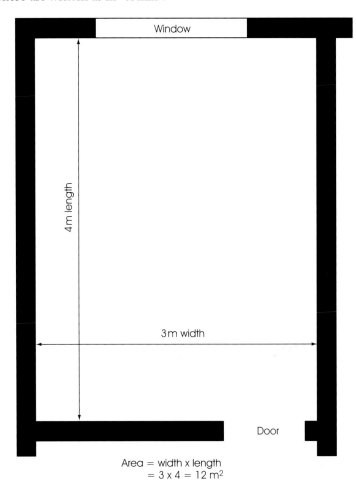

Area = width x length
= 3 x 4 = 12 m^2

▲ **Figure 8.4** Area of a room

Example

A floor of sides 3 m and 4 m has an area of 12 square metres written as 12 m^2.

Measuring volume and capacity

Volume is concerned with the space taken up by a solid object. The room illustrated in Figure 8.5 is 4 m wide by 5 m long and 3 m high, it has a volume of 60 cubic metres, written as 60 m^3.

Capacity is concerned with the amount of space taken up by liquids or the amount of liquid that can fit into a given volume (Figure 8.6):

- The **litre** is the unit of capacity. To avoid confusion with the number 1, 'litre' is best written in full or as 'L'.
- The **millilitre** (ml) is the sub-multiple.

  1 litre = 1000 ml; 1 ml = 0.001 L

Capacity is linked to volume.

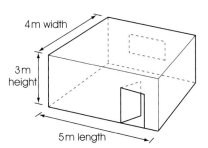

Volume = width x length x height
= 4 x 5 x 3
= 60 m^3

▲ **Figure 8.5** Volume

▲ **Figure 8.6** Capacity

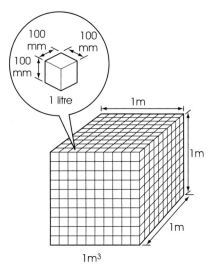

▲ **Figure 8.7** Relationship between capacity and volume

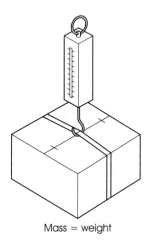

Mass = weight

▲ **Figure 8.8** Mass

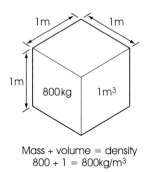

Mass ÷ volume = density
800 ÷ 1 = 800kg/m³

▲ **Figure 8.9** Density

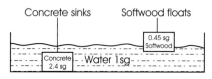

Specific gravity = Density ÷ 1000

▲ **Figure 8.10** Specific gravity

Example

A carton 100 mm × 100 mm × 100 mm encloses a volume of 0.1 m × 0.1 m × 0.1 m = 0.001 m³. This volume holds a litre of water. It would therefore take 1000 of these cartons to fill 1 m³, see Figure 8.7.

Measuring mass or weight

Mass is concerned with the weight of an object, see Figure 8.8:

- The unit of mass is the **kilogram** (kg).
- Its multiple is the **tonne** (to avoid confusion with the imperial ton, no abbreviation is used).
- The sub-multiple unit is the **gram** (g). For very small objects the sub-sub-multiple **milligram** (mg) is used.

1 tonne = 1000 kg; 1 kg = 1000 g; 1 g = 1000 mg
1 mg = 0.001 g; 1 g = 0.001 kg; 1 kg = 0.001 tonne

Measuirng density of an object

Density is a measure of a material's consistency. A more solid material is generally considered to have a higher density (more dense), but a true measure of density can only be calculated by measuring its weight and volume (Figure 8.9) since the ratio of weight (mass) for a given volume is a measure of density. Thus:

Mass ÷ Volume = Density

- The density of water is 1000 kg/m³
- The density of softwood is typically 450 kg/m³
- The density of concrete is typically 2400 kg/m³

Note that in order to compare densities of different materials, the ratio of weight to volume has to be converted to a standard unit of measurement, in this case kg/m³.

Specific gravity. The density of materials is normally expressed in smaller units than kg/m³ by dividing values by 1000. This is referred to as the specific gravity (sg).

Density ÷ 1000 = Specific gravity (sg) or relative density

- The specific gravity of water is the value 1 sg.
- The specific gravity of softwood is typically 0.45 sg
- The specific gravity of concrete is typically 2.4 sg
- Materials heavier than water have an sg greater than 1 sg and sink; materials lighter than water have an sg smaller than 1 sg and float, as illustrated in Figure 8.10.

Example

Determine the density of a sample which has a mass of 25 kg and a volume of 0.125 m³. Will it sink or float?
Desnity = 25 ÷ 0.125 = 200 kg/m³
Specific gravity = 200 ÷ 1000 = 0.2
Specific gravity of water = 1, therefore the sample is lighter than water and will float

Measuring force

Force is concerned with the effect that the earth's gravitational pull has on a mass (Figure 8.11).

- The unit of force is the Newton (N)
- Its multiple is the kilonewton (kN)
- A mass of 1 kg will exert a gravitational force of 9.81 N

Newtons are used in place of kilograms in calculations involving the force exerted by the mass of building materials in a structure.

A 25 kg bag of cement placed on firm ground will exert a force of 25 × 9.81 = 245.25 N; the earth in turn will push back with an equal and opposite force of 245.25 N called a reaction, see Figure 8.12.

As force acts over an area the following units may be encountered:
- kN/m^2 for superimposed loads over a wide area;
- N/mm^2 for stress on a localised area.

Measurement of time

Time is a measure of the continued progress of existence, the past, the present and the future:
- The second (s) is the main unit of time.
- 60 s = 1 minute (min).
- 60 min = 1 hour (hr).
- 24 hr = 1 day

Unlike the previous units, time is not based on multiples of 10 or sub-multiples of 1000. Periods of time less than one day are measured in multiples of 60. Thus extra care must be taken calculating time.

Example

A team of four carpenters takes 8 hours 48 minutes to erect a timber frame. Determine the total time worked.

Multiply the hours and minutes separately:

4 × 8 = 32 hrs and 4 × 48 = 192 min

Divide the total of minutes by 60 to convert to hours:

192 ÷ 60 = 3.2 hr

Note: this is 3.2 hr **not** 3 hrs 20 min. Multiply the decimal part of the hours by 60 to convert back to minutes;

0.2 × 60 = 12 min

Add the two totals together, keeping the hours and minutes in separate columns:

32 hr + 3 hr 12 min = 35 hr 12 min

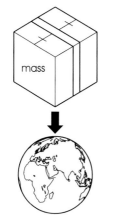

▲ **Figure 8.11** Force

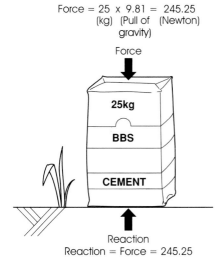

▲ **Figure 8.12** Force and reaction

Measuring angles

An angle is the amount of space between two intersecting straight lines (Figure 8.13). The angle does not depend on the length of the lines or the distance from the intersection that it is measured. The size of an angle is measured in degrees (°). See Figure 8.14 for angle properties:

- A **right angle** is 90°. They are shown on drawings by either a small square in the corners, or an arc with arrowheads and the figures 90°.
- An **acute angle** is between 0° and 90°.
- An **obtuse angle** is between 90° and 180°.
- A **reflex angle** is between 180° and 360°.
- Angles on a **straight line** always add up to 180°: A + B + C = 180°.
- Angles at a **point** always add up to 360°: A + B + C = 360°.
- Angles in a **circle** always add up to 360°: A + B = 360°.
- Angles in a **quadrilateral** (four-sided figure) always add up to 360°: A + B + C + D = 360°.
- Angles in a **polygon** (more than four sides) always add up to 360°: A + B + C + D + E = 360°.

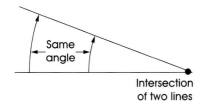

▲ **Figure 8.13** Angle

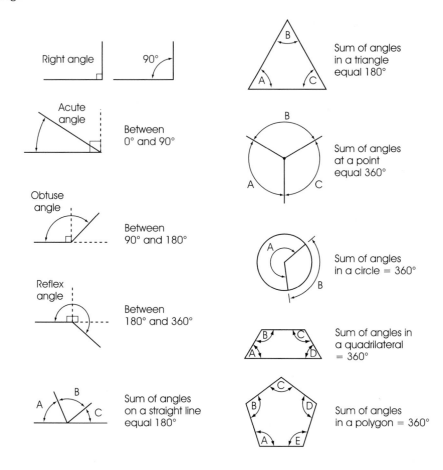

▶ **Figure 8.14** Angle properties

Imperial to metric conversion

Comparisons between imperial and metric units of measurement can be made using the following information:

- 1 inch = 25.4 mm
- 12 inches = 1 foot = 0.3048 m or 304.8 mm
- 3 feet = 1 yard = 0.9144 m or 914.4 mm
- 1 m = 1.0936 yards or 3.281 feet
- 1 yard^2 = 0.836 m^2
- 1 m^2 = 1.196 yard^2 or 10.76 feet^2
- 1 m^3 = 1.31 yard^3 or 35.335 feet^3
- 1 gallon = 8 pints
- 1 pint = 20 fluid oz (ounces)
- 1 litre = 0.22 gallons or 1.76 pints
- 1 ml = 0.035 fluid oz
- 1 ton = 2240 lb (pound)
- 112 lb = 1 cwt (hundredweight)
- 14 lb = 1 st (stone)
- 1 lb = 16 oz (ounces)
- 1 tonne = 2205 lb
- 1 kg = 2.2 lb
- 30 g = 1 oz

Examples

350 feet = 350 × 0.3048 m = 106.7 m
12 yards = 12 × 0.9144 m = 10.973 m
1 gallon = 1 ÷ 0.22 litres = 4.545 litres
10 lbs = 10 ÷ 2.2 kg = 4.545 kg

Ratio, proportion and percentages

Ratios and proportions

These are used to compare or show the relationship between similar or related quantities. **Ratios** are written in the form 5:3 and we say this as 'the ratio of five to three'.

When 8 items are shared in the ratio 5:3, this means that one share contains 5 items out of the 8 and the other share contains 3 items.

Examples

£96 is to be shared by two people in the ration of 5:3. How much will each receive?

- Add up the total number of parts or shares
- Work out what one part is worth
- Work out what the other parts are worth

Total shares = 5 + 3 = 8
One share = 96 ÷ 8 = 12
Five shares = 5 × 12 = 60
Three shares = 3 × 12 = 36

Proportions

These are written in the same way as ratios. A bricklaying mortar may be described as a 1:6 (1 in 6) mix. This means the proportion of the mix is 1 part cement and six parts fine aggregate (sand), see Figure 8.15. You could use bags, buckets or barrows as the unit of measure, providing the proportion is kept the same; they will all make a suitable mortar mix. All that changes is the amount of mortar produced.

Proportions are also used in expressing gradients, i.e. slope of roofs.

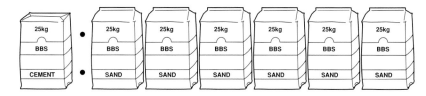

◀ **Figure 8.15** Proportion 1:6

Example

The pitched roof shown in the example has a 1:3 (1 in 3) or $\frac{1}{3}$ pitch (slope). This means that for every 3 m span, the roof rises 1 m.

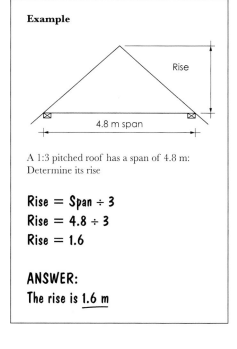

Example

Rise

4.8 m span

A 1:3 pitched roof has a span of 4.8 m:
Determine its rise

Rise = Span ÷ 3
Rise = 4.8 ÷ 3
Rise = 1.6

ANSWER:
The rise is 1.6 m

Percentages

Percentages are commonly used to determine increases and decreases of given quantities.

■ Percentage % means 'per hundred' or 'per cent'.

■ All percentages are fractions over a 100. So 1% is $\frac{1}{100}$ and 50% is $\frac{50}{100}$ etc.

As before when working with fractions it is best to turn them into decimals before undertaking any further calculations. In this case it is necessary to divide by 100.

This can be done quickly by placing an imaginary point behind the percentage and then moving it two places forward in order to divide by 100:

■ 5% becomes 0.05

■ 10% becomes 0.1

■ 75% becomes 0.75

Using percentages. There are four ways in which percentages are used:

1. Finding the percentage of a quantity:
    ■ Turn the percentage into a decimal.
    ■ Multiply the quantity by the decimal.

    ### *Example*
    Find 12% of £55.
    $12\% = \frac{12}{100} = 0.12$
    £55 × 0.12 = £6.60

2. Increasing a quantity by a percentage
    ■ Turn the percentage into a decimal.
    ■ Place a one in front of it to include the original quantity.
    ■ Multiply the quantity by the decimal.

    ### *Example*
    Increase 75 kg by 20%.
    20% = 0.2
    75 kg × 1 + 0.2 = 75 kg × 1.2 = 90 kg

3. Reducing a quantity by a percentage:
    ■ Take the percentage reduction away from 100.
    ■ Convert to a decimal.
    ■ Multiply the quantity by the decimal.

    ### *Example*
    Reduce £95 by 15%.
    100% – 15% = 85%
    85% = 0.85
    £95 × 0.85 = £80.75

4. Expressing one quantity as a percentage of another quantity:
    ■ Write the two quantities as a fraction.
    ■ Convert to a decimal by dividing the top by the bottom.
    ■ Multiply the decimal by 100 to convert to a percentage.

    ### *Example*
    Express 17 out of 20 as a percentage.
    Ratio $\frac{17}{20} = 17 \div 20 = 0.85$
    0.85 × 100 = 85%

Averages, roots and formulae

Averages

An average is the mean value of several numbers or quantities.

> Average of a set of numbers = Sum of numbers ÷ Number of quantities

Example

To find the average of these six numbers, 3, 5, 6, 7, 9 and 15, we add them together and divide by 6:

Average = (3 + 5 + 6 + 7 + 9 + 15) ÷ 6 = 45 ÷ 6 = 7.5

The average value of the six numbers is 7.5.

Mean, median and mode

These are different types of averages:

- *Mean* is the true average we mainly refer to. It equals the sum of values divided by the number of values in the range, group or set of numbers.
- *Median* is the middle value when the numbers are put in order of size.
- *Mode* is the value that occurs most frequently.

Example

The number of people late for work over a ten-day period was: 4, 2, 0, 1, 2, 3, 3, 6, 4 and 2.

Mean = (4 + 2 + 0 + 1 + 2 + 3 + 3 + 6 + 4 + 2) ÷ 10 = 27 ÷ 10 = 2.7

This means that 2.7 people are late per day on average.

To calculate the **median**, first put in size order:

0, 1, 2, 2, 2, 3, 3, 4, 4, 6

Then cross off numbers from either end to find the middle value(s):

~~0, 1, 2, 2,~~ 2, 3, ~~3, 4, 4, 6~~

The median is the middle value or the mean of the two middle values, 2 and 3 in this example = 2.5

Therefore, the median is 2.5 people late per day

The **mode** = 2 people late per day, because it occurs the most number of times.

Range of numbers. When using mode the range should be stated, to show how much the information is spread. In the above example:

Range = highest value – Lowest value = 6 – 0 = 6

Powers and roots of numbers

Powers

A simple way of writing repeated multiplication of the same number is shown below:

$10 \times 10 = 100$ or 10^2
$10 \times 10 \times 10 = 1000$ or 10^3
$10 \times 10 \times 10 \times 10 = 10,000$ or 10^4 and so on

The small raised number is called the power or index. Numbers raised to the power 2 are usually called square numbers. We say 10^2 is '10 squared', and 10^3 is '10 cubed' but we say 10^4 is '10 to the power 4', and 10^5 is '10 to the power 5', etc.

Large numbers can therefore be written in a *standard shorthand form* by the use of an index or power:

$$30,000,000 = 3 \times 10,000,000 \text{ or } 3 \times 10^7$$
$$6,600,000 = 6.6 \times 1000,000 \text{ or } 6.6 \times 10^6$$
$$990 = 9.9 \times 100 \text{ or } 9.9 \times 10^2$$

The index number is the number of places that the decimal point has to be moved to the right if the number is written in full.

Small numbers less than one can also be written in standard shorthand form, by using a negative power or index:

$$0.036 = 3.6 \times 10^{-2} \text{ or } 36 \times 10^{-3}$$
$$0.0099 = 9.9 \times 10^{-3} \text{ or } 99 \times 10^{-4}$$
$$0.00012 = 1.2 \times 10^{-4} \text{ or } 12 \times 10^{-5}$$

The negative index is the number of places that the decimal point will have to be moved to the left if the number is written in full. For example:

$$99 \times 10^{-4} = 0.0099$$

Roots

Square root. It is sometimes necessary to find a particular root of a number. Finding a root of a number is the opposite process of finding the power of a number. Hence, the square root of a number is a smaller number multiplied by itself once to give the original number:

The square of 5 is 5^2 or $5 \times 5 = 25$.
Therefore the square root of 25 is 5.

The common way of writing this is to use the square root sign $\sqrt{}$ as in $\sqrt{25} = 5$.

Cube root. This is a number multiplied by itself twice to give the number in question:

The cube of 5 is 5^3 or $5 \times 5 \times 5 = 125$.
Therefore, the cube root of 125 is 5.

The common way of writing this is to use the root sign and an index of 3, i.e. $\sqrt[3]{}$ as in $\sqrt[3]{125} = 5$

From this we can see why there is a connection between powers and roots as opposite processes.

$$10^2 = 100; \sqrt{100} = 10$$
$$10^3 = 1000; \sqrt[3]{1000} = 10$$

Calculating roots with the aid of a calculator. Roots of numbers are easily found using a calculator with square or cube root function keys: $\sqrt{}$ $\sqrt[3]{}$

Depending on the layout of your calculator you may have to press an inverse key first (INV).

Formulae

Formulae are normally stated in algebraic terms. Algebra uses letters and symbols instead of numbers to simplify statements and enable general rules and relationships to be worked out.

Transposition of formulae

When solving a problem, sometimes formulae have to be rearranged in order to change the subject of the formulae before the calculation is carried out. Basically anything can be moved from one side of the equals sign to the other by changing its symbol.

This means that on crossing the equals sign:

- plus changes to minus;
- multiplication changes to division;
- powers change to roots.

This is also true vice versa.

Example

The relationship between the length of the three sides (a, b, c) of a right-angled triangle is given by the following formula (Pythagoras' theorem):

$c^2 = a^2 + b^2$, e.g. $5^2 = 3^2 + 4^2$

This formula can be transposed in the following ways:

$b^2 = a^2 - c^2$; $c^2 = a^2 - b^2$

$b = \sqrt{(a^2 - c^2)}$; $c = \sqrt{(a^2 - b^2)}$

Cross-multiplication. This means that on crossing the equals sign anything on the top line moves to the bottom line, and conversely anything on the bottom line moves to the top line. Hence: if $V = IR$

then $I = \dfrac{V}{R}$ and $R = \dfrac{V}{I}$

Use of formulae with triangles

Area. The area of a triangle can be found by using the formula (Figure 8.16):

Area = (Base × Height) ÷ 2 or $A = \dfrac{B \times H}{2}$

Example

Area = $\dfrac{(2.5\,\text{m} \times 4.8\,\text{m})}{2}$ = $6\,\text{m}^2$

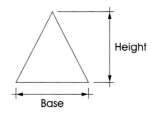

▲ **Figure 8.16**

Use of formulae with rectangles

Perimeter. The perimeter of a rectangle can be found by using the formulae (Figure 8.17):

Perimeter = 2 × (Length + Breadth)

This can be abbreviated to:

$P = 2(L + B)$

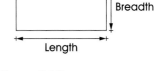

▲ **Figure 8.17**

Note that the multiplication sign (×) has been removed although it is still there in reality: the figure '2' means '2 ×' when placed in front of another part of the formula. Plus and minus signs (+ and −) cannot be 'hidden' in this way and must always be shown in the formulae. Using the rule for combined operations, L must be added to B before multiplying by 2.

Example

Find the perimeter of a rectangle having a length of 3.6 m and a breadth of 2.2 m.

Perimeter = 2(L + B) = 2(3.6 + 2.2) = 2(5.8) = 11.6 m

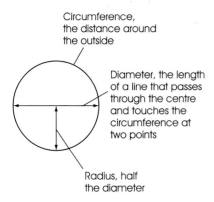

▲ **Figure 8.18** Parts of a circle

▶ **Figure 8.19** Relationship between circumference and diameter

Use of formulae with circles

Perimeter and diameter: The formula for the perimeter or circumference of a circle is (Figure 8.18):

Circumference = π × Diameter
C = πD

'π' (spoken as 'pi') is the number of times that the diameter will divide into the circumference (Figure 8.19). It is the same for any circle and is taken to be 3.142.

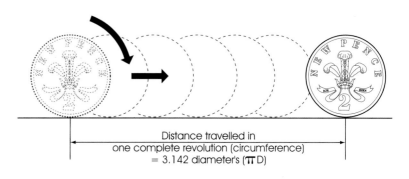

Distance travelled in one complete revolution (circumference) = 3.142 diameter's (πD)

Suppose we were given the circumference and asked to find the diameter. Since C = πD, then C ÷ π = D. Hence:

Diameter = Circumference ÷ π

Example

Find the diameter of a circle having a circumference of 7.855 m.
D = C ÷ π = 7.855 ÷ 3.142 = 2.5

Formulae for areas and volumes

Values of **common shapes** can be found by using the formulae illustrated in Figure 8.20 (following page).

■ The **perimeter** of a figure is the distance or length around its boundary, linear measurement, usually given in metres run (m).

■ The **area** of a figure is the extent of its surface, square measurement, usually given in square metres (m²).

■ The **volume** of an object can be defined as the space it takes up, cubic measurement, usually given in cubic metres (m³).

Complex areas can be calculated by breaking them into a number of recognisable areas and solving each one in turn, converting all measurements first to the same units e.g. metres (m).

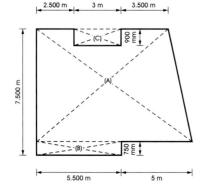

▲ **Figure 8.21** Complex areas

Example

The area of the room shown in Figure 8.21 is equal to area A plus area B minus area C. Find the total area.

Area A = (9 + 10.5) ÷ 2 × 6.75 = 65.813 m²
Area B = 0.75 × 5.5 = 4.125 m²
Area C = 0.9 × 3 = 2.7 m²
Total area = A + B + C = 67.238 m²

Many solids have a uniform cross-section and parallel edges (Figure 8.22). The volume of these can be found by multiplying the base area by the height:

Volume = Base area × Height

▲ **Figure 8.22** Volume of a solid

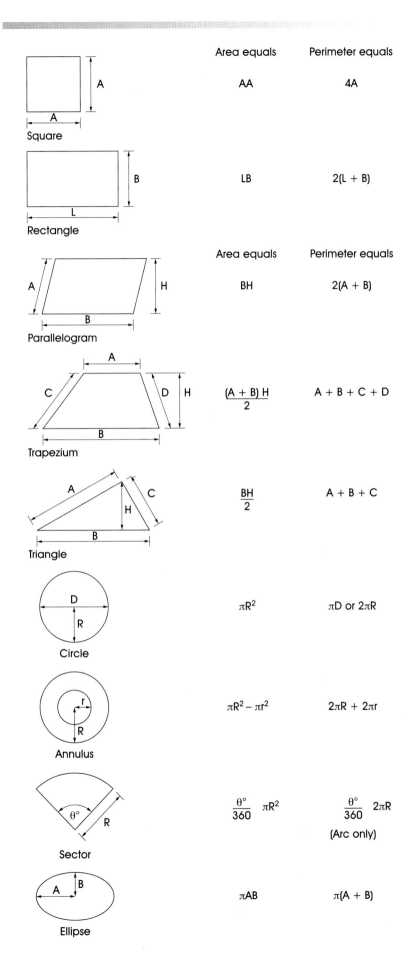

	Area equals	Perimeter equals
Square	AA	4A
Rectangle	LB	2(L + B)

	Area equals	Perimeter equals
Parallelogram	BH	2(A + B)
Trapezium	$\dfrac{(A + B) H}{2}$	A + B + C + D
Triangle	$\dfrac{BH}{2}$	A + B + C
Circle	πR^2	πD or $2\pi R$
Annulus	$\pi R^2 - \pi r^2$	$2\pi R + 2\pi r$
Sector	$\dfrac{\theta°}{360}\ \pi R^2$	$\dfrac{\theta°}{360}\ 2\pi R$ (Arc only)
Ellipse	πAB	$\pi(A + B)$

◀ **Figure 8.20** Formulae for areas and perimeters

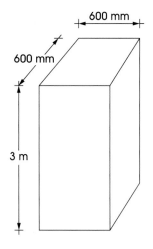

▲ Figure 8.23

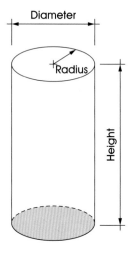

▲ Figure 8.24

Example

Find the volume of concrete required for the column illustrated in Figure 8.23.

Volume = Base area × Height = 0.6 × 0.6 × 3 = 1.08 m³

Example

A house contains 48 50 mm × 225 mm softwood joists 4.5 m long. How many cubic metres of timber are required?

Volume of 1 joist = Width × Depth × Length = 0.05 × 0.225 × 4.5
= 0.050625 m³
Total volume = 0.050625 × 48 = 2.43 m³

The **lateral surface area** of a solid with a uniform cross-section is found by multiplying its base perimeter by its height (Figure 8.24).

Lateral surface area = Base perimeter × Height

For a cylinder:

Lateral surface area = π(Diameter × Height)

The total surface area can be determined by adding the areas of ends or base to the lateral surface area. For a cylinder:

Total surface area of a cylinder = π × Diameter × Height + π(Radius × Radius) × 2
= (Lateral area) + 2(Area of each end)
Surface area = $\pi DH + 2\pi R^2$

Example

Find the total surface area of a cylinder 1.2 m diameter and 2.4 m height.

Total surface area = $\pi D \times H + (\pi R^2 \times 2)$
= (3.142 × 1.2 × 2.4) + (3.142 × 0.6 × 0.6 × 2)
= 9.04896 + 2.26224 = 11.3112 m² = 11.31 m² to 2 dec. pl.

Common solids. The formulae illustrated in Figure 8.25 (following page) can be used for calculating the volume and lateral surface area of frequently used common solids. It can be seen that the volume of any pyramid or cone will always be equal to one third of its equivalent prism or cylinder.

Complex volumes are found by breaking them up into a number of recognisable volumes and solving each one in turn. This is the same as the method used when solving for complex areas.

Example

Suppose you were asked to find the volume of concrete required for a 2.4-m high column having the plan or horizontal cross-section shown in Figure 8.26. The column can be considered as a rectangular prism (A) and half a cylinder (B).

Volume A = 0.4 × 0.6 × 2.4 = 0.576 m³
Volume B = 3.142 × 0.2 × 0.2 × 2.4 ÷ 2 = 0.151 m³
Total volume of concrete required = 0.576 + 0.151 = 0.727 m³

Example

Where a solid tapers, its volume can be found by multiplying its average cross-section by its height (Figure 8.27).

Volume = Average cross-section × Height
Volume = (0.8 × 0.8) + (0.5 × 0.5) ÷ 2 × 4.5
= 0.64 + 0.25 ÷ 2 × 4.5 = 2.003 m³

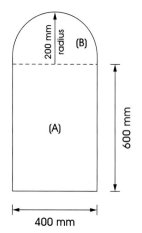

▲ Figure 8.26

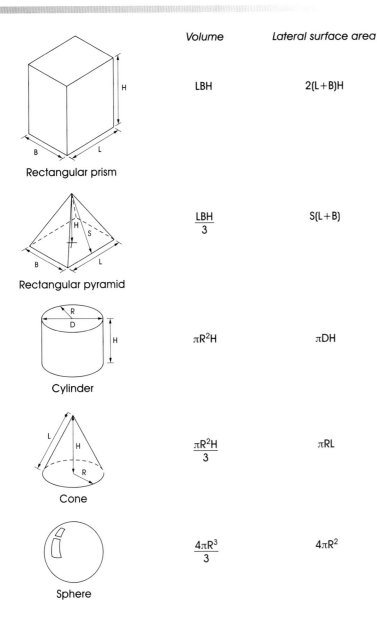

	Volume	Lateral surface area
Rectangular prism	LBH	2(L+B)H
Rectangular pyramid	$\dfrac{LBH}{3}$	S(L+B)
Cylinder	$\pi R^2 H$	πDH
Cone	$\dfrac{\pi R^2 H}{3}$	πRL
Sphere	$\dfrac{4\pi R^3}{3}$	$4\pi R^2$

◀ **Figure 8.25** Formulae for volume and surface area

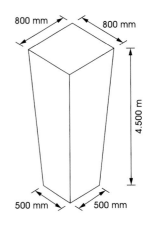

▲ **Figure 8.27**

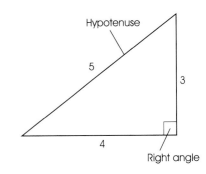

▲ **Figure 8.28** Hypotenuse is the longest side and is opposite the right angle

Pythagoras' rule

The lengths of the sides in a right-angled triangle can be found using Pythagoras' rule, which is also known as Pythagoras' theorem. He was an early Greek who discovered a unique fact about right-angled triangles. According to his rule, in a right-angled triangle the square of the length of the hypotenuse (longest side) is equal to the sum of the squares of the lengths of the other two sides (Figure 8.28).

Example

Find the length of the hypotenuse 'C' in Figure 8.29:

$A^2 + B^2 = C^2$

$6^2 + 8^2 = 100 = C^2$

$C = \sqrt{100} = 10$

The length of the hypotenuse, C, is 10.

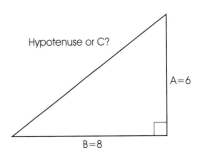

▲ **Figure 8.29** Pythagoras' rule

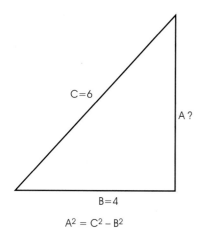

$$A^2 = C^2 - B^2$$

▲ **Figure 8.30** Rearranging Pythagoras' rule

Example

Finding the length of the unknown side 'A' in Figure 8.30:

$$A^2 + B^2 = C^2$$

As 'A' is the unknown we need to take B^2 away from both sides of the equals sign to give:

$$A^2 = C^2 - B^2 = 6^2 - 4^2 = 20$$
$$A = \sqrt{20} = 4.4721359$$

The length of the unknown side is 4.472 to three decimal places.

Intersecting chords rule

Where two chords intersect in a circle, the product (after multiplication) of the two parts of one chord will always be equal to the product of the two parts of the other chord, see Figure 8.31. The rule, which is written as $A \times B = C \times D$ is very useful for finding radius lengths.

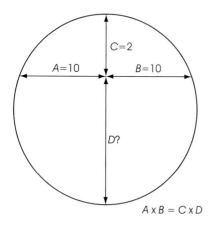

$$A \times B = C \times D$$

▶ **Figure 8.31** Intersecting chord rule

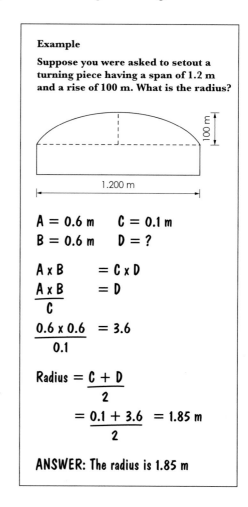

Example

Suppose you were asked to setout a turning piece having a span of 1.2 m and a rise of 100 m. What is the radius?

A = 0.6 m   C = 0.1 m
B = 0.6 m   D = ?

$$A \times B = C \times D$$
$$\frac{A \times B}{C} = D$$
$$\frac{0.6 \times 0.6}{0.1} = 3.6$$

$$Radius = \frac{C + D}{2}$$
$$= \frac{0.1 + 3.6}{2} = 1.85 \text{ m}$$

ANSWER: The radius is 1.85 m

Trigonometry

Trigonometry involves understanding the relationship between the sides and angles of right-angled triangles.

You will have to set up and solve equations to find unknown lengths or angles. There are three ratios to understand in right-angled triangles. These are sine, cosine and tangent ratios.

In trigonometry, the sides of a right-angled triangle are given temporary names in relation to the angle 'θ' being considered, see Figure 8.32.

The ratios of these sides to each other are the **trigonometrical ratios** of the angle θ:

The **sine** of θ = Opposite side/Hypotenuse (longest side), or abbreviated:

$$\sin \theta = \frac{\text{opp}}{\text{hyp}}$$

The **cosine** of θ = Adjacent side/ Hypotenuse, or abbreviated:

$$\cos \theta = \frac{\text{adj}}{\text{hyp}}$$

The **tangent** of θ = Opposite side/Adjacent side, or abbreviated:

$$\tan \theta = \frac{\text{opp}}{\text{adj}}$$

▲ **Figure 8.32** Names given to the sides of a triangle in relation to angle θ

Using trigonometry

If we know two parts of the equation we can use trigonometry to find the unknown third part.

Finding the length of an unknown side

Your calculator may have different angle modes; you should set it to degrees. With DEG showing on the display enter 30 then the sin key. If the mode is correctly set it should display 0.5:

$$\sin 30° = 0.5$$

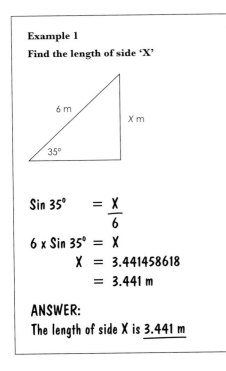

Example 1

Find the length of side 'X'

6 m   X m   35°

Sin 35° $= \dfrac{X}{6}$

6 x Sin 35° = X

X = 3.441458618

= 3.441 m

ANSWER:
The length of side X is 3.441 m

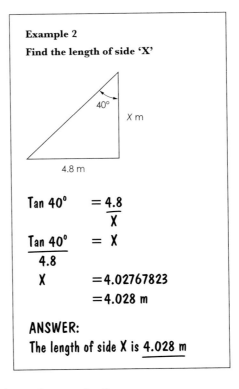

Example 2

Find the length of side 'X'

40°   X m   4.8 m

Tan 40° $= \dfrac{4.8}{X}$

$\dfrac{\text{Tan } 40°}{4.8} = X$

X = 4.02767823

= 4.028 m

ANSWER:
The length of side X is 4.028 m

In Example 1 we know the hypotenuse and the angle opposite X. Thus use the sine ratio.

Transpose formula to get **X** on its own. Enter 35 sin × 6 = into the calculator and round answer to three decimal places.

In Example 2 we know the angle adjacent to **X** and the length of the opposite side. Thus use the tangent ratio. Transpose formula to get **X** on its own. Enter 40 tan ÷ 4.8 = into the calculator and round to three decimal places.

Finding an unknown angle

On your calculator you will find keys marked $\sin^{-1}$, $\cos^{-1}$ and $\tan^{-1}$. You may have to press an inverse INV key first. $\sin^{-1}$ means 'the angle whose sin is ...'

$$\sin^{-1} 0.50 = 30°$$

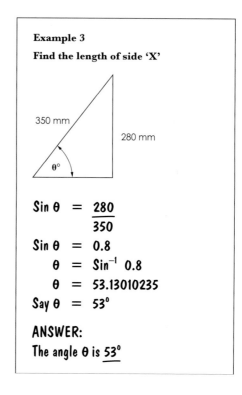

Example 3

Find the length of side 'X'

350 mm

280 mm

$\theta°$

$$\sin \theta = \frac{280}{350}$$

$$\sin \theta = 0.8$$

$$\theta = \sin^{-1} 0.8$$

$$\theta = 53.13010235$$

$$\text{Say } \theta = 53°$$

ANSWER:

The angle θ is $\underline{53°}$

In Example 3 we know the lengths of the opposite side and the hypotenuse. Thus use the sine ratio. Divide opposite by hypotenuse. Enter $0.8 \sin^{-1} =$ (or 0.8 INV sin =) into the calculator. Round the answer to nearest 0.5 of a degree.

Measuring and costing materials

Where timber is sold by the cubic metre you may be required to calculate the metres run of a particular cross-section that can be obtained from it.

To calculate the metres run of a section that can be obtained from a cubic metre, divide the cross-sectional area of a cubic metre (in square millimetres) by the cross-sectional area of the section required, see Figure 8.33.

Example

Determine the metres run of 50×200 mm that can be obtained from $1\,m^3$ of timber.

Cross-section of $1\,m^3 = 1000 \times 1000 = 1,000,000\,mm^2$

Cross-section of 50×200 mm $= 10,000\,mm^2$

Metres run $= 1,000,000 \div 10,000 = 100$

Hence, 100 m run of 50×200 boards can be obtained from $1\,m^3$.

Flooring

In order to determine the amount of floor covering materials required for an area, multiply its width by its length.

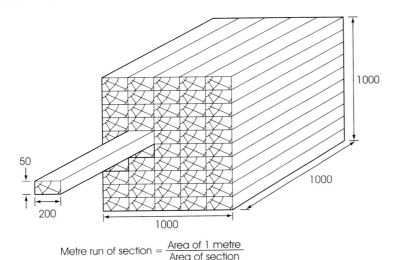

$$\text{Metre run of section} = \frac{\text{Area of 1 metre}}{\text{Area of section}}$$

Floorboards. To calculate the metres run of floorboards required to cover a floor area:

Metres run required = Area ÷ Width of board

Example

An area of $4.65\,\text{m}^2$ is to be boarded. If the floorboards have a covering width of 137 mm, see Figure 8.34, calculate the metres run required:

$4.65 \div 0.137 = 33.94\,\text{m}$, say 34 m run

It is standard practice to order an additional amount of flooring to allow for **cutting and wastage**. This is often between 10% and 15%

137 mm covering width

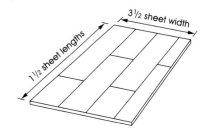

▲ **Figure 8.34**

Example

If 34 m run of floor boarding is required to cover an area, calculate the amount to be ordered including an additional 12% for cutting and wastage.

Amount to be ordered = $34 \times 1.12 = 38.08$, say 38 m run

Sheet material. In order to determine the number of sheets of plywood or chipboard required to cover a room either:

■ Divide area of room by area of sheet, or
■ Divide width of room by width of sheet and divide length of room by length of sheet. Convert these numbers to the nearest whole or half and multiply them together.

Example

To calculate the number of 600 mm × 2400 mm chipboard sheets required to cover the floor area of 2.05 m × 3.6 m shown in Figure 8.35:

Number of sheets required = Area of room ÷ Area of sheet
Area of room = $2.05 \times 3.6\,\text{m} = 7.38\,\text{m}$
Area of sheet = $0.6 \times 2.4 = 1.44\,\text{m}$
Number of sheets = $7.38 \div 1.44 = 5.125$, say 6 sheets

Alternatively:

Number of sheet widths in room width = $2.05 \div 0.6 = 3.417$, say 3.5
Number of sheet lengths in room length = $3.6 \div 2.4 = 1.5$
Total number of sheets = $3.5 \times 1.5 = 5.25$, say 6 sheets

▲ **Figure 8.35**

Joists

To determine the number of joists required and their centres for a particular area the following procedure can be used (Figure 8.36):

■ Measure the distance between adjacent walls, say 3150 mm.

■ The first and last joist would be positioned 50 mm away from the walls. The centres of 50-mm breadth joists would be 75 mm away from the wall. The total distance between end joists centres would be 3000 mm (Figure 8.36a).

■ Divide the distance between end joists centres by the specified joist spacing, say 400 mm. This gives the number of spaces between joists. Where a whole number is not achieved round up to the nearest whole number above. There will always be one more joist than the numbers of spaces so add another one to this figure to determine the number of joist (Figure 8.36b).

■ Where T&G boarding is used as a floor covering the joist centres may be spaced out evenly, i.e. divide the distance between end joist centres by the number of spaces.

■ Where sheet material is used as a joist covering to form a floor, ceiling or roof surface, the joist centres are normally maintained at 400 mm or 600 mm module spacing to coincide with sheet sizes. This would leave an undersized spacing between the last two joists (Figure 8.36c).

Number of joists = (Distance between end joist centres × Joist spacing) + 1 = (3000 ÷ 400) + 1 = 8.5, say 9

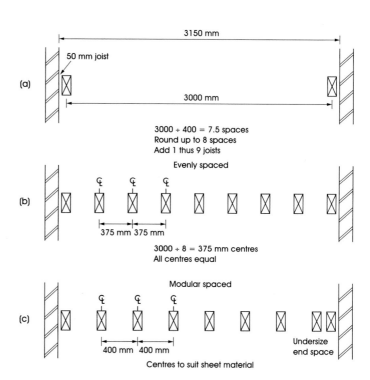

▶ **Figure 8.36**

Stud partitions

To determine the number of studs required for a particular partition the following procedure can be used as illustrated in Figure 8.37:

■ Measure the distance between the adjacent walls of the room or area which the partition is to divide, say 3400 mm.

■ Divide the distance between the walls by the specified spacing, say 600 mm. This gives the number of spaces between the studs. Use the whole number above. There will always be one more stud than the number of spaces so add one to this figure to determine the number of studs. Stud centres must be maintained to suit sheet material sizes leaving an undersized space between the last two studs.

- The lengths of head and sole plates are simply the distance between the two walls.

- Each line of noggins will require a length of timber equal to the distance between the walls.

The total length of timber required for a partition, can be determined by the following method:

> 7 studs at 2.4 mm: $7 \times 2.4 = 16.8$ m
> Head and sole plates at 3.4 m: $2 \times 3.4 = 6.8$ m
> 3 lines of noggins at 3.4 m: $3 \times 3.4 = 10.2$ m
> Metres run required $= 16.8 + 6.8 + 10.2 = 33.8$ m

An allowance must be added to this length for cutting, say 10%.

> Total metres run required $= 33.8 \times 1.1 = 37.18$ say 38 m

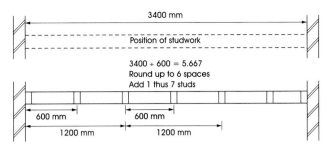

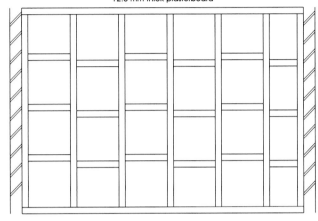

▲ **Figure 8.37**

Rafters

The **number of rafters** required for a pitched roof can be determined using the same method as used for floor joists. For example, divide the distance between the end rafter centres by the rafter spacing. Round up and add one, then double the number of rafters to allow for both sides of the roof.

> ### Example
>
> If distance between end rafter centres is 12 m and spacing is 400 mm as illustrated in Figure 8.38:
>
> Number of rafters = Distance between end rafter centres ÷ Rafter spacing $+ 1 = (12 \div 0.4) + 1 = 30 + 1 = 31$ rafters

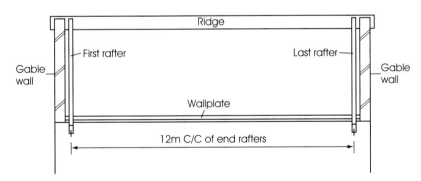

Number of rafters = C/C of end rafters ÷ rafter spacing +1
= 12 ÷ 0.4 + 1
= 30 + 1
= 31

◀ **Figure 8.38** To determine number of rafters on one side

> Therefore total number of rafters required for both sides of the roof is 62.

Where **overhanging verges** are required, noggins and an additional pair of rafters must be allowed at each end to form the gable ladders which in turn provide a fixing for the bargeboard and soffit.

To determine the **length of rafters**, Pythagoras' rule of right-angled triangles can be used.

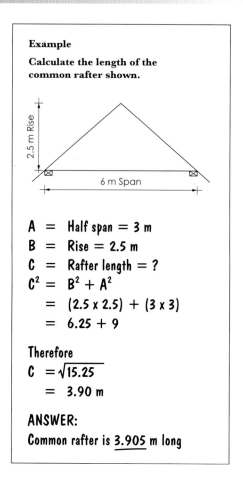

Example

Calculate the length of the common rafter shown.

2.5 m Rise

6 m Span

A = Half span = 3 m
B = Rise = 2.5 m
C = Rafter length = ?
C² = B² + A²
   = (2.5 x 2.5) + (3 x 3)
   = 6.25 + 9

Therefore
C = √15.25
  = 3.90 m

ANSWER:
Common rafter is **3.905 m long**

An allowance must be added to the length of rafters for the eaves overhang and the cutting. Say 0.5 m and 10% (Figure 8.39). In the example:

Total length of rafter = 3.905 + 0.5 + 10% = 4.405 × 1.1 = 4.845 m

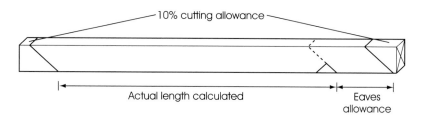

10% cutting allowance

Actual length calculated

Eaves allowance

▶ **Figure 8.39** Length of rafter

Fascia, barge and soffits

Calculating the lengths of material required for fascia boards, barge boards and soffits is often a simple matter of measuring, allowing a certain amount extra for jointing, and adding lengths together to determine total metres run.

▬▬ *Example*

A hipped-end roof requires two 4.4 m lengths and two 7.2 m lengths of ex 25 mm × 150 mm PAR softwood for its fascia boards.

Metres run required = (4.4 × 2) + (7.2 × 2) = 8.8 + 14.4 = 23.2 m

Total metres run required allowing for 10% cutting and jointing = 23.2 × 1.1 = 25.52, say 26 m

The length of timber for bargeboards may require calculation using Pythagoras' rule.

Sheet material. Where sheet material is used for fascias and soffits, the amount that can be cut from a full sheet often needs calculating. This entails dividing the width of the sheet by the width of the fascia or soffit, then using the resulting whole number to multiply by the sheet's length to give the total metres run.

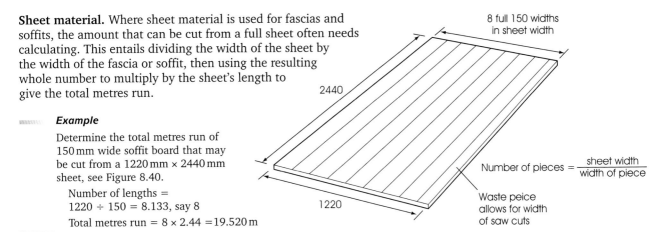

Example

Determine the total metres run of 150 mm wide soffit board that may be cut from a 1220 mm × 2440 mm sheet, see Figure 8.40.

Number of lengths =
1220 ÷ 150 = 8.133, say 8
Total metres run = 8 × 2.44 = 19.520 m

8 full 150 widths
in sheet width

$$\text{Number of pieces} = \frac{\text{sheet width}}{\text{width of piece}}$$

Waste peice allows for width of saw cuts

▲ **Figure 8.40** To determine number of pieces that can be cut from a sheet

Trim

To determine the amount of trim required for any particular task is a fairly simple process, if the following procedures are used.

Architraves. The jambs or legs in most situations can be taken to be 2100 mm long. The head can be taken to be 1000 mm. These lengths assume a standard full-size door and include an allowance for mitring the ends. Thus the length of architrave required for one face of a door lining/frame is 5200 mm or 5.2 m, see Figure 8.41.

Multiply this figure by the number of architrave sets to be fixed. This will determine the total metres run required, say 8 sets, both sides of four doors: 5.2 × 8 = 41.6 m.

Skirtings and other horizontal trim can be estimated from the perimeter. This is found by adding up the lengths of the walls in the area. The widths of any doorways and other openings are taken away to give the actual metres run required.

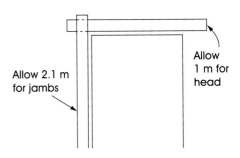

Allow 2.1 m for jambs

Allow 1 m for head

▲ **Figure 8.41**

Example

Determine the total length of skirting required for the area illustrated in Figure 8.42.

Perimeter = 2 + 3.6 + 2.5 + 1.6 + 0.5 + 2 = 12.2 m
Metres run required = 12.2 − 0.9 (door opening) = 11.3 m
Allowing 10% for cutting and jointing, total metres run required =
11.3 × 1.1 = 12.43, say 13 m

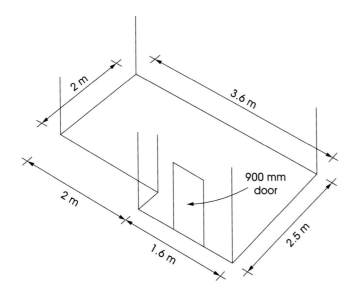

2 m

3.6 m

2 m

900 mm
door

1.6 m

2.5 m

◀ **Figure 8.42**

Sheet Materials				Price per item
Product	Size	Product code	£ inc. VAT	£ exc. VAT
Blockboard BB/CC 18 mm	2440 × 1220 mm	5022652518455	20.32	17.29
PLYWOOD				
Hardwood WBP BB/CC 4 mm	2440 × 1220 mm	5022652504588	7.00	5.96
Hardwood WBP BB/CC 6 mm	2440 × 1220 mm	5022652506605	8.50	7.23
Hardwood WBP BB/CC 9 mm	2440 × 1220 mm	5022652519629	11.55	9.83
OSB				
OSB2 11 mm	2440 × 1220 mm	5022652560386	7.41	6.31
OSB3 18 mm	2440 × 1220 mm	5014957148673	15.98	13.60
MDF				
MDF 6 mm	2440 × 1220 mm	5022652550035	6.57	5.60
MDF 12 mm	2440 × 1220 mm	5022652512231	8.48	7.22
MDF 18 mm	2440 × 1220 mm	5022652518257	10.97	9.34
HARDBOARD				
Standard Hardboard 3.2 mm	2440 × 1220 mm	5022652503314	2.60	2.21
White Faced Hardboard 3.2 mm	2440 × 1220 mm	5022652503338	5.17	4.40
CHIPBOARD				
Flooring Grade 18 mm P4	2400 × 600 mm	5014957105706	3.98	3.39
Flooring Grade 22 mm P4	2400 × 600 mm	5014957105720	5.58	4.75
Flooring Grade 18 mm P5	2400 × 600 mm MR	5014957054691	5.31	4.52
Flooring Grade 22 mm P5	2400 × 600 mm MR	5014957088320	7.98	6.79
Standard Grade 12 mm	2440 × 1220 mm	5014957054677	4.95	4.21

▲ **Figure 8.43** Extract from supplier's price list

Costing materials

This can be carried out once the required quantities of material have been calculated. It is a simple matter of finding out prices and multiplying these by the number of items required.

Example

Suppose you were asked to find the total price of four 2440 × 1220 sheets of 18 mm MDF. From the typical extract of the supplier's price list shown in Figure 8.43, these are £10.97 each including VAT.

Total price = Price per sheet × Number of sheets = £10.97 × 4 = £43.88

Example

The following materials are required for a refurbishing contract. Calculate the total cost including an allowance of 10% for cutting and wastage and 17.5% for VAT.

Sawn softwood at £258.50 per m^3:

Joists 16 pieces size 50 × 225 × 3600 mm
Strutting 10 pieces size 50 × 50 × 4200 mm
Studwork 84 pieces size 50 × 100 × 2400 mm
Battening 50 pieces size 25 × 50 × 4800 mm

18 mm flooring grade chipboard at £49 per 10 m^2:

30 sheets 2400 × 600 mm

Softwood requirement:

Joists 16 × 0.05 × 0.225 × 3.600 = 0.648 m^3
Strutting 10 × 0.05 × 0.05 × 4.200 = 0.105 m^3
Studwork 84 × 0.05 × 0.1 × 2.400 = 1.008 m^3
Battening 50 × 0.025 × 0.05 × 4.800 = 0.3 m^3

Volume = 2.061 m^3
Total volume including 10% allowance = 2.061 × 1.1 = 2.2671 m^3
Cost of softwood = 2.2671 × 258.5 = £586.04535 say £586.05

Chipboard for flooring:

18 mm chipboard = 30 × 2.4 × 0.6
Area = 43.2 m^2
Total area including 10% allowance = 43.2 × 1.1 = 47.52 m^2
Cost of flooring = 47.5 × 49 ÷ 10 = £232.848 say £232.85

(Note that dividing by 10 converts the 10 m^2 price to a 1 m^2 price.)

Cost of materials required = £586.05 + £232.85 = £818.90
Total cost including 17.5% VAT = 818.9 × 1.175 = £962.2075 say £962.20

(Note that VAT prices are always rounded down, not up, to the nearest £0.01.)

Chapter Nine

Applied geometry

This chapter covers the work of both the site carpenter and bench joiner. It is concerned with the types of geometry that you may be required to undertake. It includes the following:

- ✐ angles and lines
- ✐ shapes and solids
- ✐ basic techniques
- ✐ construction of plane figures
- ✐ solid geometry
- ✐ practical applications of geometry including arch centres, pitched roofs, louvre ventilators, mouldings, splayed linings and geometrical stairs.

Geometry is concerned with the values, relationship and measurement, of points, lines, curves and surfaces.

Knowledge of geometry is required by the woodworker in order to determine the true shapes, lengths and angles of components that are required for a range of practical woodworking tasks.

Angles and lines

An **angle** is the amount of space between two intersecting straight lines. The size of an angle is measured in degrees (°) with each degree being sub-divided into 60 seconds.

Angles are also a measure of turning (Figure 9.1): standing on a spot looking in one direction, if you were to turn completely round to look again in the same direction, you would have turned through 360° (degrees).

- ■ A three-quarter turn is 270°.
- ■ A half turn is 180°.
- ■ A quarter turn is 90°, also called a right angle.

Parallel lines and angles

Refer to Figure 9.2:

- ■ **Parallel lines** are always the same distance apart: they never meet. They are shown on a drawing using either a single or double pair of arrowheads.
- ■ **Traverse line**: a line drawn across a pair of parallel lines.
- ■ **Alternate angles** of a traverse are always equal (B = B).

▶ **Figure 9.1** Angles are a measure of turning

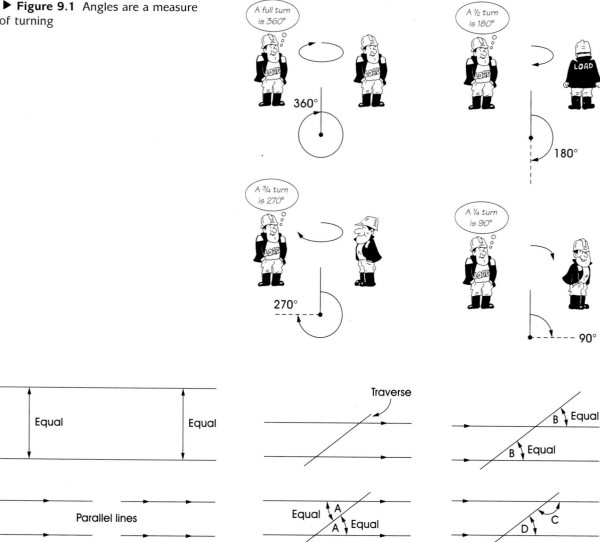

Equal Equal

Parallel lines

Traverse

Equal A Equal

B Equal

B Equal

D C

▲ **Figure 9.2** Parallel lines

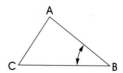

▲ **Figure 9.3** Angle ABC

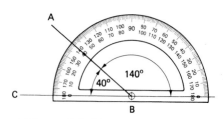

▲ **Figure 9.4** Measuring or constructing angles using a protractor

- **Corresponding angles** of a traverse are always equal (A = A).
- **Supplementary angles** of a traverse always add up to 180° (C + D = 180°).

Measuring the size of angles between lines

If asked to find the angle between AB and CB (called angle ABC), it means you have to find the angle of the middle letter, B (Figure 9.3).

Protractors can be used to measure angles, as illustrated in Figure 9.4:

- Place the protractor with its centre point over the intersection of the lines AB and CB and its baseline over one of the lines.
- Read off the angle at the edge of the protractor. It could be 40° or 140°. As it is obviously less than 90°, it must be 40° in this case.

Shapes and solids

Shapes

A **plane** is a flat surface. Its shape is formed by lines, which are known as the sides of the shape which give it length and breadth. The total distance all the way round the sides is called the perimeter. The number of sides is used to classify standard shapes.

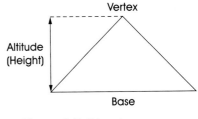

▲ **Figure 9.5** Triangle

Triangle

A plane shape, which is bounded by three straight lines is called a triangle (Figure 9.5).

■ The **vertex** is the angle opposite the base.
■ The **altitude** is the vertical height from the base to the vertex.

Triangles are classified by either the length of their sides or by the size of their angles (Figure 9.6):

■ An **equilateral triangle** has three equal length sides and three equal angles.
■ An **isosceles triangle** has two sides of equal length.
■ A **scalene triangle** has sides which are all unequal in length.
■ A **right-angled triangle** has one 90° angle.
■ In an **acute-angled triangle** all angles are less than 90°.
■ In an **obtuse-angled triangle** one of the angles is between 90° and 180°.

Quadrilaterals

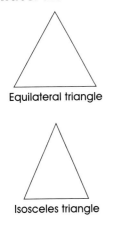

Equilateral triangle

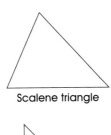

Scalene triangle

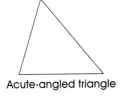

Acute-angled triangle

Isosceles triangle

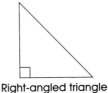

Right-angled triangle

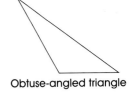

Obtuse-angled triangle

▲ **Figure 9.6** Types of triangle

A plane shape that is bounded by four straight lines (Figure 9.7) is a quadrilateral. A straight line joining opposite angles is called a diagonal; it divides the figure into two triangles.

■ A **square** has all four sides of equal length and all four angles are right angles.
■ A **rectangle** has opposite sides of equal length and all four angles are right angles.
■ A **rhombus** has four equal sides, opposite sides being parallel, but none of the angles is a right angle.
■ A **parallelogram** has opposite sides parallel and equal in length but none of its angles is a right angle.
■ A **trapezium** has two parallel sides.
■ A **trapezoid** has no parallel sides.

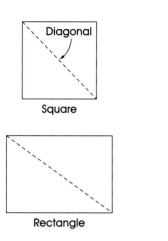

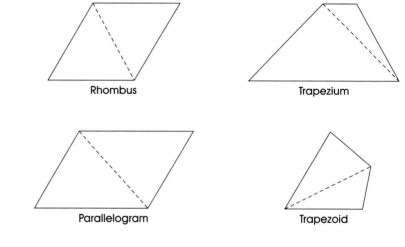

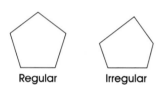

▲ **Figure 9.7** Types of quadrilateral

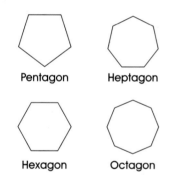

▲ **Figure 9.8** Polygons

▲ **Figure 9.9** Classification of polygon by number of sides

Polygon

A plane shape that is bounded by more than four straight lines is called a polygon. Polygons may be classified as either (Figure 9.8):

- **Regular polygon**, having sides of the same length and equal angles.
- **Irregular polygon**, having sides of differing length and unequal angles.

Both regular and irregular polygons are classified by the number of sides they have (Figure 9.9):

- A **pentagon** consists of five sides.
- A **hexagon** consists of six sides.
- A **heptagon** consists of seven sides.
- An **octagon** consists of eight sides.

Circles

A circle is a plane shape, bounded by a continuous, curved line, which at every point is an equal distance from its centre (Figure 9.10). The main elements of a circle are:

- **Centre**, the mid point. It is equidistant from any point on the circumference.
- **Circumference**, the curved outer line (perimeter) of the circle.
- **Diameter**, a straight line, which passes through the centre and is terminated at both ends by the circumference.
- **Radius**, the distance from the centre to the circumference. The radius is always half the length of the diameter.
- **Chord**, a straight line that touches the circumference at two points but does not pass through the centre.
- **Arc.** Any section of the circumference.

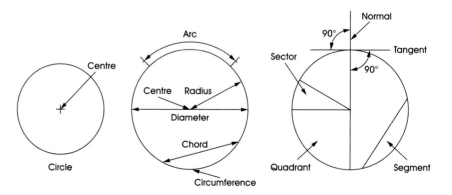

▶ **Figure 9.10** Elements of a circle

- **Normal.** Any straight line which starts at the centre and extends beyond the circumference.
- **Tangent.** A straight line that touches the circumference at right angles to the normal.
- **Sector.** The portion of a circle contained between two radii and an arc. (Radii is the plural of radius.)
- **Quadrant,** a sector whose area is equal to a quarter of the circle.
- **Segment,** the portion of a circle contained between an arc and a chord.

Other shapes related to circles:

- **Annulus.** The area of a plane shape that is bounded by two circles, each sharing the same centre but having different radii (Figure 9.11).
- **Ellipse.** A plane shape bounded by a continuous curved line drawn round two centre points called **foci.** The longest diameter is known as the *major axis* and the shortest diameter is the *minor axis* (Figure 9.12). An ellipse is derived from a section of cylinder or cone made by a cutting plane inclined to the axis of the solid.

Solids

Solid figures are three-dimensional; they have length, breadth and thickness (Figure 9.13).

Prism

A solid figure contained by plane surfaces that are parallel to each other. If cut into slices, they would all be the same shape (Figure 9.14).

All prisms are named according to the shape of their ends (Figure 9.15):

- **Cube:** all sides are equal in length and each face is a square.
- **Cuboid** or **rectangular prism:** each face is a rectangle and opposite faces are the same size. Bricks and boxes are examples of cuboids.
- **Triangular prism** the ends are triangles and other faces are rectangles.
- **Octagonal prism** the ends are octagons and other faces are rectangles.

Pyramid

A solid figure contained by a base and triangular sloping sides. The sides meet at a point called the **apex.** All pyramids are named according to their base shape (Figure 9.16):

- **Triangular pyramid**
- **Square pyramid**
- **Hexagonal pyramid**

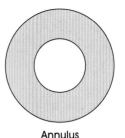

Annulus

▲ **Figure 9.11**

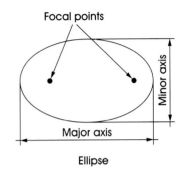

Focal points

Minor axis

Major axis

Ellipse

▲ **Figure 9.12**

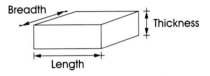

Breadth

Thickness

Length

▲ **Figure 9.13** Solid

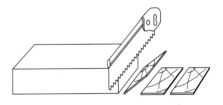

▲ **Figure 9.14** Sections of a prism

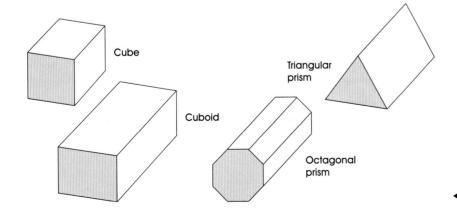

Cube

Cuboid

Triangular prism

Octagonal prism

◀ **Figure 9.15** Types of prism

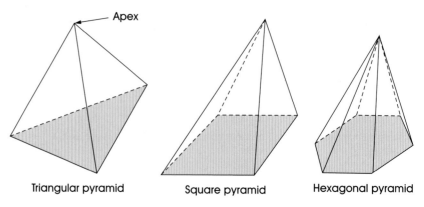

▲ **Figure 9.16** Types of pyramid

Circular solids

Refer to (Figure 9.17):

- **Cylinder** a circular prism, the ends are circular in shape. Most tins are cylindrical.
- **Cone** a circular pyramid; the base is circular in shape.
- **Sphere** a solid figure, where all sections are circular in shape. Most balls are spherical.

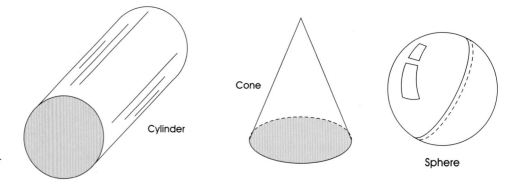

▶ **Figure 9.17** Circular solids

Net of solids

The net of a solid is the two-dimensional (2-D) shape, which can be folded to make the three-dimensional (3-D) shape, as illustrated in Figure 9.18.

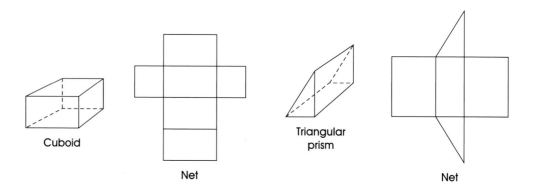

▲ **Figure 9.18** Net of solids

Basic geometrical techniques

There are two simple operations, which occur repeatedly in geometry:

- **Bisecting a line**, or dividing it into two equal parts.
- **Dividing a line** into a greater number of equal parts.

Bisecting and dividing a line

The method used to bisect a given line is shown in Figure 9.19:

1. Draw line AB.
2. With centre A and a radius greater than half AB, mark arcs above and below line AB.
3. With centre B and same radius, mark arcs above and below line AB to cross the previous ones.
4. Draw a line through the intersections of the arcs.

The method used to divide a line into a number of parts is shown in Figure 9.20:

1. Draw given line AB.
2. Draw line AC any length at a convenient angle to line AB.
3. Mark on AC the number of equal parts required: 9 in this case. Any convenient spacing is suitable.
4. Join the last numbered point to B.
5. Draw lines through each of the numbered points parallel to the line drawn in stage 4.

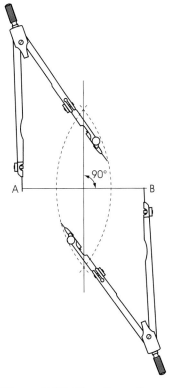

▲ **Figure 9.19** Bisecting a line

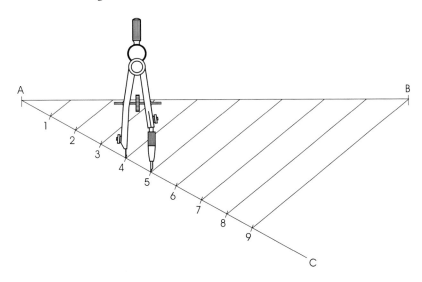

▲ **Figure 9.20** Dividing a line into a number of equal parts

Construction of angles

Angles can be drawn by a variety of methods:

- **Set squares.** A number of different angles can be formed using set squares either separately or combined. Figure 9.21 shows some of the angles that can be constructed using this method.
- **Protractor.** All angles can be set out or measured using a protractor. (Refer to Figure 9.4).

▶ **Figure 9.21** Angles using set squares

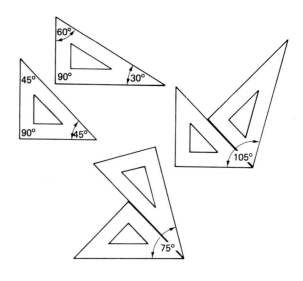

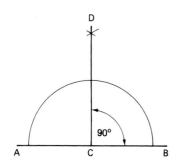

▲ **Figure 9.22** Forming a right angle with a compass

■ **Compasses.** Almost any angle can be formed using compasses. Figure 9.22 shows how a right angle is formed. This is a similar method to that used when bisecting a straight line.

■ From a **scale of chords**.

Bisecting a 90° angle to form a 45° angle, Figure 9.23:

1. Construct a right angle as before. With centre A and any convenient radius, draw arc ED.
2. With centre E and a radius of over half ED, mark a small arc over ED.
3. With centre D and same radius as stage 2, mark a small arc to cross the previous one.
4. Draw a line through the intersection of the small arcs and A.

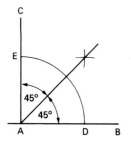

▲ **Figure 9.23** Forming a 45° angle with a compass

Forming 30° and 60° angles by bisection, Figure 9.24:

1. Construct a right angle as before.
2. With centre A and radius set to AB, draw arc BC.
3. With centre B and radius set to AB, mark point D on arc.
4. With centre C and radius set to AB, mark point E on arc.
5. Draw lines from A to D and A to E.
6. Angle DAB will be 60° and angle EAB will be 30°.

Angles other than 30°, 45° and 60° can be formed by further bisection. Figure 9.25 shows an angle of 15° formed by the bisection of a 30° angle. The 30° angle is bisected using a similar method to that used when forming the 45° angle.

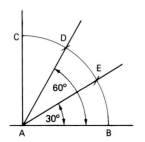

▲ **Figure 9.24** Forming 30° and 60° angles with a compass

Scale of chords. Figure 9.26 shows a scale of chords. The scale is often set out on a strip of plywood and used in the workshop for setting out various angles:

1. Draw line AB.
2. Draw line CD at right angles to AB.
3. With centre D and any convenient radius draw arc BC.
4. Mark on arc BC, 30° and 60° points. (Use the same method as shown for constructing 30° and 60° angles.)
5. Divide the three spaces 0–30°, 30–60° and 60–90° on arc BC into three equal parts using dividers.
6. With centre B, draw arcs from all points down onto AB.
7. Construct scale below AB as shown.

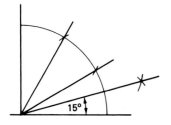

▲ **Figure 9.25** Forming a 15° angle with a compass

Note: By further subdivision of arc BC a wide range of angles can be obtained.

Using a scale of chords. In Figure 9.27 an angle of 70° has been set out using a scale of chords:

1. Draw line AB.
2. With centre A and radius set from 0 to 60° on the scale, draw arc BC.
3. With centre B and radius set from 0 to 70° on the scale, mark point C on arc BC.
4. Draw line AC. This produces the required 70° angle CAB.

Note: The radius for the arc BC is always 0 to 60°, no matter what angle is being produced.

An angle of 40°, which has been set out using a scale of chords, is illustrated in Figure 9.28. The method used to produce the angle is the same as before, except that in stage 3 the radius must be set from 0° to the required angle.

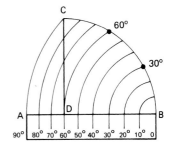

▲ **Figure 9.26** Scale of chords

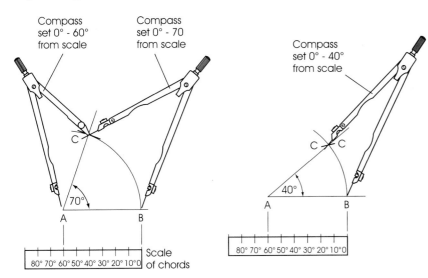

▲ **Figure 9.27** Use of a scale of chords to set out a 70° angle

▲ **Figure 9.28** Use of a scale of chords to set out a 40° angle

Construction and development of plane figures

Construction of triangles

In order to construct a triangle, any one of the following four sets of information is required:

- The lengths of the three sides.
- The lengths of two sides and the included angle.
- The length of one side and two angles.
- The length of the perimeter and the ratio of the sides.

Method 1 (Figure 9.29). To construct an equilateral triangle with a base of 100 mm:

1. Draw base AB 100 mm long.
2. From A set radius AB, mark arc BC.

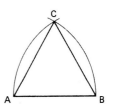

▲ **Figure 9.29** Constructing a triangle: method 1

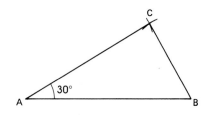

▲ **Figure 9.30** Constructing a triangle: method 2

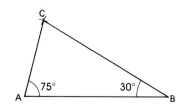

▲ **Figure 9.31** Constructing a triangle: method 3

3. From B set radius BA; mark arc AC.

4. Join points AC and BC.

Method 2 (Figure 9.30). To construct a triangle with a base of 100 mm, one side of 85 mm and an included angle of 30°:

1. Draw base AB 100 mm long.

2. From A, draw a line at 30° using a set square.

3. Mark point C 85 mm from A.

4. Draw line BC.

Method 3 (Figure 9.31). To construct a triangle with a base of 100 mm and an angle of 75° at one end, and an angle of 30° at the other end:

1. Draw base AB 100 mm long.

2. From A, draw a line at 75° using a protractor.

3. From B, draw a line at 30°. The lines cross at C.

4. Draw lines AC and BC.

Method 4 (Figure 9.32). To construct a triangle, which has a perimeter of 200 mm and sides with a ratio of 4:6:5:

1. Draw line AB 200 mm long.

2. Divide AB into 4:6:5 units using the method shown in Figure 9.20.

3. From C with radius set at length CA, mark arc.

4. From D with set radius DB, mark arc. The arcs cross at E.

5. Draw lines CE and DE.

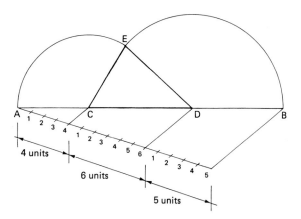

▶ **Figure 9.32** Constructing a triangle: method 4

Constructing quadrilaterals

Squares and rectangles can be constructed using setsquares, while the other four quadrilaterals are constructed using a similar method to that used in the construction of triangles.

When constructing quadrilaterals sufficient information is required:

- *Square:* the length of the base.
- *Rectangle:* the lengths of the base and a side.
- *Rhombus:* the length of the base and one angle.
- *Parallelogram:* the length of the base, the vertical height and one angle.
- *Trapezium:* the lengths of the parallel sides, the vertical height and one base angle.
- *Trapezoid:* the lengths of three sides and the angles between them.

Constructing a polygon

Figure 9.33 shows how to construct any regular polygon, given the length of one side:

1. Draw line AB to given length.
2. With centre A and radius AB, draw semicircle CB.
3. Divide semicircle CB into a number of parts equal to the number of sides required in the polygon (in this example five).
4. Draw side A2.
5. Bisect lines A2 and AB. The two bisectors cross at point 0.
6. With centre 0 and radius OA draw a circle.
7. Mark points D and E on the circle at a distance equal to AB.
8. Join points up as shown. ABED2 is the required figure.

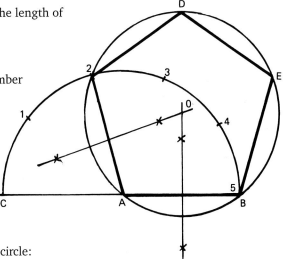

▲ **Figure 9.33** Constructing a regular polygon

Constructions with circles

Figure 9.34 shows the method used to find the centre of a given circle:

1. Draw on the circle any two chords, AB and CD.
2. Bisect AB and CD.
3. The point where the two bisections cross is the required centre.

Constructing a segment of a circle. This method can be used when setting out a turning piece or centre for a segmental arch (Figure 9.35):

1. Draw line AB equal to the span.
2. Bisect AB.
3. From C, mark the rise. Let this point be D.
4. Draw line AD and bisect it. The point where the two bisections cross is the required centre.
5. With radius set from centre to A, the arc can be drawn.

Drawing a circle that passes through any three given points. Providing these points are not in a straight line, then (Figure 9.36):

1. Connect the three points with straight lines.
2. Bisect the lines. The point where the two bisections cross is the required centre.
3. With radius set from the centre to any one of the points, the circle then can be drawn to pass through all three points.

Joining two straight lines with an arc. In Figure 9.37a, the lines are at right angles. The method used to set out the arc is:

1. Draw lines AB and CD parallel to the straight lines, at a distance equal to the radius of the arc.
2. Draw the arc. The point where the parallel lines cross is the centre required.

Where the lines are not at right angles, as in Figure 9.37b, a similar method is used to draw the arc. The difference is the addition of two extra lines to indicate the ends of the arc. **Note:** The extra lines are in fact normals and must be at right angles to the parallel lines.

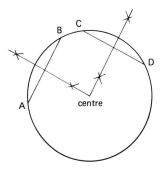

▲ **Figure 9.34** Finding the centre of a circle

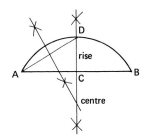

▲ **Figure 9.35** Constructing a segment of a circle

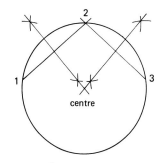

▲ **Figure 9.36** Drawing a circle to pass through three points

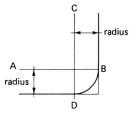 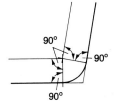

▲ **Figure 9.37** Joining two straight lines with an arc: (a) at 90°; (b) at any angle

Constructing ellipses

There are many methods by which an ellipse may be drawn, a number of which are covered in the following figures.

Drawing an ellipse using the trammel method, Figure 9.38:

1. Draw major and minor axes AB and CD.

2. Make trammel from a strip of card. Mark points so that 1–3 is equal to $\frac{1}{2}$ major axis and 2–3 is equal to $\frac{1}{2}$ minor axis.

3. Move trammel so that point 1 travels on the minor axis and point 2 travels on the major axis.

4. Mark off at suitable intervals point 3.

5. Join the point 3 marks with a smooth curve to draw the ellipse.

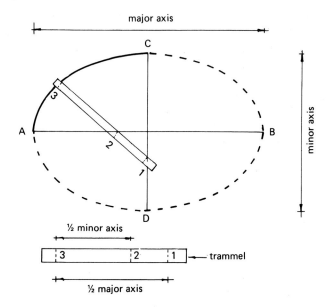

▶ **Figure 9.38** Trammel method of drawing an ellipse

Drawing an ellipse using the foci pins and string method, Figure 9.39:

1. Draw major and minor axes AB and CD.

2. With centre C and radius AE, mark two arcs on AB. These arcs give the two focal points F^1 and F^2.

3. Place pins at F^1, F^2 and C and stretch a piece of taut string around them.

4. Remove the pin at C and insert a pencil.

5. Move the pencil around keeping the string taunt to draw an ellipse.

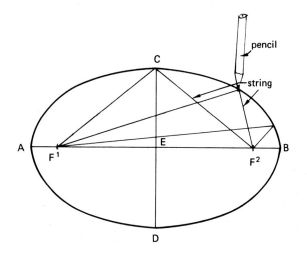

▶ **Figure 9.39** Foci pins and string method

Drawing an ellipse using the intersecting lines within a rectangle method Figure 9.40:

1. Draw a rectangle FGHI equal to the major and minor axes.
2. Draw the major and minor axes AB and CD.
3. Divide lines EA, EB, FA and GB into the same number of equal parts.
4. Join points along FA and GB to C.
5. Join points along EA and EB to D. Continue them to intersect the lines joined to C.
6. Join the intersecting points with a smooth curve.
7. Repeat the process, to draw the other half of the ellipse.

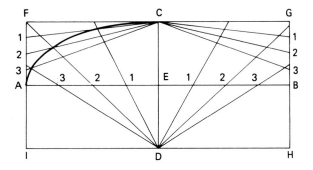

▲ **Figure 9.40** Intersecting lines method

Drawing an ellipse with the aid of two circles, Figure 9.41:

1. Draw two concentric circles with diameters equal to the major and minor axes.
2. Divide the circumference of the larger circle into twelve equal parts, using a 30°/60° set square passing through the centre of the circle.
3. Draw vertical lines from the points on the outer circle and draw horizontal lines from the points on the inner circle.
4. A smooth curve can be drawn through the intersection of the vertical and horizontal lines to form the required ellipse.

Drawing a false ellipse using compasses, Figure 9.42:

1. Draw the major and minor axes AB and CD.
2. Draw line AC.
3. With centre E and radius AE, draw arc AF.
4. With centre C and radius CF, draw arc FG.
5. Bisect line AG to give points H and I.
6. With centre E and radius EH, draw an arc to give H^1.
7. With centres H, H^1 and I, draw arcs to give the semi-ellipse.
8. Draw arcs with centres H, H^1 and I^1 to complete the other half of the ellipse below line AB.

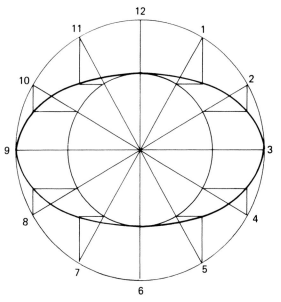

▲ **Figure 9.41** Concentric circles method

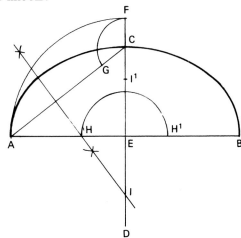

◀ **Figure 9.42** False ellipse drawn using compasses

Practical examples of geometric developments

Hexagonal column

Figure 9.43 shows a part elevation and plan of a hexagonal column which supports an inclined concrete ramp. The problem is to find the true shape of the hole required in the formwork for the soffit of the ramp.

1. Draw plan and elevation.
2. Draw lines at right-angles to the ramp where it intersects the column.
3. Mark lines 1, 2 and 3 parallel to the ramp at equal spacing to these on the plan.
4. Join intersections to give the required true shape.

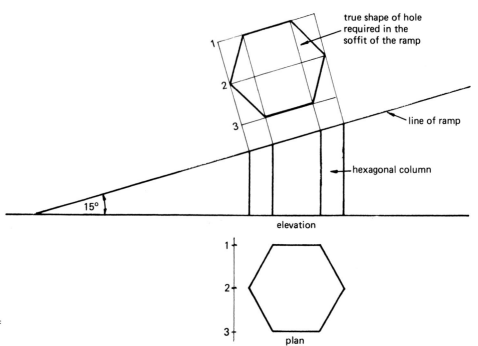

Figure 9.43 Practical development (true shape) of hexagonal column

Skew arch

Figure 9.44 shows a plan and elevation of two walls A and B, which are joined by a skew arch. The problem is to find the true shape of the soffit or lagging required for the arch centre.

1. Draw plan and elevation.
2. Divide the semicircle into six parts to give points 1–7.
3. Project points 1–7 down vertically onto the plan to give points 1^1–7^1

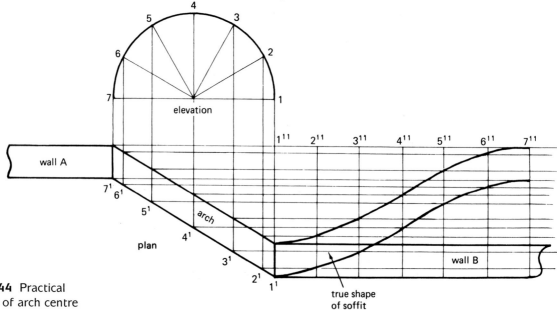

Figure 9.44 Practical development of arch centre

4. Where vertical lines touch either side of the arch, project lines across horizontally.

5. Take distances 1–2, 2–3, 3–4, etc. from elevation and mark points 1^{11}– 7^{11}

6. Project points 1^{11}–7^{11} down vertically where the same numbered lines intersect and draw the development of the soffit.

Dormer window

Figure 9.45 shows a sketch of a house with a dormer window in a pitched roof. The problem is to find the true shape of the opening in the main roof and the development of the dormer roof:

1. Draw the plan and elevations.

2. Project up points A, B and C at 90° to line of main roof.

3. Mark points 1^1, 2^1 and 3^1 equal to distances 1, 2 and 3 taken from elevation.

4. Draw lines to complete the true shape of opening.

5. With centre E and radius E D, draw arc to give point F.

6. From F, project vertically down to G and from G horizontally across on to plan.

7. Draw lines to give development of half of dormer roof.

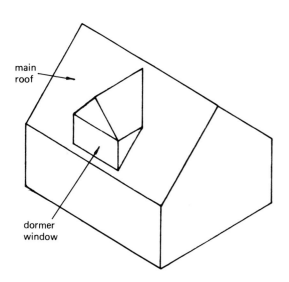

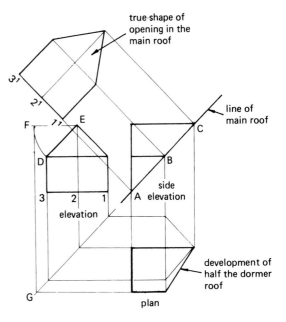

▲ **Figure 9.45** Practical development of dormer window

Segmental dormer in pitched roof

Figure 9.46 is a similar problem to the previous example except in this case the dormer window has a segmental-shaped roof.

1. Draw the elevation and side elevation.

2. Divide the segmental roof into a number of equal divisions 1, 2, 3, 4, 5 on the elevation and project these points on to the main roofline on the side elevation.

True shape of opening:

3. Project lines down and across from the two elevations and draw in the plan.

4. Project up the points of intersection on the main roofline at right angles.

5. Draw in a centre line, which will carry 5^1.

6. Mark on either side of centre line points 1^1 to 4^1 equal to distances 1^1 to 4^1 on the elevation, and draw lines from these points to form an intersecting grid with the right-angled projections from the main roofline.

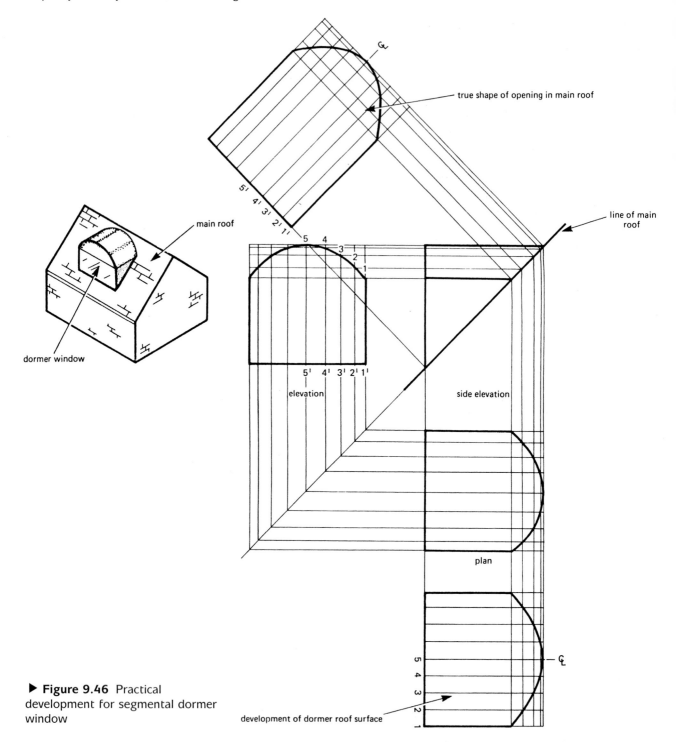

▶ Figure 9.46 Practical development for segmental dormer window

7. Draw a smooth curve through the intersecting grid to give the shape of the opening in the main roof.

Development of dormer roof surface:

8. Project points down from the plan and draw in the centre line which will carry point 5.

9. Mark on either side of the centre line points 1 to 4 equal to distances 1^1 to 4^1 on the elevation and draw lines from these points to form an intersecting grid with the projections from the main roof line.

10. Draw a smooth curve through the intersecting grid to give the development of the dormer roof surface.

Domed roof with square plan

Figure 9.47 shows a semicircular domed roof, which is square on plan.
In order to construct the roof it is necessary to develop the true shape of one surface of the roof and determine the outline of the hip rib and backing bevel.

Development of surface:

1. Draw plan and elevation of the roof. Divide half the elevation into a number of equal divisions 1, 2, 3, 4, 5, 6, 7.
2. Project lines down from these points on to the plan to give a series of points on the hips.
3. Draw a horizontal line from each point on the hips.
4. Mark on the centre line points 1^1 to 7^1 equal to distances 1 to 7 on the elevation and draw lines through these points to form an intersecting grid with the horizontal lines.
5. Draw smooth curves through the intersecting grid to give the development of the roof surface.

Outline of hip:

6. Draw XY line parallel to one hip.
7. Project points from the hip at right angles through the XY line.
8. Transfer distances A, B, C, D, E, F from the elevation and mark below the XY line.
9. Draw in a smooth curve to give the outline of the hip rib (used to make hip template).

To form backing bevel, when the hip rib is cut to shape it will still require a backing bevel. Placing the hip template on the hip and sliding it sideways equal to distance X marked on the plan can find this.

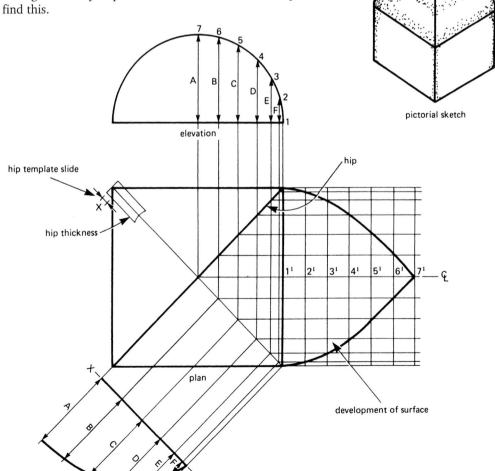

pictorial sketch

elevation

hip template slide

hip thickness

hip

plan

development of surface

outline of hip rib

◀ **Figure 9.47** Domed roof with square plan

10. Mark around the template and repeat on the other side.

11. Mark the centre line on the top edge of the hip. Chamfer the edge to these lines to form the backing bevel.

Domed roof with octagonal plan

Figure 9.48 shows another example of a domed roof, which in this case has an octagonal plan. The geometry involved is the same as that used in the previous example.

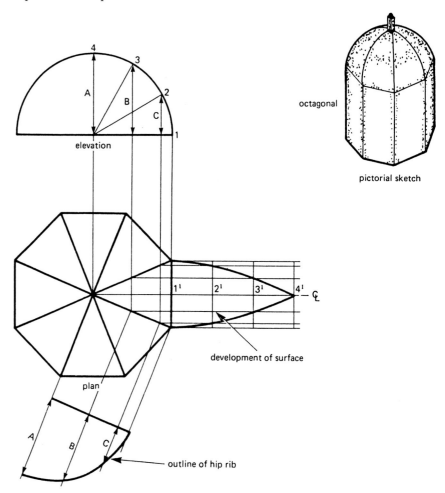

▶ **Figure 9.48** Domed roof with octagonal plan

Hemispherical domed roof

Figure 9.49 shows a hemispherical domed roof. All that is required in this example is a development of one portion of the surface. The outline of the hips will be the same as the elevation. It is not possible to develop accurately the surfaces of a hemisphere, but for practical purposes the method shown will give a close approximation:

1. Draw the plan and elevation.

2. Divide half of the elevation into a number of equal divisions 1, 2, 3, 4, 5, 6, 7.

3. Project lines down from these points on to the plan to give a series of points on the hips.

4. Draw a horizontal line from each point on the hips.

5. Mark on the centre line points 1^1 to 7^1 equal to distances 1 to 7 on the elevation and draw a line through these points to form an intersecting grid with the horizontal lines.

6. Draw smooth curves through the intersecting grid to give the approximate development of the roof surface.

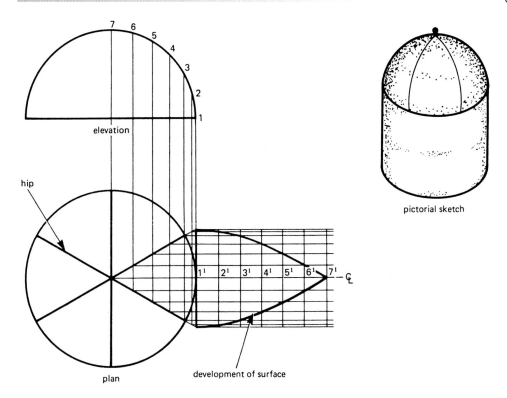

elevation

hip

plan

development of surface

pictorial sketch

▲ **Figure 9.49** Hemispherical domed roof

Arch centre geometry

In order to construct a centre for an arch (temporary support for the bricklayer) the carpenter must first set out the outline of the required arch.

Arch terminology (Figure 9.50):

- *Springing line:* The horizontal reference line at the base of an arch where the curve starts to move away from the vertical reveal (where the curve springs from).
- *Span:* The horizontal distance between the reveals or sides of the opening.
- *Rise:* The vertical distance measured on the centre line from the springing line to the highest point on the intrados.
- *Intrados:* The underside or soffit of an arch.
- *Extrados:* The upper edge of the arch bricks or stones.
- *Crown:* The highest point on the extrados.
- *Voussoirs:* The wedge-shaped bricks or stones used to build an arch.
- *Key or keystone:* The large central voussoir at the crown of an arch.
- *Centre point of an arch:* The pivoting point from which the curve is struck.
- *Radius of an arch:* The distance from the centre point to the intrados, which is used to strike the curve.

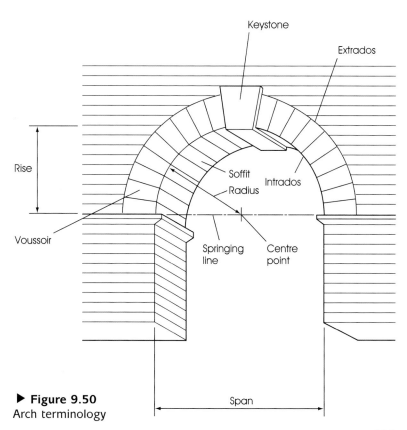

▶ **Figure 9.50**
Arch terminology

379

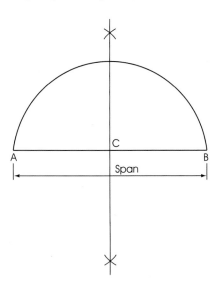

▲ **Figure 9.51** Setting out a semicircular arch

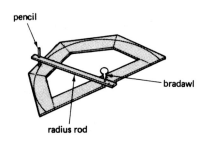

▲ **Figure 9.52** Use of radius rod

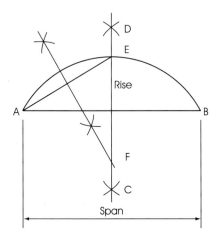

▲ **Figure 9.53** Setting out a segmental arch

▶ **Figure 9.54** Use of triangular frame or taper board

Setting out arches

Semicircular arch, Figure 9.51:

1. Draw line AB equal to the span.
2. Bisect line AB to determine the centre point C.
3. With centre C and radius AC draw the semicircle from A to B.

In the practical situation the setting out of a semicircular arch can be done with the aid of a radius rod as illustrated in Figure 9.52.

Segmental arch, Figure 9.53:

1. Draw line AB equal to the span.
2. Bisect line AB.
3. Draw a line through the intersections of the arcs to give the centre line CD.
4. Mark the required rise (less than half of the span) on the centre line to give point E.
5. Draw line AE and bisect.
6. Draw a line through the intersections of the arcs to give point F on the centre line.
7. With centre F and radius AF draw the segment from A to B.

In the practical situation the setting out of a segmental arch can be done with the aid of a triangular frame or shaped taper board depending on the rise, as illustrated in Figure 9.54.

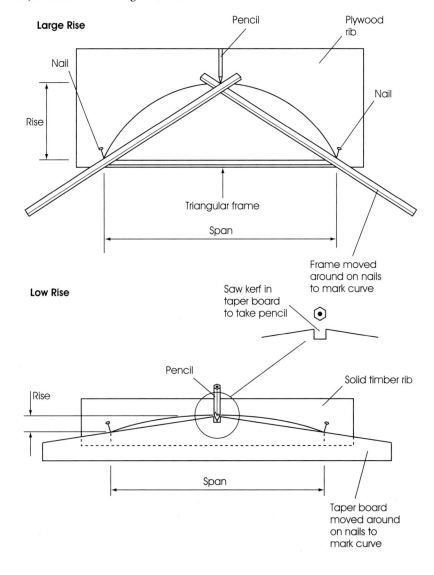

Equilateral Gothic arch, Figure 9.55:
1. Draw line AB equal to span.
2. With centres A and B and radius AB, draw two arcs to intersect at C.
3. Triangle ACB is an equilateral triangle.

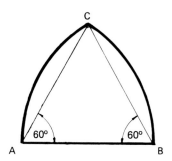

▲ **Figure 9.55** Equilateral Gothic arch

Drop Gothic arch, Figure 9.56:
1. Draw lines AB and CD equal to the required span and rise.
2. Bisect line AC to give point O^1.
3. With centre D and radius DO^1, draw arc to give point O^2.
4. With centres O^1 and O^2 and radius O^1A, draw arcs to intersect at C.

Lancet arch, Figure 9.57:
1. Draw lines AB and CD equal to the required span and rise.
2. Bisect line AC to give point O^1.
3. With centre D and radius DO^1 draw arc to give point O^2.
4. With centres O^1 and O^2 and radius O^1A, draw arcs to intersect at C.

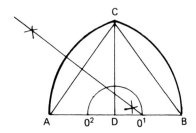

▲ **Figure 9.56** Drop Gothic arch

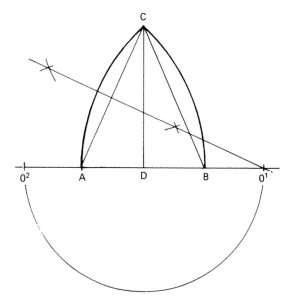

▲ **Figure 9.57** Lancet arch

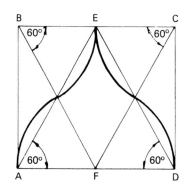

▲ **Figure 9.58** Ogee arch

Ogee arch, Figure 9.58:
1. Draw equilateral triangle AED with base equal to the required span.
2. Complete rectangle ABCD.
3. Draw lines from B and C at 60° to give point F.
4. With centres F, B and C, draw arcs.

Tudor arch, Figure 9.59, also known as a **four-centred arch**:
1. Draw rectangle ABCD equal to the required span and rise.
2. Draw vertical line FE from the centre of AD.
3. Divide AB into three equal parts. Join point 2 to E.
4. Draw line EO^2 at right angles to line 2E.
5. Make EG equal to A2.
6. With centre A and radius A2, draw arch to give O^1.
7. Draw line O^1G and bisect to give point O^2.
8. With centre F and radius FO^1, draw arc to give point O^3.
9. With centre H and radius HO^2, draw arc to give point O^4.
10. With centres O^1, O^2, O^3 and O^4 draw arcs.

▶ **Figure 9.59** Tudor arch

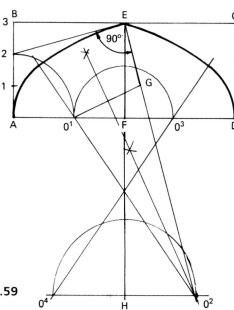

Pointed parabolic arch, Figure 9.60:

1. Draw rectangle ABCD equal, to the required span and rise.
2. Draw vertical line FE from the centre of AD.
3. Divide lines AB, BE, CE and DC into the same number of equal parts.
4. Join points on BE to A and points on CE to D.
5. Draw horizontal lines from points on lines AB and DC.
6. Draw a smooth curve through points where same numbered lines intersect.

Elliptical arches

Three-centred elliptical arch, Figure 9.61, also known as an *approximate or mock semi-elliptical arch*:

1. Draw the major and minor axes AB and CD.
2. Draw line AC.
3. With centre E and radius AE, draw arc AF.
4. With centre C and radius CF, draw arc FG.
5. Bisect line AG to give points H and I.
6. With centre E and radius EH, draw an arc to give H^1
7. With centres H, H^1 and I, draw arcs to give the semi-ellipse.

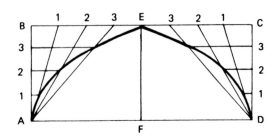

▲ **Figure 9.60** Pointed parabolic arch

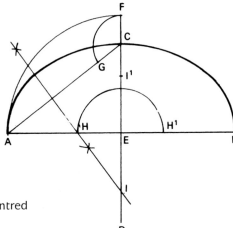

▶ **Figure 9.61** Three-centred elliptical arch

A true semi-elliptical arch outline, although rarely used, can be marked out using either the foci pins and string method or the trammel method covered previously.

Roofing geometry

Pitched roof geometry can be divided into three sections:

- The development of roof surfaces.
- Finding the true lengths of rafters, etc.
- Finding the required angles for the cuts to the rafters and other components.

In the following section the geometry required for gable roofs, hipped-end roofs, double roofs and roofs with valleys are covered. (Refer also to Book 2, Chapter 2.)

The scale drawing shown in Figure 9.62 is of a part plan and section of a hipped-end roof with purlins. Indicated on this scale drawing are all the developments, angles and true lengths required to set out and construct the roof.

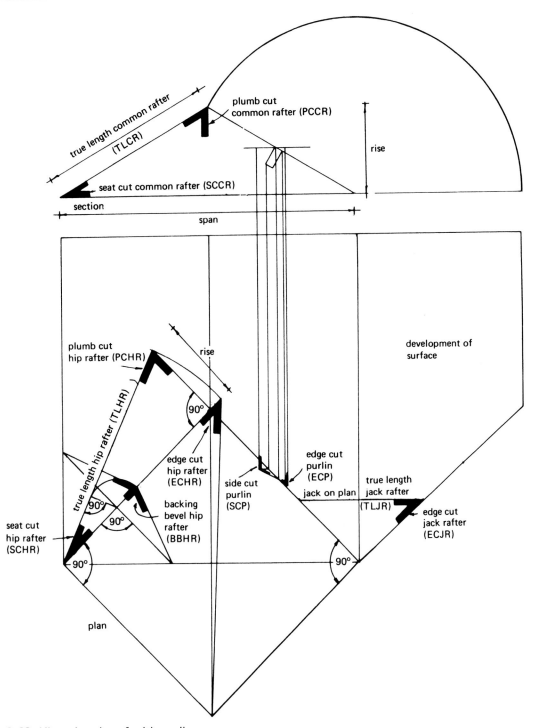

▲ **Figure 9.62** Hipped-end roof with purlins

Common abbreviations, which may be used, have been included in brackets.

The geometry for each of these developments, angles and true lengths are considered separately in the following figures.

Developments of roof surfaces

The developments of all the roofs shown are found using a similar method (Figure 9.63). The following list refers to the development of the lean-to roof (detail a):

1. Draw the plan and elevation of the roof.

383

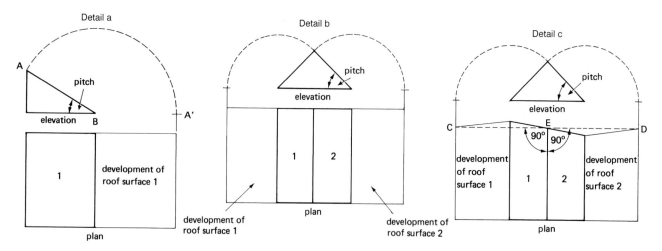

▲ **Figure 9.63** Plan views of: (a) lean-to roof; (b) gable-end roof, rectangular; (c) gable-end roof, splayed

2. With centre B and radius AB swing the sloping surface of the roof down onto the horizontal plane to give point A^1.
3. Draw a vertical line down from point A^1 to meet the horizontal lines that have been extended from the plan. This gives the development of surface 1.

When developing the surfaces of the splayed-end roof (detail c) the points C and D must be extended horizontally from the end of the ridge, point E.

Angles and true lengths for the common and hip rafters (Figure 9.64):

1. Draw to a suitable scale, the plan and section of the roof. On regular plan roofs, the hip rafters will be at 45°. On irregular plan roofs the angle will have to be bisected.
2. Indicate on the section the following:
   - the true length of the common rafter (TLCR);
   - the plumb cut for the common rafter (PCCR);
   - the seat cut for the common rafter (SCCR).
3. At right angles to one of the hips on the plan, draw line from A and mark on it the rise of the roof AB taken from the section.
4. Join B to C and indicate the following:
   - the true length of the hip rafter (TLHR);
   - the plumb cut for the hip rafter (PCHR);
   - the seat cut for the hip rafter (SCHR).

Dihedral angle or backing bevel for the hip rafter, Figure 9.65.
The dihedral angle is the angle of intersection between the two sloping roof surfaces:

1. Draw a plan of the roof and mark on TLHR as before.
2. Draw a line at right angles to hip on the plan at D to touch wall plates at E and F.
3. Draw a line at right angles to TLHR at G to touch point D.
4. With centre D and radius DG, draw an arc to touch the hip on the plan at H.
5. Join point E to H and H to F. This gives the required backing bevel (BBHR).

The backing bevel is rarely used today in hipped roofing work for economic reasons. Instead the edge of the hip rafter is usually left square.

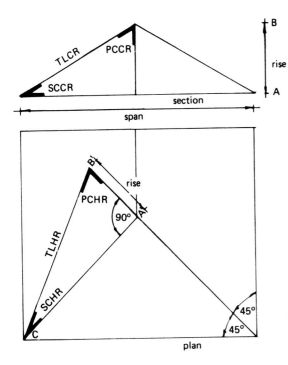

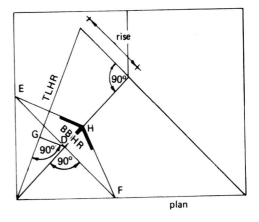

◀ **Figure 9.64** Hipped-end roof: rafter angles and lengths

▲ **Figure 9.65** Hip rafter backing bevel

Edge cut to the hip rafter, Figure 9.66. This is applied to both sides to allow it to fit up to the ridge board between the crown and common rafters:

1. Draw a plan of the roof and mark on TLHR as before.
2. With centre I and radius IB, swing TLHR down to J. (This makes IJ, TLHR.)
3. Draw lines at right angles from the ends of the hips and extend the ridge line. All three lines will intersect at K.
4. Join K to J. Angle IJK is the required edge cut (ECHR).

Development of the roof surfaces, to obtain the true length of the jack rafters (TLJR) and the edge cut for the jack rafters (ECJR), Figure 9.67. The edge cut allows the jack rafters to sit up against the hip. The plumb and seat cuts for the jack rafters are the same as those for the common rafters:

1. Draw the plan and section of roof. Mark on the plan the jack rafters.
2. Develop roof surfaces by swinging TLCR down to L and project down to M¹.
3. With centre N and radius NM¹ draw arc M¹O. Join points M¹ and O to ends of hips as shown.
4. Continue jack rafters on to development.
5. Mark the true length of jack rafter (TLJR) and edge cut for jack rafter (ECJR).

Side and edge cut for the purlin, Figure 9.68:

1. Draw a section of the common rafter with purlin and plan of hip.
2. With centre B and radii BA and BC draw arcs onto a horizontal line to give points D and E.
3. Project D and E down on to plan.
4. Draw horizontal lines from A¹ and C¹ to give points D¹ and E¹.
5. Angle DD¹B¹ is the side cut purlin (SCP) and angle B¹E¹E is the edge cut purlin (ECP).

The hip cut is rarely developed, due to its size and is normally simply cut as required.

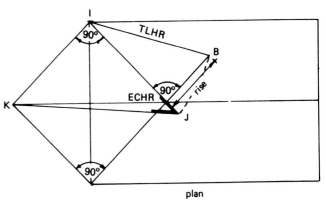

▲ **Figure 9.66** Hip rafter edge cut

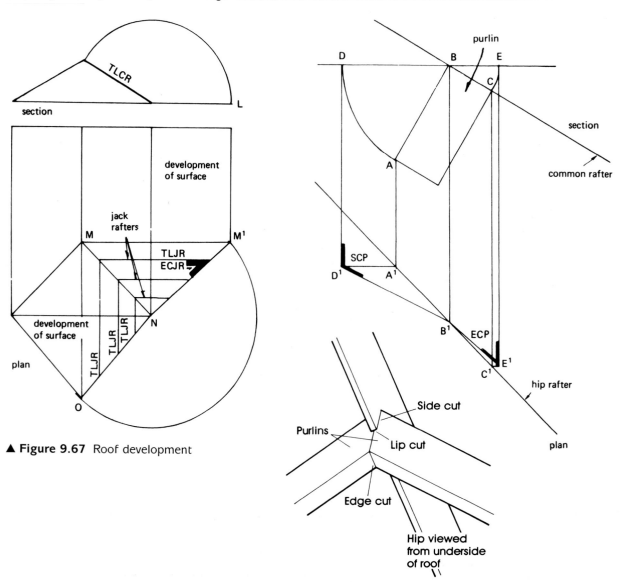

▲ **Figure 9.67** Roof development

▲ **Figure 9.68** Purlin cuts showing purling lip cut

Roofs with valleys

Where two sloping roof surfaces meet at an internal angle, a valley is formed. Figure 9.69 shows the following:

1. True length of valley rafter (TLVR).
2. Plumb and seat cuts for valley rafter (PCVR and SCVR).
3. Dihedral angle for valley rafter (DAVR).
4. Edge cut for valley rafter (ECVR).
5. Part development of one roof surface.
6. True length and edge cut for cripple rafters (TLCrR and ECCrR).

These true lengths and angles can be found by using similar methods as those used for the hip and jack rafters.

Pitch line. The single line used in roof geometry represents the pitch line, which is a line marked up from the underside of the common rafter, $\frac{1}{3}$ of its depth. As the hip and valley rafters are usually of deeper section, the pitch line on these is marked down form the top edge at a distance equal to $\frac{2}{3}$ the depth of the common rafter (Figure 9.70)

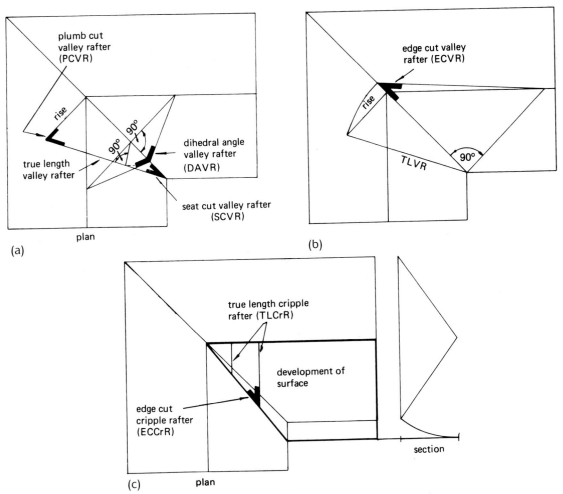

▲ **Figure 9.69** Valley rafter: (a) true length and angles; (b) edge cut; (c) cripple rafter, true lengths and edge cut

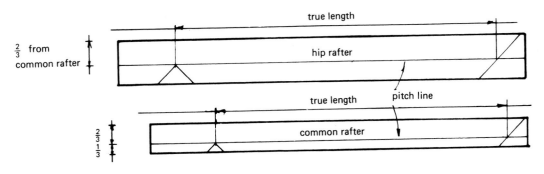

▲ **Figure 9.70** Setting out rafters

True length of rafters. The true length of the common and hip rafters is measured on the pitch line from the centre line of the ridge to the outside edge of the wall plate. For jack rafters, it is from the centre line of the hip to the outside edge of the wall plate; and for the cripple rafter it is from the centre line of the ridge to the centre line of the valley. Therefore when marking out the true lengths of the roofing components from the single line drawing, an allowance in measurement must be made. This allowance should be an addition for the eaves overhang and a reduction to allow for the thickness of the components. If this reduction is not apparent, it may be found by drawing the relevant intersecting components.

These reductions are illustrated in Figure 9.71, which shows the intersection between the ridge, common, crown and hip rafters. The reduction of the hip rafters is as shown; the reduction for the common rafters is always half the ridge thickness and for the crown rafter half the common rafter thickness. These reductions should be marked out at right angles to the plumb cut. Where a saddle board is used an additional allowance must be made for its thickness.

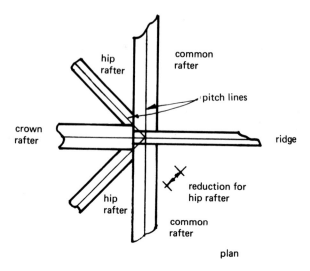

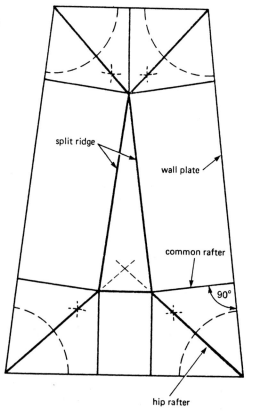

▶ **Figure 9.71** Reduction for thickness of materials

Irregular plan roofs

A hipped-end, equal-pitch roof on an irregular-shaped building is illustrated in Figure 9.72. The required true lengths and angles for this roof can be found using the methods covered in the previous examples after first establishing the positions of the hips and split ridge.

1. Draw to a suitable scale the plan of the roof.
2. Bisect external angles and mark on line of hip rafters.
3. From the point where hips intersect at the narrow end, draw lines parallel to the wall plate to indicate split ridge.
4. Common and jack rafters can be indicated by drawing lines at right angles to the wall plates.

▶ **Figure 9.72** Positioning members in an irregular plan roof

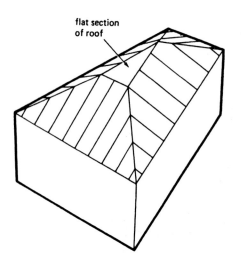

Louvre ventilator geometry

Triangular louvre ventilator

Refer to Figure 9.73:

1. Draw the elevation and vertical section of the ventilator.
2. Draw a rectangle to represent the development of inside surface of frame by projecting points on frame side over at right angles to the slope; the width of the rectangle equals the width of the frame.

To obtain bevels for housing:

3. Project points ABC and D across from the vertical section on to the development of inside surface of frame to give $A^1 B^1 C^1$ and D^1.
4. Draw lines $B^1 C^1$ and $A^1 D^1$ to give width and bevel for housing.
5. The edge bevel for the housing is the same as the slope of the frame.

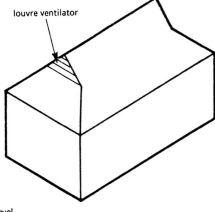

louvre ventilator

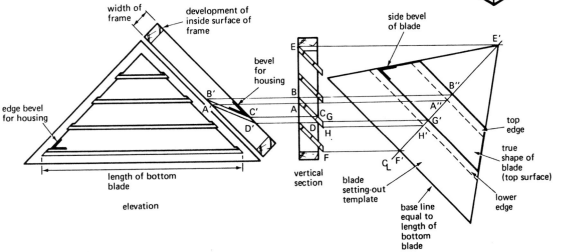

▲ **Figure 9.73** Louvre ventilator in gambrel roof end

To obtain true shape and side bevel of blade:

6. Project lines across from points E and F in the vertical section.
7. Draw centre line at same angle as the slope of the blades (45°). This gives points E^1 and F^1.
8. Draw base line at right angles to centre line to pass through F^1.
9. Make base line equal to length of bottom blade in elevation.
10. Draw lines from end of base line to E^1. This gives outline for blade setting-out template.
11. Project lines from points A, B, G and H in the vertical section across to the centre line to give points A^{11}, B^{11}, G^{11} and H^{11}.
12. Draw lines at right angles to the centre line from points A^{11}, B^{11}, G^1 and H^1 across width of template to give the true shape of blade and side bevel.

Circular louvre ventilator

The method used to obtain the true shape of the blades is similar to that used in the previous example, except that the template will be elliptical with its minor axis equal to the internal diameter of the frame, plus the depth of the housing on either side (Figure 9.74). The major axis is found by projecting lines across from the elevation onto the centre line, drawn at the same angle as the slope of the blades (45°). The ellipse can be drawn using any of the true methods previously covered.

The method of marking out the housings in the circular frame is illustrated in Figure 9.75. This entails the use of a temporary square frame made to fit over the circular one. The positions of the housings can be taken from the vertical section and marked onto the sides of the square frame. A straight edge is placed over these marks, which can then be transferred to the circular frame. A bevelled block and pencil are used to mark the housings on the inside of the circular frame.

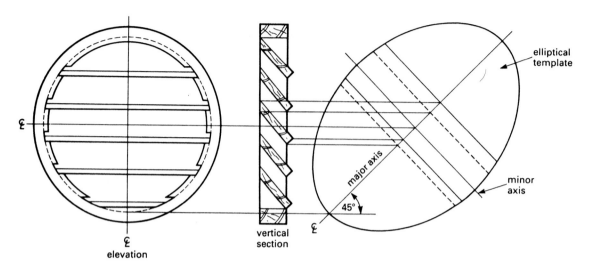

▲ **Figure 9.74** Circular louvre ventilator

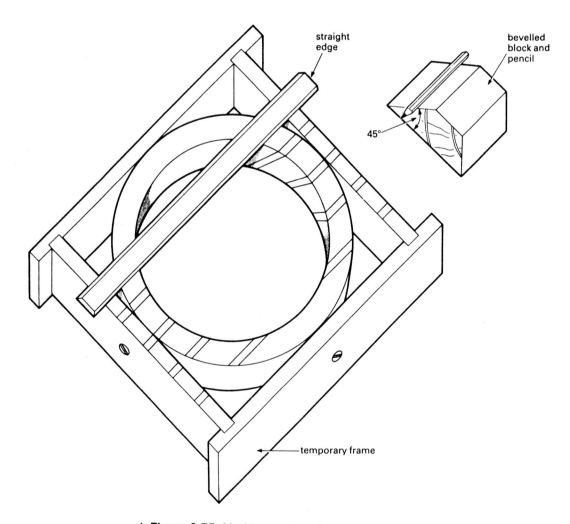

▲ **Figure 9.75** Marking out housing in louvre frame

Curved mouldings

Traditional curved mouldings can be one of two types:

- **Roman mouldings**, which are formed from arcs of circles;
- **Grecian mouldings**, which are formed from elliptical, parabolic or hyperbolic curves.

The following figures show various common mouldings on pieces of timber with Roman on the left and Grecian on the right. The setting out should be self-explanatory from the drawings. The Grecian mouldings are normally thought to be more aesthetically pleasing than the Roman mouldings. The proportions of the moulding and type of curve are usually left to the designer.

- *Ovolo moulding* (Figure 9.76) is a quarter of a circle and is the reverse shape of the cavetto. Where fillets are not used the moulding is normally called a *quadrant*.
- *Scotia moulding* (Figure 9.77) is formed by two quadrants of different radii giving a concave curve.
- *Torus moulding* (Figure 9.78) sometimes called a bull-nose or half-round moulding.
- *Cavetto moulding* (Figure 9.79) is the reverse shape of the ovolo.
- *Cyma-recta moulding* (Figure 9.80) often called an ogee moulding.
- *Cyma-reversa moulding* (Figure 9.81) also known as a reverse ogee mould.
- *Astragal or bead moulding* (Figure 9.82). In its simplest form it is a torus with side fillets, but the more elaborate mouldings include cavetto moulds on both sides of the fillets.

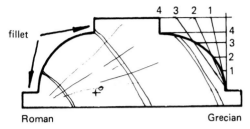

▲ **Figure 9.76** Ovolo moulding

▲ **Figure 9.77** Scotia moulding

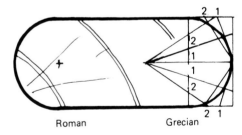

▲ **Figure 9.78** Torus moulding

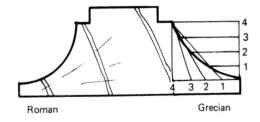

▲ **Figure 9.79** Cavetto moulding

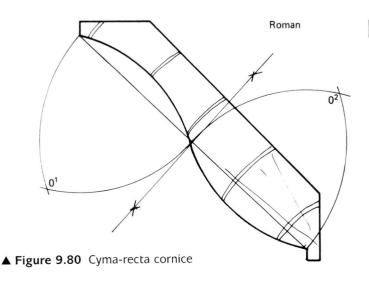

▲ **Figure 9.80** Cyma-recta cornice

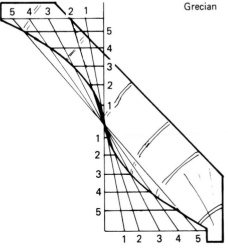

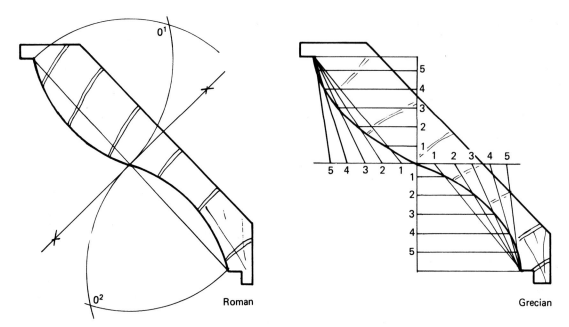

▲ **Figure 9.81** Cyma-reversa cornice

Roman

Grecian

Intersection of moulding

Where two mouldings intersect, the intersection is known as a mitre. When both of the mouldings are the same size, the mitre will be a bisection (half) of the angle of intersection. Two examples of this are shown in Figure 9.83. The 90° angle has 45° mitres and the 120° angle has 60° mitres.

Enlargement and reduction of mouldings

The procedure to enlarge or reduce a given moulding in width only, e.g. for use where the architrave over the head of a door is required to be larger or smaller than the jambs, is illustrated in Figure 9.84:

1. Draw the given mould and the outline rectangle of the required mould.
2. Project lines to determine the mitre line.

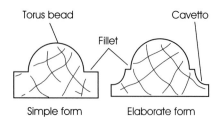

Torus bead   Cavetto

Fillet

Simple form   Elaborate form

▲ **Figure 9.82** Astragal moulding

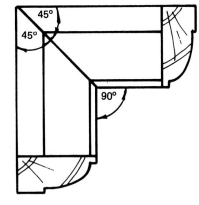

▶ **Figure 9.83** Mitring mouldings

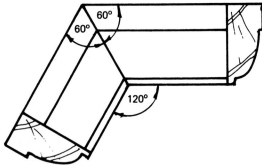

3. Select a number of suitable points on the given mould and project these onto the mitre line.

4. Draw vertical lines down from the mitre onto the required section.

5. With centre O draw arcs as shown from points on given mould to points on required mould.

6. Where lines intersect, draw the outline of the required mould.

Proportional enlargement

The procedure to enlarge and reduce proportionally a given mould is illustrated in Figure 9.85. This occurs when mouldings such as architraves, dado rails and picture rails are required in different sizes, but with their mouldings similarly proportioned:

1. Draw the given mould on base line AB.

2. Draw line AA^1 touching the top of the given mould at D.

3. Draw vertical heights of the required enlarged and reduced moulding GH and JK.

4. Select a number of suitable points on the given mould and project these both horizontally and vertically onto DC and DF.

5. Draw lines through these points radiating from point A.

6. Where the radiating lines intersect with lines GH and GI on the reduced moulding and lines JK and JL on the enlarged moulding, draw lines horizontally and vertically. This then gives the points from which the outlines can be drawn.

Intersection of curved and straight mouldings

When curved and straight mouldings of the same section intersect, there will be a curved mitre (Figure 9.86):

1. Draw the section of the straight moulding and select on it a number of suitable points: 0 to 7.

2. Draw the section of the curved moulding and mark on it the same points 0 to 7.

3. From these points, draw lines parallel to the direction of each moulding.

4. Where the lines intersect draw a smooth curve to give the mitre.

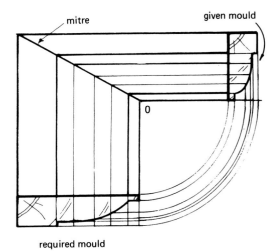

required mould

▲ **Figure 9.84** Enlargement or reduction of a given mould

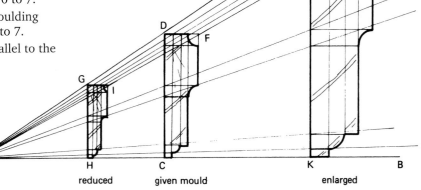

reduced   given mould   enlarged

▲ **Figure 9.85** Proportional enlargement and reduction of a given mould

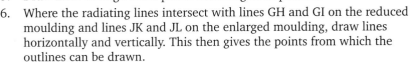
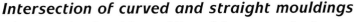

Grecian cyma-recta cornice

◀ **Figure 9.86** Intersection of a curved and straight moulding

Raking mouldings

When a raking mould and a level mould intersect at a mitre, the sections of the moulds will be different. The plan and elevation of a brick pier around which a dado height moulding is to be fixed is shown in Figure 9.87. The moulding on the face is inclined or raking at 30°. The two return moulds are level. Given the section of the raking mould, the method used to determine the sections of the two level moulds is as follows:

1. Draw the plan and elevation of the raking mould. Mark the given raking mould section on the elevation.
2. Select suitable points on the given mould 1, 2, 3 and 4. Transfer these on to the required moulds as shown.
3. Where lines intersect, draw the required sections for the top and bottom moulds.

Book 2, Chapter 8, page 278, also refers.

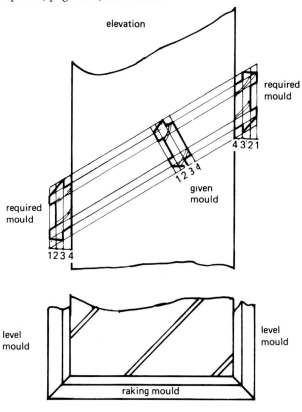

▶ **Figure 9.87** Raking mouldings

Splayed linings

The part plan and elevation of a splayed timber reveal lining to a door or window opening is illustrated in Figure 9.88. In order to construct the lining, two of three angles are required. If the corners of the lining are to be mitred, the face bevel and the edge bevel are required. If the corners of the lining are to be housed or butt jointed then the face bevel and shoulder bevel are required.

To obtain the side or face bevel (detail A):

1. Draw the plan and elevation.
2. With centre A and radius AB draw an arc to give B′ on the line drawn horizontally from A.
3. Project B′ vertically upwards to give B″ on the line drawn horizontally from B‴.
4. Join A′ to B″ to give the required side or face bevel.

To obtain the edge and shoulder bevel (detail B):

1. Draw the plan and elevation.

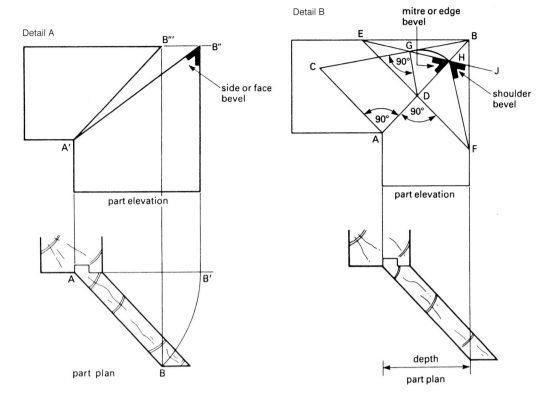

▲ **Figure 9.88** Development of bevels for splayed linings

2. At right angles to mitre line AB, draw line AC and mark on it the depth of the lining taken from the plan.
3. Join C to B.
4. Draw a line at right angles to AB at D (this can be anywhere along AB) to touch the edges of the lining at E and F.
5. Draw a line at right angles to CB at G to touch point D.
6. With centre D and radius DG, draw an arc to touch AB at H.
7. Join point E to H and extend on to J. Angle EHD is the required edge bevel.
8. Join point H to F. Angle FHJ is the required shoulder bevel.

Geometrical stairs

The construction of wreath stair strings and handrails is a highly specialised section of joinery, requiring an extensive knowledge of applied geometry. See also Book 2, Chapter 5.

Wreathed string quarter turn stair

The geometry required to develop the wreathed portion of a string for a quarter-turn geometric stair is illustrated in Figure 9.89:

1. Draw the plan of the quarter turn and set out the riser positions.
2. Draw a line at 60° through A to give points A^1 and C on lines projected horizontally across and vertically down from B.
3. From C draw lines through the ends of the risers to give points D and E on line A^1B.
4. Project points A^1, D, E, and B upwards to give positions of the riser faces in the development.

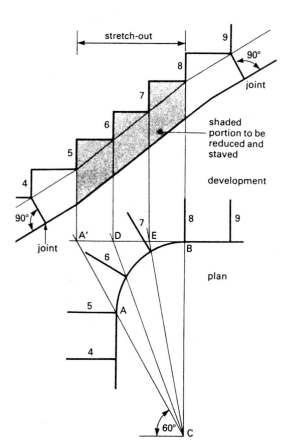

▲ **Figure 9.89** Wreathed string geometry (quarter turn)

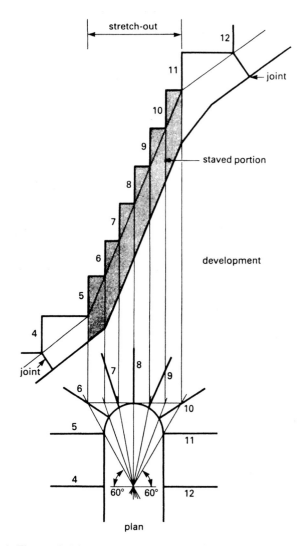

▲ **Figure 9.90** Wreathed string geometry (half turn)

5. Mark the riser positions of the two end steps (make the tread the same width as those on the straight part of the flight).

6. Draw in the treads making all the rises equal. Draw straight lines to join the lower corners of the steps. Draw lines parallel to these to give the required margin.

7. Draw the joint lines at either end of the string 'stretch-out'. Make these 90° to the underside of the string development.

This development gives the shape of the board or laminate required and also the set out of the string before bending.

Wreathed string half turn stair

The geometry required to develop the wreathed portion of a string for a half-turn stair is illustrated in Figure 9.90. The method used is similar to the quarter stair, except that two 60° lines are required to give the 'stretch-out' of the steps.

Handrail wreath: The geometry required to obtain the face moulds and bevels for a simple quarter turn rake to a level wreath is shown in Figure 9.91. This is required for a continuous handrail between a straight flight and landing:

1. Draw the plan and side elevation of the handrail and top two steps. The joint lines have been positioned a short distance past the curve to ease their jointing.
2. Draw in the rectangular section of the level-landing rail where the side elevation and plan centre lines intersect. This gives the minimum thickness of material required to form the wreath.
3. Draw VT line above the side elevation parallel to the pitch of the stair.
4. Project lines vertically down and horizontally across from the joint lines in plan to give AB.
5. Mark on line AB a series of points C, D, E, F, G, and H.
6. Project these points and AB vertically onto the VT line and continue at right angles beyond.
7. Transfer distances A, A′, A″, C, C′, C″, etc. from plan and mark above the VT line.
8. Draw in a smooth curve to complete the face mould.

Application of face mould. Figure 9.92 illustrates how the plywood face mould is applied to the timber blank and the rectangular end sections of the handrail marked out. The twist bevel is equal to the pitch of the stairs. The joint faces should be accurately prepared at right angles to the timber's face and at right angles to the centre lines prior to marking out.

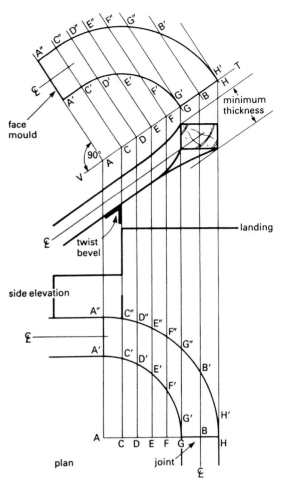

▲ **Figure 9.91** Wreathed handrail geometry

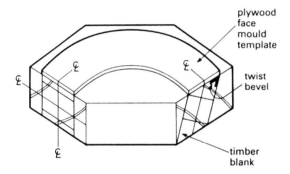

▲ **Figure 9.92** wreath blank and face mould template

Slabbed wreath

Figure 9.93 illustrates the blank after the perpendicular cuts have been made (slabbed wreath). This is often carried out on the band saw with the blank pitched at the same angle as the stairs. The falling top and bottom face lines are marked on as a smooth curve, taking care that they start at right angles to the joint surfaces. This determines the final shaping of the rectangular wreath; the surplus timber is normally removed by hand.

Prior to the hand moulding of the wreath, it should be end jointed with handrail bolts and dowels to the moulded straight sections, as illustrated in Figure 9.94. This eases the moulding operation and ensures its smooth, proper continuity.

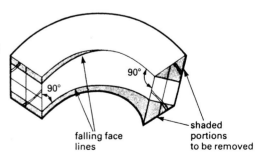

▲ **Figure 9.93** Slabbed wreath

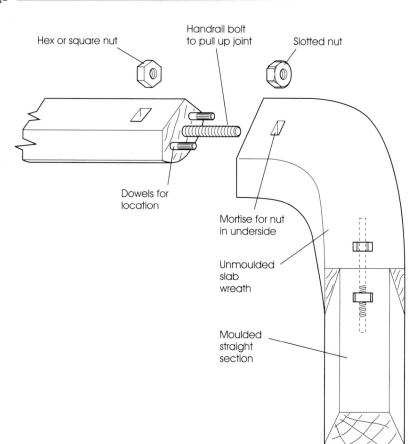

▶ **Figure 9.94** Means of joining
slabbed wreath to straight sections

Chapter Ten

Building controls

This chapter is intended to provide the site carpenter and bench joiner with an overview of the Building Controls system; it is concerned with the various statutory requirements that may be encountered, as well as general safe housekeeping practices that should be observed on a daily basis.

It includes the following:

- Planning permission and development controls
- Building regulations controls
- Health and safety controls
- General safety guidance.

Planning controls restrict the type, position and use of a building or development in relation to the environment, whereas the building regulations provide functional requirements for the design and construction of buildings to ensure the health and safety of people in and around them, promote energy efficiency and contribute to meeting the needs of disabled people. Health and safety controls, on the other hand, are concerned with the health and safety of all persons at their place of work and protecting other people, such as visitors and the general public, from risks occurring through work activities.

Planning permission and development controls

All development work is controlled by planning laws. These exist to control the use and development of land in order to obtain the greatest possible environmental advantages with the least inconvenience, both for the individual and society as a whole. The submission of a planning application provides the local authority and the general public with an opportunity to consider the development and decide whether or not it is in the general interest of the locality. The key word in planning is **development.** This means all building work, and related operations such as the construction of a driveway, and the **change of use** of land or buildings, such as using agricultural land as a garden or running a business from your home.

Certain developments are known as **permitted developments** where no formal planning approval is required. These permitted developments include limited extensions to buildings and the erection of boundary fences and walls within certain height limits.

Other considerations linked to planning permission include:

- conservation areas
- listed buildings
- tree preservation orders
- advertisement controls.

Making a planning application

To consider a planning application, the local authority will require the submission of drawings and other documents. The local authority will be concerned with the use and general appearance of the development as well as its impact on the surrounding area. The authority's highway department will be involved where the application requires vehicle access or will result in increased road traffic.

The local authority will have a written policy that sets out guidelines on the sort of building that may be built in a particular area and may also define the style of building permitted. Whether an application is approved or rejected will depend largely on their local guidelines. However in general it is government policy that development is permitted unless there is a good reason to refuse it.

It is possible to make either an outline application or a full application. The actual process in each case are similar and information will be required, concerning the nature of the proposed development, its position, size and general appearance.

Outline planning permission

This enables the owner or prospective owner to obtain approval of the proposed development in principle without having to incur the costs involved with the preparation of full working drawings, thus leaving certain aspects of the development for later approval. The granting of outline planning permission will in due course require a further application for the **approval of reserved matters**, such as site access, precise location of building and external appearance, etc., before any development can start. If the proposed development differs in any way from the outline planning permission granted, or conflicts with any conditions imposed by the local authority, a full planning application rather than an application for the approval of reserved matters should be made.

Full planning permission

After obtaining outline planning permission and when full details of the development have been decided, an application for full planning permission can be made. Alternatively an outline application can be dispensed with and full planning permission sought at the outset.

Full planning permission should always be sought at the outset in the following circumstances:

- to change the use of land or a building;
- to erect a building where the outline permission procedure has not been used;
- to erect an extension, outbuilding, garage, boundary fence or wall or create a new access;
- to gain approval for any unauthorised development;
- to carry out any development in a conservation area or involving any listed building;
- to demolish a building where no other development works is planned.

An application for full permission must normally include the following (Figure 10.1):

1. A completed **application form.**

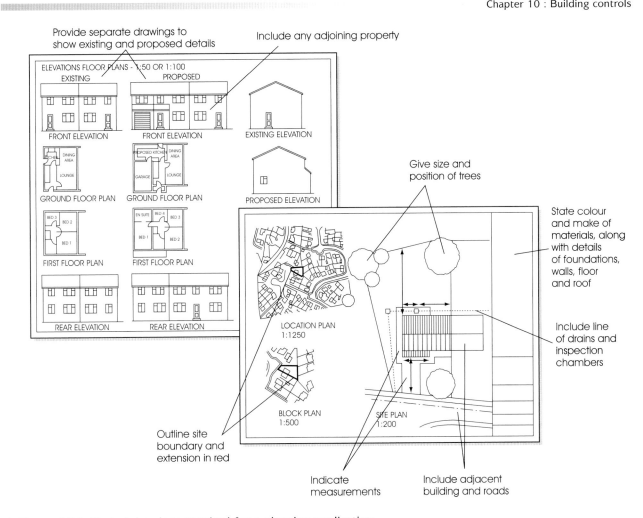

Provide separate drawings to show existing and proposed details

Include any adjoining property

Give size and position of trees

State colour and make of materials, along with details of foundations, walls, floor and roof

Include line of drains and inspection chambers

Outline site boundary and extension in red

Indicate measurements

Include adjacent building and roads

▲ Figure 10.1 Typical drawings required for a planning application

2. A completed **certificate of ownership** (stating who owns the land where the development is proposed).
3. Copies of the following **plans and drawings**.

Typical plans and drawings

- A **location plan** at a scale of 1:1250 or 1:2500, which shows the application site centred on the sheet shaded in red and its relationship to adjacent properties and roads adjoining the site. This is normally based on an Ordnance Survey plan, which is available from the local authority or an approved supplier.

- A **block plan** at a scale of 1:500 identifies the proposed site in relation to the surrounding area.

- A **site plan** at a scale of 1:500 or 1:200 shows the proposals in relation to the site boundaries and other buildings on the site. This drawing should also include the general layout of services and drainage on the proposed site as well as adjacent buildings and roads adjoining the site. In addition the position and size of all trees within or just outside the site should be shown.

- **Elevations and floor plans**, at a scale of 1:50 or 1:100 should be produced to show all side views of the proposal and position of windows, doors and internal walls. Where the proposal is attached to another property this relationship should also be shown. Floor plans should show the amount of floor space to be used for each purpose. In addition the types, colours and maker (if known) of materials for the external walls and roof should be indicated. Where the proposal is for an extension or alterations to an existing building, separate details for the existing building and the proposed alterations should be provided.

■ A **notice of the application** will be published in the local press and displayed at the property. Neighbours and any other interested parties are able to view the plans and make comments which the local authority must take into account.

After considering the application and any comments, the planning officer/ department may adopt the following options:

■ request further information or seek to negotiate changes in the proposal to make the application acceptable;

■ approve the application, where it considers the proposals to be acceptable and no objections have been received;

■ refuse the application, where it considers the proposals to be unacceptable and objections received cannot be resolved;

■ refer the application to the planning committee, where it considers the proposals to be acceptable but objections have been received.

Certain types of application are always referred to the **planning committee**. The planning officer will prepare a report recommending a course of action to the planning committee. The report to the committee is also available for public inspection.

After considering the case the planning committee can either:

■ grant permission;

■ grant permission with certain conditions;

■ refuse permission.

Once full planning permission is granted, development work may be started without applying for any further planning consent. However this is always subject to obtaining other approvals if applicable, such as building regulations control, listed building consent and complying with health and safety controls, etc.

Where permission is refused or given conditionally the committee must give their reasons for the decision. Applicants can then modify the proposed development and resubmit their application, or appeal against the decision.

Permitted developments

The *Town and Country Planning Act* sets out certain classes of development known as permitted developments where no planning approval is required as the Secretary of State has already granted automatic permission for them. These include:

■ limited extensions and certain alterations to dwellings;

■ temporary uses of land;

■ erection of boundary fences and walls within certain height limits.

Extensions to detached and semi-detached domestic houses are classified as permitted developments providing they do not exceed $70\,m^3$ or 15% of the original volume of the house up to a maximum of $115\,m^3$

Terraced houses are also restricted, however, to $50\,m^3$ or 10% of the original volume up to a maximum of $115\,m^3$.

In all cases these extensions must:

■ not project in front of the house if it faces a public highway (except small porches of $2\,m^2$ maximum floor area and less than $3\,m$ high);

■ not cover more than 50% of the garden;

■ not be higher than the house;

■ not be within $2\,m$ of the boundary if the extension is higher than $4\,m$.

Conservation areas

These are defined as 'areas of special architectural or historic interest the character and appearance of which it is desirable to preserve or enhance'. As the definition suggests, they are concerned with the maintenance of the local environment in its entirety and not just the buildings and their boundaries.

The purpose of designating a conservation area is to provide the local authority with an additional control measure over a cherished area which it considers to be of special architectural or historic interest.

Development work within a conservation area will require **conservation area consent** before it can be undertaken. This is a similar process to normal planning permission and can be done as part of the normal planning application process.

In addition to the documentation required for the granting of planning permission, the local authority will also require a design philosophy statement; a description of the intended works and the effect that they may have on the character of the conservation area.

Listed buildings

Certain buildings of historic or special architectural interest may be included in a list of buildings drawn up by the Secretary of State. Laws protect buildings included in this list. It is an offence to pull down, alter or extend a listed building unless authorised by listed building consent.

Listed building consent is required for any works which alter the style or character of a listed building either internally or externally. This includes any works, which do not require planning permission, and, as well as the listed building, it also includes any out-buildings, walls and railings etc. within the boundary.

- *Buildings of outstanding national importance* are given the top listing, being classified as Grade I.
- *Buildings of outstanding regional importance* are given a Grade II* classification.
- *Buildings of ordinary regional importance* are classified as Grade II.

Owners of listed buildings are required to preserve them in good order. If they fail to do so, the local authority may serve on the owner a **repairs notice**. This notice lists the repairs that the owner must carry out immediately; failure to do so can result in the local authority purchasing the building compulsorily, or carrying out the repairs themselves and recovering the costs from the owner. In certain cases owners of listed buildings may be eligible for special grants or loans from the local authority towards the costs of repairs.

Tree preservation orders

Trees are an essential part of our environment, both in the town and country areas. Many individual trees or groups of trees are protected by tree preservation orders (TPOs) issued by the local authority. Their purpose is to protect mature and healthy trees: owners are required to obtain the local authority's approval to lop, top, prune or fell a protected tree. Felling or damaging a protected tree without prior consent can result in prosecution. Where approval to fell is granted, the owner is normally required to plant a suitable replacement in the same position.

The majority of trees in conservation areas are automatically protected by TPOs and anyone wishing to lop, top, prune or fell a tree in such an area must give the local authority six weeks' notice of their intent.

Advertisement control

Town and country planning regulations control the display of advertisements such as shop signs, hoardings and banners etc. Certain advertisements are free of control and do not require the express approval of the local authority. Others such as illuminated advertisements and advertisements at, or above, first-floor level do generally require local authority approval.

The two main considerations taken into account when considering an advertisement application are highway safety and amenity. No advertisement will be permitted if, in the view of the local authority, it will distract drivers, be mistaken for a road sign, or spoil the appearance of a building or area. Like general planning decisions, in the event of a refusal there is right of appeal.

Enforcement of planning controls

Where work has been carried out without planning permission or when conditions attached to planning permission have not been complied with, the local authority has the power to serve an **enforcement notice**. This will state what has to be done to rectify the situation within a set time limit. If an enforcement notice is ignored, substantial fines can be imposed by subsequent court proceedings. In the case of listed buildings, the fine may be of an unlimited amount and or imprisonment.

Where an enforcement notice has been served, the local authority may, at its discretion, also serve a **stop notice**. Anyone served with enforcement or stop notices has the right of appeal to the Secretary of State.

Building regulations

The need for building regulations control arose at the time of the Industrial Revolution, when society and the economy were changing from being mainly agricultural to mainly industrial. Development at this time was very rapid but largely uncontrolled by local authorities and resulted in the appalling conditions of working-class housing.

The *Public Health Act, 1875,* allowed local authorities to make local bylaws to control the construction of new streets, the general layout of buildings, the chimneys, and external space requirements. In addition, the Act also gave local authorities powers for closing down dwellings they deemed to be unfit for human habitation.

In 1966, national building regulations replaced local bylaws to ensure that buildings were safe and healthy and that all building work was designed and implemented correctly.

In 1984 a new building act was approved by parliament. The *Building Regulations, 1985, 1991,* were applied to all building work carried out in England and Wales. The current regulations are the *Building Regulations 2000,* and the *Building (Amendment) Regulations, 2003.* Separate but similar controls and regulations exist for building work in Scotland and Northern Ireland.

The *Building Act, 1984,* is the primary legislation relating to building work. It gives the Secretary of State for the Environment powers to produce building regulations for the following broad purposes:

- Securing and maintaining the health, safety, welfare and convenience of people using or visiting buildings.
- Promoting the comfort of the occupants of a building whilst furthering the conservation of fuel and the efficient use of energy.
- Preventing or controlling waste, undue consumption of resources and misuse or contamination of water.

The *Building Regulations 2000* contain the minimum performance standards expected of buildings. They are supported by a series of **approved documents**, which are issued by the Office of the Deputy Prime Minster, ODPM. They are written in a clear but technical style and contain illustrations intended to give practical guidance on how to comply with the regulations. This guidance is non-statutory and often refers to British Standards, Building Research Establishment Papers, British Board of Agreement Certificates and other authoritative reference documents. When designing a building, you are free to use the solutions given in the approved documents or devise your own solutions providing you show that you meet the requirements of the Regulations.

Approved documents

The following approved documents are available:

- *Approved Document A: Structure* requires that all buildings are to be designed, constructed or altered so as to be safe and robust and not impair the structural stability of other buildings. It includes design standards for all types of building and in addition contains numerous tables giving timber sizes and wall thicknesses, etc., for traditional domestic buildings.

- *Approved Document B: Fire Safety* deals with fire precautionary measures required to ensure safety in the event of a fire for occupants, persons in the vicinity of buildings, and fire fighting crews. It includes requirements and guidance on means of escape in the event of a fire; fire detection and warning systems; fire resistance of the structural elements; compartmentation, separation and protection to prevent the spread of fire; restrictions on the use of flammable materials; facilities and access to buildings for fire fighting crews.

- *Approved Document C: Site Preparation and Resistance to Moisture* covers weather resistance and water tightness of buildings; site preparation and subsoil drainage; measures to deal with the erection of buildings on contaminated land, such as radon or landfill gases, methane and carbon dioxide and any other site-related dangerous or hazardous substances.

- *Approved Document D: Toxic Substances* controls against the use of specified toxic materials in buildings.

- *Approved Document E: Resistance to the Passage of Sound* deals with the ability of a building to prevent the passage of unwanted sound from internal sources. Details also apply to dwellings and rooms in buildings used for residential purposes such as hotel bedrooms, and similar rooms in hostels and residential homes for the elderly. **Pre-completion testing** (PCT) or the use of standard robust details is required to prove compliance. Guidance is also given on how to improve the acoustic standards of common areas (stairs and corridors) in these buildings as well as flats and schools.

- *Approved Document F: Ventilation* covers the means of ventilation and standards for air quality in all buildings including the requirements for preventing condensation occurring in roof spaces.

- *Approved Document G: Hygiene* sets out standards for the provision of sanitary and washing facilities, bathrooms and hot water. Also covered are safety requirements concerning the use of unvented hot water systems.

- *Approved Document H: Drainage and Waste Disposal* covers all above and below ground drainage, including sanitary pipework, foul drainage, rainwater drainage and disposal, wastewater treatment and discharge, cesspools, refuse storage and buildings close to or over public sewers.

- *Approved Document J: Combustion Appliances and Fuel Storage Systems* covers the construction, installation and use of boilers, chimneys, flues, hearths and fuel storage systems.

- *Approved Document K: Protection from Falling, Collision and Impact* sets out standards for the safety of stairs, ramps and ladders, along with requirements for balustrading, windows and vehicle barriers to prevent falling. In addition, requirements are provided for guarding or warning of hazards created by the position or use of windows and doors.

- *Approved Document L: Conservation of Fuel and Power* controls the insulation values of building elements, the allowable size of windows, doors and other openings; the air tightness of the structure; the heating efficiency of boilers and their control and installation.
- *Approved Document M: Access to and Use of Buildings* sets out requirements to ensure that buildings are accessible and usable by all people, regardless of disability, age or gender.
- *Approved Document N: Glazing Safety in relation to Impact, Opening and Cleaning* covers safety requirements relating to the use, operation, and cleaning of windows. In addition it states requirements for the use of safety glazing to avoid impact damage.
- *Approved Document to Regulation 7: Materials and Workmanship* states that all materials, products, fittings and components must be fit for their purpose and that all workmanship must be of an adequate standard.
- *Approved Document P: Electrical Safety,* a new (2005) approved document that introduces regulatory control over domestic electrical installations; also includes a self-certification scheme to prove compliance.

Building regulations and approved documents are constantly being updated, usually by adding more, at the time of writing the following are proposed:

Conservatories: Currently these and similar small buildings and extensions are exempt from building regulations control. Their re-inclusion is being considered, with particular concern being given to energy use and conservation, in order to bring them in line with AD L.

Dwellings: The introduction of a new approved document dealing specifically with all regulatory aspects of small dwellings is being considered.

Sustainable and secure buildings: A Bill is passing through Parliament at present that looks to widen the application of building regulations. Its intended aim is to introduce regulations that facilitate sustainable development, control waste and offer wider protection of the environment, and in addition provide crime prevention and detection measures related to buildings.

Application of regulations

Whenever anyone wishes to erect a new building, extend or alter an existing one, or change the use of an existing building, they will probably require building regulations approval, in addition to any planning controls applicable. Typical examples of work requiring building regulations approval are:

- All new building work.
- An extension to existing property.
- Internal structural alteration.
- The provision, extension or alteration of services or fittings, e.g.
    - sanitary conveniences and washing facilities;
    - bathrooms in dwellings;
    - hot water storage systems;
    - drainage and waste disposal systems;
    - fixed heat-producing appliances burning solid fuel, oil or gas and incinerators;
    - in dwellings, replacement windows, doors and roof lights; space heating or boilers and hot water storage;
    - in non-domestic buildings, heating and hot water systems; lighting, air conditioning and mechanical ventilation systems.

- Change of building use, e.g.
  - a building is changed into a dwelling;
  - a building contains a flat or residential room for the first time;
  - a building is changed for use as a hotel, boarding house or institution, where previously it was not;
  - a building is changed to a public building;
  - a building, which was previously exempt, but is no longer exempt;
  - a dwelling is altered to provide more or less than before;
  - a building containing a room for residential purposes is altered to provide more or less than before.
- Insertion of insulating material into a cavity wall.
- Underpinning of a building's foundations.

Exemptions to building regulations

Certain classes of building are exempted from the regulations. In addition to Crown buildings, buildings in the following classes are also exempt:

Class 1. Buildings controlled by other legislation, e.g. *Explosives Act, Nuclear Installations Act, Ancient Monuments* and *Archaeological Areas Act.*

Class 2. Buildings not used by people, e.g. a detached building where people cannot, or do not, normally enter.

Class 3. Greenhouses and agricultural buildings.

Class 4. Temporary buildings e.g. buildings which are intended to remain erected for less than 28 days.

Class 5. Ancillary buildings, e.g. temporary building-site accommodation; buildings used onsite in connection with sales; any building other than a dwelling used in connection with a mine or quarry.

Class 6. Small detached buildings, e.g. a detached building of up to $30\,m^2$ floor area, which does not contain sleeping accommodation (must be situated more than one metre from the boundary or be constructed substantially of non-combustible material); a detached building of up to $30\,m^2$ floor area designed to shelter people from the effects of nuclear, chemical or conventional weapons; a detached building of up to $15\,m^2$ floor area does not contain sleeping accommodation, e.g. garden sheds.

Class 7. Extensions (of up to $30\,m^2$ floor area), e.g. the ground-floor extension to a building by the addition of a green house, conservatory, porch, covered yard or covered way; a carport, which is open on at least two sides.

Certain other works although not exempt from the material requirements of the regulations, can be exempt from the need to seek local authority approval. For example, the installation of various combustion appliances; installation of specified plumbing services, fittings and drainage; installation of replacement doors, windows and roof lights in existing buildings. This is providing the people carrying out the work are registered under the appropriate scheme for carrying out the type of work.

In addition the local authority has the delegated powers to dispense with or grant a relaxation to any regulation requirement, if in special circumstances the terms of a requirement cannot be fully complied with.

Application procedure for building regulations approval

When building regulations approval is required, the building control section of the relevant local authority must be notified of your intentions by one of the following three methods (Figure 10.2):

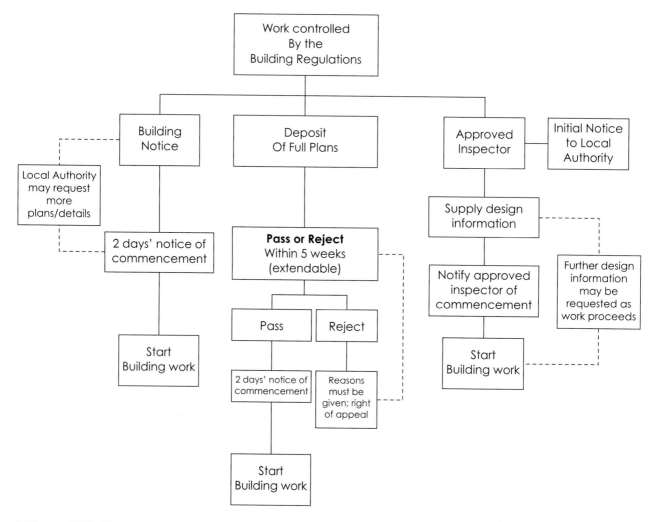

▲ **Figure 10.2** Alternative application procedures for Building Regulations

- deposit of full plans
- issue of a Building Notice
- appointment of an Approved Inspector

Providing full plans

Application using the full plans method can be made by depositing full plans of the proposed works similar to that required for planning permission. These shall consist of:

- The name and address of the person intending to carry out the work.
- A statement that the plans are deposited in accordance with the relevant building regulation(s).
- A full description of the proposed work (the specification).
- Details of the surrounding area (the block plan).
- The intended use of the proposed building.
- Drawings to a scale of not less than 1:1250 showing the size and position of the building, its boundaries and relationship to adjoining boundaries (site plan and general location plans).
- The number of storeys in the building.
- The provision made for drainage.
- Details of any cavity wall insulation and its installer.
- Details of any unvented hot-water system and its installer.
- Any other details or plans if required showing that the work will comply with the regulations.

These plans and details will be examined to see if they comply with the regulations. A decision will be made within five weeks or two months if you agree to an extension of time.

The plans may be rejected on any of the following grounds:
- The plans show a contravention of the regulations.
- The plans are defective (they fail to prove compliance with the regulations).
- They contravene or show insufficient detail with regards to one of the local authority's functions under the Building Act (e.g. drainage, water supply, public buildings and local legislation).

Where an application is refused or the applicant and the local authority are in dispute, there is an **appeals procedure** to the Secretary of State for the Environment.

Issuing a Building Notice

Application using the Building Notice method can be made by depositing a Building Notice and limited accompanying information, such as a specification, block plan, site plan and general location plans.

In addition, the local authority may request further information as the work proceeds, in order to show compliance of specific items which cannot be inspected on site including structural calculations, material specifications, etc.

Appointing an Approved Inspector

Application using the Approved Inspector method can be made by you and the inspector jointly by depositing an initial notice, limited plans and evidence of the insurance cover to the local authority. The local authority must accept or reject this initial notice within ten working days. Once accepted, their powers to enforce the Regulations are suspended and the Approved Inspector will carry out the building control function and issue a final certificate to you and the local authority when the work has been completed satisfactorily.

Inspection of building work

When either the full plans method or the Building Notice method has been adopted, the local authority's Building Control Officer will inspect the work as it proceeds, unless you are using an Approved Inspector.

The builder must give the local authority notice of the following building stages:
- At least two days before the commencement of work.
- At least one day before the covering up of any excavation for a foundation, any foundation, any damp-proof course or any concrete or other material laid over a site.
- At least one day before covering up any drain or private sewer.
- Not more than five days after laying, concreting or backfilling a drain.
- Not more than five days after completion of building work.

These periods of notice commence on the day after the day on which the notice is served. A day is defined as a 24-hr period starting at midnight but does not include Saturdays, Sundays and Bank or public holidays.

Where builders fail to notify the local authority of any stage as required, the local authority has the power to require them to 'open up' or 'pull down' part of the work at a later date to enable inspection. After inspection by the Building Control Officer, the local authority may require modifications or additional work to be carried out in order to comply with the regulations.

Where an Approved Inspector has been appointed they will be responsible for inspecting the work as it proceeds. The Inspector may also require the builder to notify the commencement and/or particular stages of building work.

The local authority will charge a set fee for considering an application and inspecting the work as it proceeds. If an Approved Inspector is appointed, they will negotiate the fee with you.

Health and safety controls

In the mid-1970s the *Health and Safety at Work etc. Act* was introduced. The HSW Act was seen as an enabling umbrella. It introduced the main statutory legislation, completely covering the health and safety of all persons at their place of work and protecting other people from risks occurring through work activities. It has overseen the gradual replacement of previous piecemeal health and safety requirements by revised and up-to-date measures prepared in consultation with industry and its workers. As a member of the construction industry you are required to know about your responsibilities with regards to various safety legislations.

The Health and Safety at Work etc. Act

The four main objectives of the HSW Act are as follows:
1. To secure the health, safety and welfare of all persons at work.
2. To protect the general public from risks to health and safety arising out of work activities.
3. To control the use, handling, storage and transportation of hazardous substances.
4. To control the release of noxious or offensive substances into the atmosphere.

These objectives can be achieved only by involving everyone in health and safety matters. This includes:
- employers and management
- employees (and those undergoing training)
- self-employed workers
- designers, manufacturers and suppliers of equipment and materials.

Employers' and management duties

Employers have a general duty to ensure the health and safety of their employees, visitors and the general public.

This means that the employer must:
1. Provide and maintain a safe working environment.
2. Ensure safe access to and from the workplace.
3. Provide and maintain safe machinery, equipment and methods of work (refer to Chapter 6).
4. Ensure the safe handling, transport and storage of all machinery, equipment and materials.
5. Provide their employees with the necessary information, instruction, training and supervision to ensure safe working.
6. Prepare, issue to employees and update as required a written statement of the firm's safety policy.
7. Involve trade union safety representatives (where appointed) with all matters concerning the development, promotion and maintenance of health and safety requirements.

Safety policy. A typical construction company's 'Safety Policy' and 'Site Safety Induction' procedure is illustrated in Figure 10.3.

BBS Construction Services

STATEMENT OF COMPANY POLICY ON HEALTH AND SAFETY

The Directors accept that they have a legal and moral obligation to promote health and safety in the workplace and to ensure the co-operation of employees in this. This duty of care extends to all persons who may be affected by any operation under the control of BBS Construction Services.

Employees also have a statutory duty to safeguard themselves and others and to co-operate with management to secure a safe work environment.

The directors shall ensure, so far as reasonably practicable, that:

- Adequate resources and competent advice are made available in order that proper provision can be made for health and safety.
- Safe systems of working are devised and maintained.
- All employees are provided with all information, instructions, training and supervision. required to secure the safety of all persons.
- All plant, machinery and equipment is safe and without risk to health.
- All places of work are maintained in a safe condition with safe means of access and egress.
- Arrangements are made for safe use, handling, storage and transport of all articles and substances.
- The working environment is maintained in a condition free of risks to health and safety and that adequate welfare facilities are provided.
- Assessment of all risks are made and control measures put in place to reduce or eliminate them.
- All arrangements are monitored and reviewed periodically.

These statements have been adopted by directors of the company and form the basis of our approach to health and safety matters.

Ivor Carpenter, Finance Director

Christine Whiteman, Human Resource Director

James Brett, Managing Director

The Director with responsibility for Health and Safety

Peter Brett, Chief Executive

BBS Construction Services

SITE SAFETY INDUCTION

As part of the company's commitment to safety the following checklist is provided for management, when employing new staff or on the transfer of existing staff to your site.

MANAGEMENT MUST:

- Issue and explain the company's Safety Policy.
- Introduce your site safety advisor controller.
- Discuss and record any previous safety training and experience.
- Issue and discuss appropriate safety method statement.
- Emphasis the following points:
  - Emergency and first aid procedures applicable to the site
  - Personal safety responsibilities, house keeping, hygiene and PPE
  - Need to report accidents 'near misses' and unsafe conditions
  - Need for authorisation and training, before use of all plant machinery and powered hand tools
  - Location of all welfare facilities
- Explain the procedure to be followed in the event of a health and safety dispute (consult safety advisor/controller in first instances).
- Show the site notice board (in the rest room) where safety notices and information are displayed.
- Finally inform of company's key phrase for all matters 'IF IN DOUBT ASK' then invite questions.

▲ **Figure 10.3** Contractor's safety policy

Employees' duties

An employee is an individual who offers his or her skill and experience, etc., to his or her employer in return for a monetary payment. It is the duty of all employees while at work to comply with the following:

1. Take care at all times and ensure that their actions do not put 'at risk' themselves, their workmates or any other person.
2. Co-operate with their employers to enable them to fulfil the employer's health and safety duties.
3. Use the equipment and safeguards provided by the employer.
4. Never misuse or interfere with anything provided for health and safety.

Self-employed duties

The self-employed person can be thought of as both the employer and employee; therefore their duties under the Act are a combination of those of the employer and employee.

Designers', manufacturers' and suppliers' duties

Under the Act, designers, manufacturers and suppliers as well as importers and hirers of equipment, machinery and materials for use at work have a duty to:

1. Ensure that the equipment machinery or material is designed, manufactured and tested so that when it is used correctly no hazard to health and safety is created.
2. Provide information or operating instructions as to the correct use, without risk, of their equipment, machinery or material. Employers should ensure this information is passed on to their employees.
3. Carry out research so that any risk to health and safety is eliminated or minimised as far as possible.

Enforcement of safety legislation

Under the HSW Act a system of control was established, aimed at reducing death, injury and ill health. This system of control is through the Health and Safety Executive (HSE). The Executive is divided into a number of specialist inspectorates or sections that operate from local offices situated throughout the country. From the local office, inspectors visit individual workplaces.

The Health and Safety Executive inspectors have been given wide powers of entry, examination and investigation in order to assist them in the enforcement of the HSW Act and other safety legislation. In addition to giving employers advice and information on health and safety matters, an inspector can do the following:

- *Enter premises in order to carry out investigations*, including the taking of measurements, photographs, recordings and samples. The inspector may require the premises to be left undisturbed while the investigations are taking place.
- *Take statements*. An inspector can ask anyone questions relevant to the investigation and also require them to sign a declaration as to the truth of the answers.
- *Check records*. All books, records and documents required by legislation must be made available for inspection and copying.
- *Give information*. An inspector has a duty to give employees or their safety representative information about the safety of their workplace and details of any action he/she proposes to take. This information must also be given to the employer.
- *Demand*. The inspector can demand the seizure, dismantling, neutralising or destruction of any machinery, equipment, material or substance that is likely to cause immediate serious personal injury.
- *Issue an improvement notice*. This requires the responsible person (employer or manufacturer, etc.) to put right within a specified period of time any minor hazard or infringement of legislation.

■ *Issue a prohibition notice.* This requires the responsible person to stop immediately any activities likely to result in serious personal injury. This ban on activities continues until the situation is corrected. An appeal against an improvement or prohibition notice may be made to an industrial tribunal.

■ *Prosecute.* All persons, including employers, employees, self-employed, designers, manufacturers and suppliers who fail to comply with their safety duty may be prosecuted in a magistrates' court or in certain circumstances in the higher court system. Conviction can lead to unlimited fines, or a prison sentence, or both.

Management of Health and Safety at Work Regulations (MHSWR)

These regulations apply to everyone at work. They require your employer and the self-employed to plan, control, organise, monitor and review their work. In doing this, they must:

■ assess the risks associated with the work being undertaken;

■ have access to competent health and safety advice;

■ provide employees with health and safety information and training;

■ appoint competent persons in their workforce to assist them in complying with obligations under health and safety legislation;

■ make arrangements to deal with serious and imminently dangerous situations;

■ co-operate in all health and safety matters with others who share the workplace.

Your duties as an employee under the regulations are:

■ to use all machinery, equipment, dangerous substances, means of production, transport equipment safely in accordance with the training and instructions given;

■ to inform your employer or named competent person of dangerous situations and/or shortcomings in the health and safety arrangements.

Risk assessment and management

This is a key part of the regulations, in order to put in place control measures. It involves employers identifying the hazards involved in their work, assessing the likelihood of any harm arising and deciding on adequate precautionary control measures.

Risk assessment is a five-step process:

■ **Step 1. Looking for the hazards.** Consider the job to be undertaken:
  – How will it be done?
  – Where is it done?
  – What equipment and materials will be used?

■ **Step 2. Decide who might be at risk and how.** Consider:
  – employees;
  – the self-employed;
  – other companies working on the job;
  – visitors to the job;
  – the general public who may be on or near the job.

■ **Step 3. Evaluate the risks and decide on the action to be taken.** Typical questions are:
  – Can the hazard be completely removed?
  – Can the job be done in another safer way?
  – Can a different, less hazardous material be used?

If any of the answers to Step 3 are 'Yes', change the job to eliminate the risk. If risks cannot be eliminated:

- Can the hazards be controlled?
- Can protective measures be taken?

■ **Step 4. Record the findings.** Employers should make a record of the risk assessment and pass it on to their employees. This should include details of significant risks involved and the measures taken to remove or control them.

■ **Step 5. Review the findings.** Periodic reviews of the findings are required to ensure that they are still effective. New assessments will be required:

- when the risks or conditions change;
- when new risks or conditions are encountered for the first time.

Checklist. A typical construction company's risk assessment checklist is illustrated in Figure 10.4.

The Manual Handling Operations Regulations

These regulations require employers and the self-employed to avoid the need to undertake manual handling operations that might create a risk of injury. Where avoidance is not reasonably practical, they have to make an assessment with the aim of removing hazards and minimising potential risk of injury by:

■ Avoiding all unnecessary manual handling;

■ Mechanising or automating handling tasks e.g. by the use of cranes, hoists, forklift trucks and conveyor belts, etc.;

■ Arranging for heavy or awkward loads to be shared when finally lifting or moving into position by hand;

■ Ordering materials in easily handled sizes, e.g. bagged sand, cement and plaster, etc., are all available in 25 kg bags;

■ Positioning all loads mechanically as near as possible to where they will be used in order to reduce the height they have to be manually lifted and the distance they have to be carried;

■ Providing employees with advice and training in safe lifting techniques and sensible handling of loads.

As an employee, you are required to make full and proper use of anything put into place by your employer to reduce the risk of injury during manual handling operations

The Personal Protective Equipment at Work Regulations

Personal protective equipment (PPE) means all pieces of equipment, additions or accessories designed to be worn, used or held by a person at work to protect against one or more risks. Typical items of PPE are:

■ safety footwear
■ waterproof clothing
■ safety helmets
■ gloves
■ high visibility clothing
■ eye protection
■ dust masks
■ respirators
■ safety harnesses.

The use of PPE is seen as the last not the first resort. The first consideration is to undertake a risk assessment with a view to preventing or controlling any risk at its source, by making machinery or work processes safer.

BBS Construction Services

RISK ASSESSMENT

Activity covered by assessment: _____

Location of activity: _____

Persons involved: _____

Date of assessment: _____

Tick appropriate box ☑

	YES	NO
• Does the activity involve a potential risk?	☐	☐

	YES	NO
• If YES can the activity be avoided?	☐	☐

	LOW	MEDIUM	HIGH
• If NO what is the level of risk?	☐	☐	☐

• What remedial action can be taken to control or protect against the risk?

1 _____

2 _____

3 _____

4 _____

5 _____

MANAGEMENT SUMMARY:

Priority for action:	LOW	MEDIUM	HIGH
	☐	☐	☐

Action to be taken: _____

Date action to be taken by: _____

Date for reassessment: _____

Assessor's name and signature: _____

ASSESS THE RISK – PUT IN CONTROLS – CHECK THEY WORK

▲ **Figure 10.4** Risk assessment checklist

PPE requirements:
- All items of PPE must be suitable for the purpose it is being used for and provision must be made for PPE maintenance, replacement and cleaning.
- Where more than one item is being worn, they must be compatible.
- Training must be provided in the correct use of PPE and its limitations.
- Employers must ensure that appropriate items are provided and are being properly used. This also applies to the self-employed.
- Employees and the self-employed must make full use of PPE provided and in accordance with the training given. In addition any defect or loss must be reported to their employers.

The Health and Safety (Safety Signs and Signals) Regulations

Safety signals and signals legislation requires employers to provide safety signs in a variety of situations that do or could affect health and safety. There are four types of safety signs in general use. Each of these signs have a designated shape and colour, to ensure that health and safety information is presented to employees in a consistent, standard way, with the minimum use of words.

Details of these signs and typical examples of use are given in Figure 10.5 (opposite page).

In addition the following points, which are of particular concern to construction work, are highlighted in the regulations:

- In order to avoid confusion, too many signs should not be placed together.
- Signs should be removed when the situation they refer to ceases to exist.
- Fire fighting equipment and its place of storage, must be identified by being red in colour.
- Traffic routes should be marked out using yellow.
- Acoustic fire evacuation signals must be continuous and sufficiently loud to be heard above other noises on site.
- Anyone giving hand signals must wear distinctive brightly coloured clothing and use the standard arm and hand movements (Figure 10.6).

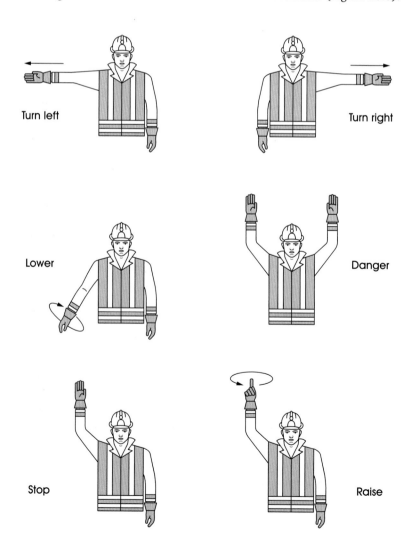

▶ **Figure 10.6** Hand signs

Purpose	Sign	Definition	Examples for use
Prohibition	white red	A sign prohibiting certain behaviour	No smoking Smoking and naked flames prohibited Do not extinguish with water Not drinking water Pedestrians prohibited
Caution	yellow black	A sign giving warning of certain hazards	Caution, risk of fire Caution, toxic hazard Caution, corrosive substance General warning caution, risk of danger Caution, risk of electric shock Perimeter of hazard
Safe condition	green	A sign providing information about safe conditions	First aid Indication of direction Indication of direction
Mandatory	blue	A sign indicating that a special course of action is required	Head protection must be worn Eye protection must be worn Hearing protection must be worn Foot protection must be worn Hand protection must be worn Respiratory protection must be worn
Supplementary	white or colour of sign it is supporting	A sign with text. Can be used in conjunction with a safety sign to provide additional information	IMPORTANT REPORT ALL ACCIDENTS IMMEDIATELY SCAFFOLDING INCOMPLETE SAFETY HELMETS ARE PROVIDED FOR YOUR SAFETY AND MUST BE WORN PETROLEUM MIXURE HIGHLY FLAMMABLE NO SMOKING OR NAKED LIGHTS WARNING HIGH VOLTAGE CABLES OVERHEAD EYE WASH BOTTLE

▲ **Figure 10.5** Safety signs

Construction industry safety legislation

The Construction (Design & Management) Regulations (CDM)

Design and management legislation requires that health and safety is taken into account and managed during all stages of a construction project, from its conception, design and planning, throughout the actual construction process and afterwards during maintenance and repair.

These regulations apply to all construction projects that:

- include any demolition work, or
- will last for more than 30 days, or
- will involve more than 500 person days of work, or
- will involve more than 5 workers on site at any one time.

Health and safety plan and file. The regulations require the client, the designers and the building contractors to play their part in improving on-site health and safety. In doing this they will have to draw up a two-stage **health and safety plan**:

- *Stage 1. A design plan*, which highlights any particular risks of the project and the equipment and the level of health and safety competence that a prospective contractor will require.
- *Stage 2. A construction plan* that sets out how health and safety will be managed during the project.

They will also have to draw up a **health and safety file**. This should be produced at the end of a project and passed on to the client or building user. It should contain details of health and safety risks that will have to be managed during future maintenance and cleaning work.

Client responsibilities:

- Appoint a planning supervisor to draw up the first stage safety plan. Co-ordinate with the principal contractor with regards the second stage. Compile the safety file.
- Appoint the principal contractor.
- Determine that the planning supervisor, principal contractor, designers and any nominated subcontractors are competent to deal with the health and safety aspects of the project.
- Ensure that construction work does not start until a suitable safety plan is in place.
- Keep the health and safety file available for inspection.

Designer responsibilities: The term 'designer' is used to describe everyone who prepares drawings or specifications for a product and thus includes architects, structural engineers, surveyors, the planning supervisor and other designers in health and safety matters.

- Consider at the design stage, the foreseeable health and safety risks associated with the project. Not only during construction, but also during later maintenance and cleaning.
- Provide information with their design, on any aspect, which might affect the health and safety of contractors, cleaners or anyone else who might be affected by their work.

Principal contractor responsibilities. This is the main contractor who has been awarded the contract to undertake the construction work. They may appoint other contractors and subcontractors, to undertake specific parts of the construction work.

- Prepare and maintain the second stage of the safety plan and supply all relevant information to the planning supervisor for inclusion in the safety file.
- Ensure co-operation between all contractors on health and safety matters.

- Ensure all contractors and employees comply with the requirements of the health and safety plan.
- Ensure that only authorised persons are permitted to enter areas where construction work is being carried out.

Contractor responsibilities. The term 'contractor' is used to include subcontractors and self-employed persons working on site.

- Co-operate with the principal contractor and other contractors in order to achieve safe and healthy site conditions.
- Provide health and safety information for inclusion as required, into both the health and safety plan and file.

The Construction (Health, Safety and Welfare) Regulations (CHSWR)

The main objective of CHSWR is to promote the health and safety of employees, the self-employed and others who may be affected by construction activities. The issues covered in the regulations include:

- provision of welfare facilities, Figure 10.7;
- provision of working platforms;
- prevention of falls;
- support of excavations;
- provision of guard rails and barriers;
- procedures in the event of fire or other emergency;
- use of vehicles and transport routes on site;
- inspections and reports.

The main requirements of CHSWR are as follows:

- **Toilets.** No specific number, but must be clean, with adequate ventilation and lighting. Men and women can use the same toilet provided each is in a separate lockable room.
- **Washing facilities.** Wash hand basins with hot and cold or warm water, to be provided in the immediate vicinity of toilets and changing rooms. These must include soap and towels or other drying facility. Where the

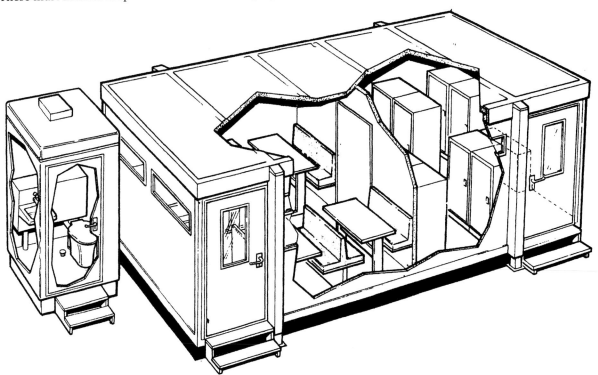

▲ **Figure 10.7** Typical welfare facilities

work is particularly dirty or involves exposure to toxic or corrosive substances, showers may be required. All rooms containing washing facilities must have adequate ventilation and lighting. Unisex facilities are suitable for washing of hands, faces and arms, otherwise they should be in a separate room for use by one person at a time and can be locked from the inside.

- **Drinking water** (wholesome) to be readily accessible in suitable places and clearly marked. Cups or other drinking containers must be provided, unless the water is supplied via a drinking fountain.

- **Storage and changing of clothing.** Secure accommodation must be provided for normal clothing not worn at work and for protective clothing not taken home. Separate lockers may be required where there is a risk of protective clothing contaminating normal clothing. This accommodation should include changing facilities and a means of drying wet clothing.

- **Rest facilities.** Accommodation must be provided for taking breaks and meals. These facilities must include tables and chairs, a means of boiling water and a means of preparing food.

- **Working platforms.** Where it is possible for a person to fall two metres or more, the working platform, must be inspected by a competent person: before its first use; after alteration; after strong winds or other events likely to affect its stability; at least once every seven days.

- **Prevention of falls.** Edge protection is required to all working platforms and other exposed edges where it is possible to fall two metres or more, Figure 10.8. They should be:
  - sufficiently rigid for the purpose;
  - include a guardrail at least 910 mm above the edge;
  - include a toe board at least 150 mm high;
  - subdivided with intermediate guardrails, additional toeboards or brick guards, etc., so that the maximum vertical unprotected gap is 470 mm;
  - other types of barrier may be used to protect edges, provided that they give the equivalent standard of protection against falls of persons and rolling or kicking of materials over the edge.

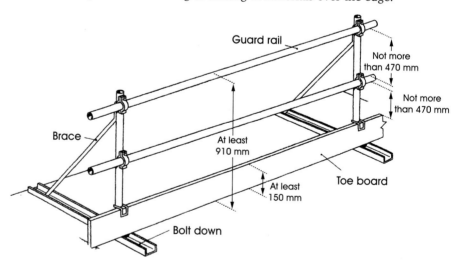

▶ **Figure 10.8** Protection to exposed edges

- **Support of excavations.** Measures must be taken to prevent injury by collapsing excavations, falling materials or contact with buried underground services, Figure 10.9. Support for excavations is to be provided at an early stage. Sides of excavations must either be battered back to a safe angle or be supported with timbering or a proprietary system. All support work is to be carried out or altered by or under the supervision of a competent person. Measures must be taken to prevent people, materials or vehicles falling into excavations, for example: by the use of edge protection guard rails; not storing materials, waste or plant items near excavations; keeping traffic routes clear of excavations.

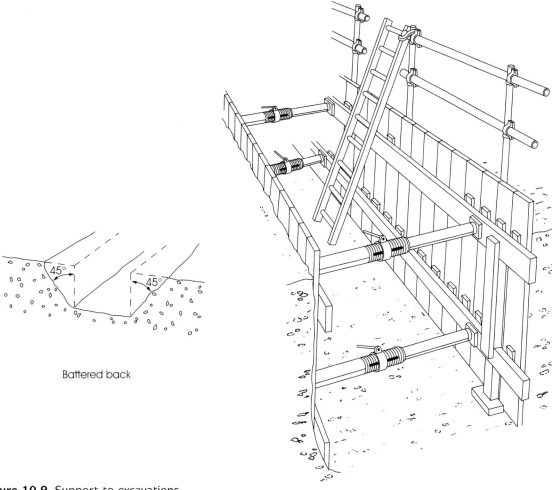

Battered back

Shoring

▲ Figure 10.9 Support to excavations

■ **Emergency procedures.** These are the arrangements made to deal with any unforeseen emergency, including fire, flooding, explosion and asphyxiation. These procedures should be co-ordinated by a trained person, who will take responsibility and control and must include the following:
- Provision of emergency signals, routes and exits for evacuation. These should be kept clear, be marked and illuminated.
- Provision to notify the emergency services.
- Provision of first aid and other facilities for treating and recovering injured persons.

■ **Use of vehicles:**
- All site drivers must be trained.
- Visiting drivers must be informed of site transport rules.
- Suitable traffic routes must be provided and clearly marked, avoiding sharp bends and blind corners including safe entry and exit points.
- Pedestrians and vehicles should be separated as far as possible.
- Reversing should be avoided wherever possible. Audible alarms are advisable where reversing is necessary.
- Provide trained signallers, wearing high visibility clothing to assist drivers.

■ **Inspections and reports.** Competent persons must carry out the following inspections and make written reports Figure 10.10:
- *Working platforms.* Inspect before use, after alteration, after any event, which may have affected its stability and at least once every seven days.

Construction (Health, Safety and Welfare) Regulations 1996

INSPECTION REPORT

Report of results of every inspection made in pursuance of regulation 29(1)

1. Name and address of person for whom inspection was carried out.

2. Site address.   3. Date and time of inspection.

4. Location and description of workplace (including any plant, equipment or materials inspected)

5. Matters which give rise to any health and safety risks.

6. Can work be carried out safely?   Y / N

7. If not, name of person informed.

8. Details of any other action taken as a result of matters identified in 5 above.

9. Details of any further action considered necessary.

10. Name and position of person making the report.

11. Date report handed over.

Construction (Health, Safety and Welfare) Regulations 1996

INSPECTION REPORTS: NOTES

Place of work requiring inspection	Timing of frequency of inspection					
	Before being used for the first time.	After substantial addition, dismantling or alteration.	After any event likely to have affected its strength or stability.	At regular intervals not exceeding 7 days.	Before work at the start of every shift.	After accidental fall of rock, earth or any material.
Any working platform or part thereof or any personal suspension equipment.	✓	✓	✓	✓		
Excavations which are supported in pursuit of paragraphs (1), (2) or (3) of regulation 12.			✓		✓	✓
Cofferdams and caissons.			✓		✓	

NOTES
General
1. The inspection report should be completed before the end of the relevant period.
2. The person who prepares the report should, within 24 hours, provide either the report or a copy to the person on whose behalf the inspection was carried out.
3. The report should be kept on site until work is complete. It should then be retained for three months at the office of the person for whom inspection was carried out.

Working platforms only
1. An inspection is only required where a person is liable to fall more than 2 metres from a place of work.
2. Any employer or any other person who controls the activities of persons using a scaffold shall ensure that it is stable and of sound construction and that the relevant safeguards are in place before his employees or persons under his control first use the scaffold.
3. No report is required following the inspection of any mobile tower scaffold which remains in the same place for less than 7 days.
4. Where an inspection of a working platform or part thereof or any personal suspension equipment is carried out.
    i.   before it is taken into use for the first time; or
    ii.   after any substantial addition, dismantling or other alterations;
    not more than one report is required for any 24 hour period.

Excavations only
1. The duties to inspect and prepare a report apply only to any excavation which needs to be supported to prevent any person being trapped or buried by an accidental collapse, or dislodgement of material from its sides, roof or area adjacent to it. Although an excavation must be inspected at the start of every shift, only one report of such inspections is required every 7 days. Reports must be completed for all inspections carried out during this period for other purposes, e.g. after accidental fall material.

Checklist of typical scaffolding faults

Footings	Standards	Ledgers	Bracing	Putlogs and transoms	Couplings	Bridles	Ties	Boarding	Guard-rails and toe-boards	Ladders
Soft and uneven	Not plumb	Not level	Some missing	Wrongly spaced	Wrong fitting	Wrong spacing	Some missing	Bad boards	Wrong height	Damaged
No base plates	Jointed at same height	Joints in same bay	Loose	Loose	Loose	Wrong couplings	Loose	Trap boards	Loose	Insufficient length
No safe plates	Wrong spacing	Loose	Wrong fittings	Wrongly supported	Damaged	No check couplers	Not enough	Incomplete	Some missing	Not bad
Undermined	Damaged	Damaged	–	–	No check couplers	–	–	Insufficient supports	–	–

▲ **Figure 10.10** Inspections and reports

- *Excavations.* Inspect at the start of each shift before work commences and after any fall of material, rock or earth.
- No work is to commence unless the competent person is satisfied that work can be carried out safely.
- The workplace must not be used until defects have been put right.

■ **Notifications reports and records.** These are also required for the following actions and incidents and are usually submitted on standard forms obtainable from the relevant authority. A record can be kept on-site by making a photocopy of the completed form before submitting it.

NOTES

Notification of project

1 This form can be used to notify any project covered by the Construction (Design and Management) Regulations 1994 which will last longer than 30 days or 500 person days. It can also be used to provide additional details that were not available at the time of initial notification of such projects (any day on which construction work is carried out (including holidays and weekends) should be counted, even if the work on that day is of short duration. A person day is one individual, including supervisors and specialists, carrying out construction work for one normal working shift.)

2 The form should be completed and sent to the HSE area office covering the site where construction work is to take place. You should send it as soon as possible after the planning supervisor is appointed to the project.

3 The form can be used by contractors working for domestic clients. In this case only parts 4–8 and 11 need to be filled in.

HSE - For official use only

| Client | V | PV | NV | | Planning supervisor | V | PV | NV |
| Focus serial number | | | | Principal contractor | V | PV | NV |

1 Is this the initial notification of this project or are you providing additional information that was not previously available?

Initial notification ☐   Additional notification ☐

2 **Client:** name, full address, postcode and telephone number *(if more than one client, please attached details on separate sheet)*

Name:
Address: Telephone number:

Postcode:

3 **Planning Supervisor:** name, full address, postcode and telephone number

Name:
Address: Telephone number:

Postcode:

4 **Principal Contractor:** *(or contractor when project for domestic*

Name:
Address:

Postcode:

5 **Address of site:** where construction is to be carried out

Address:

Postcode:

F10 (rev0.3 95)

6 **Local Authority:** name of the local government district council or island council within whose district the operations are to be carried out

7 **Please give your estimates on the following:** Please indicate if these estimates are original ☐ revised ☐ *(tick relevant box)*

a. The planned date for the commencement of the construction work

b. How long the construction work is expected to take *(in weeks)*

c. The maximum number of people carrying out construction work on site at any one time

d. The number of contractors expected to work on site

8 **Construction work:** give brief details of the type of construction work that will be carried out

9 **Contractors:** name full address and postcode of those who have been chosen to work on the project *(if required continue on a separate sheet). (Note this information is only required when it is known at the time notification is first made to HSE. An update is not required)*

Declaration of planning supervisor

10 I hereby declare that...(name of organisation) has been appointed as planning supervisor for the project

Signed by or on behalf of the organisation.......................................(print name)..

Date...

Declaration of principal contractor

11 I hereby declare that...(name of principal contractor) has been appointed as principal contractor (or contractor undertaking project for domestic client)

Signed by or on behalf of the organisation.......................................(print name)..

Date...

▲ **Figure 10.11** Notification of a project

– *Notification of a construction project,* which will last more than 30 days or 500 person days or have more than 5 workers on site at a time, Figure 10.11.

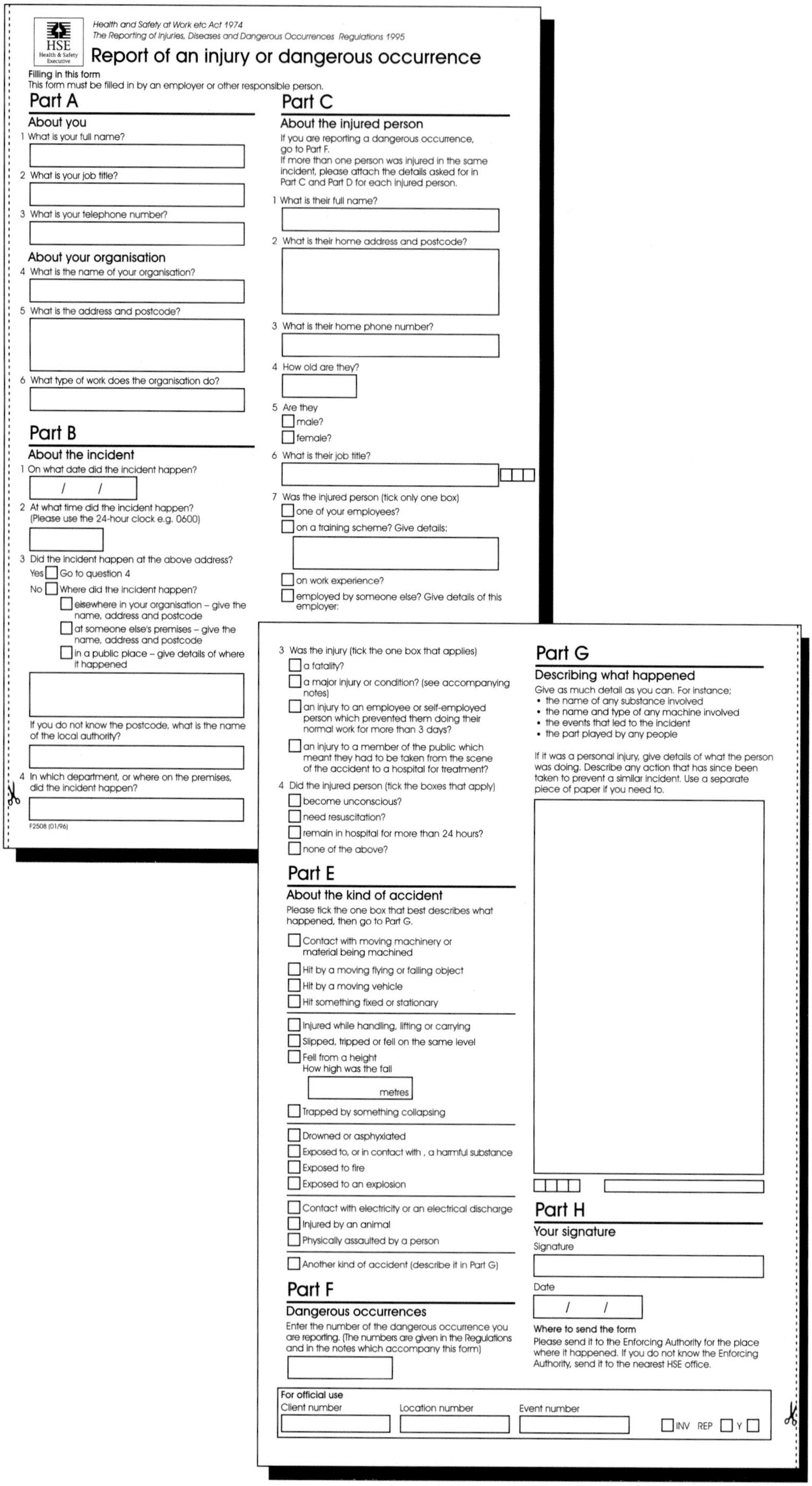

HSE
Health & Safety Executive

Health and Safety at Work etc Act 1974
The Reporting of Injuries, Diseases and Dangerous Occurrences Regulations 1995

Report of an injury or dangerous occurrence

Filling in this form
This form must be filled in by an employer or other responsible person.

Part A

About you
1 What is your full name?

2 What is your job title?

3 What is your telephone number?

About your organisation
4 What is the name of your organisation?

5 What is the address and postcode?

6 What type of work does the organisation do?

Part B

About the incident
1 On what date did the incident happen?

/ /

2 At what time did the incident happen?
(Please use the 24-hour clock e.g. 0600)

3 Did the incident happen at the above address?
Yes ☐ Go to question 4
No ☐ Where did the incident happen?
☐ elsewhere in your organisation – give the name, address and postcode
☐ at someone else's premises – give the name, address and postcode
☐ in a public place – give details of where it happened

If you do not know the postcode, what is the name of the local authority?

4 In which department, or where on the premises, did the incident happen?

F2508 (01/96)

Part C

About the injured person
If you are reporting a dangerous occurrence, go to Part F.
If more than one person was injured in the same incident, please attach the details asked for in Part C and Part D for each injured person.

1 What is their full name?

2 What is their home address and postcode?

3 What is their home phone number?

4 How old are they?

5 Are they
☐ male?
☐ female?

6 What is their job title?

7 Was the injured person (tick only one box)
☐ one of your employees?
☐ on a training scheme? Give details:

☐ on work experience?
☐ employed by someone else? Give details of this employer:

3 Was the injury (tick the one box that applies)
☐ a fatality?
☐ a major injury or condition? (see accompanying notes)
☐ an injury to an employee or self-employed person which prevented them doing their normal work for more than 3 days?
☐ an injury to a member of the public which meant they had to be taken from the scene of the accident to a hospital for treatment?

4 Did the injured person (tick the boxes that apply)
☐ become unconscious?
☐ need resuscitation?
☐ remain in hospital for more than 24 hours?
☐ none of the above?

Part E

About the kind of accident
Please tick the one box that best describes what happened, then go to Part G.

☐ Contact with moving machinery or material being machined
☐ Hit by a moving flying or falling object
☐ Hit by a moving vehicle
☐ Hit something fixed or stationary

☐ Injured while handling, lifting or carrying
☐ Slipped, tripped or fell on the same level
☐ Fell from a height
How high was the fall

metres

☐ Trapped by something collapsing

☐ Drowned or asphyxiated
☐ Exposed to, or in contact with , a harmful substance
☐ Exposed to fire
☐ Exposed to an explosion

☐ Contact with electricity or an electrical discharge
☐ Injured by an animal
☐ Physically assaulted by a person

☐ Another kind of accident (describe it in Part G)

Part F

Dangerous occurrences
Enter the number of the dangerous occurrence you are reporting. (The numbers are given in the Regulations and in the notes which accompany this form)

Part G

Describing what happened
Give as much detail as you can. For instance;
• the name of any substance involved
• the name and type of any machine involved
• the events that led to the incident
• the part played by any people

If it was a personal injury, give details of what the person was doing. Describe any action that has since been taken to prevent a similar incident. Use a separate piece of paper if you need to.

Part H

Your signature
Signature

Date
/ /

Where to send the form
Please send it to the Enforcing Authority for the place where it happened. If you do not know the Enforcing Authority, send it to the nearest HSE office.

For official use
Client number   Location number   Event number

☐ INV REP ☐ Y ☐

▲ **Figure 10.12** Notification of an injury or dangerous occurrence

- *Notification and record of accidents* resulting in death or major injuries or notifiable dangerous occurrences, or more than three days absence from work, or for a specified disease associated with the work, Figure 10.12. Major injuries can be defined as most fractures, amputations, loss of sight or any other injury involving a stay in hospital. Many incidents can be defined as notifiable dangerous occurrences but in general they include the collapse of a crane, hoist, scaffolding or building, an explosion or fire, or the escape of any substance that is liable to cause a health hazard or major injury to any person.
- *A record of all accidents and first aid treatments* (see later topic on first aid).

The Construction (Head Protection) Regulations

This legislation places a duty on employers and the self-employed to provide and ensure that suitable head protection is worn on site. Employees and the self-employed are obliged to wear them. Employers must ensure that suitable head protection is properly worn at all times on site, unless there is no foreseeable risk or injury to the head. Site rules should be set down giving guidance to employees and the self-employed and for site visitors.

The Provision and Use of Work Equipment Regulations (PUWER)

The main objectives of PUWER are to ensure that all equipment used in the workplace is:
- suitable for its intended purpose;
- properly maintained;
- provided with all appropriate safety devices and warning notices;
- all users and supervisors of equipment are given health and safety information, training and written instructions.

Also refer to Chapter 6.

The Control of Substances Hazardous to Health Regulations (COSHH)

Where people use or are exposed to hazardous substances, the COSHH regulations require:
- the assessment of risks involved;
- the prevention of exposure to risk;
- measures taken to adequately control it;
- monitoring of the effectiveness of the measures taken.

Identification. People may be exposed to risks either because they handle a hazardous substance, or because during their work a hazardous substance is created. Manufacturers and suppliers of such substances are required to provide safety data sheets for reference purposes Figure 10.13.

Assessment. Must be undertaken by employers (see under risk assessment). They must look at the ways people may be exposed to hazardous substances in the particular type of work undertaken. For example:
- breathing in dust, fumes or vapours;
- swallowing or eating contaminated materials;
- contact with skin or eyes.

Prevention. Where harm from a substance is likely, the first course of action should be prevention of exposure by:
- doing the job in a different way so that the substance is removed or not created, e.g. rodding out blocked drains rather than using hazardous chemicals;
- using a less hazardous substitute substance, e.g. using water-based paints rather than more hazardous spirit-based ones.

BBS: Panel Products
33 Stafford Thorne Street
Nottingham NG22 3RD
Tel. 011594000

SAFETY DATA SHEET

Chemical Name:	Interior Medium Density Fibreboard (MDF)	
Trade Name:	MeDFit	
Chemical Family:	Wood Based Panel Product	
Formula:	Mixture	
Ingredients:	Mixed Softwoods	82%
	Urea Formaldehyde Resins	8–10%
	Paraffin Wax	0.5%
	Water	6–8%
	Silica	<0.05%
	Free Formaldehyde	<0.04%

Physical and Chemical Characteristics:

Specific Gravity:	0.65–0.99
Appearance/colour:	Cream to light brown, solid wood texture.

Fire and Explosion Hazard:

Extinguisher Media:	Water
Explosion Hazard:	None for the sheet material. However airbourne dust produced during re-manufacturing operations may cause an explosion hazard. Dust should be continuously removed from processing machinery. Smoking should not be permitted in the working area.

Health Hazards: During re-manufacture wood dust may:
- increase mucosal output
- cause reddening and itching of the skin
- irritation of the throat and eyes.

Most of the effects are readily reversible after the end of exposure.

Personal Protection: During re-manufacturing operations:
- wear dust mask and eye protection
- apply a barrier cream (replenish after washing)
- wash before eating, drinking, smoking and going to the toilet.

Special controls: During re-manufacturing operations the use of high efficiency dust collection equipment is strongly recommended to ensure compliance with the COSHH regulations.

First Aid:

Inhalation of dust:	Clean nasal passages, take in plenty of fresh air
Contamination of eyes:	Flush with an approved eye wash solution for a prolonged period.

▶ **Figure 10.13** Typical safety data sheet

Control. Where the substance has to be worked with because either there is no choice, or because alternatives also present equal risks, exposure must be controlled by:

- using the substance in a less hazardous form, e.g. use a sealed surface glass fibre insulation quilt rather than an open fibre one, to reduce the risk of skin contact or the inhalation of fine strands;
- using a less hazardous method of working with the substance, e.g. wet rubbing down of old lead based painted surfaces rather than dry rubbing down which causes hazardous dust; or applying spirit-based products by brush or roller rather than by spraying;
- limiting the amount of substance used;
- limiting the amount of time people are exposed;
- keeping all containers closed when not in use;
- providing good ventilation to the work area: mechanical ventilation may be required in confined spaces;
- using tools when cutting or grinding fitted with exhaust ventilation or water suppression to control dust.

Typical **hazardous substances used in construction** are listed in Table 10.1, along with their potential health risks and suggested controls.

Protection. If exposure cannot be prevented, or adequately controlled using any of the above, also use personal protective equipment (PPE):

- Always wear protective clothing when stipulated: overalls, gloves (for protection and anti-vibration), boots, helmets, ear protection, eye protection goggles or visors and dust masks or respirators as appropriate.
- The use of barrier and after-work creams is recommended to protect skin from contact dermatitis.
- Ensure items of PPE are kept clean so that they do not themselves become a source of contamination.

Table 10.1 Hazardous substances in construction

Substances	Health risk	Jobs	Controls
DUSTS:			
Cement (Also when wet)	SK I ENT	Masonry, rendering	Prevent spread. Protective clothing, respirator when handling dry, washing facilities, barrier cream.
Gypsum	SK I ENT	Plastering	
Man-made mineral fibre	I SK ENT	Insulation	Minimise handling/cutting, respirator, one piece overall, gloves, eye protection.
Silica	I	Sand blasting, grit blasting: scrabbling granite, polishing	Substitution – e.g. with grit, silica-free sand; wet methods; process enclosure/extraction; respirator.
Wood dust (Dust from treated timber e.g. with pesticide may present extra hazards)	I SK ENT	Power tool use in carpentry, especially sanding	Off-site preparation; on-site – enclosures with exhaust ventilation; portable tools – dust extraction; washing facilities; respirator.
Mixed dusts (Mineral and biological)	I SK ENT	Demolition and refurbishment	Minimise dust generation; use wet methods where possible; segregate or reduce number of workers exposed; protective clothing, respirator; good washing facilities/showers. Tetanus immunisation.
FUMES/GASES:			
Various welding fumes from metals or rods	I	Welding/cutting activities	Mechanical ventilation in enclosed spaces; air supplied helmet; elsewhere good general ventilation.
Hydrogen sulphide	I ENT	Sewers, drains, excavations, manholes	All work in confined spaces – exhaust and blower ventilation; self contained breathing equipment confined space procedures.
Carbon monoxide/nitrous oxide	I	Plant exhausts	Position away from confined spaces. Where possible maintain exhaust filters; forced ventilation and extraction of fumes.
SOLVENTS: In many construction products – paints, adhesives, strippers, thinners, etc.	I SK SW	Many trades, particularly painting, tile fixing. Spray application is high risk. Most brush/roller work less risk. Regulation exposure increases risks	Breathing apparatus for spraying, particularly in enclosed spaces; use of mistless/airless methods. Otherwise ensure good general ventilation. Washing facilities, barrier cream.
RESIN SYSTEMS:			
Isocynates (MDI:TDI)	I ENT SK SW	Thermal insulation	Mechanical ventilation where necessary; respirators; protective clothing, washing facilities. Skin checks, respirators checks.
Polyurethane paints	I ENT SK SW	Decorative surface coatings	Spraying – airline/self contained breathing apparatus; elsewhere good general ventilation. One-piece overall, gloves, washing facilities.
Epoxy	I SK SW	Strong adhesive applications	Good ventilation, personal protective equipment (respirator; clothing) washing facilities, barrier cream.
Polyester	I SK ENT SW	Glass fibre claddings and coatings	As above.
PESTICIDES: (e.g. timber preservatives, fungicides, weed killers)	I SK ENT SW	Particularly in-situ timber treatment. Handling treated timber	Use least toxic material. Mechanical ventilation, respirator, impervious gloves, one-piece overall and head cover. In confined spaces – breathing apparatus. Washing facilities, skin checks. If necessary biological checks. Handle only dry material.
ACIDS/ALKALIS:	SK ENT	Masonry cleaning	Use weakest solutions. Protective clothing, eye protections. Washing facilities (first aid including eye bath and copious water for splash removal).
MINERAL OIL:	SK I	Work near machines, compressors, etc. Mould release agents	Filters to reduce mist. Good ventilation. Protective clothing. Washing facilities; barrier creams. Skin checks.
SITE CONTAMINANTS: e.g. Arsenic. Phanols; heavy metals; Micro organisms etc. e.g. Wells disease, tetanus, hepatitis B	I SK SW	Site re-development of industrial premises or hospitals – particularly demolition ground work and drain/sewers	Thorough site examination and clearance procedures. Respirators, protective clothing. Washing facilities/showers. Immunisation for tetanus.

Health risk
SK = skin; I = inhalation; ENT = irritant eyes, nose, throat; SW = ingestion
Table extracted from Control of Substances Hazardous to Health (COSHH) regulations

- All items of PPE should be regularly maintained, checked for damage and stored in clean dry conditions.
- Replacement items of PPE and spare parts must be available for use when required.

Personal hygiene. Protection does not stop with PPE. Hazardous substance can be easily transferred from contaminated clothing and unwashed hands and face:

- always wash hands and face before eating, drinking or smoking and also at the end of the working shift;
- never eat, drink or smoke near to the site of exposure;
- change out of contaminated work wear into normal clothes before travelling home;
- ensure contaminated work wear is regularly laundered.

Monitoring and health surveillance. This must be carried out to ensure exposure to hazardous substances is being adequately controlled:

- *Monitoring of the workplace* is required to ensure exposure limits are not being exceeded, e.g. regular checks on noise levels and dust or vapour concentrations.
- *Health surveillance is a legal duty* for a limited range of work exposure situations (for example exposure to asbestos dust). However many employers operate a health surveillance programme for all their employees. This gives medical staff the opportunity to check the general health of workers, as well as giving early indications of illness, disease and loss of sensory perception. Simple checks can be made on a regular basis including blood pressure, hearing, eyesight and lung peak flow. Any deterioration over time indicates the need for further action.

Information. Employers must provide their employees who are or may be exposed to hazardous substances with the following information (Figure 10.14):

BBS: Shopfitting Services
33 Stafford Thorne Street
Nottingham NG22 3RD
Tel. 0115 94000

SAFETY METHOD STATEMENT

Process:

The re-manufacture of MDF panel products. During this process a fine airborne dust is produced. This may cause skin, eye, nose and throat irritation. There is also a risk of explosion. The company has controls in place to minimise any risk. However, for your own safety and the protection of others, you must play your part by observing the following requirements.

General Requirements: At all times observe the following safety method statements and the training you have received from the company.
- Manual Handling
- Use of Woodworking Machines
- Use of Powered Hand Tools
- General House Keeping

Specific Requirements:

- When handling MDF, always wear gloves or barrier cream as appropriate. Barrier cream should be replenished after washing.
- When sawing, drilling, routing or sanding MDF, always use the dust extraction equipment and wear dust masks and eye protection.
- Always brush down and wash thoroughly to remove all dust, before eating, drinking, smoking, going to the toilet and finally at the end of the shift.
- Do not smoke outside the designated areas.
- If you suffer from skin irritation or other personal discomfort seek first aid treatment or consult the nurse.

IF IN DOUBT ASK

▶ **Figure 10.14** Typical safety method statement

- *Safety information and training* for them to know the risks involved.
- *A safe working method statement*, including any precautions to be taken or PPE to be worn.
- *Results* of any monitoring and health surveillance checks.

Accident statistics

Each year there are over 14,000 accidents reported to the Health and Safety Executive that occur during construction-related activities in the United Kingdom.

Reported accidents are those, which result in death, major injury, more than three days absence from work or are caused by a notifiable dangerous occurrence.

At the time of writing the latest figures show that 1% of the reported accidents were fatal, on average 6 deaths a month; 34% resulted in major injuries, nearly 20 a week and 65% resulted in absence from work for more than three days, an average of 40 each working day. These percentage distributions along with a further breakdown by cause of accident are illustrated in Figure 10.15.

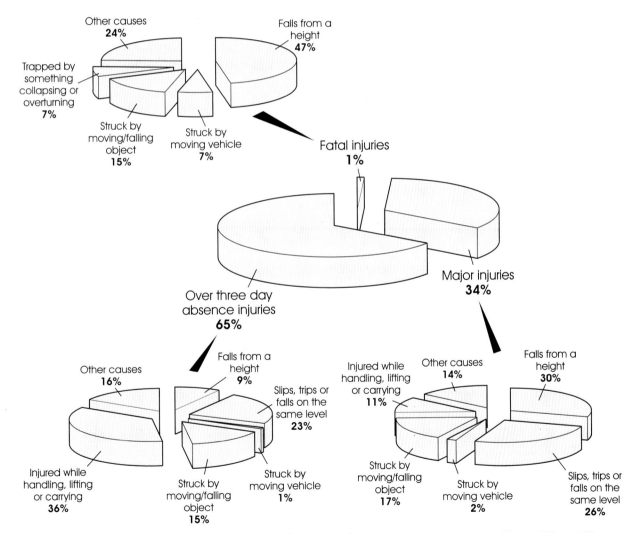

▲ **Figure 10.15** Distribution of reported accidents by type and cause

General safety guidance

It should be the aim of everyone to prevent accidents. Remember, you are required by law to be aware and fulfil your duties under the *Health and Safety at Work Act* and other regulations.

The main contribution an operative can make towards the prevention of accidents is to work in the safest possible manner at all times, thus ensuring that their actions do not put at risk the operator, workmates or the general public. Refer to Chapter 6 for woodworking accident statistics.

Safety: on site and in the workshop

A safe working area is a tidy working area. All unnecessary obstructions, which may create a hazard should be removed, e.g. off-cuts of material, unwanted materials, disused items of plant, and the extraction or flattening of nails from discarded pieces of timber. Therefore:

- Clean up your work bench/work area periodically as off-cuts and shavings are potential tripping and fire hazards.
- Learn how to identify the different types of fire extinguishers and what type of fire they can safely be used on. Staff in each work area should be trained in the use of fire extinguishers, Figure 10.16.
- Careful disposal of materials from heights is essential. They should always be lowered safely and not thrown or dropped from scaffolds and window openings, etc. Even a small bolt or fitting dropped from a height can penetrate a person's skull and almost certainly lead to brain damage or even death.
- Ensure your tools are in good condition. Blunt cutting tools, loose hammer heads, broken or missing handles and mushroom heads must be repaired immediately or the use of the tool discontinued, Figure 10.17.
- When moving materials and equipment always look at the job first; if it is too big for you to tackle then get help. Look out for splinters, nails and sharp or jagged edges on the items to be moved. Always lift with your back straight, elbows tucked in, knees bent and feet slightly apart. When putting an item down ensure that your hands and fingers will not be trapped, Figure 10.18.

Fire Classification

Class A:
Wood, paper, textiles and any other carbonaceous materials

Class B:
Flammable gases such as petrol, oils, fats and paints

Class C:
Flammable gases such as propane, butane and natural gas

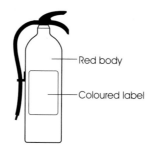

— Red body
— Coloured label

TYPE OF FIRE RISK	White label Water	Cream label Foam	Black label Carbon dioxide	Blue label Dry powder	Red Fire blanket
USE OF FIRE EXTINGUISHERS					
Class A	✔	✔	✘	✔	Can be used for smothering all types of fire. Also for use where clothing is alight since it does not pose a risk to skin or to breathing as some extinguishers do
Class B	✘	✔	✔	✔	
Class C	✘	✘	✔	✔	
Electrical	✘	✘	✔	✔	
Vehicle	✘	✔	✘	✔	

▶ **Figure 10.16** Fire classification and types of extinguisher

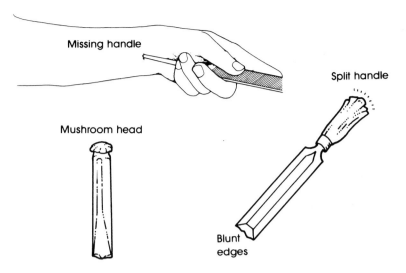

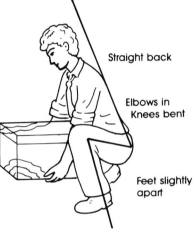

▲ **Figure 10.17** Unsafe tool conditions

■ Materials must be stacked on a firm foundation; stacks should be of reasonable height to allow easy removal of items. They should also be bonded to prevent collapse and battered to spread the load. Pipes, drums, etc., should be wedged or chocked to prevent rolling. Never climb on a stack or remove material from its sides or bottom, Figure 10.19.

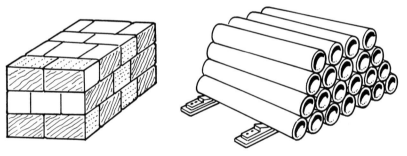

▲ **Figure 10.18** Safe manual handling

▲ **Figure 10.19** Safe stacking of materials

■ Excavations and inspection chambers should be either protected by a barrier or covered over completely to prevent people carelessly falling into them, Figure 10.20.

Extra care is needed when working at heights. **Ladders** should be of sufficient length for the work in hand and should be in good condition and not split, twisted or with rungs missing. They should also be used at a working angle of 75° and securely tied at the top (Figure 10.21). This angle is a slope of four vertical units to one horizontal unit. Where a fixing at the top is not possible, an alternative is the stake-and-guy rope method. Otherwise, arrange for someone to stand on the foot of the ladder. (The 'footer' must wear the appropriate headgear and pay attention to the task at all times and not 'watch the scenery'.)

▲ **Figure 10.20** Warning sign and covering to make safe small holes, excavations, etc.

Wooden ladders must not be painted as this may hide defects. Ensure that extension ladders have sufficient overlap for strength (at least two rungs for short ladders and up to four for longer ones and that the latching hook is engaged. Never overreach when working on a ladder: always take the time to stop and move it closer to the work position. Ladders should be lowered and locked away at night.

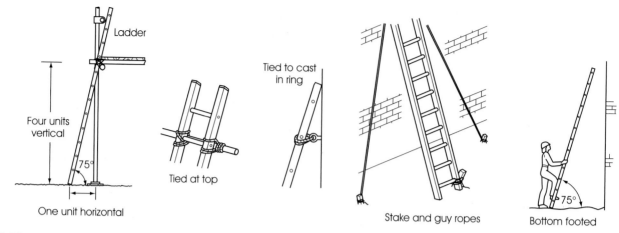

▲ **Figure 10.21** Safe working angle and security of ladders

Scaffolds should be inspected before working on them. Check to see that all components are there and in good condition, not bent, twisted, rusty, split, loose or out of plumb and are level (Figure 10.22). Also ensure that the base has not been undermined or is too close to excavations. If in doubt do not use, and have it looked at by an experienced scaffolder or report it to your supervisor. Never remove any part from a scaffold: you may be responsible for its total collapse. Never block a scaffold with your tools, equipment or materials and always clear up any mess made as you go (don't leave it to form a hazard).

You may be required to erect and use other **working platforms:** hop-ups, split-head type working platforms, stepladders and trestle scaffolds during the execution of your work (Figure 10.23). This type of equipment is only generally suitable for internal use and at restricted heights. Ensure all the equipment is in good order and that it is only erected on a flat level surface.

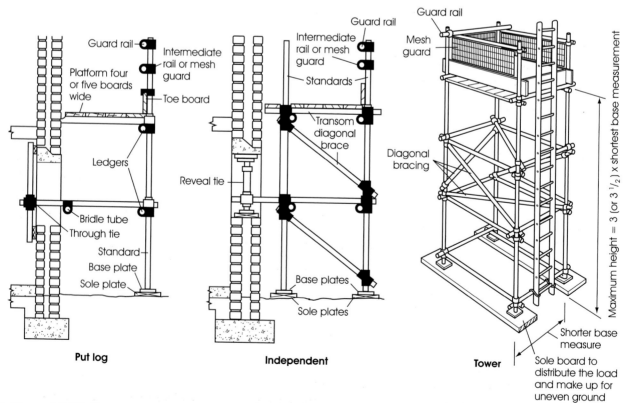

▲ **Figure 10.22** Types of scaffold with safety features

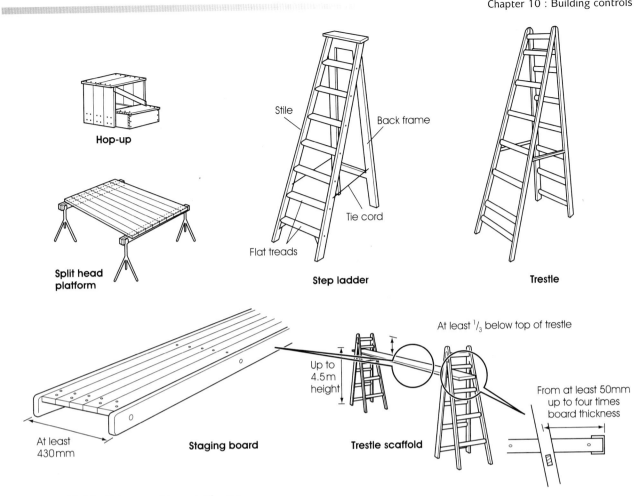

Hop-up

Split head platform

Stile

Back frame

Tie cord

Flat treads

Step ladder

Trestle

At least $^1/_3$ below top of trestle

Up to 4.5m height

From at least 50mm up to four times board thickness

At least 430mm

Staging board

Trestle scaffold

▲ **Figure 10.23** Other working platforms

Working on roofs, **roofing ladders** or **crawl boards** should be used to provide safe access and/or to avoid falling through fragile coverings, Figure 10.24.

Working with **electrical and compressed air equipment** brings additional hazards, as they are both potential killers (see Chapter 5). Qualified personnel should check installations and equipment regularly. If anything is incomplete, damaged, frayed, worn or loose, do not use it, but return it to stores for attention. Ensure cables and hoses are kept as short as possible and routed safely out of the way to prevent risk of tripping and damage, or in the case of electric cables, from lying in damp conditions.

Protective clothing and equipment

Always wear the correct personal protective equipment (PPE) for the work in hand:

- *Safety helmets*, *safety footwear* and a *high visibility vest* should be worn at all times.
- Wear *ear protectors* when carrying out noisy activities, and *safety goggles* when carrying out any operation that is likely to produce dust, chips or sparks, etc.
- *Dust masks* or *respirators* should be worn where dust is being produced or fumes are present.
- Wear *gloves* when handling materials.
- *Wet weather clothing* is necessary for inclement conditions.

Many of these items must be supplied free of charge by employers. See Chapter 5 for further information on PPE.

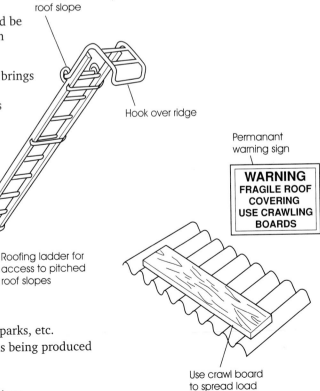

Wheels to slide up roof slope

Hook over ridge

Permanant warning sign

WARNING FRAGILE ROOF COVERING USE CRAWLING BOARDS

Roofing ladder for access to pitched roof slopes

Use crawl board to spread load

▲ **Figure 10.24** Safe access to roofs

▶ **Figure 10.25** Typical warning signs (hazard classifications) shown on packaging

HIGHLY FLAMMABLE (F)

HARMFUL (X)

TOXIC (T)
VERY TOXIC (T+)

OXIDISING (O)

EXTREMELY FLAMMABLE (F+)

DANGEROUS FOR THE ENVIRONMENT (N)

CORROSIVE (C)

IRRITANT (XI)

Personal hygiene

Care should be taken with personal hygiene, which is just as important as physical protection. Some building materials have an irritant effect on contact with the skin. Some are poisonous if swallowed, while others can result in a state of unconsciousness (narcosis) if their vapour or powder is inhaled. These harmful effects can be avoided by wearing the appropriate PPE and by taking proper hygiene precautions:

- follow the manufacturer's instructions;
- avoid inhaling fumes or powders;
- wear a barrier cream;
- thoroughly wash your hands before eating, drinking, smoking and after work.

Typical warning signs that are displayed on packaging by manufacturers are illustrated in Figure 10.25. ·

First aid

First aid is the treatment of persons with the purpose of preserving life until medical help is obtained and also the treatment of minor injuries for which no medical help is required.

In all cases only a trained first-aider should administer first aid. Take care not to become a casualty yourself. Send for the nearest first-aider and/or medical assistance (phone 999) immediately. A record should be made of all accidents and first aid treatments (Figure 10.26). Even minor injuries where you may have applied a simple plaster or sterilised dressing could become infected and require further attention. You are strongly recommended to seek medical attention if a minor injury becomes inflamed, painful or festered.

It is recommended that you read the first aid guidance leaflet, which should be found in every first aid box. A typical example is illustrated in Figure 10.27.

▼ **Figure 10.26** Record of accidents and first aid treatment

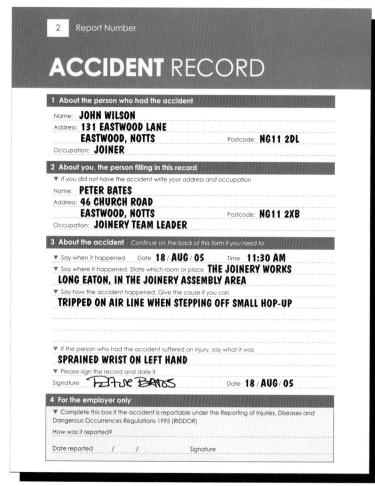

| 2 | Report Number |

ACCIDENT RECORD

1 About the person who had the accident

Name: JOHN WILSON
Address: 131 EASTWOOD LANE
EASTWOOD, NOTTS   Postcode: NG11 2DL
Occupation: JOINER

2 About you, the person filling in this record

▼ If you did not have the accident write your address and occupation
Name: PETER BATES
Address: 46 CHURCH ROAD
EASTWOOD, NOTTS   Postcode: NG11 2XB
Occupation: JOINERY TEAM LEADER

3 About the accident   Continue on the back of this form if you need to

▼ Say when it happened   Date 18 / AUG / 05   Time 11:30 AM
▼ Say where it happened. State which room or place THE JOINERY WORKS
LONG EATON, IN THE JOINERY ASSEMBLY AREA
▼ Say how the accident happened. Give the cause if you can
TRIPPED ON AIR LINE WHEN STEPPING OFF SMALL HOP-UP

▼ If the person who had the accident suffered an injury, say what it was
SPRAINED WRIST ON LEFT HAND
▼ Please sign the record and date it
Signature Peter Bates   Date 18 / AUG / 05

4 For the employer only

▼ Complete this box if the accident is reportable under the Reporting of Injuries, Diseases and Dangerous Occurrences Regulations 1995 (RIDDOR)
How was it reported?

Date reported   /   /   Signature

Health and safety testing in construction

In order to raise safety standards in construction, employers and their clients are demanding that all workers and visitors to sites have current membership of a competence-based health and safety at work registration scheme to construction sites.

The **Construction Skills Certification Scheme** (CSCS), which is administered by the Construction Industry Training Board (CITB), has the greatest uptake of all recognised schemes. CSCS cards (Figure 10.28) provide evidence that the cardholders are competent and have up-to-date

(b) *Chemical burns* Remove any contaminated clothing which shows no sign of sticking to the skin and flush all affected parts of the body with plenty of clean, cool water ensuring that all the chemical is so diluted as to be rendered harmless. Apply a sterilised dressing to exposed, damaged skin and clean towels to damaged areas where the clothing cannot be removed. (N.B. Take care when treating the casualty to avoid contamination).

(c) *Foreign bodies in the eye* If the object cannot be removed readily with a clean piece of moist material, irrigate with clean, cool water. People with eye injuries which are more than minimal must be sent to hospital with the eye covered with an eye pad from the container.

(d) *Chemical in the eye* Flush the open eye at once with clean, cool water; continue for at least 5 to 10 minutes and, in any case of doubt, even longer. If the contamination is more than minimal, send the casualty to hospital.

(e) *Electric shock* Ensure that the current is switched off. If this is impossible, free the person, using heavy duty insulating gloves (to BS 697/1977) where these are provided for this purpose near the first aid container, or using something made of rubber, dry cloth or wood or a folded newspaper; use the casualty's own clothing if dry. *Be careful* not to touch the casualty's skin before the current is switched off. If breathing is failing or has stopped, start resuscitation and continue until breathing is restored or medical, nursing or ambulance personnel take over.

(f) *Gassing* Move the casualty to fresh air but make sure that whoever does this is wearing suitable respiratory protection. If breathing has stopped, start resuscitation and continue until breathing is restored or until medical, nursing or ambulance personnel take over. If the casualty needs to go to hospital make sure a note of the gas involved is sent with him.

General

(a) *Hygiene* When possible, wash your hands before treating wounds, burns or eye injuries. Take care in any event not to contaminate the surfaces of dressings

(b) *Treatment position* Casualties should be seated or lying down while being treated

(c) *Record-keeping* An entry must be made in the accident book (for example B1 510 Social Security Act Book) of each case

(d) *Minor injuries* Casualties with minor injuries, of a sort they would attend to themselves if at home, may wash their hands and apply a small sterilised dressing from the container

(e) *First aid materials* Each article used from the container should be replaced as soon as possible

Health and Safety (First Aid) Regulations 1981

General first aid guidance for first aid boxes

Note: Take care not to become a casualty yourself while administering first aid. Be sure to use protective clothing and equipment where necessary. If you are not a trained first-aider, send immediately for the nearest first-aider where one is available.

Advice on treatment

If the assistance of medical or nursing personnel will be required, send for a doctor or nurse (where they are employed at the workplace) or ambulance immediately. When an ambulance is called, arrangements should be made for it to be directed to the scene without delay.

Priorities

(1) *Breathing* If the casualty has stopped breathing, resuscitation must be started at once *before any other treatment is given* and should be continued until breathing is restored until medical, nursing or ambulance personnel take over.

(3) *Unconsciousness* Where the patient is unconscious, care must be taken to keep the airway open. This may be done by clearing the mouth and ensuring that the tongue does not block the back of the throat. Where possible, the casualty should be placed in the recovery position.

Recovery position

(4) *Broken bones* Unless the casualty is in a position which exposes him to further danger, do not attempt to move a casualty with suspected broken bones or injured joints until the injured parts have been supported. Secure so that the injured parts cannot move.

(5) *Other injuries*

(a) *Burns and scalds* Small burns and scalds should be treated by flushing the affected area with plenty of clean cool water before applying a sterilised dressing or a clean towel. Where the burn is large or deep, simply apply a dry sterile dressing. (N.B. Do not burst blisters or remove clothing sticking to the burns or scalds).

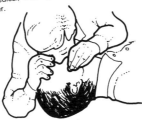

Mouth-to-mouth resuscitation

(2) *Bleeding* If bleeding is more than minimal, control it by direct pressure – apply a pad of sterilised dressing or, if necessary, direct pressure with fingers or thumb on the bleeding point. Raising a limb if the bleeding is sited there will help reduce the flow of blood (unless the limb is fractured).

▲ **Figure 10.27** Extract from a first aid guidance leaflet knowledge of health and safety legislation concerning construction site activities associated with their work.

To obtain a CSCS card you will have to attend a Test Centre and answer 35 to 40 random multiple-choice questions taken from a substantial bank of questions relating to all aspects of health and safety in construction.

 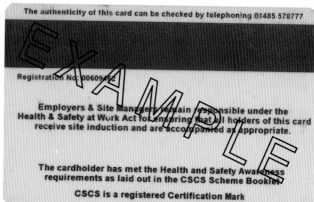

▲ **Figure 10.28** Example of a CSCS safety card

Chapter Eleven

Building principles

This chapter is intended to provide the site carpenter and bench joiner with an overview of the principles of building construction. It is concerned with the types of building and construction methods used.

It includes the following:

- ᖯ types of building
- ᖯ structural forms
- ᖯ building elements and components
- ᖯ building services and thermal insulation

A building encloses space and in doing so creates an internal environment. The actual structure of a building is termed the external envelope. This protects the internal environment from the outside elements, known as the external environment.

Types of building

The protective role of the building envelope is to provide the desired internal conditions for the building's occupants with regard to security, safety, privacy, warmth, light and ventilation (Figure 11.1).

A **structure** or **construction** can be defined as an organised combination of connected elements (components), which are constructed and interconnected to perform some required function, e.g. a bridge. The term 'building' takes this idea a step further and is used to define structures that include an external envelope.

Buildings may be divided into three types according to their height as illustrated in Figure 11.2.

- ■ *Low-rise buildings:* from one to three stories.
- ■ *Medium-rise buildings:* from four to seven stories.
- ■ *High-rise buildings:* those above seven stories.

Structural form

There are many differing structural forms in present-day use, each changing from time to time, in order to make the best possible use of new materials and developing techniques. These differing forms may be grouped together under three main categories: solid structures; framed structures; surface structures.

- ■ **Solid structures,** also known as *mass wall construction,* are constructed of brickwork, blockwork or concrete see Figure 11.3.

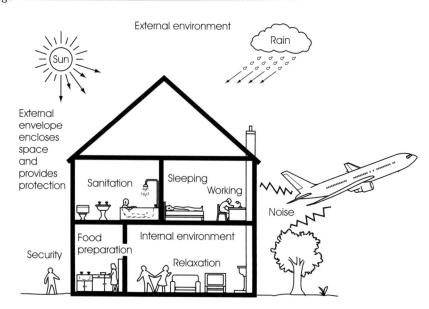

▶ **Figure 11.1** Internal/external environment

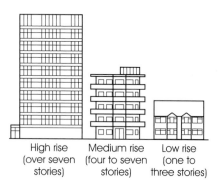

▲ **Figure 11.2** Height of buildings

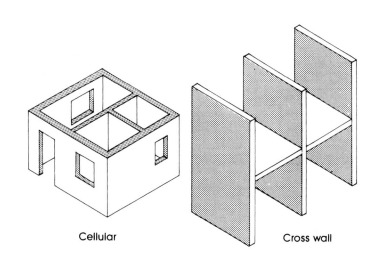

▲ **Figure 11.3** Solid structures

They form a stable box-like structure, but are normally limited to low-rise, short-span buildings.

■ **Framed structures,** also termed *skeleton construction,* consist of an interconnected framework of members having a supporting function, see Figure 11.4. Either external cladding or infill walls are used to provide the protecting external envelope. Frames made from steel, concrete or timber are often pre-made in a factory as separate units, which are simply and speedily erected on site. Framed construction is suitable for a wide range of buildings and civil engineering structures from low to high rise.

■ **Surface structures** consist of a thin material that has been curved or folded to obtain strength, or alternatively a very thin material that has been stretched over supporting members or medium. Surface structures are often used for large clear span buildings with a minimum of internal supporting structure.

Structural parts

All structures consist of two main parts: that below ground and that above ground, see Figure 11.5.

■ **Substructure** comprises all of the structure below ground and that up to and including the ground floor slab and damp-proof course.

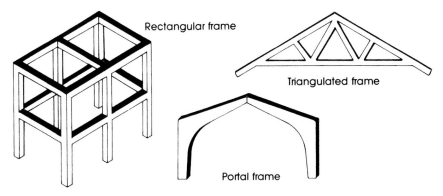

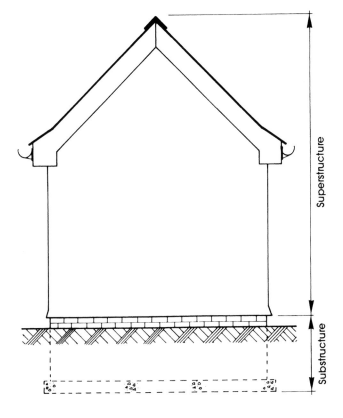

◄ Figure 11.5 Structural parts

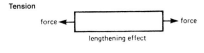

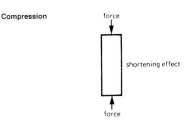

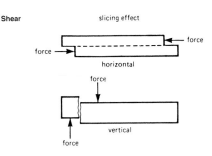

Its purpose is to receive the loads from the main building superstructure and its contents and transfer them safely down to a suitable load-bearing layer of ground.

■ **Superstructure** comprises all of the structure above the substructure both internally and externally. Its purpose is to enclose and divide space, and transfer loads safely on to the substructure.

Although classified as separate parts, the substructure and the superstructure should be designed to operate as one structural unit.

Structural members and loading

The main parts of a structure which themselves carry a load are said to be in a state of stress (a body subjected to a force). There are three types of stress, see Figure 11.6:

■ **Compression.** This causes squeezing, pushing and crushing; it has a shortening effect.

■ **Tension.** This tends to pull or stretch a material; it has a lengthening effect.

▲ Figure 11.6 Types of stress

■ **Shear.** This occurs when one part of a member tends to slip or slide over another part; it has a slicing effect.

The two main **types of load** are defined as:

■ *Dead loads:* the self-weight of the building materials used in the construction, service installations and any permanent built-in fitments, etc.

■ *Imposed loads:* the weight of any movable load, such as the occupants of a building, their furniture and other belongings (goods and chattels), and any visitors to the property and their belongings. Also included are any environmental forces exerted on the structure from the external environment (wind, rain and snow).

The three main types of **load-bearing structural members** are illustrated in Figure 11.7:

1. *Horizontal members.* Their purpose is to carry and transfer a load back to its point of support. Horizontal members include beams, joists, lintels and floor or roof slabs, etc. When a load is applied to a horizontal member, bending will occur, resulting in a combination of tensile, compressive and shear stresses. This bending causes compression in the top of the member, tension in the bottom and shear near its supports and along its centre line. Bending causes the member to sag or deflect. (For safe design purposes deflection is normally limited to a maximum of 3 mm in every 1 m span.) In addition, slender members, which are fairly deep in comparison with their width, are likely to buckle unless restrained (e.g. strutting to floor joists).

2. *Vertical members.* Their purpose is to transfer the loading of the horizontal members down onto the substructure. Vertical members include walls, columns, stanchions and piers. Vertical members are in compression when loaded. Buckling tends to occur in vertical members if they are excessively loaded or are too slender.

3. *Bracing members.* They are used mainly to triangulate rectangular frameworks in order to stiffen them. These can be divided into two types: struts and ties:
   - **strut:** a bracing member that is mainly in compression.
   - **tie:** a bracing member that is mainly in tension.

At certain times bracing members may, depending on their loading conditions, act as either struts or ties. In these circumstances they may be termed as **braces**.

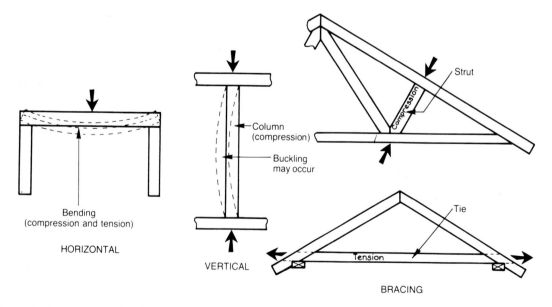

Bending
(compression and tension)

HORIZONTAL

Column
(compression)

Buckling
may occur

VERTICAL

Strut

Compression

Tie

Tension

BRACING

▲ **Figure 11.7** Types of load bearing members

Examples of loading. An example of dead and imposed loads and how they are transferred down through the structural members to the soil is illustrated in Figure 11.8.

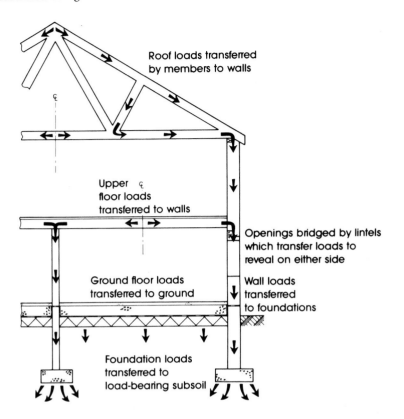

Roof loads transferred by members to walls

Upper ₵ floor loads transferred to walls

Openings bridged by lintels which transfer loads to reveal on either side

Ground floor loads transferred to ground

Wall loads transferred to foundations

Foundation loads transferred to load-bearing subsoil

◀ **Figure 11.8** Transfer of loads

Building elements

An element can be defined as a constructional part of either the substructure or superstructure having its own functional requirements. These include the foundations, walls, floors, roof, stairs and the structural framework or skin. Elements may be further classified into three main groups: primary elements, secondary elements and finishing elements.

Primary elements

These are named because of the importance of their supporting, enclosing and protection functions. In addition, they have mainly internal roles of dividing space and providing floor-to-floor access. Typical examples of primary elements are shown in Figure 11.9.

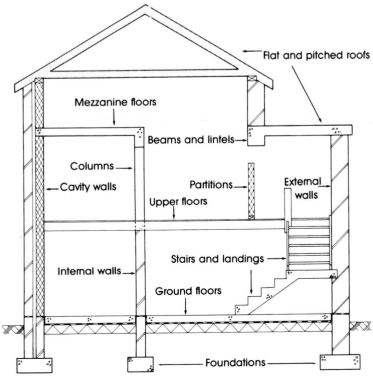

Flat and pitched roofs

Mezzanine floors

Beams and lintels

Columns

Cavity walls

Partitions

External walls

Upper floors

Internal walls

Stairs and landings

Ground floors

Foundations

▶ **Figure 11.9** Primary elements

441

Foundations

This is the part of the structure (normally in-situ concrete) that transfers the dead weight and imposed loads of the structure safely onto the ground. The width of a foundation is determined by the total load of the structure exerted per square metre on the foundation and the safe load-bearing capacity of the ground. Wide foundations are used for either heavy loads or weak ground and narrow foundations for light loading or high load-bearing ground. The load exerted on foundations is spread to the ground at an angle of 45 degrees. Shear failure leading to building subsidence (sinking) will occur if the thickness of the concrete is less than the projection from the wall/column face to the edge of the foundation. Alternatively, steel reinforcement may be included to enable the load to spread across the full width of the foundation (see Figure 11.10).

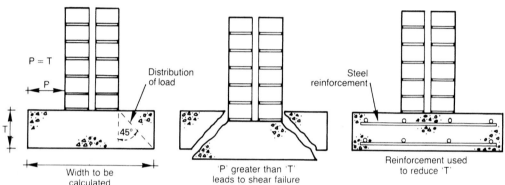

▶ **Figure 11.10**
Foundation proportions

Foundations are taken below ground level to protect the structure from damage resulting from ground movement. The actual depth below ground level is dependent on a number of factors: load-bearing capacity of the ground, need to protect against ground movement and tree roots, etc. In most circumstances, a depth of 1 m to the bottom of the foundation is considered to be the minimum.

Ground movement is caused mainly by the shrinkage and expansion of the ground near the surface owing to the wet and dry conditions. Compact granular ground suffers little movement whereas a clay (cohesive) ground is at high risk. Frost also causes ground movement when the water in the ground expands on freezing. This is known as frost heave and is limited to about 600 mm in depth. The main problem with tree roots is shrinkage of the ground owing to the considerable amounts of water they extract from it. Tree roots can extend out in all directions further than its height.

The four most common **types of foundations** are strip, pad, raft and pile (see Figure 11.11). For most small-scale building works **strip foundations** are commonly used for solid structures and **pad foundations** for frame structures, except where the subsoil is of a poor unstable quality. In these circumstances, a **raft or pile foundation** would be more suitable.

Walls

The walls of a building may be classed as either load bearing or non-load bearing. In addition, external walls have an enclosing role and internal walls a dividing one. Thus load-bearing walls carry out a dual role of supporting and enclosing or dividing. Internal walls, both load and non-load bearing are normally termed **partitions**. Openings in load-bearing walls (windows and doors) are bridged by either arches or lintels, which support the weight of the wall above.

STRIP FOUNDATIONS

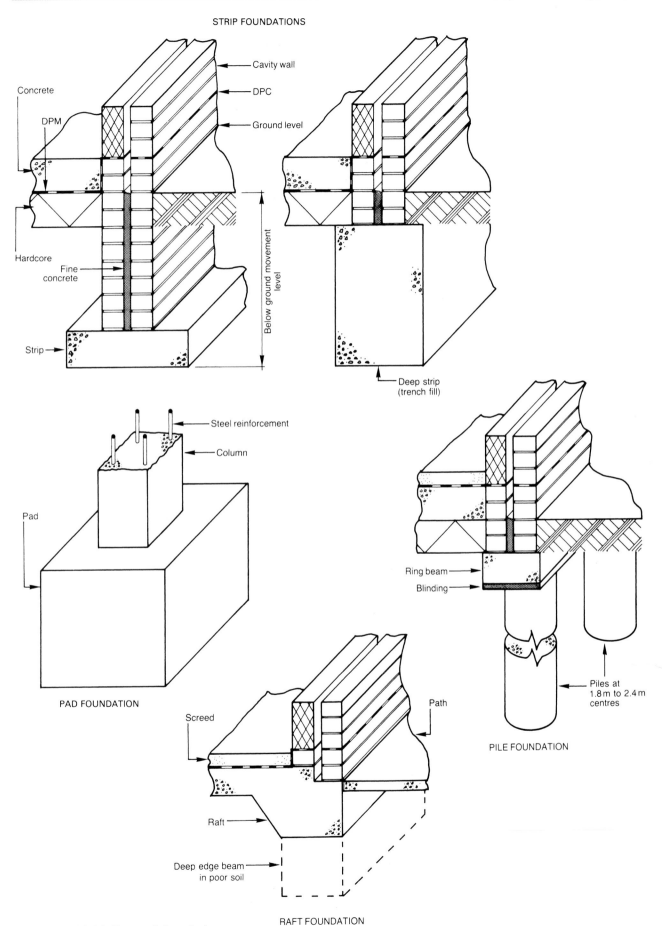

Cavity wall

DPC

Ground level

Concrete

DPM

Hardcore

Fine concrete

Strip

Below ground movement level

Deep strip (trench fill)

Steel reinforcement

Column

Pad

PAD FOUNDATION

Ring beam

Blinding

Piles at 1.8 m to 2.4 m centres

PILE FOUNDATION

Screed

Path

Raft

Deep edge beam in poor soil

RAFT FOUNDATION

▲ **Figure 11.11** Types of foundation

Walls may be divided into three main groups according to their method of construction (see Figure 11.12). These are solid, cavity and framed:

■ **Solid walls.** These are made from bricks, blocks or concrete. When used externally, very thick walls are required (450 mm or over) in order to provide sufficient thermal insulation. This thickness also prevents rain being absorbed through the wall to the inside causing internal dampness before heat and air circulation can evaporate it from the outside. Because of the costs involved this method is now rarely used. An alternative method is to use thinner external solid walls (normally insulating blocks) and apply an impervious (waterproof) surface finish to the outside, e.g. cement rendering.

■ **Cavity walls.** These consist of two relatively thin walls or 'leaves' (about 100 mm each) separated by a 50 mm to 75 mm cavity. The cavity prevents the transfer of moisture from the outside to the inside and also improves the wall's thermal insulation properties. Cavity walls are in common use for enclosing walls of low to medium-rise dwellings. The standard form of the wall is a brick outer leaf and an insulating block inner leaf, or as an alternative, a timber-framed inner leaf. To reduce heat transfer through the wall, it is fairly common practice to fill the cavity with a **thermal insulating material** (mineral wool, fibreglass, foam or polystyrene, etc.).

■ **Framed walls.** These are normally of timber construction and are made up in units called panels. They may be either load or non-load bearing and also for use externally or internally. They consist of vertical members, which are called **studs** and horizontal members, the top and bottom of which are called the head and sole plates whilst any intermediates are called **noggins**. Their thermal and sound insulating properties are greatly improved by filling the spaces between the members with mineral wool or fibreglass, etc. Sheathing may be fixed to one or both sides of the panel to improve strength. This type of wall panel is used in the majority of present-day timber-frame house construction as the internal 'leaf' of the cavity wall. Refer to Book 2, Chapters 3 and 4.

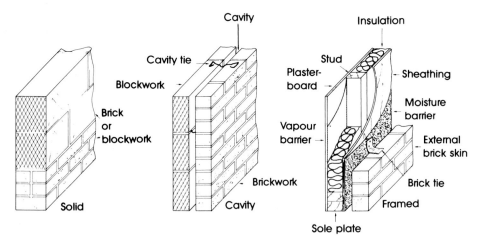

▶ **Figure 11.12**
Types of wall

Bonding of solid and cavity walls. The strength of brickwork is dependent on its bonding (overlapping of vertical joints). This is necessary to spread any loading evenly throughout the wall (see Figure 11.13). The actual overlap or bonding pattern will vary depending on the type of wall and the decorative effect required, typical examples are illustrated in Figure 11.14

■ Solid walls are normally built in either English or Flemish bond. **English bond** consists of alternate rows (courses) of bricks laid lengthways along the wall (stretchers) and bricks laid widthways across the wall (headers). **Flemish bond** consists of alternate stretchers and headers in the same course. In both, the quarter lap is formed by placing a queen closer (brick reduced in width) next to the quoin (corner brick).

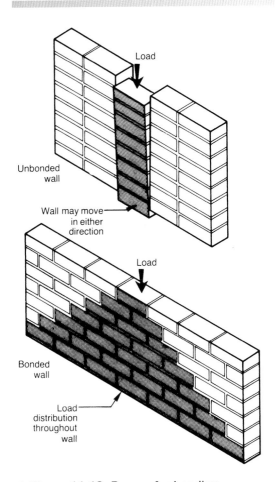

▲ **Figure 11.13** Reason for bonding

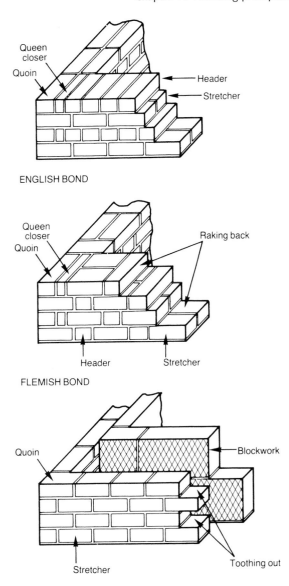

▶ **Figure 11.14** Types of bond

- Cavity walls are built in **stretcher bond** where all bricks show their stretcher faces, although adjacent courses overlap by half a brick. To ensure sufficient strength, the inner and outer leaves are tied together across the cavity at intervals with cavity ties.

Openings in walls for doors and windows are spanned by steel or concrete lintels. These bridge the opening and transfer loads to the reveal on either side (Figure 11.15).

▼ **Figure 11.15** Lintel

Jointing of walls. Brickwork and blockwork walls are jointed by means of mortar (mixture of sand and cement and/or lime forming an adhesive). Horizontal joints are known as **bed joints** and vertical ones as **perpends** or perps. The face of these joints may be finished in a variety of profiles intended to improve the weathering resistance and appearance of the work (Figure 11.16).

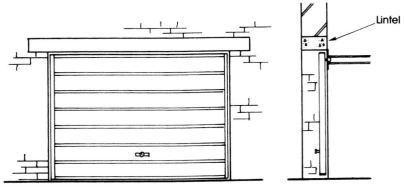

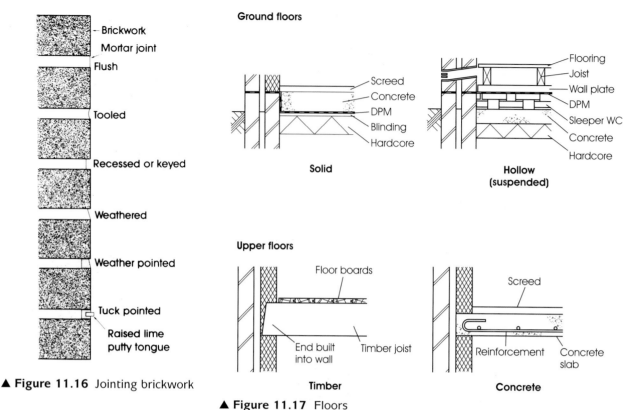

▲ **Figure 11.16** Jointing brickwork

▲ **Figure 11.17** Floors

Floors

These are the horizontal internal surfaces at ground and upper levels (Figure 11.17). Their main functions are to provide a level surface, standard of insulation and carry and transfer any loads imposed upon it. In addition, ground floors are also required to prevent moisture penetration and weed growth.

■ Ground floors are either solid or hollow (suspended).

■ Upper floors are suspended: timber construction is mainly used for house construction and concrete for other works.

Refer to Book 2, Chapter 1.

Roofs

These are part of the external envelope that spans the building at high level and has weathering and insulation functions. They are classified according to their **roof pitch** (slope of the roof surface) and also their shape, the most common of which are illustrated in Figure 11.18. Refer to Book 2, Chapter 2.

Stairs

Stairs provide floor-to-floor access. They can be defined as a series of steps (combination of tread and riser), each continuous set of steps being called a flight. Landings may be introduced between floor levels, to break up a long flight giving rest points, or to change the direction of the stair. Stairs can be classified according to their plan shape as shown in Figure 11.19 or by the material from which they are made. Timber stairs are common in dwelling houses, whilst concrete is most common for other works. Refer to Book 2, Chapter 5.

Secondary elements

The non-essential elements of a structure, having mainly a completion role around openings in primary elements and within the building in general, see Figure 11.20.

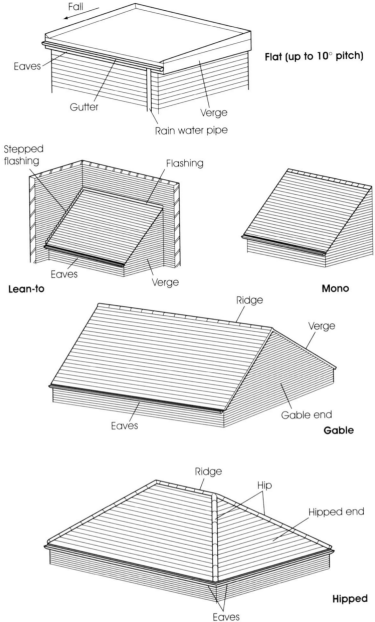

◀ **Figure 11.18** Common roof shapes

Fall

Eaves

Gutter

Verge

Rain water pipe

Flat (up to 10° pitch)

Stepped flashing

Flashing

Eaves

Verge

Lean-to

Mono

Ridge

Verge

Eaves

Gable end

Gable

Ridge

Hip

Hipped end

Eaves

Hipped

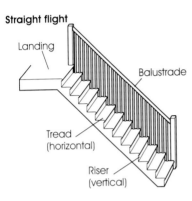

▼ **Figure 11.19** Classification of stairs by plan shape

Straight flight

Landing

Balustrade

Tread (horizontal)

Riser (vertical)

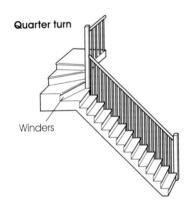

Quarter turn

Winders

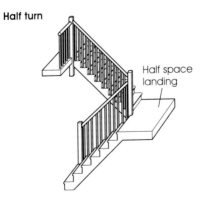

Half turn

Half space landing

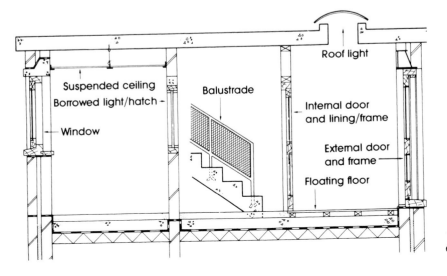

Roof light

Suspended ceiling

Borrowed light/hatch

Window

Balustrade

Internal door and lining/frame

External door and frame

Floating floor

◀ **Figure 11.20** Secondary elements

■ **Doors** are moveable barriers used to cover an opening in a structure. Their main function is to allow access in a building and passage between its interior spaces. Other functional requirements include weather protection, fire resistance, sound and thermal insulation, security, privacy, ease of operation and durability. They may be classified by their method of construction and method of operation. The surround to the wall opening on which doors are hung may be either a frame or lining. Refer to Book 2, Chapter 6.

■ **Windows** are glazed openings in a wall used to allow daylight and air in and give occupants a view outside. Windows are normally classified by their method of opening and the material from which they are made. Windows that project beyond the face of a building at ground level are known as a bay windows; those which project from an upper storey are known as oriel windows; those with a continuous curve are bow windows; those that contain a pair of casements for giving access to a garden or balcony are called French windows. Refer to Book 2, Chapter 7.

Finishing elements

A finish is the final surface of an element, which can be a self-finish as with face brickwork and concrete or an applied finish such as plaster, wallpaper and paint. Typical examples of finishing elements are shown in Figure 11.21. Included in this category are internal trims (skirting, architraves and coving or cornices), which mask the joint between adjacent elements, external flashings that weatherproof the joint and cladding, cement rendering and tile hanging, which are all used to either weatherproof or give a decorative finish to external walls.

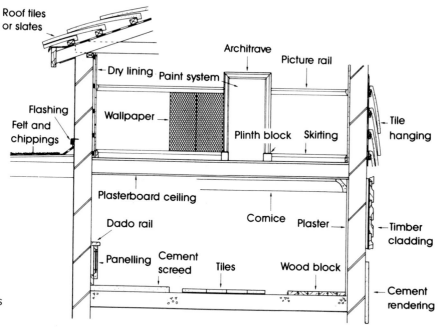

▶ **Figure 11.21** Finishing elements

Building components and services

The primary elements, secondary elements and services of a building are made up invariably from a number of different parts or materials; these are known as components. Examples of three main types of components are shown in Figure 11.22:

■ *Section components.* A section is a material that has been processed to a definite cross-sectional size but of an unspecified or varying length, e.g. a length of timber.

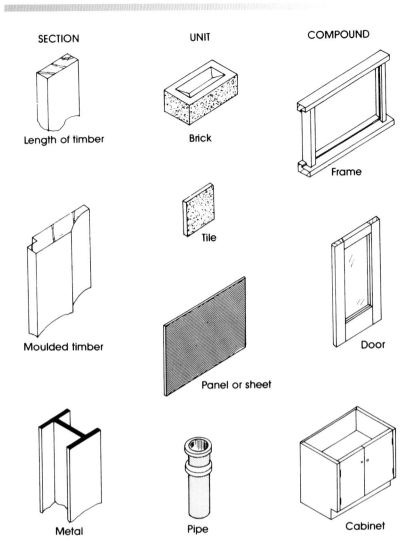

SECTION   UNIT   COMPOUND

Length of timber

Brick

Frame

Moulded timber

Tile

Panel or sheet

Door

Metal

Pipe

Cabinet

◀ **Figure 11.22** Building components

- *Unit components.* A unit is a material that has been processed to a definite cross-sectional size and length, e.g. a brick.
- *Compound components.* These are combinations of sections or units put together to form a complex article, e.g. a window frame.

The various components are combined to form the elements of a building.

Building services

Certain basic services are considered as essential requirements for all buildings:

- **Water** for drinking, washing, heating, cooking, waste/soil disposal and industrial processes (Figure 11.23).
- **Drainage** for the disposal of wastewater and sewage (Figure 11.24).
- **Electricity** for lighting, heating, cooking, cleaning, air conditioning, entertainment, telecommunications and industrial processes (Figure 11.25).
- **Gas** for heating, cooking and industrial processes (Figure 11.26).

These basic services consist of systems of pipes or wires, which are either fixed within, or on, the surface of the elements. They are connected to the distribution system, usually via a meter, which records the amount used. Each supply company has its own set of regulations concerning the supply, use of, and any alterations to, its services. These regulations must be complied with when provision is made within a building for connection to the particular service.

449

(a)

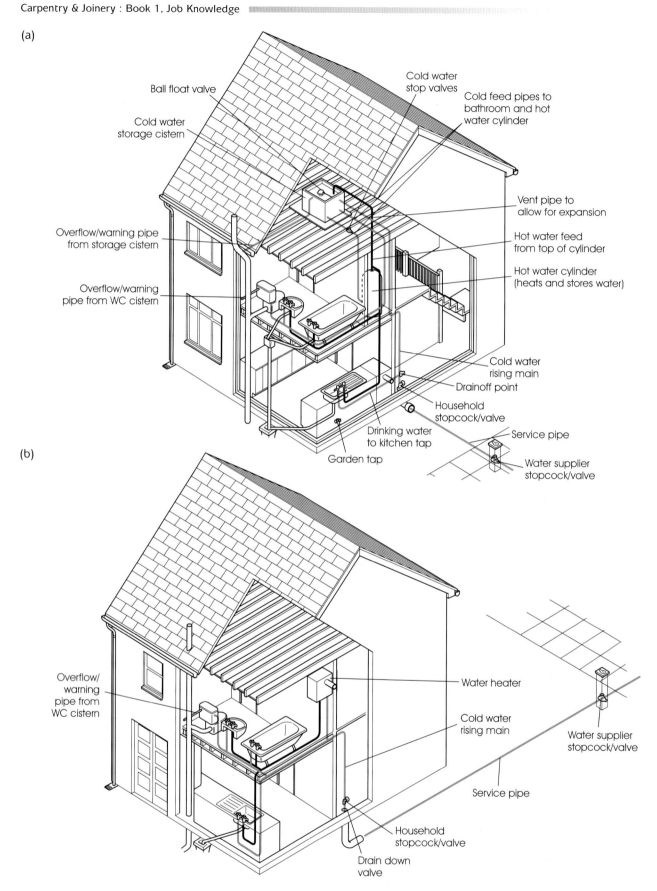

Ball float valve

Cold water storage cistern

Overflow/warning pipe from storage cistern

Overflow/warning pipe from WC cistern

Cold water stop valves

Cold feed pipes to bathroom and hot water cylinder

Vent pipe to allow for expansion

Hot water feed from top of cylinder

Hot water cylinder (heats and stores water)

Cold water rising main

Drainoff point

Household stopcock/valve

Service pipe

Water supplier stopcock/valve

Drinking water to kitchen tap

Garden tap

(b)

Overflow/warning pipe from WC cistern

Water heater

Cold water rising main

Water supplier stopcock/valve

Service pipe

Household stopcock/valve

Drain down valve

▲ **Figure 11.23** Domestic water systems: a) stored water system; b) direct water system

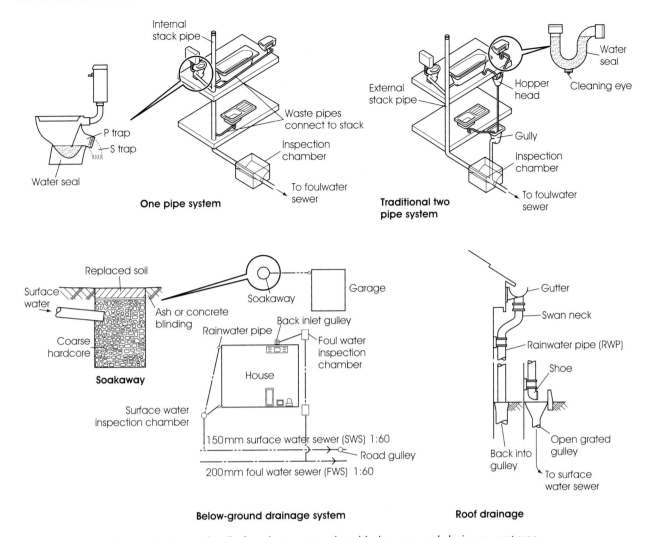

▲ Figure 11.24 Typical drainage details for above-ground and below-ground drainage systems

The installation and maintenance of services should be undertaken only by *competent persons*.

Thermal insulation

Thermal insulation is essential if heating costs are not to be wasted. Figure 11.27 illustrates the percentage heat losses through the various building elements of a typical house.

Thermal insulation standards for new buildings are controlled by the building regulations, for reasons of energy conservation. These regulations stipulate minimum requirements in the form of **U-values** (heat transfer per metre square though an element of construction, e.g. wall, roof or floor). Present values for a domestic building are illustrated in Figure 11.28.

Prior to the early 1970s only very modest standards for thermal insulation were stipulated for new buildings; earlier buildings may be totally uninsulated. Improved levels of thermal insulation in these buildings will reduce the rate of heat loss, reduce the amount of energy used and increase the standards of comfort for the buildings' occupants (Figure 11.29).

- **Roof spaces** are insulated using fibreglass, mineral wool, expanded polystyrene or Vermiculite between the ceiling joists. Laying additional insulation over the ceiling joists can top up existing insulation.

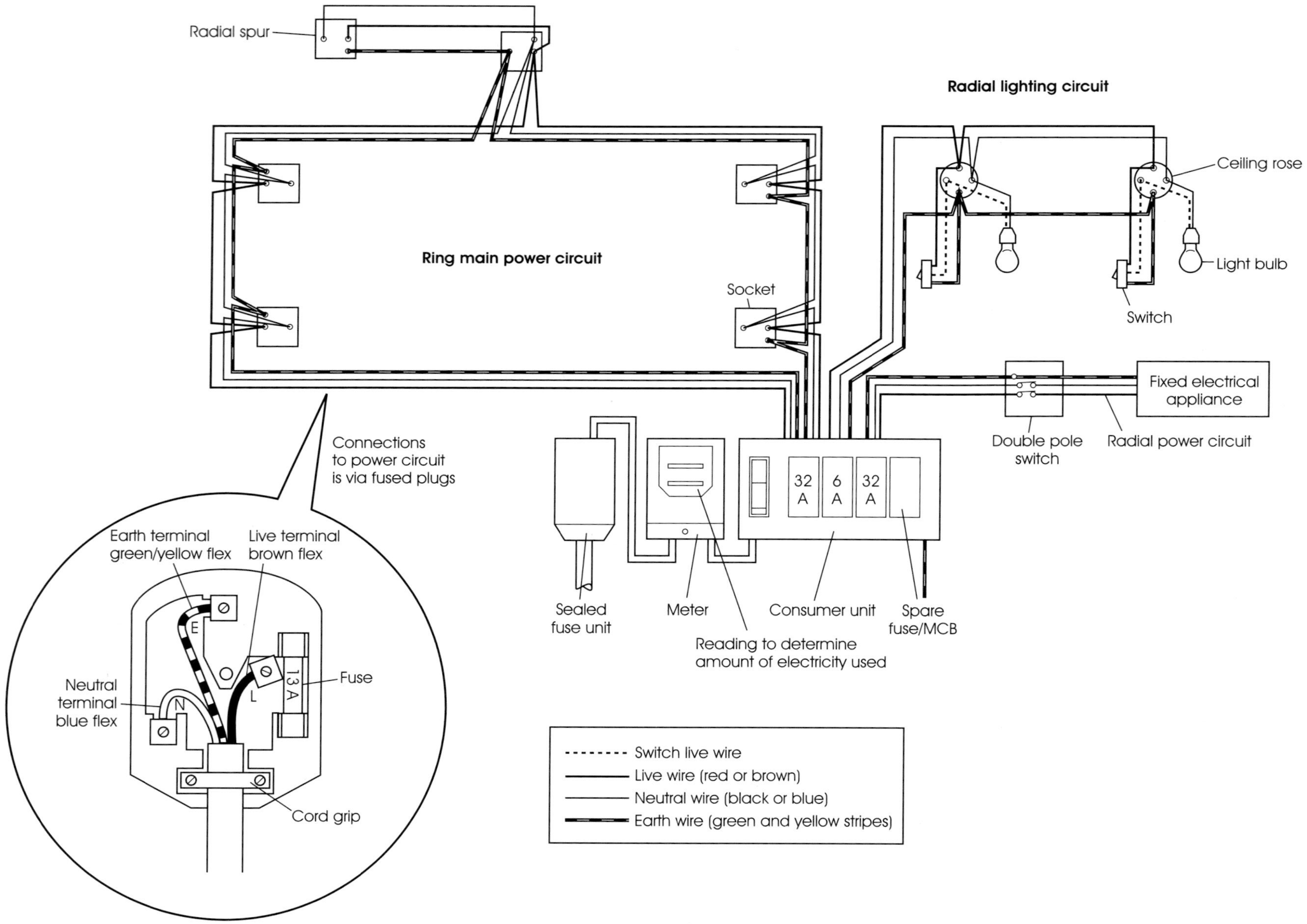

Radial lighting circuit

Ceiling rose

Light bulb

Switch

Fixed electrical appliance

Radial power circuit

Double pole switch

Spare fuse/MCB

32 A | 6 A | 32 A

Consumer unit

Reading to determine amount of electricity used

Socket

Meter

Sealed fuse unit

Ring main power circuit

Radial spur

Connections to power circuit is via fused plugs

Switch live wire
Live wire (red or brown)
Neutral wire (black or blue)
Earth wire (green and yellow stripes)

Earth terminal green/yellow flex

Live terminal brown flex

Fuse

13 A

Cord grip

E

L

N

Neutral terminal blue flex

▲ **Figure 11.25** Domestic electric circuits

452

■ **Cold-water storage tanks** should be lagged to prevent them freezing in the winter. As an extra precaution, the tank will be kept warmer if the ceiling insulation is not continued under them.

■ **Cavity walls** where not already filled with insulation during the building process can have insulating material injected into them. This may be plastic foam, polystyrene granules or mineral wool fibre.

■ **Solid walls** can be improved by adding an internal insulating layer. This may take the form of a cavity by lining the walls with foil-backed plasterboard on battens. Extra benefit is achieved by filling the space between battens with fibreglass or mineral wool.

■ **Sealed-unit double-glazing** is most effective against heat loss through windows and glazed doors.

■ **Secondary double-glazing** is normally not so good as sealed units as its primarily use is for sound insulation.

■ **Ground floors.** Hollow ground floors may be improved by laying fibreglass or mineral wool between the joists and suspended on wire mesh. A floating floor finish with insulation between the battens may be used to improve solid ground floors.

■ **Sealing strips.** A simple and fairly cheap method of cutting down on heat loss through draughts around windows and doors is to fix flexible sealing strips around their joints and over postal flaps.

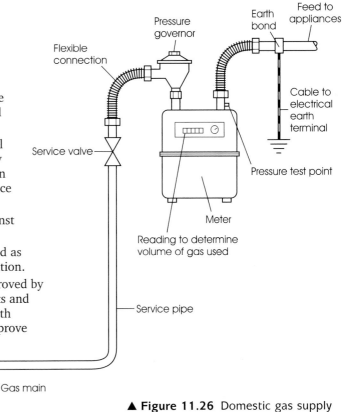

▲ **Figure 11.26** Domestic gas supply

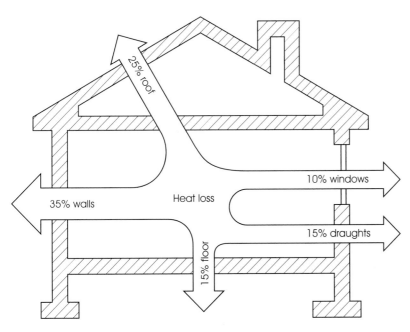

▲ **Figure 11.27** typical heat loss through building elements

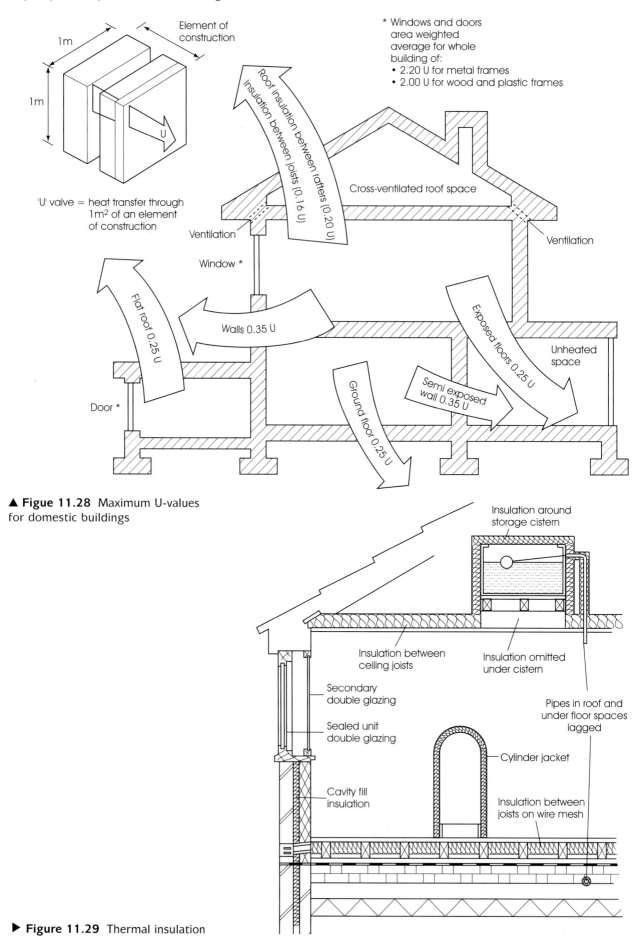

* Windows and doors area weighted average for whole building of:
• 2.20 U for metal frames
• 2.00 U for wood and plastic frames

1m

1m

Element of construction

'U' valve = heat transfer through 1m² of an element of construction

Roof insulation between rafters (0.20 U)

Insulation between joists (0.16 U)

Cross-ventilated roof space

Ventilation

Window *

Ventilation

Flat roof 0.25 U

Walls 0.35 U

Exposed floors 0.25 U

Unheated space

Door *

Ground floor 0.25 U

Semi exposed wall 0.35 U

▲ **Figue 11.28** Maximum U-values for domestic buildings

Insulation around storage cistern

Insulation between ceiling joists

Insulation omitted under cistern

Secondary double glazing

Sealed unit double glazing

Cavity fill insulation

Pipes in roof and under floor spaces lagged

Cylinder jacket

Insulation between joists on wire mesh

▶ **Figure 11.29** Thermal insulation

Chapter Twelve

Building procedures

This chapter is intended to provide the site carpenter and bench joiner with an overview of the building industry and its administrative procedures. It is concerned with the personnel, procedures and terms that may be encountered on a daily basis:

- the building industry and its team
- types of building contractor and the work role of personnel
- the design process
- contract documents
- the construction process
- contract planning
- site and personal communications.

The building industry

Government statistics divide the building industry into five main areas, by the type of firm or organisation undertaking the work:

- *Construction and repair of buildings.* The construction, improvement and repair of both residential and non-residential buildings, including specialist organisations bricklaying, carpentry, general building maintenance, roofing, scaffolding and the erection of framed structures for buildings.

- *Civil engineering.* The construction and maintenance of roads, car parks, bridges, railways, airport runways, and works associated with dams, reservoirs, harbours, rivers, canals, irrigation and land drainage. In addition the laying of pipelines or cables for sewers, gas and water mains and electricity, including overhead lines and supporting structures.

- *Installation of fixtures and fittings.* The installation of fixtures and fittings for electricity, gas, plumbing, heating and ventilation, sound and thermal insulation.

- *Building completion work:* Firms specialising in work having a completion role to a building, including on-site carpentry and joinery, painting and decorating, glazing, plastering, tiling, and flooring, etc.

- *General construction and demolition work.* A general classification to include firms and other organisations engaged in building and civil engineering whose work is not sufficiently specialised to be classified in one of the previous four. Demolition work and direct labour establishments of local authorities and government departments are also included under this heading.

The value of this work in the United Kingdom is about £85 billion pounds per year, at the time of writing. About 54% of the annual total is for new work and 46% for repairs and maintenance. Figure 12.1 shows in pie chart form the further percentage breakdown of these statistics.

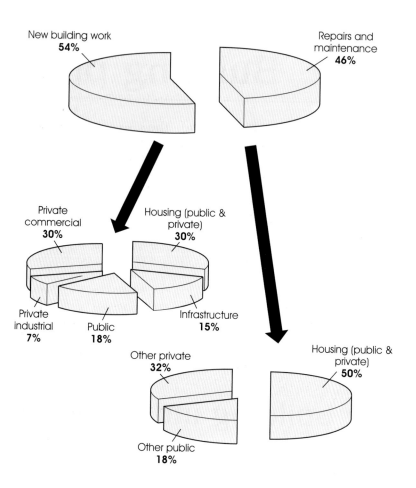

▶ **Figure 12.1** Distribution of building work

The work of the building industry can also be divided into work carried out either from the public or private sectors.

■ *Public work* is for any public authority such as local and central government departments, public utilities, nationalised industries, universities, new town corporations and housing associations, etc.

■ *Private work* is for a private owner, organisation or developer and includes all work undertaken by businesses on their own initiative. In addition work carried out under the government's PFI (Private Financial Initiative) is included. PFI is work where an individual firm or consortium take on the responsibility for providing a public service, typical examples being in health care and education where the private sector is actively engaged in building, equipping and maintaining the infrastructure.

Building and construction companies

At the time of writing there are over 166,000 registered firms in the United Kingdom of varying size undertaking building work. The vast majority, 93%, are small firms who employ 1 to 13 people, 6% are medium size firms employing 14 to 79 people. Large companies employing 80 or more people account for the remaining 1%. See Figure 12.2.

Many of the smaller firms are self-employed people or small organisations, which are employed by the medium and larger firms on a subcontract basis.

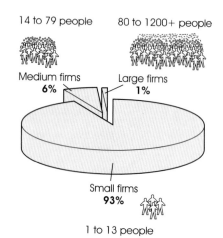

▲ **Figure 12.2** Classification of building firms by number of employees

In total the building industry employs almost 2 million workers, this represents 7% of the country's working population. Of these, 60% are directly employed and 40% self-employed. The top occupations are carpenters and joiners with over 10% of the total; plumbers 7.5%; electricians 7%; managers 6.5%; painters and decorators 6% and bricklayers 5%.

The building team

The construction of a building is a complex process, which requires a team of professionals working together to produce the desired results. Collectively they are known as the building team and are a combination of the following parties:

■ **Client.** This is the building owner; the person or persons who have an actual need for building work, e.g. the construction of a new house, office block, factory, etc. or extensions, repairs and alterations to existing buildings. The client is the most important member of the building team, without whom the work would simply not exist. The client is responsible for commissioning and the overall financing of the work and in effect employs either directly or indirectly the entire team. In particular they have the specific responsibility under the Construction (Design and Management) Regulations of appointing the **planning supervisor** (normally the architect and nominating the **principal contractor**. Refer to Chapter 10.

■ **Architect.** The architect is the client's agent and is considered to be the leader of the building team. The role of an architect is to interpret the client's requirements, translate them along with other specialist designers into a building form and generally supervise all aspects of the work until it is completed. In addition architects are often appointed by the client under the Construction (Design and Management) Regulations as the planning supervisor who is responsible for drawing up a safety plan, coordinating with the principal contractor with regards on-site **health and safety** and the compilation of a health and safety file. Refer to Chapter 8.

■ **Quantity surveyor.** In effect, the quantity surveyor (QS) is the client's economic consultant or accountant. This specialist surveyor advises during the design stage as to how the building may be constructed within the client's budget, and measures the quantity of labour and materials necessary to complete the building work from drawings and other information prepared and supplied by the architect. These quantities are incorporated into a document known as the **bill of quantities**, which is used by building contractors when pricing the building work. During the contract, the quantity surveyor will measure and prepare valuations of the work carried out to date to enable interim payments to be made to the building contractor and at the end of the building contract they will prepare the final account for presentation to the client. In addition, the quantity surveyor will advise the architect on the cost of any additional work or variations.

■ **Consulting (specialist) engineers.** These are engaged as part of the design team to assist the architect in the design of the building within their specialist fields, e.g. civil engineers, structural engineers and service engineers. They will prepare drawings and calculations to enable specialist contractors to quote for these areas of work. In addition, during the contract the specialist engineers will make regular inspections to ensure the installation is carried out in accordance with the design.

■ **Clerk of works.** Appointed by the architect or client to act as their on-site representative. On large contracts they will be resident on-site whilst on smaller ones will only visit periodically. The clerk of works or COW is an 'inspector of works' and as such will ensure that the contractor carries out the work in accordance with the drawings and other contract documents. This includes inspecting both the standard of workmanship and the quality of materials. The COW will make regular reports back to the architect, keep a diary in case of disputes, make a daily record of the weather, and of personnel employed on-site and any stoppages. He or she will also agree general matters directly with the building contractor. However the architect must confirm them to be valid.

- **Local authority.** The local authority normally has the responsibility of ensuring that proposed building works conform to the requirements of relevant planning and building legislation. For this purpose, they employ planning officers and building control officers to approve and inspect building work. In some areas, building control officers are known as building inspectors or district surveyors (DS). Alternatively the client may appoint an approved inspector acting with the local authority, to approve the work in accordance with the Building Regulations and supervise during construction.

- **Health and safety inspector.** The health and safety inspector, also known as the factory inspector, has the duty to ensure that the government legislation concerning health and safety is fully implemented by the building contractor.

- **Principal building contractor.** The building contractor enters into a contract with the client to carry out, in accordance with the contract documents, certain building works. Each contractor will develop their own method and procedures for tendering and carrying out building work which in turn, together with the size of the contract, will determine the personnel required. In addition they have the specific responsibility under the Construction (Design and Management) Regulations for health and safety on site. See heading under Building Control.

- **Subcontractors.** The building contractor may call upon a specialist firm to carry out a specific part of the building work; for this they will enter into a subcontract, hence the term subcontractor. Contractor-appointed subcontractors may also be known as **domestic subcontractors**. The client or architect often names or nominates a specific subcontractor in the contract documents for specialist construction or installation work. These must be used for the work and are known as **nominated subcontractors.**

- **Suppliers.** Building materials, equipment and plant are supplied by a wide range of merchants, manufacturers and hirers. The building contractor will negotiate with these to supply their goods in the required

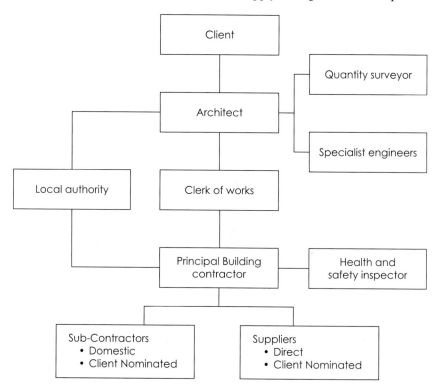

▶ **Figure 12.3** The building team

quantity and quality, at the agreed price, and finally in accordance with the building contractor's delivery requirements. The client or architect often nominates specific suppliers who must be used and are therefore termed **nominated suppliers.**

There is a recognised pattern by which the building team operates and communicates. This is illustrated in the form of a line diagram in Figure 12.3.

The building team may also be divided into a number of smaller teams each with their specific interests and overlapping roles in the total building process, see Figure 12.4.

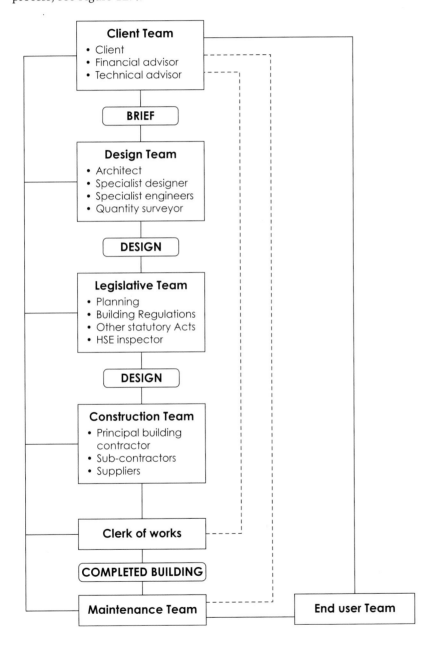

◀ **Figure 12.4** Sub-division of the building team

The design process

The design of a building is usually in the hands of a team of specialists brought together for this specific purpose. This design team, all of whom are employed directly by the client, will consist of the:

■ client (or specialist advisor);

■ architect (leader);

■ quantity surveyor (economic consultant);
■ specialist engineers (specialist consultants).

In addition, depending on the nature of the work, it may also include:
■ interior designers;
■ landscape designers;
■ town planners.

The first step in building projects is for a prospective building client to appoint an architect to act for them in the construction or alteration of a building. On being appointed the architect will obtain a **client brief**, consisting of full details of their requirements and the proposed site. Having inspected the site and assessed the feasibility of the client's requirements, the architect will prepare sketch designs and submit them to the client for approval and apply for outline planning permission.

When approval is obtained, a **design team** consisting of the architect, a structural engineer, a services engineer and a quantity surveyor is formed. This team will consider the brief and sketch designs and come up with proposals that will form the basis of the structure. Location drawings, outline specifications and preliminary details of costs are then produced and submitted to the client for approval. If these details are acceptable, applications will be made for full planning permission and building regulations approval. When these approvals have been obtained, contract documents will be prepared and sent to a number of building contractors for them to produce and submit **tenders**. The quantity surveyor will advise the architect and client of the most suitable contractor, after considering the returned tenders. The client will then engage a contractor and draw up and sign the contract.

Building contractors

Building contractors may operate as: **traditional companies** directly employing the majority of their workforce to undertake the construction or **management companies** who take on a co-ordinating role during the construction using only subcontractors and suppliers, with many operating someway between the two.

Traditionally building contractors only subcontracted out certain work such as structural steelwork, formwork, mechanical services and electrical installations, plastering, tiling and often painting. Today the move is towards a greater use of subcontractors for both the main and specialist operations leaving the main contractor to perform a management role.

Subcontractors may be labour-only where they contract to fit the building contractor's material, or they may contract to supply and fix their own material.

Choice of main contractor

The size, type and complexity of construction along with the resources to undertake the work in the required timescale, play an important role in the selection of a building contractor. Members of both the client and design teams may further influence this selection, by expressing a preference or dislike for a particular builder with whom they have had past experience.

Types of contractor

Building contractors may operate anywhere on the scale between traditional companies directly employing the majority of their workforce to undertake the construction with the use of subcontractors for specialist work, through to **management companies** who take on a co-ordinating role during the construction using subcontractors and suppliers to complete all the work. Although complicated by this fact, building contractors can be divided into the following three main types:

■ **General builder.** Mainly these are the smaller firms undertaking a wide range of work, but concentrating on a particular form of construction, geographical area or specialist service. The smallest are the **self-employed** craft operatives, working as subcontractors, moving on to **jobbing builders** who are mainly concerned with maintenance of existing buildings, adaptations and small extensions. The principal workers are often **craft operatives** (carpenters or bricklayers), with other specialist tasks being subcontracted out as required. The majority of these general builders operate in a more formal organisational structure, with work being divided between office and on-site activities as illustrated in Figure 12.5.

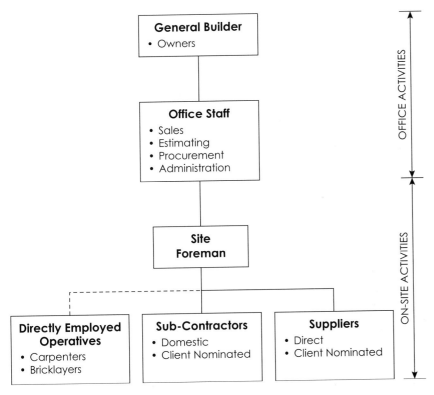

▲ **Figure 12.5** Small to medium general builders: organisational structure

■ **Building contractor.** These are the medium to larger firms undertaking a wide range of building work, civil engineering work or specialist activities. There will be a clear distinction between office and site activities, as illustrated in Figure 12.6.

■ **Design and build contractor.** These companies combine the responsibilities of the building design and construction, see Figure 12.7. They may be employed directly by the client to provide a 'package deal' or 'one-stop shop'. Alternatively they may work as a **speculative builder** where the building contractor on their own or as a part of a consortium design and build a project for later sale, mainly private housing, but also offices, shops and industrial units for sale or lease.
The building contractor is speculating or taking a chance that they will find a buyer or occupier (client) for the building on completion.

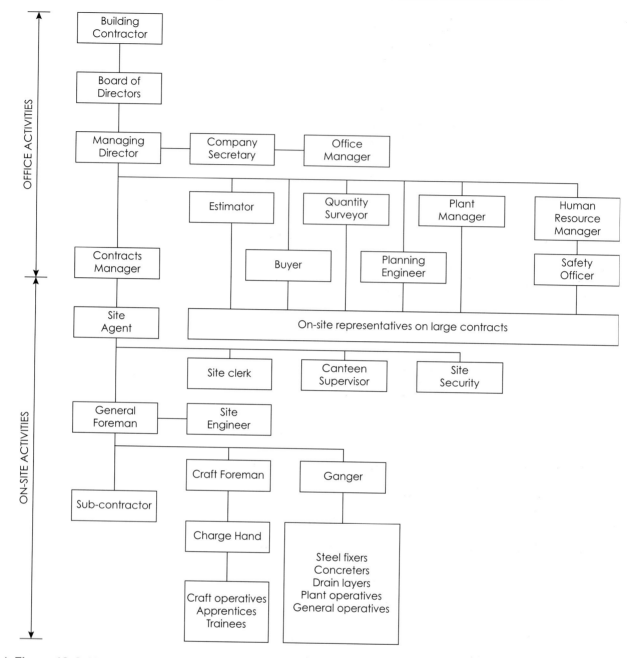

▲ **Figure 12.6** Medium to large building contractors: organisational structure

Contract documents

These documents will vary depending on the nature of the work, but will normally consist of (Figure 12.8):

- working drawings;
- specification;
- schedules;
- bill of quantities;
- conditions of contract.

Working drawings

These are scale drawings showing the plans, elevations, sections, details and locality of the proposed construction. These drawings can be divided into a number of main types as covered in Chapter Four.

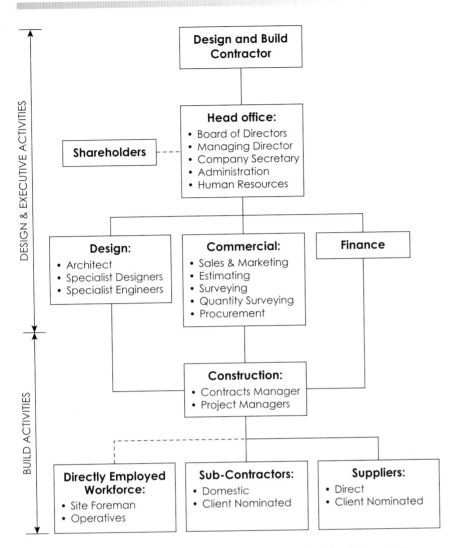

▲ **Figure 12.7** Design and build contractors: organisational structure

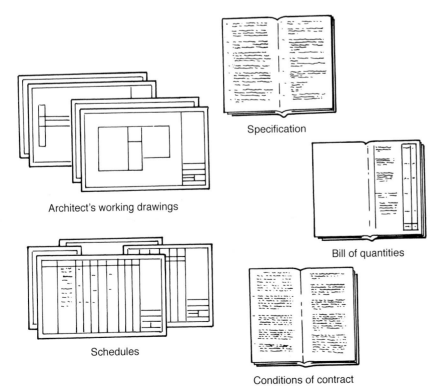

Specification

Architect's working drawings

Bill of quantities

Schedules

Conditions of contract

◄ **Figure 12.8** Contract documents

Specification

Except in the case of very small building works the drawings cannot contain all the information required by the builder, particularly concerning the required standards of materials and workmanship. For this purpose the architect will prepare a document, called the specification, to supplement the working drawings. The specification is a precise description of all the essential information and job requirements that will affect the price of the work but cannot be shown on the drawings. Typical items included in specifications are:

- site description;
- restrictions (limited access and working hours etc.);
- availability of services (water, electricity, gas, telephone);
- description of materials, quality, size, tolerance and finish;
- description of workmanship, quality, fixing and jointing;
- other requirements: site clearance; making good on completion; nominated suppliers and subcontractors; who approves the work, etc.

Various clauses of a typical specification are shown in Figure 12.9.

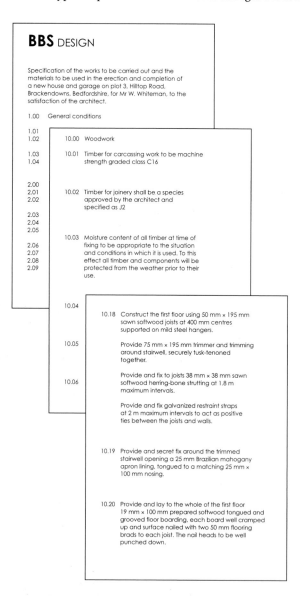

BBS DESIGN

Specification of the works to be carried out and the materials to be used in the erection and completion of a new house and garage on plot 3, Hilltop Road, Brackendowns, Bedfordshire, for Mr W. Whiteman, to the satisfaction of the architect.

1.00 General conditions

1.01
1.02
1.03
1.04

2.00
2.01
2.02
2.03
2.04
2.05
2.06
2.07
2.08
2.09

10.00 Woodwork

10.01 Timber for carcassing work to be machine strength graded class C16

10.02 Timber for joinery shall be a species approved by the architect and specified as J2

10.03 Moisture content of all timber at time of fixing to be appropriate to the situation and conditions in which it is used. To this effect all timber and components will be protected from the weather prior to their use.

10.04

10.05

10.06

10.18 Construct the first floor using 50 mm × 195 mm sawn softwood joists at 400 mm centres supported on mild steel hangers.

Provide 75 mm × 195 mm trimmer and trimming around stairwell, securely tusk-tenoned together.

Provide and fix to joists 38 mm × 38 mm sawn softwood herring-bone strutting at 1.8 m maximum intervals.

Provide and fix galvanized restraint straps at 2 m maximum intervals to act as positive ties between the joists and walls.

10.19 Provide and secret fix around the trimmed stairwell opening a 25 mm Brazilian mahogany apron lining, tongued to a matching 25 mm × 100 mm nosing.

10.20 Provide and lay to the whole of the first floor 19 mm × 100 mm prepared softwood tongued and grooved floor boarding, each board well cramped up and surface nailed with two 50 mm flooring brads to each joist. The nail heads to be well punched down.

▶ **Figure 12.9** Extracts from a specification

Schedules

These are used to record repetitive design information about a range of similar components. The main areas where schedules are used includes:

- doors, frames, linings;
- windows;
- ironmongery;
- joinery fitments;
- sanitary ware, drainage;
- heating units, radiators;
- finishes, floor, wall, ceiling;
- lintels;
- steel reinforcement.

A typical window schedule along with its related range drawing is illustrated in Figure 12.10.

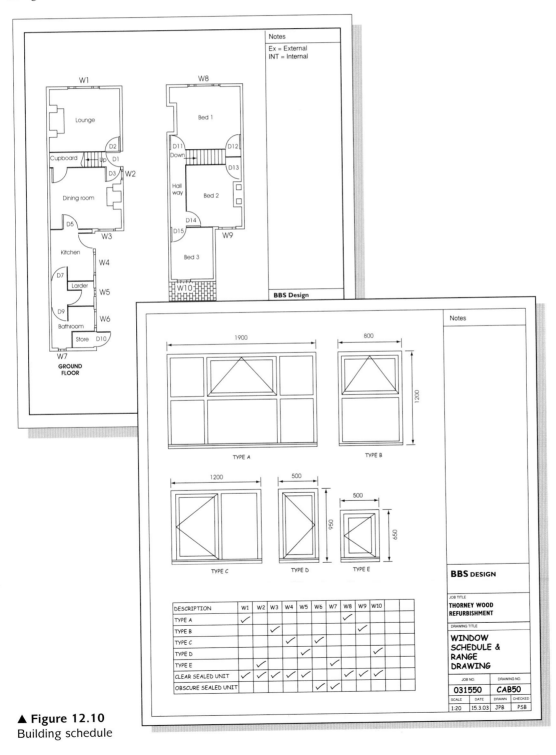

▲ **Figure 12.10**
Building schedule

Bill of quantities

The bill of quantities (BOQ) is prepared by the client's quantity surveyor. This document gives a complete description and measure of the quantities of labour, material and other items required to carry out the work based on drawings, specification and schedules. Its use ensures that all estimators prepare their tender on the same information. An added advantage is that as each individual item is priced in the tender they can be used for valuing the work in progress and also form the basis for valuing any variation to the contract.

All bill of quantities will contain the following information:

- *Preliminaries.* These deal with the general particulars of the work, such as the names of the parties involved, details of the works, description of the site and conditions of the contract, etc.
- *Preambles.* These are introductory clauses to each trade covering descriptions of the material and workmanship similar to those stated in the specifications.
- *Measured quantities.* A description and measurement of an item of work, the measurement being given in metres run, metres square, kilograms, etc., or just enumerated as appropriate.
- *Provisional quantities.* Where an item cannot be accurately measured, an approximate quantity to be allowed for can be stated. Adjustments will be made when the full extent of the work is known.
- *Prime cost sum (PC sum).* This is an amount of money to be included in the tender for work services or materials provided by a nominated subcontractor, supplier or statutory body.
- *Provisional sum.* A sum of money to he included in the tender for work which has not yet been finally detailed or for a 'contingency sum' to cover the cost of any unforeseen work.

Extracts from a typical bill of quantities are shown in Figure 12.11.

Conditions of contract

Most building work is carried out under a 'standard form of contract' such as one of the Joint Contractors Tribunal (JCT) forms of contract The actual standard form of contract used will depend on the following:

- type of client (local authority, public limited company or private individual);
- size and type of work (subcontract, small or major project, package deal);
- contract documents (with or without quantities or approximate quantities).

A building contract is basically a legal agreement between the parties involved in which the contractor agrees to carry out the building work and the client agrees to pay a sum of money for the work. The contract should also include the rights and obligations of all parties and details of procedures for variations, interim payments, retention, liquidated and ascertained damages and the defects liability period:

- *Variations.* A modification of the specification by the client or architect. The contractor must be issued with a written variation order or architect's instruction. Any cost adjustment as a result of the variation must be agreed between the quantity surveyor and the contractor.
- *Interim payment.* A monthly or periodic payment made to the contractor by the client. It is based on the quantity surveyor's interim valuation of the work done and the materials purchased by the contractor. On agreeing the interim valuation the architect will issue an interim certificate authorising the client to make the payment.
- *Final account.* Final payment on completion. The architect will issue a certificate of practical completion when the building work is finished. The quantity surveyor and the contractor will then agree the final account less the retention.

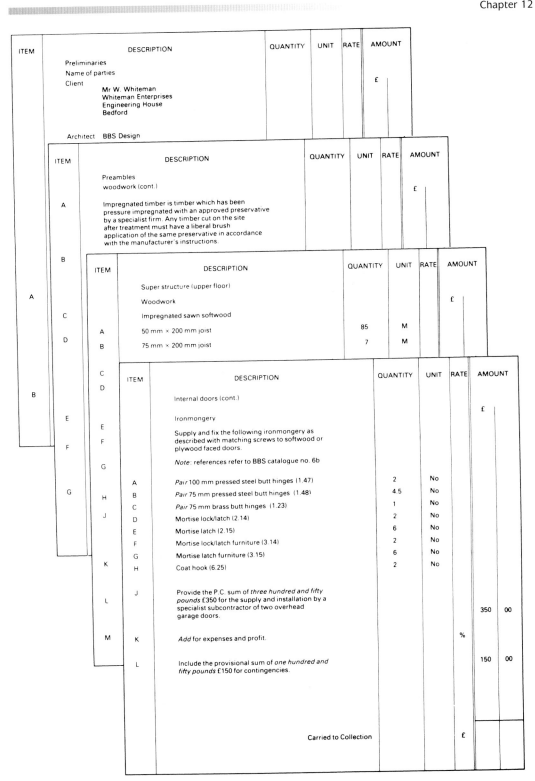

▲ **Figure 12.11** Bill of quantities

- *Practical completion.* The time at which the building work has been completed to the clients/architects satisfaction. A certificate of practical completion will be issued and the contractor will be entitled to the remainder of monies due, less any retention.
- *Main contractors discount.* A sum of money, which may be included in a subcontractor's quotation, often 2.5%, to cover the main contractor's administration costs associated with a subcontract.

- *Retention.* A sum of money, which is retained by the client until the end of an agreed defects liability period.
- *Liquidated and ascertained damages (LADs).* A sum of money payable on a daily or weekly basis, agreed in advance in the contract, as being fair and reasonable payment for damages or loss of revenue to the client as a result in a delay of practical completion.
- *Defects liability period.* A period of normally six months after practical completion to allow any defects to become apparent. The contractor will be entitled to the retention after any defects have been rectified to the architect's satisfaction.

Tendering for contract

The main way a building contractor obtains work is via the preparation and submission of tenders. There are three methods of tendering in common use: open tendering; selective tendering; negotiated contracts:

- *Open tendering.* Architects place advertisements in newspapers and construction journals inviting contractors to tender for a particular project. Interested contractors will apply for the contract documents, and prepare and submit a tender within a specific time period. At the close of this tender period the quantity surveyor will open all the tenders and make recommendations to the architect and client as to the most suitable contractor, bearing in mind the contractor's expertise and his tender price.

- *Selective tendering.* Architects establish a list of contractors with the expertise to carry out a specific project and will ask them to submit tenders for it. The architect may make up this list, simply from their experience of various contractors' expertise. Alternatively, advertisements may be placed in the newspapers and construction journals inviting contractors to apply to be included in a list of tenderers. From these applications, the architect will produce a shortlist of the most suitable contractors and ask them to tender. Again the quantity surveyor will open the returned tenders and make their recommendation to the architect and client.

- *Negotiated contracts.* Here the architect selects and approaches suitable contractors and asks them to undertake the project. If the contractor is willing to undertake the project they will negotiate with the quantity surveyor to reach an agreed price.

The construction process

The construction stage in the total building process can itself be subdivided into two stages:

- *The pre-construction stage,* which involves tendering and contract planning.
- *The on-site construction stage,* which consists of the actual physical tasks and the administration processes.

The construction process can be illustrated best in the form of a **flow chart** to show the varied range of tasks and the order in which they are carried out. Figure 12.12 shows a typical flow chart of the construction process for a detached house and the contract documents.

Contract planning and control

On obtaining a contract for a building project a contractor will prepare a programme that shows the sequence of work activities. In some cases an architect may stipulate that the contractor submits a programme of work at the time of tendering; this gives the architect a measure of the contractor's organising ability.

The work programme will show the interrelationship between the different tasks and also when and for what duration resources such as materials, equipment and workforce are required.

Once under way the progress of the actual work can be compared with the target times contained in a programme. If the target times are realistic, a programme can be a source of motivation for the site management who will make every effort to stick to the programme and retrieve lost ground when required. There are a number of factors, some outside the management's control, which could lead to a programme modification. These factors include: bad weather; labour shortages; strikes; late material deliveries; variations to contract; lack of specialist information; bad planning and bad site management, etc. Therefore when determining the length of a contract the contractor will normally make an addition of about 10% to the target completion date to allow for such eventualities.

Bar charts. The most widely used and popular type of work programme is the bar or **Gantt chart**. This is probably the most simple control system to use and understand. They are drawn up either manually or on a computer using a spreadsheet or specialised project planning software.

The individual tasks are listed in a vertical column on the left-hand side of the sheet and a horizontal time scale along the top. A horizontal bar shows the target start and end times of the individual tasks. A second horizontal bar is shaded in to show the work progress and the actual time taken for each task. Plant and labour requirements are often included along the bottom of the sheet. A typical bar/Gantt chart is shown in Figure 12.13. In addition to their use as overall contract management, bar charts can be used for short-term, weekly and monthly plans.

Site communication

No building site or construction contract could function effectively without a certain amount of day-to-day **paperwork** and form filling. Methods of communication, which enable information to flow, include:

■ **Time sheets.** Each employee completes these on a weekly basis, on which they give details of their hours worked and a description of the job or jobs carried out (Figure 12.14). Time sheets are used by the employer to determine wages and expenditure, gauge the accuracy of target programmes, provide information for future estimates and form the basis for claiming daywork payments. The foreman and timekeeper, especially on larger sites where a time clock is used, may complete these sheets.

■ **Daywork sheets.** A common misconception is that daywork sheets are the same as time sheets: they are not. Daywork is work carried out without an estimate. This may range from emergency or repair work carried out by a jobbing builder to work that was unforeseen at the start of a major contract, for example, repairs, replacements, demolition, extra ground work, late alterations, etc. (Figure 12.15). Daywork sheets should be completed by the contractor, authorised by the clerk of works or architect and finally passed on to the quantity surveyor for inclusion in the next interim payment. This payment is made from the provisional contingency sum included in the bill of quantities for any unforeseen work. Details of daywork procedures should be included in the contract conditions. A written architect's instruction is normally required before any work commences.

■ **Confirmation notice.** Where architects issue verbal instructions for daywork or variations, written confirmation of these instructions should be sought by the contractor from the architect before any work is carried out (Figure 12.16). This does away with any misunderstanding and prevents disputes over payment at a later date.

■ **Daily report/site diary.** This is used to convey information back to head office and also to provide a source for future reference, especially should a problem or dispute arise later in the contract regarding verbal

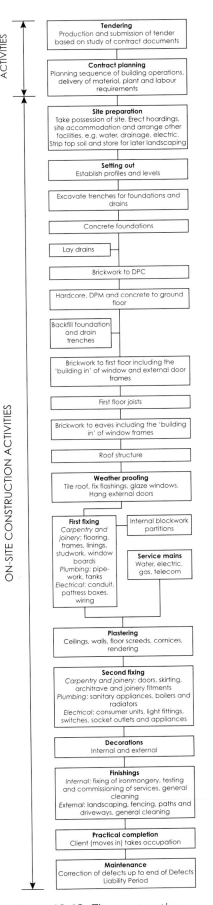

Figure 12.12 The construction process

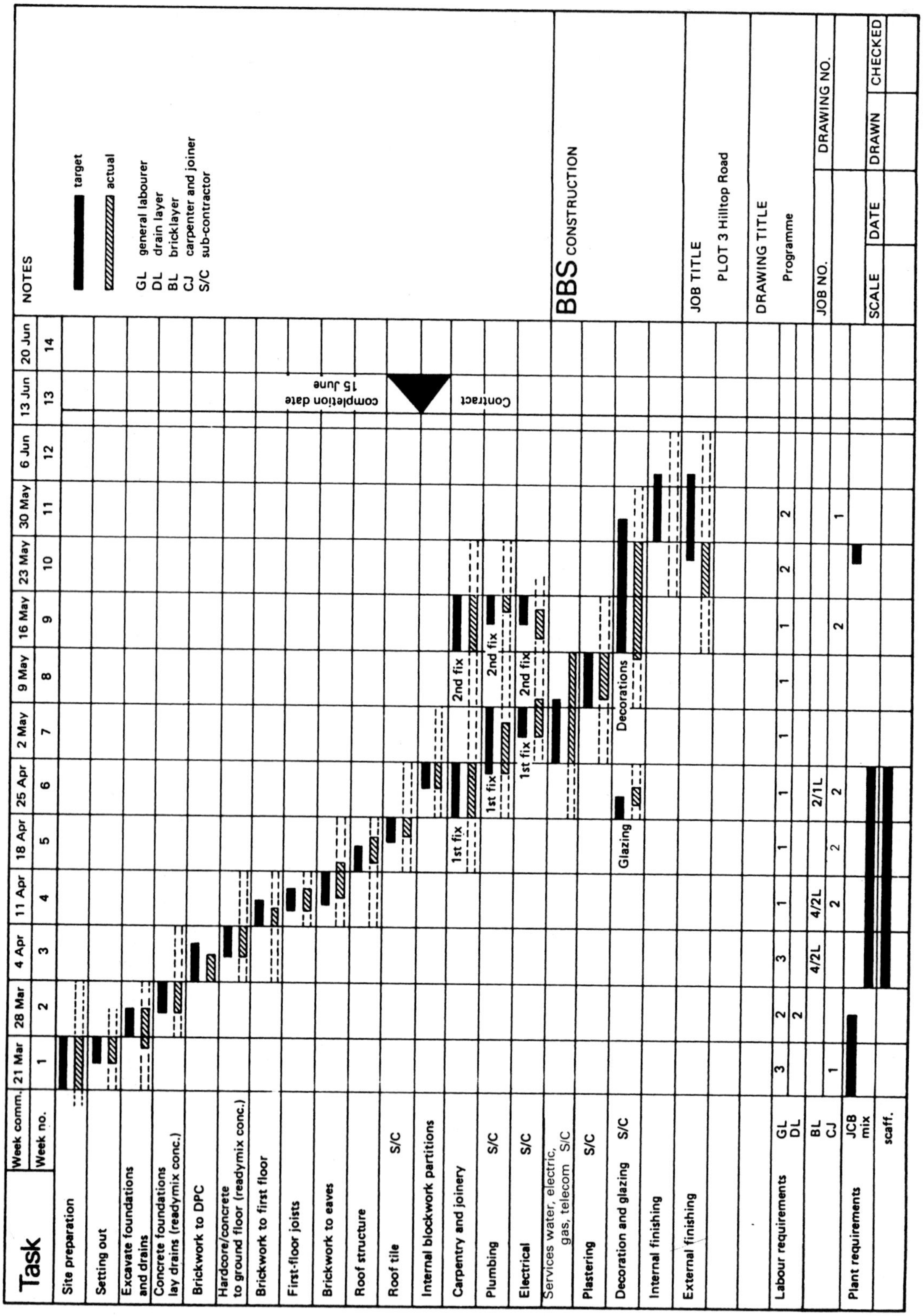

▲ **Figure 12.13** Bar/Gantt chart

instructions, telephone promises, site visitors, delays or stoppages due to late deliveries, late starts by subcontractors or bad weather conditions (Figure 12.17). Like all reports its purpose is to disclose or record facts; it should therefore be brief and to the point. Many contractors use a duplicate book for the combined daily report and site diary. After filling in, the top copy is sent to head office, the carbon copy being retained on site. Some firms use two separate documents to fulfil the same function.

▲ Figure 12.15 Daywork sheet

BBS CONSTRUCTION
DAYWORK SHEET

Registered office
BRETT HOUSE
1 HAGELY ROAD
BIRMINGHAM
B11 N4

Sheet no. __201/05__
Job title __EAST WOOD LANE__
Week commencing __15 MAR 05__

Description of work
GENERAL SNAGGING TO PLOT 2. INC. NEW FRONT DOOR FITTED AS A RESULT OF VANDALISM

Labour	Name	Craft	Hours	Gross rate	Total	
201/05	J. BROWN	C & J	2.5	£20	£50	00
			Total labour		£50	00

Materials	Quantity	Rate	% Addition		
NEW DOOR	1	£75	20	£90	00
		Total materials		£90	00

Plant	Hours	Rate	% Addition		
		Total plant			

Sub total	00	00
VAT (where applicable) N/A %	£140	–
Total claim	£140	00

Note Gross labour rates include a percentage for overheads and profit as set out in the contract conditions

Site manager/foreman ____[signature]____
Architect _____

▲ Figure 12.14 Time sheet

BBS CONSTRUCTION
WEEKLY TIME SHEET

Registered office
BRETT HOUSE
1 HAGELY ROAD
BIRMINGHAM
B11 N4

Name __JAMES BROWN__
Craft __CARPENTER__
Week commencing __15 MAR 05__

	Job title	Description of work	Time: start/finish	total
MON	EAST WOOD LANE	SECOND FIXING PLOT 10	8.00 TO 5.00	8.5
TUE	EAST WOOD LANE	SECOND FIXING PLOT 10	8.00 TO 5.00	8.5
WED	EAST WOOD LANE	SECOND FIXING PLOT 11	8.00 TO 5.00	8.5
THUR	EAST WOOD LANE	HANGING EXTERNAL DOORS PLOT 17	8.00 TO 5.00	8.5
FRI	EAST WOOD LANE	FIXING DOOR FURNITURE AND GENERAL SNAGGING FOR HANDOVER PLOT 2	8.00 TO 3.00	6.5
SAT				
SUN				

Details of expenses (attach receipts) __N/A__

Authorized by ____[signature]____ Position __PROJECT MANAGER__

For office use only
Standard hours _____ at _____ =
Overtime hours _____ at _____ =
Overtime hours _____ at _____ =
Overtime hours _____ at _____ =
TOTAL =

Figure 12.17 — BBS Construction Daily Report/Site Diary

BBS CONSTRUCTION
DAILY REPORT/SITE DIARY

No. __24/1005__   Date __24 OCT 05__

Registered office
**BRETT HOUSE
1 HAGELY ROAD
BIRMINGHAM
B11 N4**

Job title __PROJECT MANAGER__

Labour force on site			Labour force required		
Our employ	Subcontract		Our employ	Subcontract	
6	22		6	24	2 EXTRA JOINERS REQUESTED

Materials Received (state delivery no.)	N/A	Required by (state requisition no.) POD 24/205	Information Received N/A	Required by N/A
Plant Received (state delivery no.)	N/A	Required by (state requisition no.) PRD 32/205	Information Received N/A	Required by CASE GOODS LAYOUT 30 OCT 05

Telephone calls
To
From

Site visitors
**J. BIRD HAVELOCK P. M.
PRE-START MEETING**

Accidents
NONE

Stoppages
NONE

Weather conditions
WET AND OVERCAST ALL DAY

Temperature   a.m. __6 °C__   p.m. __9 °C__

Brief report of progress and other items of importance
**GOOD PROGRESS IN LINE WITH BAR CHART
HAVELOCK'S CONFIRMED THAT CASEGOODS TO LEVEL 1 WILL BE
UNDERTAKEN ON 31 OCT.**

Site manager/foreman _B Brett_

Note Send top copy daily to head office and retain carbon copy as an on-site record. _____

▲ **Figure 12.17** Daily report/site diary

Figure 12.16 — BBS Construction Confirmation Notice

BBS CONSTRUCTION
CONFIRMATION NOTICE

No. __106/05__   Date __10 MAR 05__

Registered office
**BRETT HOUSE
1 HAGELY ROAD
BIRMINGHAM
B11 N4**

Job title __EAST WOOD LANE__

From __PETER BRETT__

To __A J ARCHITECTURAL__

I confirm that today I have been issued with *verbal/written instructions from __A. J. POWELL__

Position __ARCHITECT__

to carry out the following *daywork/variation to the contract _____

Additions

**SUPPLY AND INSTALL NEW FRONT DOOR TO PLOT 2
(VANDALISM DAMAGE)**

Omissions

N/A.

Please issue your official *confirmation/variation order

Copies to head office _____

Signed _B Brett_

Position __PROJECT MANAGER__

*Delete as appropriate

▲ **Figure 12.16** Confirmation notice

BBS SUPPLIES
DELIVERY NOTE

Registered office
BRETT HOUSE
1 HAGELY ROAD
BIRMINGHAM
B11 N4

No. **8914**

Date **15 MARCH 2002**

Delivered to
T. JOYCEE
25 DAWNCRAFT WAY
STENSON DERBY D. 70

Invoice to
FELLOWS, MORTON PLC
JOSHER STREET
BIRMINGHAM B21

Please receive in good condition the undermentioned goods

SAWN, TREATED SOFTWOOD
50 OFF   25 x 50 x 3600
50 OFF   50 x 50 x 4.800

KILN SEASONED HARDWOOD
25 OFF   25 x 150 x 2400 (REBATED WINDOW SILLS)

(SHRINK-WRAPPED IN PLASTIC)

Received by _B Brett_
Remarks **ONLY 48 LENGTHS OF 50 x 50 RECEIVED**

Note Claims for shortages and damage will not be considered unless recorded on the sheet

▲ **Figure 12.19** Delivery note

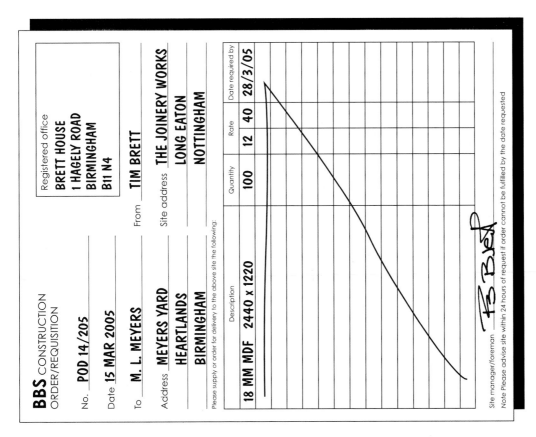

BBS CONSTRUCTION
ORDER/REQUISITION

Registered office
BRETT HOUSE
1 HAGELY ROAD
BIRMINGHAM
B11 N4

No. **POD 14/205**

Date **15 MAR 2005**

To **M. L. MEYERS**   From **TIM BRETT**

Address **MEYERS YARD**   Site address **THE JOINERY WORKS**
HEARTLANDS   **LONG EATON**
BIRMINGHAM   **NOTTINGHAM**

Please supply or order for delivery to the above site the following:

Description	Quantity	Rate	Date required by
18 MM MDF 2440 x 1220	100	12 40	28/3/05

Site manager/foreman _B Brett_

Note Please advise site within 24 hours of request if order cannot be fulfilled by the date requested

▲ **Figure 12.18** Order/requisition

- **Orders/requisitions.** The majority of building materials are obtained through the firm's buyer, who at the estimating stage would have sought quotes from the various suppliers or manufacturers in order to compare prices, qualities and discounts. It is the buyer's responsibility to order and arrange phased deliveries of the required materials to coincide with the contract programme. Each job would be issued with a duplicate order/requisition for obtaining sundry items from the firm's central stores or, in the case of a smaller builder, direct from the supplier (Figure 12.18). Items of plant would be requisitioned from the plant manager or plant hirers using a similar order/requisition.

- **Delivery notes.** When materials and plant are delivered to the site, the foreman is required to sign the driver's delivery note (Figure 12.19). A careful check should be made to ensure all the materials are there and undamaged. Any missing or damaged goods must be clearly indicated on the delivery note and followed up by a letter to the supplier. Many suppliers send out an **advice note** prior to delivery which states details of the materials and the expected delivery date. This enables the site management to make provision for its unloading and storage.

- **Delivery record.** This forms a complete record of all the materials received on site and should be filled in and sent to head office along with the delivery notes on a weekly basis (Figure 12.20). This record is used to check deliveries before paying suppliers' invoices and also when determining the interim valuation.

T. Joycee Construction
DELIVERIES RECORD

Registered office
**RIDGE HOUSE
NORTON ROAD
CHELTENHAM
GL59 1DB**

Week no. **P3 WK3**   Date **MARCH 02**

Job title **STENSON FIELDS**

Delivery note no.	Date	Supplier	Description of delivery	For office use only	
				Rate	Value
241	14/3/02	G. BLOGGS	SANITARY WARE		
1535	14/3/02	I. BLUNDER	READY-MIX CONCRETE		
					Total

Site manager/foreman
Note: Send weekly to head office with delivery notes

▲ **Figure 12.20** Delivery record

■ **Memorandum (memo).** This is a printed form on which internal communications can be carried out (Figure 12.21). It is normally a brief note about the requirements of a particular job or details of an incoming inquiry (representative/telephone call) while a person was unavailable.

Internal memos written between site staff should be friendly but not frivolous, brief but factual. Carbon copies should be kept on file in case of contractural queries.

■ **Letters.** Letters provide a permanent record of communication between organisations and individuals. They can be hand-written, but formal business letters give a better impression of the organisation if they are printed (Figure 12.22).

Letters should be written using simple concise language. The tone should be polite and business-like, even if it is a letter of complaint. They must be clearly constructed with each fresh point contained in a separate paragraph for easy understanding. When you write a letter for any reason, remember the following basic rules:

■ Your own address should be written in full, complete with the post code.

■ Include the recipient details. This is the title of the person (plus name if known) and the name and address of the organisation you are writing to in full, for future reference. This should be the same as appears on the envelope.

■ Write the date in full, e.g. 30 January 2006.

■ For greetings, use 'Dear Sir/Madam' if you are unsure of name and gender of the person you are writing to, otherwise 'Dear Sir' or 'Dear Madam' as applicable but use the person's name if you know it.

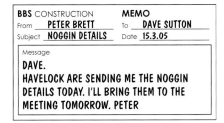

▲ **Figure 12.21** Memorandum (memo)

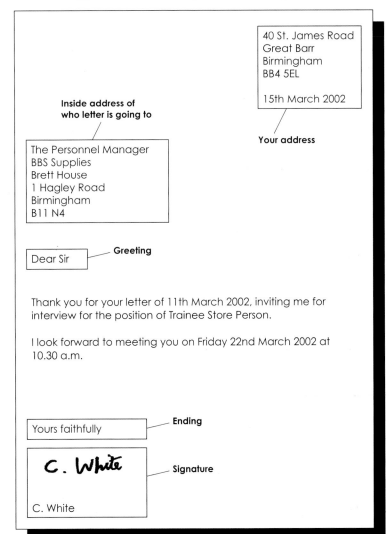

◀ **Figure 12.22** Typical letter

■ For endings, use 'Yours faithfully' unless you have used the person's name in the greeting, in which case use 'Yours sincerely'.

■ Sign below the ending. Your name should also be printed below the signature for clarity with your status if appropriate.

Telecommunications

■ **Facsimile transmission or FAX.** Fax is a method of sending images, both text and pictures, by a telecommunications link (Figure 12.23). Most fax transmissions are via the normal telephone network. Fax machines can send and receive hand-written notes, drawings, diagrams, photographs or printed text from one fax machine to another, anywhere in the world. Often used as a fast means of sending letters, which should normally be followed up with a postal copy.

■ **E-mail.** E-mail is the Internet's version of the postal service. Instead of faxing a message, you send a message from your computer down a telephone line. Copies of e-mails can be sent to other people who have

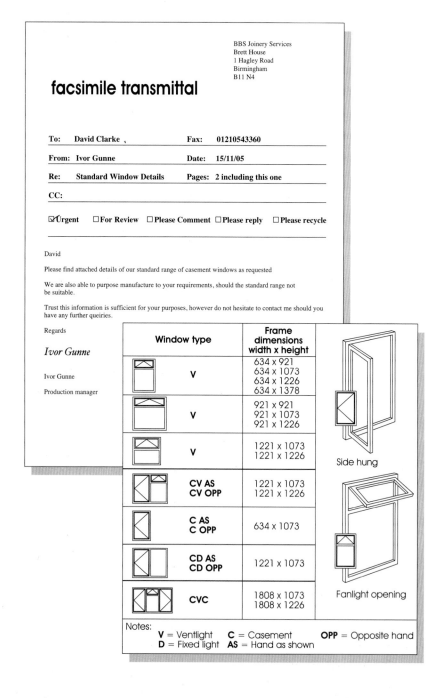

▶ **Figure 12.23** Example of a fax

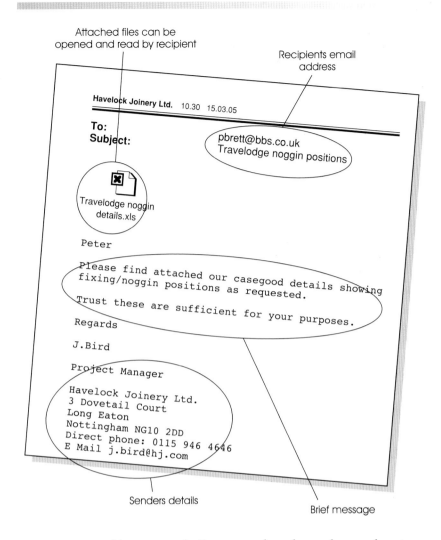

Attached files can be
opened and read by recipient

Recipients email
address

Havelock Joinery Ltd. 10.30 15.03.05

To:
Subject:

pbrett@bbs.co.uk
Travelodge noggin positions

Travelodge noggin
details.xls

Peter

Please find attached our casegood details showing
fixing/noggin positions as requested.

Trust these are sufficient for your purposes.

Regards

J.Bird

Project Manager

Havelock Joinery Ltd.
3 Dovetail Court
Long Eaton
Nottingham NG10 2DD
Direct phone: 0115 946 4646
E Mail j.bird@hj.com

Senders details

Brief message

◀ **Figure 12.24** Example of
an e-mail

a computer with access to the Internet and can be used as an almost
instant form of memo or letter (Figure 12.24). It is advisable to print out
important e-mails to retain file copies in paper form for others to refer to.
Electronic copies should be stored and backed up for contractual
purposes.

■ **Telephone.** Telephones play an important communication role both
within an organisation and to customers and suppliers. It is useful to keep
a record of incoming and outgoing telephone calls in the form of a log.

Telephone manner is important. Remember that you cannot be seen and
there are no facial expressions or other forms of body language to help make
yourself understood. The tone, volume and pace of your voice are
important. Speak clearly and loud enough to be heard without shouting;
sound cheerful, with a 'smile in your voice', speaking at a speed which the
recipient can take down any message, key words or phrases that you are
trying to relay.

When making telephone calls, if you initiate a call you are more likely to be
in control of the conversation and when you have achieved your objective
you will be in the best position to end the call without causing offence.
Make notes before you begin. Have times, dates and other necessary
information ready including words you find difficult to spell 'from your
head'. The call may take the following form (Figure 12.25):

■ 'Good morning' or 'Good afternoon'.

■ 'This is (your name), of (organisation) speaking'.

■ Give the name of the person you wish to speak to, if a specific individual
is required.

■ State the reason for your call.

▶ **Figure 12.25** Making a telephone call

▲ **Figure 12.26** Receiving a telephone call

■ Keep the call brief but courteous.

■ Thank the recipient, even if the call did not produce the results required.

When receiving a telephone call, a good telephone manner is as vital as when making a call. The call may take the following form (Figure 12.26):

■ 'Good morning' or 'Good afternoon'.

■ State your organisation and your name.

■ Ask 'How can I help you?'

■ If the call is not for you and the person required is unavailable ask if you can take a message.

Telephone messages. It is important that you understand what someone is saying to you on the telephone, and you may need to make notes of the conversation. When the message is not for you it is essential that you make written notes during the call, even though you may be seeing the person soon and be able to give the message verbally. Always make sure that the message contains all the necessary details e.g. Who, Where? When? What? and How? Any vagueness or omission of details could lead to problems later. Do not guess how to spell names and other details: ask the caller to spell them for you. A typical telephone message is illustrated in Figure 12.27.

Telephone Message

Date __15-3-05__   Time __10.15 AM__

Message for __JOHN BIRD__

Message from (Name) __PETER BRETT__

(Address) __PBRETT@BBS.CO.UK (EMAIL)__

(Telephone) __0121 478 478__

Message __CAN YOU SEND HIM AS SOON AS POSSIBLE A COPY OF THE TRAVELODGE NOGGIN DETAILS__

Message taken by __EMMA B.__

◀ **Fgure 12.27** Telephone message pad

Personal communications

Between yourself and work colleagues. It is necessary, in order for companies to function effectively, that they establish and maintain **good working relationships** within their organisational structure. This can be achieved by co-operation and communication between the various sections and individual workers: good working conditions (pay, holidays, status, security, future opportunities and a pleasant safe working environment) are important. But so is nurturing a good team spirit, where people are motivated, appreciated and allowed to work on their own initiative under supervision for the good of the company as a whole. Your working relationships with your immediate colleagues are important to the team spirit and overall success of the company. Remember: always plan your work to ensure ease of operation, co-ordination and co-operation with other members of the workforce.

With customers. Ultimately it is the customer who pays your wages; they should always be treated with respect. You should be polite at all times, even with those who are difficult. Listen carefully to their wishes and pass to a higher authority in the company anything you cannot deal with to the customer's satisfaction.

Remember, when working on their property you are a guest and you should treat everything accordingly:

- Always treat customers' property with the utmost care.
- Use dustsheets to protect carpets and furnishings when working indoors.
- Clean up periodically and make a special effort when the job is complete.
- If any problems occur contact your supervisor.

Personal hygiene

Ensure good standards of personal hygiene especially when working in occupied customers' premises. A smelly, dirty work person will make the customer assume the work will be poor. This may cause them to withdraw their offer of employment or may result in further work being given to another company.

- Wash frequently.
- Use deodorant if you have a perspiration problem.
- Wear clean overalls and have them washed at least once a week.
- Take off your muddy boots when working on customers' premises.

Index